DRIVER ACCEPTANCE OF NEW TECHNOLOGY

Human Factors in Road and Rail Transport

Series Editors

Dr Lisa Dorn
*Director of the Driving Research Group, Department of Human Factors,
Cranfield University*

Dr Gerald Matthews
Professor of Psychology at the University of Cincinnati

Dr Ian Glendon
*Associate Professor of Psychology at Griffith University, Queensland,
and President of the Division of Traffic and Transportation Psychology
of the International Association of Applied Psychology*

Today's society confronts major land transport problems. Human and financial costs of road vehicle crashes and rail incidents are increasing, with road vehicle crashes predicted to become the third largest cause of death and injury globally by 2020. Several social trends pose threats to safety, including increasing vehicle ownership and traffic congestion, advancing technological complexity at the human-vehicle interface, population ageing in the developed world, and ever greater numbers of younger vehicle drivers in the developing world.

Ashgate's Human Factors in Road and Rail Transport series makes a timely contribution to these issues by focusing on human and organisational aspects of road and rail safety. The series responds to increasing demands for safe, efficient, economical and environmentally-friendly land-based transport. It does this by reporting on state-of-the-art science that may be applied to reduce vehicle collisions and improve vehicle usability as well as enhancing driver wellbeing and satisfaction. It achieves this by disseminating new theoretical and empirical research generated by specialists in the behavioural and allied disciplines, including traffic and transportation psychology, human factors and ergonomics.

The series addresses such topics as driver behaviour and training, in-vehicle technology, driver health and driver assessment. Specially commissioned works from internationally recognised experts provide authoritative accounts of leading approaches to real-world problems in this important field.

Driver Acceptance
of New Technology
Theory, Measurement and Optimisation

Edited by

MICHAEL A. REGAN
University of New South Wales, Australia

TIM HORBERRY
University of Queensland, Australia, and University of Cambridge, UK

ALAN STEVENS
Transport Research Laboratory (TRL), UK

CRC Press
Taylor & Francis Group
Boca Raton London New York

CRC Press is an imprint of the
Taylor & Francis Group, an **informa** business

CRC Press
Taylor & Francis Group
6000 Broken Sound Parkway NW, Suite 300
Boca Raton, FL 33487-2742

First issued in paperback 2017

Version Date: 20160226

ISBN 13: 978-1-138-07703-4 (pbk)
ISBN 13: 978-1-4094-3984-4 (hbk)

Visit the Taylor & Francis Web site at
http://www.taylorandfrancis.com

and the CRC Press Web site at
http://www.crcpress.com

Contents

PART V: OPTIMISING DRIVER ACCEPTANCE

PART VI: CONCLUSIONS

List of Figures

List of Tables

About the Editors

Michael A. Regan is currently a Professor in the Transport and Road Safety Research group in the School of Aviation at the University of New South Wales, in Sydney, Australia. Before that he held research appointments with the French Institute of Science and Technology for Transport, Development and Networks (IFFSTAR) in Lyon, France, and the Monash University Accident Research Centre in Melbourne, Australia. Mike's current research interests focus on human interaction with, and acceptance of, intelligent transport systems, driver distraction and inattention, use of instrumented vehicles for naturalistic observation of driver and pilot behaviour, and aviation safety. He sits on the Editorial Boards of five peer-reviewed journals, including *Human Factors*, is the author of more than 200 publications, including two books, and sits on numerous expert committees on traffic safety. He is the 25th President of the Human Factors and Ergonomics Society of Australia.

Tim Horberry is Associate Professor of Human Factors at the University of Queensland, Australia. He is also a Senior Research Associate at the University of Cambridge, UK, and before that he was at the UK's Transport Research Laboratory. Tim has published his work widely, including four books published either by Ashgate or CRC press: 'The Human Factors of Transport Signs' (2004) and 'Human Factors in the Maritime Domain' (2008), 'Understanding Human Error In Mine Safety' (2009) and 'Human Factors for the Design, Operation and Maintenance of Mining Equipment' (2010). Tim has undertaken many applied Human Factors research projects in Australia, the UK and Europe for organisations such as the European Union, Australian Research Council and the UK Department for Transport. Currently Tim is leading several projects in the minerals industry that are examining acceptance of new technology for mining vehicles – including collision detection systems and shovel automation.

Alan Stevens is Chief Research Scientist and Research Director, Transportation, at the Transport Research Laboratory TRL, in the UK, where he has been working on the application of new technology to transport for 25 years. He is an internationally recognised expert in 'Human–Machine Interaction' (HMI) in the driving environment and was co-author of the 'European Statement of Principles on HMI' through his work within the iMobility initiative, where he co-chairs the HMI Working Group. He was also an active member of the responsible international standards committee, regularly participating in meetings with European, US, Canadian and Japanese colleagues. He was recently appointed to the EU–US Working Group on Driver Distraction following the EU–US

High Level Cooperation agreement and continues to be involved in the international IHRA (International Harmonized Research Agenda) group and on the Management Committee of IBEC (International Benefit Evaluation and Costs) group. Alan's consultancy activities focus on providing advice on policy and interoperability issues to Government, developing research programs and carrying out specific technical and Human Factors studies in Intelligent Transportation Systems. He participates in university teaching at MSc level, supervises PhD students and is Editor in Chief of an international peer-review journal of Intelligent Transport Systems.

List of Contributors

Adell, Emeli, Trivector Traffic, Sweden

Belin, Matts-Åke, Swedish Transport Administration, Vision Zero Academy, Borlänge, Sweden, and School of Health, Care and Social Welfare, Mälardalen University, Västerås, Sweden

Brookhuis, Karel, Delft University of Technology, the Netherlands and University of Groningen, the Netherlands

Brusque, Corinne, Institut Français des Sciences et Technologies des Transport, de l'aménagement et des Réseaux (IFSTTAR), Bron, France

Burnett, Gary, Human Factors Research Group, Faculty of Engineering, University of Nottingham, Nottingham, UK

Cooke, Tristan, Minerals Industry Safety and Health Centre, University of Queensland, Australia

Diels, Cyriel, Coventry School of Art and Design, Department of Industrial Design, Coventry University, Coventry, UK

Dorn, Lisa, Cranfield University, UK

Edmunds, Robert, Cranfield University, UK

Ghazizadeh, Mahtab, Department of Industrial and Systems Engineering, University of Wisconsin-Madison, USA

Green, William S., University of Canberra, Australia

Hollnagel, Erik, University of Southern Denmark, Denmark

Horberry, Tim, Minerals Industry Safety and Health Centre, University of Queensland, Australia, and Engineering Design Centre, University of Cambridge, UK

Huth, Véronique, Institut Français des Sciences et Technologies des Transport, de l'aménagement et des Réseaux (IFSTTAR), Bron, France

Jordan, Patrick W., University of Surrey, UK

Källhammer, Jan-Erik, Autoliv Development AB, Sweden

Keinath, Andreas, BMW Group, Germany

Labeye, Elodie, Institut Français des Sciences et Technologies des Transport, de l'aménagement et des Réseaux (IFSTTAR), Bron, France

Lee, John D., Department of Industrial and Systems Engineering, University of Wisconsin-Madison, USA

Maguire, Martin C., Loughborough Design School, Loughborough University, UK

Marchau, Vincent, Radboud University, Nijmegen School of Management, the Netherlands

Meleckidzedeck, Khayesi, World Health Organization (WHO), Department of Violence and Injury Prevention and Disability, Geneva, Switzerland

Mitsopoulos-Rubens, Eve, Monash University Accident Research Centre, Monash University, Australia

Nilsson, Lena, Swedish National Road and Transport Research Institute (VTI), Sweden

Reed, Nick, Transport Research Laboratory, UK

Regan, Michael A., Transport and Road Safety Research, University of New South Wales, Australia

Rudin-Brown, Christina M., Monash University Accident Research Centre, Australia

Skrypchuk, Lee, Jaguar Land Rover, UK

Smith, Kip, Naval Postgraduate School, USA

Stevens, Alan, Transport Research Laboratory, UK

Tingvall, Claes, Swedish Transport Administration, Borlänge, Sweden, and Department of Applied Mechanics, Chalmers University, Gothenburg, Sweden

Van der Pas, Jan-Willem, Delft University of Technology, Faculty of Technology, Policy and Management, the Netherlands

Várhelyi, András, Lund University, Sweden

Vedung, Evert, Institute for Housing and Urban Research, Uppsala University, Uppsala, Sweden

Vilimek, Roman, BMW Group, Germany

Vlassenroot, Sven, Ghent University, Belgium, and Flanders Institute for Mobility, Belgium

Walker, Warren E., Delft University of Technology, Faculty of Technology, Policy and Management, and Faculty of Aerospace, the Netherlands

Young, Kristie L., Monash University Accident Research Centre, Australia

Acknowledgements

The editors wish to thank the following organisations and individuals for the important roles they played in enabling this book to be completed:

- The anonymous reviewers, recruited by Ashgate, for recommending that development of the book proceed;
- Guy Loft and the editorial team at Ashgate for their professional guidance, trust and patience;
- The invaluable help provided by Mei Regan – for the many hours she spent sub-editing the whole manuscript, chasing copyright agreements and permissions, and generally supporting us in keeping the entire process running smoothly;
- The support of the Institut Français des Sciences et Technologies des Transport, de l'aménagement et des Réseaux (IFSTTAR), the University of New South Wales, the University of Queensland, the University of Cambridge and the Transport Research Laboratory. Dr Horberry also acknowledges the support of an EC Marie Curie Fellowship 'Safety in Design Ergonomics' (project number 268162); and
- All the authors, for their insightful contributions, patience and goodwill in adhering to the requirements of the editorial process.

Michael A. Regan
Tim Horberry
Alan Stevens

PART I
Introduction

Chapter 1

Driver Acceptance of New Technology: Overview

Michael A. Regan
University of New South Wales, Australia

Alan Stevens
Transport Research Laboratory, UK[1]

Tim Horberry
University of Queensland, Australia, and University of Cambridge, UK

Introduction

Anyone who has worked in the area of driver acceptance of new vehicle technologies will know the frustrations of doing so: there are many definitions of driver acceptance; the methods and metrics for measuring acceptance vary enormously across studies; terms like 'driver acceptability' and 'driver acceptance', although seemingly different, are often used interchangeably; and even if acceptance is measured and quantified, the data yielded by the methods used may not be in a form that is practically useful for informing system design. These issues, and the need for a single volume that pulls the field together, were the primary factors motivating development of this book. In this, the introductory chapter, we set the scene for what is to come.

The Changing Motor Vehicle

Since its advent, the motor vehicle has undergone some significant transformations: engines have become more efficient and reliable; vehicle bodies and interior cockpit structures have become more crashworthy; and mechanical linkages have been replaced increasingly by electronic connections. Until quite recently the vehicle cockpit remained largely unchanged; driving was the central focus of activity, and the driver remained completely in control of the vehicle. However, all of that is changing, rapidly.

The last decade has witnessed an explosion in the availability of new vehicle technology; a general term referring to the application of mechanical, electronic, information and communication systems and new materials in the driving environment. Some technology has been built into the vehicle by manufacturers, some has been added within aftermarket products and other technologies have been brought into the vehicle by drivers (e.g., mobile phones). 'Infotainment' systems have emerged to keep drivers informed and entertained; communication systems, such as phone, fax and email, allow the driver to stay connected with the outside world; and driver assistance systems, such as collision warning and adaptive cruise control, support the driver to drive safely, efficiently and comfortably. In doing so, technologies automate – partially, highly or fully – aspects of vehicle control, or even all aspects of control. Driverless vehicles are starting to be driven in some parts of the world as this book goes to press. The advent of new propulsion systems (electric and hydrogen vehicles) have contributed further to this revolution in vehicle technology, changing the face of driving, and the human machine interfaces and interactions through which driving is accomplished (e.g., Labeye et al. 2013).

Paralleling these developments has been an explosion in the application of technology to modern roadways: technologies that inform drivers (e.g., variable message signs) and technologies that support them to drive safely and efficiently (e.g., ramp meters, speed cameras, red light cameras, etc.). The advent of cooperative intelligent transport systems (C-ITS), which enable wireless vehicle-to-vehicle (V2V), vehicle-to-infrastructure (V2I) and vehicle-to-nomadic device (V2N) communication (NTC 2012), open up an almost unlimited new world of technology applications to improve the comfort, efficiency and safety of drivers.

The Importance of Driver-Centred Design and Deployment

The rapid development of new technology has resulted in many new systems for drivers being deployed without them having been designed systematically, integrated into work environments and evaluated from a driver-centred perspective. Typical issues that arise without a driver-centric approach to technology design include information overload from multiple information and warning systems, inadequate driver training and support, drivers being outside the system control loop, over-reliance on technology by drivers, de-skilling of drivers, negative behavioural adaptation to the technology and, ultimately, low acceptance or even misuse of the new technology after introduction (Lee and Seppelt 2009). Human factors are, thus, of great importance during the design and introduction of new technologies, but often are not considered in sufficient detail.

A Spotlight on Driver Acceptance

The technologies that drivers use to inform, entertain, communicate, comfort and protect themselves are no different from other technologies: unless they are accepted by drivers, they will not deliver the benefits intended by those who designed them. If they are not accepted, drivers will not buy them; and even if they do, they may disable them out of frustration or use them in a manner unintended by designers. This is especially salient for vehicle safety technologies. There is much evidence that Advanced Driver Assistance Systems (ADAS) have huge potential to save lives and reduce serious injury (US DOT 2008). Yet if they are not accepted by drivers, their potential to save lives and deliver economic benefits to society will never be realised.

Human Factors and ergonomics professionals have been interested for a long time in identifying and understanding the determinants of user acceptance of technology, in order to support engineers in ensuring new systems and products are designed and deployed to minimise resistance and maximise uptake (Dillon 2001). This interest was spawned in part by a realisation within the IT industry that some investments in information technology were not producing the intended benefits because the technologies themselves were not accepted by users. There has been a long and learned preoccupation by those in the IT industry with user acceptance; what it means, how it is measured and how it can be optimised. Similarly, defence and other complex occupational domains such as health care and nuclear power have long had an interest in integrating new technologies into existing work systems. More recently, this interest has spilled over into transportation Human Factors (e.g., Young, Regan and Mitsopoulos 2004). A range of measures, from initial design to user-centred deployment, can be implemented to improve driver acceptance of new technologies.

At its most basic level, acceptance of new technology can simply be aligned with use of that technology: if it is acceptable to people, they will use it. So there might be interest, for example, in how many drivers use their cruise control, under what circumstances, and how often. However 'acceptance equals use' is simplistic at best, and does not help system designers to develop and deploy successful products. A more fundamental decomposition of acceptance is necessary to set the scene for this book and to illustrate why different authors (implicitly or explicitly) think about acceptance in different ways.

Defining 'Acceptability' and 'Acceptance'

As a scientific construct, acceptance has been variously defined. In the information technology domain, it has been defined as 'the demonstrable willingness within a user group to employ information technology for the tasks it is designed to support' (Dillon and Morris 1996: 4). The determinants of user acceptance, however, are complex and derive from the technology itself, from those who use it and from the context in which it is implemented. The characteristics of technology that determine its level of acceptance include such characteristics as relative advantage over other available tools, compatibility with social practices and norms, complexity in ease of use and learning, 'trial-ability' of the technology before use, and 'observability' – or the extent to which the benefits of the technology are obvious (Dillon 2001, Rogers 1995).

There is debate in the literature about the psychological variables that distinguish users who accept or reject technologies: cognitive style, personality, demographic variables (e.g., age and education) and user-situational variables are among those that have been cited as variables that influence user acceptance of technologies (Dillon 2001, Alavi and Joachimsthaler 1992). Acceptance of technology is also influenced by the social, legal, cultural, political and organisational context in which the technology is implemented, and by the amount and type of exposure the user has had to the technology. Some attempts have been made to link these kinds of variables into a unified, predictive, theory of acceptance (e.g., Davis, Bagozzi and Warshaw 1989, Venkatesh et al. 2003).

The terms 'acceptance' and 'acceptability' are used, often interchangeably, in the literature. Driver reactions to technology can be studied at different times in the technology lifecycle: before it exists; when it exists in prototype form and when it is commercially mature. In advance of actually experiencing a new product, individuals will invariably have a view about it, although most researchers would not yet ascribe the term 'acceptance' to this judgement; at this point most talk about 'acceptability' as a 'prospective judgement of measures to be introduced in the future' (Schade and Schlag 2003: 45–61). Product designers are very interested in characterising acceptability (potential acceptance) even though it is a personal judgement about a product yet to be experienced. No objective measures are available but opinions can be sought and designers will probably also want to know how certain are individuals about their likely future reactions and whether there are important variables that are important to them.

As well as a focus on individual drivers – on their behaviour and their acceptance of technology – it is possible to research acceptance of technology at an organisational, cultural or societal level. Policy makers are very interested not only in the impact that driver acceptance of new technology has on transport outcomes (such as safety), but in how desirable outcomes can be supported by promoting acceptance of new technology more generally. Here, we can identify concepts such 'early adopters' (Rogers 1962) and look at how the use of new technology spreads through organisations and society. Issues include how acceptance of individuals should be amalgamated in order to represent acceptance at a group or society level and whether (as suggested by Van der Laan, Heino and De Waard 1997) social acceptance is a concept distinct from user acceptance requiring a more holistic evaluation of the consequences of adoption of the new technology.

Purpose and Structure of This Book

The purpose of this book is to bring together into a single volume a body of accumulated scientific and practical knowledge that can be used to optimise driver acceptance and uptake of new technologies in cars and other vehicles. The book has four main parts:

- In Part II, the chapters focus on theories and definitions of acceptance and related concepts, and review a number of different models of driver acceptance.
- The chapters in Part III look at the scientific and practical issues around measurement of driver acceptance with a description of some of the main tools, techniques and metrics available and used.
- Part IV presents case studies involving the measurement of driver acceptance of new technology, providing empirical data and findings on user acceptability and acceptance of a range of new technologies, and drawing also on experience from wider domains and perspectives.
- In Part V, the chapters turn to the issue of how driver acceptance of new technology can be optimised, both through design and by considering the wider context of use.

Finally, in the concluding chapter, we bring together and discuss the key themes that have emerged and identify future research, design and deployment needs in the area.

This book aims to provide a balanced treatment of driver acceptance of new technology, with contributions from experts in their field from around the world. All contributions have been peer-reviewed. Contributors represent a range of stakeholders including academics, vehicle manufacturers, road and transport safety authorities, equipment manufacturers and injury prevention researchers, providing multiple perspectives on the issue. While the main focus of the book is on driver acceptance, several chapters broaden the scope to consider also the optimisation of user/operator acceptance in other areas (e.g., consumer products, mining equipment and motorcycle technologies).

There is much that can be done to improve driver acceptance of new technologies – and in turn, to increase the safety, efficiency and comfort of driving. We hope that the information, insights and advice contained in this volume will help to guide and facilitate this process.

References

Alavi, M. and Joachimsthaler, E.A. 1992. Revisiting DSS Implementation Research: A Meta-Analysis of the Literature and Suggestions for Researchers. *MIS Quarterly*, 16(1): 95–116.

Davis, F., Bagozzi, R. and Warshaw, P. 1989. User Acceptance of Computer Technology: A Comparison of Two Theoretical Models. *Management Science*, 35(8): 982–1003.

Dillon, A. 2001. User Acceptance of Information Technology. In *Encyclopaedia of Human Factors and Ergonomics*. Edited by W. Karwowski. London: Taylor and Francis.

Dillon, A. and Morris, M.G. 1996. User Acceptance of Information Technology: Theories and Models, *Annual Review of Information Science and Technology*, 31: 3–32.

Labeye, E., Adrian, J., Hugot, M., Regan, M.A. and Brusque, C. 2013. Daily Use of an Electric Vehicle: Behavioural Changes and Potential for ITS Support. *IET Intelligent Transport Systems*, 17(2): 210–14.

Lee, J.D. and Seppelt, B.D. 2009. Human Factors in Automation Design. In *Springer Handbook of Automation*. Edited by S.Y. Nof. New York: Springer Publishing Company.

National Transport Commission (NTC). 2012. Cooperative ITS Regulatory Policy Issues: Discussion Paper. Melbourne, Australia: NTC.

Rogers, E.M. 1962. *Diffusion of Innovations*. Glencoe: Free Press.

———. 1995. *Diffusion of Innovations*. New York: Free Press.

Schade, J. and Schlag, B. 2003. Acceptability of Urban Transport Pricing Strategies, *Transportation Research Part F: Traffic Psychology and Behaviour*, 6(1): 45–61.

US Department of Transportation (DOT). 2008. *Intelligent Transport Systems Benefits, Costs, Deployment, and Lessons Learned*: 2008 Update. Report No. FHWA-JPO-08-032. Washington, DC: US DOT.

Van der Laan, J.D., Heino, A. and De Waard, D. 1997. A Simple Procedure for the Assessment of Acceptance of Advanced Transport Telemetics. *Transportation Research Part C*, 5(1): 1–10.

Venkatesh, V., Morris, M.G., Davis, G.B. and Davis, F.D. 2003. User Acceptance of Information Technology: Toward a Unified View. *MIS Quarterly*, 27(3): 425–78.

Young, K., Regan, M.A. and Mitsopoulos, E. 2004. Acceptability to Young Drivers of In-Vehicle Intelligent Transport Systems. *Road & Transport Research*, 13(2): 6–16.

PART II
Theories and Models
of Driver Acceptance

Chapter 2

The Definition of Acceptance and Acceptability

Emeli Adell
Trivector Traffic, Sweden

András Várhelyi
Lund University, Sweden

Lena Nilsson
Swedish National Road and Transport Research Institute (VTI), Sweden

Abstract

Despite the recognised importance of the concept of acceptance, how and why new technologies are actually accepted by drivers is not well understood. While many studies claim to have measured acceptance, few have explicitly defined what it is. This chapter points out the importance of defining acceptance and categorises definitions that have been used according to their 'essence'. Distinctions between different types of acceptance as well as between acceptance and acceptability are also described. A proposal for a common definition of acceptance is then presented and discussed.

Introduction

Acceptance has often been pointed out as a key factor for successful introduction and intended use of new technology in the vehicle context and elsewhere. The literature also contains some statements on the purpose of investigating acceptance. Najm et al. (2006: 5-1) claim that 'driver acceptance is the precondition that will permit new automotive technologies to achieve their forecasted benefit levels' and that there is a need to determine whether drivers will accept and use the new technologies as intended. Further, Najm et al. (2006: 5-1) state that 'driver acceptance measurement also provides a means to estimate drivers' interest in purchasing and using new technologies as a basis for estimating the safety benefit associated with its use'. Van der Laan, Heino and De Waard (1997) see acceptance as the link to usage, thereby materialising the potential safety effects, whereas Van Driel (2007) sees acceptance as a predictor of the willingness to buy a system.

As can be seen, there are different ways of viewing acceptance and acceptability. Common to all of them is that acceptance and acceptability are recognised to be important and are based on the individual's judgement of, for example, the driver assistance system.

Despite the recognised importance of the concept of acceptance, how and why new technologies, like driver assistance systems, are accepted by drivers is not well understood. While many studies claim to have measured acceptance of these systems, few have explicitly defined what it is. As Regan et al. (2002: 9) put it, 'While everyone seems to know what acceptability is, and all agree that acceptability is important, there is no consistency across studies as to what "acceptability" is and how to measure it'.

The definition of acceptance is one of the three elements of the acceptance concept (Figure 2.1). It is the fundamental foundation upon which both assessment structure and acceptance models rest. Without a definition it is not possible to examine the validity and reliability of any assessment methods and/or models. Although, there is no common and established definition of acceptance, various definitions can be found in the literature as well as descriptions of different types of acceptance.

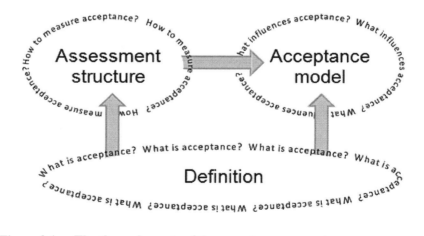

Figure 2.1   The three elements of the acceptance concept

Five Different Ways of Defining Acceptance

A recent literature review (Adell 2009) shows that acceptance definitions identified in the literature can be classified into five categories. The first category simply uses the word *accept* to define acceptance: for example, 'acceptance is the degree to which a law, measure or device is accepted' (Risser, Almqvist and Ericsson 1999: 36). The second category is concerned with the satisfaction of the needs and requirements of users (and other stakeholders) and may be interpreted as the *usefulness of the system*. For example, Nielsen (1993: 24) describes acceptance as 'basically the question of whether the system is good enough to satisfy all the needs and requirements of the users and other potential stakeholders'. The third category sees acceptance as *the sum of all attitudes*, implying that other, for example more emotionally formed, attitudes are added to the more 'rational' evaluation of the usefulness of the system (as in Category 2). For example Risser and Lehner (1998: 8) write 'Acceptance refers to what the objects or contents for which acceptance is measured are associated to; what do those objects or contents imply for the asked person'. The fourth category focuses on *the will to use the system*. For example, Chismar and Wiley-Patton (2003) state that acceptance is the intention to adopt an application. This can be based on either theoretical knowledge of the application or real experience. This definition of acceptance aims for a behavioural change and may be seen as being based on the earlier categories; in that the will to use a system is based on a driver's assessment of the usefulness of the system (as in Category 2) as well as on all other attitudes to the system and its effects (as in Category 3). This fourth category stresses the will to act as a consequence of a positive attitude towards the system in question. The fifth category of acceptance emphasises *the actual use of the system*; for example, Dillon and Morris (1996: 5) define acceptance as 'the demonstrable willingness within a user group to employ information technology for the task it is designed to support'. This is presumably influenced by the will to use it (as in Category 4).

Viewing the acceptance categories in this way, they may to some extent be seen as a progression from assessing the usefulness of a system towards the actual use of that system, with the latter categories including the earlier ones (see Figure 2.2). This progression perspective, however, cannot include Category 1, which uses the word *accept* to define acceptance, but does not provide any information about what is implied by acceptance or accept.

Definition categories

1	2	3	4	5
Using the word "accept"	Satisfying needs and requirements	Sum of attitudes	Willingness to use	Actual use

Figure 2.2   The five categories of acceptance, based on definitions found in a literature review by Adell (2009)

Different Types of Acceptance

There are also different *types* of acceptance described in the literature. Authors have made distinctions between attitudinal and behavioural acceptance (Kollmann 2000, Franken 2007), between social and practical acceptance (Nielsen 1993) and between different levels of problem awareness of the individual (Katteler 2005).

Attitudinal acceptance is, according to Franken (2007), based on emotion and experience and provides a basis for accepting a system. *Behavioural acceptance* is displayed in the form of observable behaviour (Franken 2007). Relating to the definition categories described above, attitudinal acceptance is comparable to the 'sum of attitudes' (Category 3) and behavioural acceptance to the 'actual use' (Category 5). Similar to this, Kollmann (2000) describes acceptance as consisting of three levels: the general connection of inner assessment and expectation (*the attitude level*, Category 3), the acquisition or purchase of the product (*the action level*) and its voluntary use with a frequency greater than that of other traffic participants (*the utilisation level*, Category 5).

Slightly later, Katteler (2005) defined different types of acceptance of driver assistance systems depending on the driver's awareness of the problem the assistance system is aimed at tackling. The *well-founded, firm acceptance* indicates, apart from a positive attitude towards the system that the individual is aware of the problem the system is designed to tackle. *Opportunistic acceptance* indicates low problem awareness and is, according to Katteler (2005), likely to be less stable and more sensitive to changes in the system design, the terms of using the system, the opinions of others about the system and so on.

There is also discussion about 'conditional' and 'contextual' acceptance in the literature. *Conditional acceptance* indicates that acceptance is dependent on certain preconditions (Saad and Dionisio 2007); for example, 'I will use the system if I am free to turn it off when I want to' or 'I will use the system if everybody else does'. Similarly, *contextual acceptance* indicates that acceptance depends on the situational context (Saad 2004); for example, 'I will use the system on roads with speed cameras' or 'I won't use the system in rush hour'.

Goldenbeld (2003) makes a distinction between acceptance and support, where *acceptance* is defined as the willingness to be subjected to something (e.g., pay taxes) while *support* is the liking for doing so. Vlassenroot et al. (2006: 1) further claim that (public) support is a precondition for acceptance since it 'defines the degree of acceptance or intentions people have to adapt or not to adapt to the desired behaviour'. According to Vlassenroot and De Mol (2007), the sum of the individuals' acceptance indicates whether there is public support. By the reasoning of Vlassenroot et al. (2006), the willingness to do something has to be preceded by liking to do it.

Acceptability vis-à-vis Acceptance

Some scientists stress the importance of making a distinction between *acceptability* and *acceptance*. Schade and Schlag (2003: 47) define *acceptability* as the 'prospective judgement of measures to be introduced in the future'. *Acceptability* is measured when the subject has no experience of the system, and is therefore an attitude construct. *Acceptance*, on the other hand, consists of attitudes and behavioural reactions after the introduction of a technology. According to Jamson (2010) *acceptability* is how much a system is liked, while *acceptance* is how much it would be used. In addition she defines *uptake* of a system as how likely it is that someone would buy it.

Pianelli, Saad and Abric (2007) differentiate between two types of acceptability: priori and posteriori acceptability. *Priori acceptability* is acceptability without experience of the system while *posteriori acceptability* is the acceptability after having tried the system. The posteriori acceptability includes experience of the system, but does not necessarily include behavioural reactions, making it different from the acceptance definition described by Schade and Schlag (2003).

Another related concept, *social acceptability*, is described by an example provided by Nielsen (1993: 24):

> Consider a system to investigate whether people applying for unemployment benefits are currently gainfully employed and thus have submitted fraudulent applications. The system might do this by asking applicants a number of questions and searching their answers for inconsistencies or profiles that are often indicative of cheaters. Some people may consider such a fraud-preventing system highly socially desirable, but others may find it offensive to subject applicants to this kind of quizzing and socially undesirable to delay benefits for people fitting certain profiles.

Comparably, a driver might find it socially unacceptable for a government to impose a driver assistance system on a user, even if it results in a reduction in road trauma. *Practical acceptability* includes dimensions like cost, compatibility, reliability, usefulness and so on (Nielsen 1993).

In summary, there are, today, many different ways of viewing acceptance and acceptability. Common to all of them is that acceptance and acceptability are based on the individual's judgement of, for example, a driver assistance system. The fact only matters when believed by the individual. Further, one has to remember that any assistance system only gives the expected effects if the system is used by the driver. From a traffic safety perspective this means that it is important that the system is used – emphasising the acceptance definitions in categories 4 and 5 (willingness to use and usage), behavioural acceptance and utilisation level. In this perspective it is less important if the drivers support the use of the system.

For other perspectives – for example, estimations of willingness to pay – other aspects might be more relevant. It is, however, according to our view, questionable whether willingness to pay is comparable to acceptance.

Defining Acceptance

The situation with many different ways of viewing and defining acceptance is problematic. If acceptance is not defined, it is not possible to validate different measuring tools and build models to understand how acceptance is formed. The definition of acceptance is the foundation upon which both assessment structure and acceptance models rest. Therefore it is vital to come to an agreement on what acceptance is.

Below follows a discussion regarding important aspects that should be included in the definition of acceptance in order to make it a usable and effective construct when designing and evaluating driver assistance systems.

The Driver's Understanding of the System

Working on driver acceptance of new technologies makes it essential to understand the importance of a driver-centred view, as it is the driver who makes the decision to use or not use a system, at least for non-mandatory systems. Since acceptance is individual, it can only be based on an individual's personal attitudes, expectations, experiences and subjective evaluation of the system, and the effects of using it (Schade and Baum 2007). The effects of a certain system (e.g., reduction in accident risk) can only influence acceptance if they are known, understood, believed and valued by the driver. A misunderstanding of the system will influence acceptance as much as a correct conception. This also implies that trust in the system, on an individual level, is important for acceptance. Trust is an important determinant in the operator's choice to use automation (see e.g., Muir 1994), and – in the same way as the concept of acceptance – needs more research for establishing a coherent framework for modelling and measuring. Here, we conclude that trust, like acceptance, is based on the individual's perception of the system. This is, of course, among other things, influenced by how the driver experiences the system.

The Gain for the Driver

It is also important to remember that to achieve acceptance and use of new technologies/systems, the personal importance to the users has to be valued more highly than the degree of innovation (Ausserer and Risser 2005). However, policies and political goals are often confused with the driver's personal goals. Societal goals and individual goals do not necessarily coincide. For example, the policy goal behind ISA (Intelligent Speed Adaptation; a system which warns the drivers when they exceed the speed limit, and may even prevent them from doing so) could be to increase traffic safety or to increase speed limit compliance. These goals might not be relevant to some drivers, for example, due to their feeling that safety measures are redundant because of their own personal driving skills (Brookhuis and Brown 1992) or because speeding is not seen as a 'real crime' (Corbett 2001). Nevertheless, they might find that the system helps them to avoid speeding tickets or they want to use the system simply because they have a general interest in innovative systems.

The multidimensional definition of acceptance proposed by Katteler (2005) and the approach chosen by Vlassenroot et al. (2006) are interesting but focus on the societal gains of drivers using the system. Katteler (2005) studied ISA and defines speeding as the 'problem awareness' dimension. However, this might not be the 'problem' for which drivers wish to use the ISA system. Similar systems are marketed as problem-solvers for speeding tickets. Similarly, in the approach of Vlassenroot et al. (2006), the drivers have to agree that high speeds are a problem and that ISA is a good way of reducing them.

It is the Use of the System that Gives Results

The actual use of technology is vital in striving to improve traffic safety by deploying driver assistance systems. It is the use of the system that will materialise its potential and hopefully produce benefits for the driver and the society. Neither attitudinal acceptance (Franken 2007) nor support (Goldenbeld 2003) requires any impact on the actual use of a system. Hence, the main aim and focus when working with acceptance should, in our view, be on behavioural acceptance (Franken 2007), the utilisation level as described by Kollmann (2000) and acceptance definition Category 5 – 'actual use' (described above), which all emphasise the use of the system. From this perspective, the second and third categories of acceptance definitions ('usefulness' and 'all attitudes'), attitudinal acceptance (Franken 2007) and the attitude level described by Kollmann (2000) influence the will to use and the actual usage of the system, and should not to be seen as acceptance per se.

Usable in the Whole Development/Implementation Process

It is desirable to accurately predict user acceptance as early as possible in the design process to be able to evaluate different alternatives and identify obstacles to overcome. Furthermore, for technologies that are available to drivers, the use of them has to be seen as part of a process, including the will to use as a step towards usage.

Sometimes the term *acceptability* is used when a driver has no practical experience of the system, for example, in the development phase. Differentiating between acceptance and acceptability is however not always easy due to the problematic situation of defining experience. Can testing the system in a driving simulator be considered as experience? Or must the system be used in real life? And for how long? Can we consider using a mock-up as experience of a system?

And so on. Due to this, it is advised not to rely on the term *acceptability* to describe the experience the driver has with the system; but instead, to explicitly describe the situation.

To summarise, it is important to define acceptance as something that is based on the driver's understanding of the system and focus his/her gain of using the system rather than societal/political gains. The acceptance should be connected to the use of the system since it is the use that creates the expected effects. Preferably the acceptance definition should also be appropriate in different stages of the idea – development – implementation process of a driver assistance system.

Proposal for a Common Definition of Driver Acceptance

Building on these aspects, Adell (2009: 31) proposes a definition of driver acceptance focusing on a system's potential to realise its intended benefits (e.g., traffic safety potential); that is, the drivers' incorporation of the technology into their driving:

> Acceptance is the degree to which an individual incorporates the system in his/
> her driving, or, if the system is not available, intends to use it.

This definition has the advantages of focusing on the *individual* perspective, both regarding the subjective evaluation of the system and the gains of using the system. The system must both address an aspect that is important to the driver (e.g., not being fined for speeding or not falling asleep while driving) and its solution to the problem/way to attain the gain must be known, understood and believed by the driver.

Further, this definition stresses the importance of *using* the system. In this way acceptance is tightly connected to demonstration of the judgement of the system. A general liking of a system is thereby not acceptance of the system; to accept the system, the individual has to incorporate the system in his/her driving. This provides the potential of realising the expected effects of the system.

This definition also provides an opening for assessing a system in development by addressing the *intention to use the system* if the system was available. This can be seen as potential acceptance, but should not be confused with acceptability. In this definition the term *acceptability* is avoided due both to the diversity of meaning put into the term (see the section 'Acceptability vis-à-vis Acceptance' in this chapter) and due to the problematic situation of defining experience.

The proposed definition also states clearly that acceptance is of a *continuous nature* and not limited to acceptance/nonacceptance (nominal scale). Of course, the degree of acceptance could also be zero when the driver does not use the system and/or has no intention to do so.

By this definition it follows that the driver does not necessarily have to like to use the system to demonstrate acceptance. To show high acceptance, it is enough that the driver decides to use the system, which, under the given circumstances, he/she sees as the best option. In this way, tolerating the use of the system can be seen as part of acceptance: for example, by the driver who would not normally choose to use an ISA system but decides to do so due to a number of speeding fines, or by the driver who agrees to use the system since it is required by law. The driver accepts the system as the best option in a given situation.

Conclusion

In this chapter we have illustrated how acceptance is defined and used in different studies and we have put forward a new definition.

If acceptance has not been defined, then we cannot be sure that the tool we use to measure it will give valid results. Without knowing how acceptance is defined, it is impossible to understand how driver experiences influence it. The wide variety of acceptance definitions and corresponding measurement methods, and thereby the diversity of results, present a breeding ground for misinterpretations and misuse of the results. What is more, this variety makes comparisons between technologies, systems and settings almost impossible to achieve.

Acknowledgements

This chapter draws on the dissertation 'Driver experience and acceptance of driver assistance systems – a case of speed adaptation' (Adell 2009).

References

Adell, E. 2009. *Driver experience and acceptance of driver assistance systems – a case of speed adaptation.* Bulletin 251. PhD thesis, Lund University, Sweden.

Ausserer, K. and Risser, R. 2005. *Intelligent Transport systems and services – chances and risks.* Proceedings of the 18th ICTCT workshop, Helsinki, Finland.

Brookhuis, K. and Brown, I.D. 1992. Ergonomics and road safety. *Impact of science on society*, 165: 35–40.

Chismar, W.G. and Wiley-Patton, S. 2003. *Does the Extended Technology Acceptance Model Apply to Physicians.* Proceedings of the 36th Hawaii International Conference on System Sciences (HICSS'03), Hawaii, USA.

Corbett, C. 2001. The social construction of speeding as not 'real' crime. *Crime Prevention and Community Safety: An International Journal*, 2(4): 33–46.

Dillon, A. and Morris, M. 1996. User acceptance of new information technology – theories and models, *in Annual Review of Information Science and Technology*, Vol. 31: 3–32. Edited by M. Williams. Medford, NJ: Information Today.

Franken, V. 2007. *Use of navigation systems and consequences for travel behaviour.* ECTRI-FEHRL-FERSI Young Researcher Seminar, Brno, Czech Republic.

Goldenbeld, C. 2003. Publiek draagvlak voor verkeersveiligheid en veiligheidsmaatregelen. SWOV, Leidschendam. In *Defining the public support: what can determine acceptability of road safety measures by a general public?* Edited by S. Vlassenroot, J. de Mol, T. Brijs and G. Wets. 2006. Proceedings of the European Transport Conference 2006, Strasbourg, France.

Jamson, S. 2010. *Acceptability data – what should or could it predict?* Presented at the International seminar on Acceptance, Paris, France, November 2010.

Katteler, H. 2005. Driver acceptance of mandatory intelligent speed adaptation. *EJTIR*, 5(4): 317–36.

Kollmann, T. 2000. Die Messung der Akzeptanz bei Telekommunikations-systemen, *JBF – Journal für Betriebswirtschaft*, 50(2): 68–78. In *User requirements for a multimodal routing system to support transport behavior.* Edited by V. Franken and B. Lenz. 2007. Proceedings of the Sixth European Congress and Exhibition on Intelligent Transport Systems and Services, Aalborg, Denmark.

Muir, B.M. 1994. Trust in automation: Part I. Theoretical issues in the study of trust and human intervention in automated systems. *Ergonomics*, 37(11): 1905–22.

Najm, W.G., Stearns, M.D., Howarth, H., Koopmann, J. and Hitz, J. 2006. Evaluation of an automotive rear-end collision avoidance system (chapter 5). US Department of Transportation, National Highway Traffic Safety Administration. Report no. DOT HS 810 569. Research and Innovative Technology Administration, Volpe National Transportation Systems Centre, Cambridge, USA.

Nielsen, J. 1993. *Usability engineering.* San Diego, CA: Academic Press.

Pianelli, C., Saad, F. and Abric, J-C. 2007. *Social representations and acceptability of LAVIA (French ISA system).* Proceedings of the Fourteenth World Congress and Exhibition of Intelligent Transport Systems and Services, Beijing, China.

Regan, M.A., Mitsopoulos, E., Haworth, N. and Young, K. 2002. *Acceptability of in-vehicle intelligent transport systems to Victorian car drivers.* Report no. 02/02. Monash University Accident Research Centre, Melbourne, Australia.

Risser, R., Almqvist, S. and Ericsson, M. 1999. *Fördjupade analyser av acceptansfrågor kring dynamisk hastighetsanpassning.* Bulletin 174, Lund University, Sweden (In Swedish).

Risser, R. and Lehner, U. 1998. *Acceptability of speeds and speed limits to drivers and pedestrians/cyclists.* Deliverable D6, MASTER-project.

Saad, F. 2004. *Behavioural adaptations to new assistance systems – some critical issues.* Proceedings of the IEEE International Conference on Systems, Man, and Cybernetics, 288–93.

Saad, F. and Dionisio, C. 2007. *Pre-evaluation of the "mandatory active" LAVIA: Assessment of usability, utility and acceptance.* Proceedings of the 14th World

Congress and Exhibition of Intelligent Transport Systems and Services, Beijing, China.

Schade, J. and Baum, M. 2007. Reactance or acceptance? Reactions towards the introduction of road pricing. *Transportation Research Part A,* 41(1): 41–8.

Schade, J. and Schlag, B. 2003. Acceptability of urban transport pricing strategies. *Transportation Research Part F*, 6(1): 45–61.

Van der Laan, J.D., Heino, A. and De Waard, D. 1997. A simple procedure for the assessment of acceptance of advanced transport telemetics. *Transportation research Part C*, 5(1): 1–10.

Van Driel, C. 2007. *Driver support in congestion – an assessment of user needs and impacts on driver and traffic flow.* PhD thesis, Thesis Series, T2007/10, TRAIL Research School, the Netherlands.

Vlassenroot, S. and De Mol, J. 2007. *Measuring public support for ISA: development of a unified theory.* Proceedings of the 14th World Congress and Exhibition of Intelligent Transport Systems and Services, Beijing, China.

Vlassenroot, S., de Mol, J., Brijs, T. and Wets, G. 2006. *Defining the public support: what can determine acceptability of road safety measures by a general public?* Proceedings of the 6th European Congress and Exhibition of Intelligent Transport Systems and Services, Strasbourg, France.

Chapter 3

Modelling Acceptance of Driver Assistance Systems: Application of the Unified Theory of Acceptance and Use of Technology

Emeli Adell
Trivector Traffic, Sweden

András Várhelyi
Lund University, Sweden

Lena Nilsson
Swedish National Road and Transport Research Institute (VTI), Sweden

Abstract

This chapter provides a brief overview of acceptance models used within the area of information technology. One particular model, the Unified Theory of Acceptance and Use of Technology (UTAUT), is then discussed, and a study is reported in which the model was used to assess driver acceptance of a particular driver assistance system. The key findings of that study are reported, and suggestions are made for refining UTAUT to make it more suitable for assessing acceptance of driver assistance systems.

Introduction

To understand how acceptance of driver assistance systems is formed, what factors influence it and what stimulates acceptance, there is a need for an acceptance model. Driver assistance systems are technology-based systems to help the driver in the driving process. They integrate sensors, information processing, communication and control technologies to constantly monitor the vehicle surroundings as well as driving behaviour to detect critical situations. These systems continuously support the driver by informing, warning and/or intervening to avoid any dangerous situations. In this chapter, acceptance models derived in other domains within information technology are reviewed, and the Unified Theory of Acceptance and Use of Technology (UTAUT) (Venkatesh et al. 2003) is used to assess driver acceptance of a driver assistance system. Based on this assessment, suggestions are made for refining UTAUT to make it more suitable for assessing acceptance of driver assistance systems.

Frameworks for Assessing Acceptance of Driver Assistance Systems

There are only a few frameworks for understanding acceptance discussed in the literature on driver assistance systems. The National Highway Traffic Safety Administration's (NHTSA) strategic plan, 1997–2002, states that driver acceptance should be understood in terms of ease of use, ease of learning, adaptation and perception of the system in question (Najm et al. 2006). Measurement of these aspects of driver assistance systems should show whether the system satisfies the needs and requirements of drivers (corresponding to the second acceptance definition category discussed in the previous chapter). The NHTSA framework was revised in 2001, to include ease of use, ease of learning, perceived value, driving performance and advocacy of the system or willingness to endorse it (Stearns, Najm and Boyle 2002, Najm et al. 2006). Regan et al. (2002) state that acceptability, as it relates to driver assistance systems, is a function of usefulness, ease of use, effectiveness, affordability and social acceptability. For Regan et al. (2002), these constructs define acceptance.

When studying Intelligent Speed Adaptation (ISA), Molin and Brookhuis (2007) showed, by means of a Structural Equation Model (SEM), that acceptability of the system was related to the 'belief that speed causes accidents', whether the system can 'contribute to personal or societal goals' and 'if one prefers an ever limiting mandatory ISA'. Molin and Brookhuis (2007) did not define acceptance; nevertheless, in a questionnaire, using closed questions, they measured it by questions with the following content: 'intention to buy ISA if it is for free', 'wants to possess ISA' and 'support for policy to impose ISA on all cars'. These indicators do not clearly fit into any of the five acceptance definition categories described in the previous chapter. However, the first two indicators seem to be consistent with the acceptance definitions in Category 4 (willingness to use) and the third indicator might be connected to categories 2 or 3 (satisfying needs and requirements or sum of attitudes) (see the discussion regarding the connection between measurements and definitions in later chapters of this book).

Neither Najm et al. (2006) nor Regan et al. (2002) have shown if and how the attributes of acceptance they put forward influence the actual acceptance of a system by drivers, which limits the use of these frameworks for understanding how acceptance is formed and how to influence it. The SEM model regarding acceptance of ISA described by Molin and Brookhuis (2007) points to the importance of the perceived usefulness of the system, but the model is too specialised to describe what stimulates acceptance in a wider perspective. In conclusion, there is a need for a model to satisfactorily describe what influences acceptance vis-à-vis driver assistance systems.

Acceptance Models Within the Area of Information Technology

Following the rapid development of new technologies and software in computer science, interest in the acceptance and use of these technologies has increased significantly. A number of different models are used in the information technology area to understand, for example, the reasons for (not) using different computer programs, how to improve computer programs to increase usage of them and reasons for (not) Internet shopping. The information technology area includes today one of the most comprehensive research bodies on acceptance and use of new technology and the models used have provided assistance in understanding what factors either enable or hinder technology acceptance and use. These models are discussed later in the chapter.

Differences Between Information Technology and Driver Assistance Systems

Applications of information technology and driver assistance systems share many important features: the user interacts with a technology that is often too complex to fully understand; new applications are incorporated into an existing interaction between the user and the technology; and both information technology and driver assistance systems seek to facilitate an ongoing task.

Despite the similarities, there are important differences between the settings in which information technology applications and driver assistance systems are used, particularly at the operational level. One important difference between computer use and car driving is the *time* aspect. When using a computer, the user normally has the possibility of pausing and pondering, and even asking for *help* with a process or decision. Continuous decision-making or execution is usually not required. It is different when driving a car. The car driver normally has a short time span in which a decision (and action) has to be made and normally does not have the possibility of acquiring assistance with a long process or decision. Car driving also demands *continuous decision-making and execution* of tasks. When using a computer the user normally does not have to interact with other humans, while a car driver must interact with other road users, making the *social dimension* of the two settings very different. When a computer user makes a *mistake* it is often repairable; the consequence is usually irritating and sometimes time-consuming, but seldom dangerous. When a car driver makes a mistake it could end in severe physical damage or fatality both for the driver him/herself (user) and others. The working *environment* when using a desktop computer is imaginary, while the use of a car takes place in the real world.

These differences are important to recognise and address. Nevertheless, the work on acceptance of driver assistance systems should be able to make use of the knowledge from the area of information technology, albeit with some modifications and caveats.

Frequently Used Acceptance Models within the Area of Information Technology

In the area of information technology a number of different models have been used. Some of the models were developed in the area of information technology while other models incorporate well-known theories developed in a broader context. The models have been developed over quite a long time span. One of the more recently developed models, the UTAUT (Venkatesh et al. 2003), integrates eight of the most used models of individual acceptance in the area of information technology (in bold in the list below):

- The Pleasure, Arousal and Dominance paradigm (Mehrabian and Russell 1974)
- **Theory of Reasoned Action (Ajzen and Fishbein 1980)**
- Expectation Disconfirmation Theory (Oliver 1980)
- Social Exchange Theory (Kelley 1979, Emersson 1987)
- **Technology Acceptance Model (TAM) (Davis 1989)**
- **Theory of Planned Behaviour (TPB) (Ajzen 1991)**
- **The Model of PC Utilisation (Thompson, Higgins and Howell 1991)**
- Social Influence Model (Fulk, Schmitz and Steinfield 1990, Fulk 1993)
- **Motivational Model (Davis, Bagozzi and Warshaw 1992)**
- **A combined model of TAM and TPB (Taylor and Todd 1995)**
- **Social Cognitive Theory (Compeau and Higgins 1995)**
- **Innovation Diffusion Theory (Rogers 1995)**
- Task Technology Fit (Goodhue and Thompson 1995)
- System Implementation (Clegg 2000)
- Technology Readiness (Parasuraman 2000)
- IS Continuance (Bhattacherjee 2001)
- Three-Tier Use Model (Liaw et al. 2006)
- Motivation variable of LGO (Saadé 2007)
- Social Identity Theory (e.g., Yang, Park and Park 2007)

The Unified Theory of Acceptance and Use of Technology – UTAUT

The UTAUT is based on an extensive literature review and empirical comparison of the models included (for references, see Venkatesh et al. 2003). The key element in all these models is the behaviour; that is, the use of a new technology.

The UTAUT model was validated for use in understanding acceptance and use of computer software by computer users in the USA. It outperformed the eight individual models, accounting for 70 per cent of the variance (adjusted R2) in use. It was concluded that the UTAUT is a useful tool for assessing the likelihood of success for new technology introduction and provides knowledge of what stimulates acceptance, which can be used to proactively design interventions (including training, marketing, etc.) targeted at populations of users that may be less inclined to adopt and use new systems (Venkatesh et al. 2003).

In UTAUT, Venkatesh et al. (2003) postulate two direct determinants of use: 'behavioural intention' and 'facilitating conditions'. 'Behavioural intention' is in turn influenced by 'performance expectancy', 'effort expectancy' and 'social influence'. Gender, age, experience and voluntariness of use act as moderators.

The items used in assessing the constructs were selected from the eight investigated models. Through empirical evaluation, using a seven-point scale from 'strongly disagree' (1) to 'strongly agree' (7), the four most significant items for each construct were chosen as indicators for the specific constructs in the UTAUT model (see Table 3.1). Behavioural intention was assessed through three items and use was measured as the duration of use via system logs (Venkatesh et al. 2003).

It was found that 'performance expectancy' is a determinant of 'behavioural intention' in most situations. The strength of the relationship is, however, moderated by age and gender, being more significant for men and younger workers. The effect of 'effort expectancy' on behavioural intention is also moderated by gender and age, but contrary to 'performance expectancy', is more significant for women and older workers. The effect of 'effort expectancy' decreases with experience. The effect of 'social influence' on behavioural intention is conditioned by age, gender, experience and voluntariness such that the authors found it to be non-significant when the data were analysed without the inclusion of moderators. The effect of 'facilitating conditions' is only significant when examined in combination with the moderating effects of age and experience; that is, they only matter for older workers with more experience (Venkatesh et al. 2003).

The Use of the UTAUT Model in Other Areas

The UTAUT model has also been utilised in areas other than information technology, such as for adoption of mobile services among consumers (Carlsson et al. 2006) and in the health sector. Application examples from the health sector include examination of the viability of motes (tiny, wireless sensor devices) as health monitoring tools, health professionals' reluctance to accept and utilise information and communication technologies, physicians' acceptance of a pharmacokinetics-based clinical decision support system and physician adoption of electronic medical records technology (e.g., Lubrin et al. 2006, Chang et al. 2007, Hennington and Janz 2007, Schaper and Pervan 2007).

The studies largely support the appropriateness of the UTAUT model in these areas. However, social influence was not found to be as strong a predictor as suggested by the model when investigating information/communication technologies and decision support in the health sector (Chang et al. 2007, Schaper and Pervan 2007). Extensions/modifications of the model were recommended both in the adoption of mobile services and within the health sector (Carlsson et al. 2006, Lubrin et al. 2006).

Using the UTAUT Model in the Context of Driver Assistance Systems

A first proposal to use the UTAUT model for understanding acceptance of driver assistance systems was made by Adell (2007), and a pilot test of the model in the area of driver assistance systems was undertaken in 2008 (Adell 2009).

Data were collected during a field trial evaluating a prototype driver assistance system. The purpose of the pilot was to explore the potential of using UTAUT in the context of driver assistance systems. Thus, the original model was applied as far as the experimental design allowed it. Additional questions to the already designed field trial questionnaires allowed data collection for examination of the interrelationships of 'performance expectancy', 'effort expectancy' and 'social influence', with 'behavioural intention', including gender and age, as moderators. A summary of the field trial is given below and reported in full by Adell, Várhelyi and Dalla Fontana (2009).

The SASPENCE system is a driver assistance system that assists the driver to keep a safe speed (according to road and traffic conditions) and a safe distance to the vehicle ahead. The 'Safe Speed and Safe Distance' function informs/warns the driver when (a) the car is too close to the vehicle in front, (b) a collision is likely due to a positive relative speed, (c) the speed is too high considering the road layout and (d) the car is exceeding the speed limit. The driver receives information and feedback from the system by means of an external speedometer display located on the instrument panel, haptic feedback in the accelerator pedal, or in the seat belt, and an auditory message. For further information about the system see Adell et al. (2009).

Two test routes were used: one in Italy, and one in Spain. Both routes were approximately 50 km long and contained urban, rural and motorway driving. The test drivers drove the test route twice, once with the system on and once with the system off, thus serving as their own controls. The order of driving was altered to minimise bias due to learning effects. At each site, 20 randomly selected inhabitants, balanced according to age and gender, participated in the trial. Prior to using the SASPENCE system, the participants were given a brief explanation of the system. The questions regarding the UTAUT assessment were given to the drivers as part of the questionnaire after the second drive.

The items for assessing 'behavioural intention', 'performance expectancy', 'effort expectancy' and 'social influence' were adopted from Venkatesh et al. (2003). Some of the items had to be adapted to fit the context of driver assistance systems (see Table 3.1). Each item was rated using a seven-point scale, ranging from 'strongly disagree' (1) to 'strongly agree' (7) (identical to Venkatesh et al. 2003).

Table 3.1   The original UTAUT items and the modified items used in the pilot test to assess acceptance of a driver assistance system (Adell et al. 2009)

Original items (Venkatesh et al., 2003)		Modified items
Behavioural intention to use the system (BI):		
		Imagine that the system was on the market and you could get the system in your own car.
BI1	I intend to use the system in the next <n> months.	I *would* intend to use the system in the next 6 months.
BI2	I predict I would use the system in the next <n> months.	I *would* predict I would use the system in the next 6 months.
BI3	I plan to use the system in the next <n> months.	I *would* plan to use the system in the next 6 months.
Performance expectancy (PE):		
PE1	I would find the system useful in my job.	I would find the system useful in my *driving.*
PE2	Using the system enables me to accomplish tasks more quickly.	Using the system enables me to *react to the situation more quickly.*
PE3	Using the system increases my productivity.	Using the system increases my *driving performance.*
PE4	If I use the system, I will increase my chances of getting a raise.	If I use the system, I will *decrease my risk of being involved in an accident.*
Effort expectancy (EE):		
EE1	My interaction with the system would be clear and understandable.	My interaction with the system would be clear and understandable.
EE2	It would be easy for me to become skilful at using the system.	It would be easy for me to become skilful at using the system.
EE3	I would find the system easy to use.	I would find the system easy to use.
EE4	Learning to operate the system is easy for me.	Learning to operate the system is easy for me.
Social influence (SI):		
		Imagine that the system was on the market and you could get the system in you own car.
SI1	People who influence my behaviour would think that I should use the system.	People who influence my behaviour would think that I should use the system.
SI2	People who are important to me would think that I should use the system.	People who are important to me would think that I should use the system.
SI3	The senior management of this business has been helpful in the use of the system.	*Authorities would* be helpful in the use of the system.
SI4	In general, the organization has supported the use of the system.	In general, *authorities would* support the use of the system.

Factor analysis confirmed, on the whole, the similarity of the items within the four constructs ('behavioural intention to use the system' [BI], 'performance expectancy' [PE], 'effort expectancy' [EE] and 'social influence' [SI]). However, items PE3 and PE4 did not show high loadings on performance expectancy. PE3 showed more resemblance to social influence while item PE4 did not show any clear resemblance to any of the four constructs. These two items were excluded and the remaining items were represented by four summated scale variables (averages of item scores).

The relationships between the independent constructs (PE, EE, SI) and behavioural intention to use the SASPENCE system (BI) were examined by applying linear regression analysis. First, the unadjusted effects – that is, the crude effects (meaning that there was only one independent variable in the model) – and then the adjusted effects of variables (by simultaneously entering other independent variables into the model) were analysed.

The results highlighted the importance of 'performance expectancy' and 'social influence' for 'behavioural intention' but did not verify the significance of 'effort expectancy'. This may be a consequence of limitations in the pilot test. However, the context of computer use, for which the UTAUT model was developed, differs from the context of using driver assistance systems (driving). Car driving demands interactions with other road users and is therefore by its nature a task with a strong social dimension. The importance of 'social influence' as a predictor of 'behavioural intention' in the context of a driver assistance system could be a consequence of this. Further, the effort associated with the use of, for example, a computer program, and the use of a driver assistance system may be different. Employing a computer program normally demands actions by the user, while a driver assistance system normally runs without requiring input from the driver, informing/warning the driver only when there is a need to do so. The results of testing the UTAUT model on a driver assistance system are summarised in Figure 3.1.

The inclusion of the moderators 'gender' and 'age' did not affect the results, regardless of whether 'effort expectancy' was included in the analysis.

The relatively low explanatory power of the UTAUT model in the pilot test (20 per cent) led to further investigation of the significance of the individual items comprising the constructs used in the model. This suggested that some individual items used in assessing the constructs were better predictors of 'behavioural intention' than the constructs themselves. The items 'usefulness' (PE1), 'driving performance' (PE3), 'accident risk' (PE4) and 'important people' (SI2) had significant crude effects on 'behavioural intention'. The significant crude effects of the items 'driving performance' and 'accident risk' indicated that they, although not clearly belonging to the construct 'performance expectancy', touched on important aspects for explaining the 'behavioural intention' of using the system. Still, there seemed to be a considerable overlap between these items and the items 'usefulness' and 'important people'. When including 'driving performance' and 'accident risk' in the model and using backwards elimination, only the items

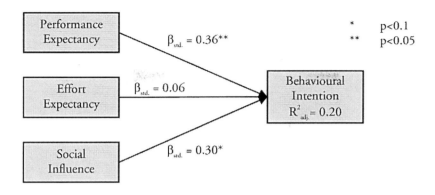

Figure 3.1   Regression coefficients and explanatory power for the UTAUT model when applied to acceptance of the driver assistance system SASPENCE

'usefulness' and 'important people' were left in the model. The model using only 'usefulness' and 'important people' explained 33 per cent of the behavioural intention, which is more than the originally tested model.

Proposed Modifications to the Model
Three modifications to the UTAUT model are suggested to improve its use in the context of assessing acceptance of driver assistance systems: (1) adding a new construct to include the emotional reactions of the driver, such as driving enjoyment, irritation, stress, feeling of being controlled and image of the system, (2) weighting the constructs by their perceived importance and (3) including reliability issues in the model (Adell 2009). Further, the need to identify items that can assess the 'essence' of the constructs in a driver assistance system perspective is highlighted.

Conclusion

Following the definition of acceptance proposed in the previous chapter, the intention to use and the actual usage of a driver assistance system is central. It is the use of a system that enables the potential benefits of the system to materialise. Therefore, increased knowledge of factors that influence the acceptance (and hence, use) of driver assistance systems and their interrelationship is crucial if driver assistance systems are to play a major role in achieving a better traffic system. Better understanding of what influences acceptance would give us valuable insights into the causes for (not) using the driver assistance system and how to improve the system to increase use of it.

The most extensive research bodies dealing with the acceptance and use of new technology today are found in the information technology area, where a number of different acceptance models are used and developed. The UTAUT model was deemed interesting to apply to the context of driver assistance systems since it summarises eight of the most significant models applied in the information technology area, and has already been used in other contexts outside information technology.

The UTAUT model, using age and gender as moderators, was examined in a pilot test with in-vehicle technologies. The results support to some extent the use of UTAUT as a framework to assess acceptance of a driver assistance system, but the explanatory power of the original model (from the IT domain) was only 20 per cent. Both 'performance expectancy' and 'social influence' indicated a relationship with the drivers' intention to use the system. 'Performance expectancy' was found to be the strongest predictor of the behavioural intention to use the system. Adell (2009) suggests some modifications to the UTAUT model to make it more suitable for driver assistance systems. However, the same model of acceptance can have different explanatory power in assessing the acceptance of specific technologies.

These modifications are promising and, if further validated, could help to better understand driver acceptance of new technologies.

Acknowledgements

This chapter draws on content from the dissertation 'Driver experience and acceptance of driver assistance systems – a case of speed adaptation' (Adell 2009).

References

Adell, E. 2007. *The concept of acceptance.* Proceedings of the 20th ICTCT workshop, Valencia, Spain.

Adell, E. 2009. *Driver experience and acceptance of driver assistance systems – a case of speed support.* Bulletin 251. PhD thesis, Lund University, Lund, Sweden.

Adell, E., Várhelyi, A. and Dalla Fontana, M. 2009. The effects of a driver assistance system for safe speed and safe distance – a real-life field study. *Transportation Research Part C, 19*: 145–55.

Ajzen, I. 1991. The theory of planned behavior. *Organizational Behavior and Human Decision Processes*, 50: 179–211.

Ajzen, I. and Fishbein, M. 1980. *Understanding Attitudes and Predicting Social Behavior.* Englewood Cliffs, NJ: Prentice-Hall.

Bhattacherjee, A. 2001. Understanding information systems continuance: an expectation-confirmation model. *MIS Quarterly* 25(3): 351–70.

Carlsson, C., Carlsson, J., Hyvönen, K., Puhakainen, J. and Walden, P. 2006. *Adoption of Mobile Devices/Services – Searching for Answers with the UTAUT.* Proceedings of the 39th Hawaii International Conference on System Sciences, Hawaii, USA.

Chang, I-C., Hwang, H-G., Hung, W-F. and Li, Y-C. 2007. Physicians' acceptance of pharmacokinetics-based clinical decision assistance systems. *Expert Systems with Applications*, 33: 296–303.

Clegg, C.W. 2000. Sociotechnical principles for system design. *Appl. Ergon.* 31: 463–77.

Compeau, D.R. and Higgins, C.A. 1995. Computer self-efficacy: development of a measure and initial test. *MIS Quarterly*, 19(2): 189–211.

Davis, F.D. 1989. Perceived usefulness, perceived ease of use, and user acceptance of information technology. *MIS Quarterly*, 13: 319–40.

Davis, F.D., Bagozzi, R.P. and Warshaw, P.R. 1992. Extrinsic and intrinsic motivation to use computers in the workplace. *Journal of Applied Social Psychology*, 22(14): 1111–32.

Emersson, R.M. 1987. Toward a theory of value in social exchange, In *Social exchange theory.* Edited by K.S. Cook. Newbury Park: Sage: 11–46.

Fulk, J. 1993. Social construction of communication technology. *Academy of Management Journal*, 36(5): 921–50.

Fulk, J., Schmitz, J.A. and Steinfield, C.W. 1990. A social influence model of technology use. In *Organizations and communication technology.* Edited by J. Fulk and C. Steinfield. Newbury Park, CA: Sage: 117–40.

Goodhue, D.L. and Thompson, R.L. 1995. Task-technology fit and individual performance, *MIS Quarterly* 19(2): 213–36.

Hennington, A.H. and Janz, B.D. 2007. Information systems and healthcare XVI: Physician adoption of electronic medical records: applying the UTAUT model in a healthcare context. *Communications of the Association for Information Systems*, 19: 60–80.

Kelley, H.H. 1979. *Personal relationships: their structures and processes.* Hillsdale, NJ: Lawence Erlbaum.

Liaw, S.S., Chang, W.C., Hung, W.H. and Huang, H.M. 2006. Attitudes toward search engines as a learning assisted tool: approach of Liaw and Huang's research model. *Computers in Human Behavior*, 22(2): 177–90.

Lubrin, E., Lawrence, E., Zmijewska, A., Navarro, K.F. and Culjak, G. 2006. *Exploring the benefits of using motes to monitor health: an acceptance survey.* Proceedings of the International Conference on Networking, International Conference on Systems and International Conference on Mobile Communications and Learning Technologies, Washington DC, USA.

Mehrabian, A. and Russell, J.A. 1974. *An approach to environmental psychology.* Cambridge, MA: MIT Press.

Molin, E. and Brookhuis, K. 2007. Modelling acceptability of the intelligent speed adapter. *Transportation Research Part F*, 10(2): 99–108.

Najm, W.G., Stearns, M.D., Howarth, H., Koopmann, J. and Hitz, J. 2006. *Evaluation of an automotive rear-end collision avoidance system (chapter 5)*. US Department of Transportation, National Highway Traffic Safety Administration, DOT HS 810 569; Research and Innovative Technology Administration, Volpe National Transportation Systems Centre, Cambridge, MA, USA.

Oliver R.L. 1980. A cognitive model for the antecedents and consequences of satisfaction. *Journal of Marketing Research*, 17: 460–69.

Parasuraman, A. 2000.Technology readiness index (TRI): a multiple-item scale to measure readiness to embrace new technologies. *Journal of Service Research*, 2: 307–20.

Regan, M.A., Mitsopoulos, E., Haworth, N. and Young, K. 2002. *Acceptability of in-vehicle intelligent transport systems to Victorian car drivers*. Report no. 02/02 Monash University Accident Research Centre, Australia.

Rogers, E. 1995. *Diffusion of innovations*. New York: Free Press.

Saadé, G.R. 2007. Dimensions of perceived usefulness: toward enhanced assessment. *Decision Sciences Journal of Innovative Education*, 5(2): 289–310.

Schaper, L.K. and Pervan, G.P. 2007. ICT and OTs: a model of information and communication technology acceptance and utilisation by occupational therapists. *International Journal of Medical Informatics*, 76: 212–21.

Stearns, M., Najm, W. and Boyle, L. 2002. *A methodology to evaluate driver acceptance*. Presentation at the Transportation Research Board Annual Meeting, Washington DC.

Taylor, S. and Todd, P. 1995. Assessing IT usage: the role of prior experience. *MIS Quarterly*, 19(4): 561–70.

Thompson, R.L., Higgins, C.A. and Howell, J.M. 1991. Personal computing: towards a conceptual model of utilization. *MIS Quarterly*, 15(1): 124–43.

Venkatesh, V., Morris, M.G., Davis, G.B. and Davis, F.D. 2003. User acceptance of information technology: toward a unified view. *MIS Quarterly*, 27(3): 425–78.

Yang, S., Park, J. and Park, J. 2007. Consumers' channel choice for university-licensed products: exploring factors of consumer acceptance with social identification. *Journal of Retailing and Consumer Services*, 14(3): 165–74.

Chapter 4

Socio-Psychological Factors That Influence Acceptability of Intelligent Transport Systems: A Model

Sven Vlassenroot
Ghent University, Belgium
Flanders Institute for Mobility, Belgium

Karel Brookhuis
Delft University of Technology, the Netherlands
University of Groningen, the Netherlands

Abstract

A success factor in the future implementation of new in-vehicle technologies is in understanding how users will experience and respond to these devices. Although it is recognised that acceptance and acceptability are important, consistency in the definition of acceptability, and how it can be measured, is absent. In this chapter we focus on the socio-psychological factors that will influence acceptability of intelligent transport systems (ITS) by drivers who have not experienced use of the system. First, different theories and methods are described to define our concept of acceptability. This conceptual framework is described and tested in the case of Intelligent Speed Assistance (ISA) by the use of a large-scale survey. This results in a model that may be used for policymaking actions regarding the implementation of ISA.

Introduction

To increase the chances of their policies being successful, policymakers will try to obtain widespread public support. 'Acceptance', 'acceptability', 'social acceptance', 'public support' and so on are all terms frequently used to describe a similar phenomenon: how will (potential) users act and react if a certain measure or device is implemented? The interest in defining support can be seen in the light of a growing awareness that policymaking has to be considered as a two-way phenomenon between the authorities and the public, in which interaction, transaction and communication are the key elements (Nelissen and Bartels 1998).

There is no good definition of what the term 'public support' means; in most cases it has been related to acceptability, commitment, legitimacy and participation (Goldenbeld 2002). An important distinction that has been made is between political, public and social 'support'. To a certain extent, the terms 'acceptance' and 'support' are strongly related. Goldenbeld (2002), however, describes a nuance between support and acceptance. Acceptance may be available, but would not necessarily lead to the support of a measure. Generally, acceptance should be seen as a precondition for support.

To determine the support for a particular policy, or how the support is developing, measuring instruments are necessary. Via support measurement the expected effectiveness of the measures and opinions about the measure and possible alternatives can be made visible. In most research, the term 'support' is not used because of its vagueness. The terms 'acceptance' and 'acceptability' are mostly used in the context of defining, getting or creating support for a policy measure (Hedge and Teachout 2000, Molin and Brookhuis 2007).

Some authors (e.g., Molin and Brookhuis 2007) state that there seem to be as many questionnaires as methods to measure acceptance and acceptability. In addition to the problems in finding the right approach for measuring acceptance or acceptability, the terms 'acceptance' and 'acceptability' are defined in different ways by different researches. In the field of ITS, Ausserer and Risser (2005) define acceptance as a phenomenon that reflects the extent to which potential users are willing to use a certain system. Hence, acceptance is closely linked to usage, and the acceptance will then depend on how user needs are integrated into the development of the system. Nielsen (cited in Young et al. 2003) described acceptability as the question of whether the system is good enough to satisfy all the needs and requirements of the users and other potential stakeholders. More generally, in Rogers's diffusion of innovations (2003), acceptability research is defined as the investigation of the perceived attributes of an ideal innovation in order to guide the research and development (R&D), to create such an innovation. Schade and Schlag (2003) make a clear distinction between acceptance and acceptability. They describe acceptance as the respondents' attitudes, including their behavioural responses after the introduction of a measure, and acceptability as the prospective judgement of something that should be introduced in the future. In the last case, the respondents will not have experienced any of the measures or devices in practice, which makes acceptability an attitude construct. Acceptance is then more related to user acceptance of a device. Van der Laan, Heino and De Waard (1997) distinguished user acceptance and social acceptance. User acceptance is directed more towards evaluation of the ergonomics of the system, while social acceptance is a more indirect evaluation of the consequences of the system.

In this chapter we will focus on how the acceptability of ITS can be measured. Different theories, methods and studies are analysed to define acceptability (Vlassenroot et al. 2010) and to identify the factors that could influence the degree of acceptability. A conceptual framework is described and tested by the use of a large-scale survey. This leads to a model that can be used for policymaking actions.

The Conceptual Model

Models and Theories

The lack of a theory and definition regarding acceptance has resulted in a large number of different attempts to measure ITS acceptance, often with quite different results (Adell 2008).

As noted elsewhere in this book, one of the most frequently used frameworks to define acceptance is the Theory of Planned Behaviour (TPB). Based on the Theory of Reasoned Action (Fischbein and Ajzen 1975), the TPB assumes that behavioural intentions, and therefore behaviour, may be predicted by three components: attitudes towards the behaviour, which are individuals' evaluation of performing a particular behaviour; subjective norms, which describe the perception of other peoples' beliefs; and perceived behavioural control, which refers to peoples' perception of their own capability.

Another successful model is the Technology Acceptance Model (TAM) (Davis, Bagozzi and Warshaw 1989). TAM was designed to predict information technology acceptance and usage on the job. TAM assumes that perceived usefulness and perceived ease of use determine an individual's intention to use a system with the intention to use serving as a mediator of actual system use. TAM has been used – in the field of ITS – in the prediction of electronic toll collection (Chen, Fan and Farn 2007).

Van der Laan et al. (1996) published a simple method to define acceptance. Acceptance is measured by direct attitudes towards a system and provides a system evaluation in two dimensions. The technique consists of nine rating-scale items. These items are mapped on two scales, one denoting the usefulness of the system, and the other satisfaction.

Venkatesh et al. (2003) noted that there are several theories and models of user acceptance of information technology, which presents researchers with difficulties in choosing the proper model. Venkatesh et al. (2003) found different underlying basic concepts in acceptance models by means of a detailed description and analysis of different models such as TPB, the motivational model, TAM, innovation diffusion theory and combined models. Based on these theories, they constructed a unified model that they named the Unified Theory of Acceptance and Use of Technology (UTAUT). In the UTAUT, four constructs play a significant role as direct determinants of user acceptance: (1) performance expectancy – the degree to which an individual believes that using the system would help him or her to attain gains in job performance; (2) effort expectancy – the degree of convenience with the use of the system; (3) social influence – the importance of other people's beliefs when an individual uses the system; and (4) facilitating conditions – how an individual believes that an organisational and technical infrastructure exists to support use of the system. The supposed key moderators within this framework are gender, age, voluntariness of use and experience. Although in several models, 'attitude towards use', 'intrinsic motivations', or 'attitude towards behaviour' are the most significant determinants of intention, these are not mentioned in the UTAUT. Venkatesh et al. (2003) presumed that attitudes towards using the technology would not have a significant influence.

Stern (2000) developed the Value-Belief-Norm (VBN) theory to examine which factors are related to acceptability of energy policies. Stern and colleagues proposed the VBN theory of environmentalism to explain environmental behaviour, including the acceptability of public policies. They proposed that environmental behaviour results from personal norms; that is, a feeling of moral obligation to act pro-environmentally. These personal norms are activated by beliefs that environmental conditions threaten the individual values (awareness of consequences) and beliefs that the individual can adopt to reduce this threat (ascription of responsibility). VBN theory proposes that these beliefs are dependent on general beliefs about human-environment relations and on relatively stable value orientations (see also Steg, Dreijerink and Abrahamse 2005). VBN theory was successful in explaining various environmental behaviours, among which were consumer behaviour, environmental citizenship, willingness to sacrifice and willingness to reduce car use.

Schlag and Teubel (1997) defined the following essential issues determining acceptability of traffic measures: problem perception, important aims, mobility-related social norms, knowledge about options, perceived effectiveness and efficiency of the proposed measures, equity (personal outcome expectation), attribution of responsibility and socio-economic factors.

We want to describe the most common and relevant socio-psychological factors that influence acceptance and acceptability of ITS. The theories and methods described above have some limitations, especially when the research is focused on people who have not experienced the device. An in-depth analysis was conducted on different user acceptance models, acceptability theories and studies that were used in the ITS field. This analysis resulted in 14 factors or indicators that could possibly influence acceptability the most. These 14 factors could be categorised into three main groups:

- Indicators related to the characteristics of the device (e.g., usefulness, effectiveness);
- Indicators related to the context wherein the device is used (e.g., social norms, problem perception). These indicators can influence the specific factors and acceptability; and
- The third group are more general issues like personal information (age, gender, education) and driving information (mileage, experience, accident involvement). These background factors will influence the contextual and device-specific indicators.

A New Theoretical Model

A distinction is made above between general indicators (related to the context awareness of the system) and system-specific indicators (directly related to the characteristics of the device). The definition of every indicator is described below.

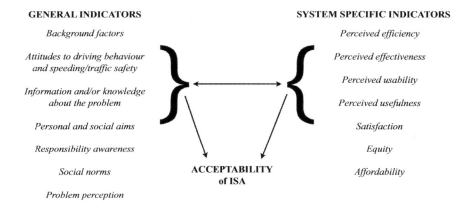

GENERAL INDICATORS

Background factors

Attitudes to driving behaviour and speeding/traffic safety

Information and/or knowledge about the problem

Personal and social aims

Responsibility awareness

Social norms

Problem perception

SYSTEM SPECIFIC INDICATORS

Perceived efficiency

Perceived effectiveness

Perceived usability

Perceived usefulness

Satisfaction

Equity

Affordability

ACCEPTABILITY of ISA

Figure 4.1   Theoretical model

General Indicators
Gender, age, level of education and (income) employment are the *individual indicators* and are considered to have an important influence on how people think about a device (Parker and Stradling 2001). On the indicator *attitudes to driving behaviour,* travel behaviour and driving style are brought into relation with the functionality of the device (Stradling et al. 2003). Schade and Schlag (2003) describe *personal and social aims* as the conflict between social or personal aims. They assume that a higher valuation of common social aims will be positively related to acceptability. Perceived *social norms* and perceived *social pressure* can be measured by quantifying the (assumed) opinions of others (peers) multiplied by the importance of others' opinions for the individual (Azjen 2002). *Problem perception* has been defined as the extent to which a certain social problem (e.g., speeding, drinking and driving, tailgating, etc.) is perceived as a problem. There is common agreement that high problem awareness will lead to increased willingness to accept solutions for the perceived problems (Goldenbeld 2002, Steg, Vlek and Rooijers 1995). *Responsibility awareness* explains how much an individual recognises responsibility for the perceived problem: is it the government (others/extrinsic) or is it the individual itself (own/intrinsic) (Schade and Baum 2007)?

The level of acceptability for the device can depend on how well informed (information *and knowledge about the problem)* the respondents are about the problem and about the (new) device that is introduced to solve the problem (Schade and Baum 2007, Steg et al. 1995).

Device Specific Indicators

The *perceived efficiency* indicates the possible benefits users expect of a concrete measure (or device) as compared to other measures. *Effectiveness* refers to the system's functioning according to its design specifications, or in the manner it was intended to function (Young et al. 2003). *Perceived usability* can be defined as the ability to use the system successfully and with minimal effort (SpeedAlert 2005). *Perceived usefulness and satisfaction* are indicators from the above-mentioned acceptance scale of Van der Laan et al. (1997). *Equity* refers to the distribution of costs and benefits among affected parties. However, from a psychological point of view, perceived justice, integrity, privacy and so on are considered basic requirements for acceptability (Schade and Baum 2007). In many ITS trials, acceptance was also defined by *willingness to pay* and *affordability* of the system (Biding and Lind 2002, Broeckx et al. 2006, Hjalmdahl and Várhelyi 2004). Giving incentives, like lower road taxes and lower insurance fees, can stimulate acceptability (Lahrmann et al. 2007, Schuitema and Steg 2008).

In our conceptual study on acceptability, based on a literature review and factor analyses of a small amount of test data (Vlassenroot et al. 2008), it was noted that these indicators had the highest potential to predict acceptability. However, not many acceptability studies were conducted, and thus not every indicator has been adequately studied in previous acceptability research (De Mol et al. 2001, Garvill, Marell and Westin 2003). In the next section, we describe how the theoretical acceptability concept has been tested for the case of Intelligent Speed Assistance (ISA).

The Conceptual Model

ISA is an intelligent in-vehicle transport system, which warns the driver about speeding, discourages the driver from speeding or prevents the driver from exceeding the speed limit (Regan et al. 2002). Most ISA-devices are categorised into three types depending on how intervening (or permissive) they are. An informative or advisory system will only give the driver feedback with a visual or audio signal. A supportive or warning ISA system will intervene when the speed limit is overruled; for example, the pressure on the accelerator pedal will increase when the driver attempts to drive faster than the speed limit. A mandatory or intervening system will totally prevent the driver from exceeding the speed limit: the driver cannot overrule these systems.

Based on theory and on an in-depth study of the factors that influence the acceptability of ISA (Vlassenroot et al. 2010), the following conceptual model was constructed (see Figure 4.2).

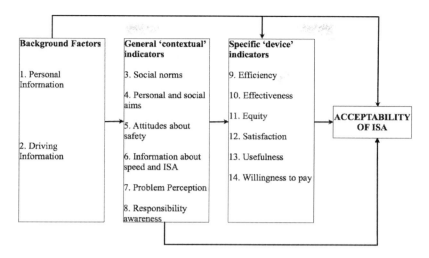

Figure 4.2   Hypothetical model of the indicators that define acceptability of ISA

In Figure 4.2, three main blocks are described that would influence acceptability. The background factors and the general contextual indicators would determine the specific device factors while the general indicators are only influenced by the background factors. It can be stated that these 14 factors may either directly or indirectly affect the acceptability of ISA and so they would influence each other as well. In the paragraphs that follow, the causal order between the factors is described. A causal order is assumed, going from the highest ranked item (1) to the lowest (15). All selected variables are assumed to directly or indirectly influence ISA acceptability.

The personal information factors (*age, gender, family situation* and *education*) are considered to be exogenous variables in the model; and, hence, not influenced by any other variables. The driving information factors (*type of car*, i.e., *company car, private vehicle*, etc., *accident involvement, mileage* and *driving experience*) are the next variables in causal rank order, only influenced by the socio-demographic variables. Both of these factors (personal and driving information) may affect any other remaining variable in the model: for example, gender and age are noted as relevant determinants in the performance of speeding behaviour; that is, speed is associated with young male drivers (Shinar et al. 2001).

The third factor, *social norms,* related to speed and speeding behaviour, may influence every contextual and device-specific factor in the model. The choice to speed or not can depend on the *personal and social aims* of people when driving. This fourth variable refers to the dilemma between social or personal aims and benefits of speeding: the hypothesis is that people who want to drive as fast as possible according to their own preferences could be less aware of the *speeding problem* and other issues that cause accidents. *Attitudes on safety* will be measured by defining which issues could cause accidents: most of the time, people will also compare the speeding problem in relation to other road safety issues (Corbett 2001), like intoxication, experience or infrastructure. Therefore the *attitudes concerning road safety* could influence the level of *problem awareness* but also the *information and knowledge* about the consequences of excessive speed. The factor *information and knowledge* refers to the assumption that people who are better informed are possibly more aware of the problem and the alternatives to tackle it. One of the main context variables is *problem perception:* in many trials (Vlassenroot et al. 2010) it was noted that the acceptability of ISA would depend on *awareness that speeding is a problem*. The last context indicator is *responsibility awareness.*

All the context factors could possibly influence the device-specific indicators. The determination of the order of the device-specific indicators was rather difficult, because most of these variables were not investigated in one and the same model. Some theories and approaches used in ISA trials formed the base to determine the causal order (Adell 2007, Agerholm et al. 2008, Biding and Lind 2002, Driscoll et al. 2007, Harms et al. 2007, Regan et al. 2006, Varhelyi et al. 2004, Vlassenroot et al. 2007).

Efficiency of ISA related to other speed management systems (e.g., speed cameras, police enforcement) can be considered as a 'gate' between the context factors and the device specific factors: it is assumed that people would compare the suggested new solution to counter the problem (speeding) with other existing measures. Efficiency, defined in this way, implies that respondents recognise that speeding is a problem, and also that they understand who is responsible for solving the problem; how the respondents get information about the solutions; and how they compare these instruments related to their own or social aims, and would possibly be influenced by their peers. If ISA is rated as efficient compared with the other measures, a next step can be to define how effective ISA is rated by the potential drivers. *Effectiveness* is first related to other ITS devices that support the driver: it is assumed that the effectiveness and acceptability of ISA will depend on how the effectiveness of other ITS is rated (Regan et al. 2006). Secondly, the effectiveness of ISA is defined by rating the effectiveness of ISA to maintain speed in different speed zones (Agerholm et al. 2008, Biding and Lind 2002). Thirdly, some secondary effects are identified, such as ISA can reduce speeding tickets, ISA is better for the environment and so on. A causal order is assumed between the effectiveness factors going from ITS *effectiveness* to ISA *effectiveness* to *secondary effects* of ISA. These three items could possibly influence the other device specific factors and the acceptability of ISA.

The third device-specific factor is *equity*. The respondents were asked to indicate when they would (*penetration level*) use a certain type of ISA and *for whom* a certain type of ISA would be the most beneficial. The assumption is made that the level of penetration would also influence for whom the system should be beneficial. Both of these factors are assumed to be influenced by the efficiency and the effectiveness parameters. The fourth and fifth device-specific factors are *satisfaction* – that is, when a certain ISA would be used – and *usefulness* of ISA to support the driver's behaviour. *Usefulness* and *satisfaction* are two parameters from the method of Van der Laan et al. (1997). *Satisfaction* will be influenced mainly by *effectiveness* and, combined with *effectiveness*, defines the level of *usefulness*. The final parameter in our model is the *willingness to pay for a certain system* that is influenced by all the parameters. *Willingness to pay* is a frequently used predictor to define the acceptability of ISA in trials (Biding and Lind 2002).

To determine *the acceptability* of ISA by drivers, the respondents had to indicate which system they preferred on a five-point scale going from no ISA, informative, warning, supportive to restrictive.

Measuring Acceptability

The Survey

A Web survey was constructed, tested and put online at the end of September 2009. The Web address of the survey was published by the Flemish and Dutch car-users organisations.

In total, 6,370 individuals responded to the Web survey in Belgium and 1,158 persons in the Netherlands. Of these 7,528 respondents, 5,599 responses of car drivers were considered useful for further analysis. Compared with the population of drivers' license owners in Belgium and the Netherlands, drivers younger than the age of 34 are under-represented and the age group 45–64 was over-represented. More male and older drivers participated. Although our sample was not representative of the whole population of drivers' license owners in the Netherlands and Belgium, both motorist organisations indicated that our results were relevant compared to their member databases, although exact data for every parameter (e.g., education level) was not available. Our research goal is mainly to define how different acceptability predictors are related to each other, rather than determining the acceptability of a certain population.

Data Analyses

It was assumed that every indicator is defined by the set of sub-questions. Factor analysis was applied to examine the structure and the dimensionality of the responses. Also Cronbach's alpha was calculated to determine the reliability of a summed scale. Not all the items of the different indicators loaded on a single factor like *problem perception, ISA effectiveness* and *equity*. Regarding the problem awareness, a main distinction could be made between low speed zones like home zones, 30 kph areas and urban areas, and higher speed zones, like outside urban areas and highways. In our model we allowed these items to correlate. The scale to define acceptability consists of five items between no intervening systems to high intervening systems (closed ISA). Therefore, it can be assumed that the acceptability of high intervening types of ISA has been measured in this model.

Cronbach's alphas of the intended scales were above 0.70, except for *responsibility awareness* and *efficiency*. It was concluded that the reliability of these scales was reasonable (e.g., Molin and Brookhuis 2007). The scale scores were constructed by summing the scores on the constituting indicator variables, equally weighting each variable.

Structural equation modelling (SEM) was used for the data analyses. SEM is a modelling approach enabling simultaneous estimation of a series of linked regression equations. SEM can handle a large number of endogenous and exogenous variables, as well as latent (unobserved) variables specified as linear combinations (weighted averages) of the observed variables (Golob 2003). SEM contains a family of advanced modelling approaches, among which is path modelling (e.g., Molin and Brookhuis 2007, Ullman 2007).

The Estimated Model

An initial model was estimated based on the causal order presented in Figure 4.2. Initially, all possible paths were drawn from factors earlier in the causal order towards all factors later in the causal order. The exogenous variables were correlated with the two variables related to speeding. The model was estimated with the program AMOS 7.

Only the variables with significant effects (p <0.05) were further used in the model. Paths that were not significant were left out of the model, which led to a total number of 139 distinct parameters in our final model to be estimated (df = 186). The probability level is 0.091 and Chi-square is 212.27. The goodness of fit (GFT) is 0.99. The probability level and the GFT indicate a good overall fit of the model. Another indication, especially when a large amount of data or cases are used to define the model fit, is the ratio between the Chi-square and the degrees of freedom: if the figure is lower than 2.0 a good fit of the model is indicated (Wijnen et al. 2002). In our estimated model the ratio is 1.141, which also indicates an acceptable fit.

More detailed information about the model and results are described in Vlassenroot et al. (2011).

Direct Effects

The effects are briefly discussed with respect to the plausibility of the significant relationships. The strength of the relationships between the variables is given between brackets. Only the most remarkable effects are described. Not every class related to age, having children, car use and mileage were kept in the model because they had no significant influence on the other variables. The different levels of education seemed to have no significant influence.

This model explains 56 per cent of the total variance in acceptability. Acceptability of ISA is directly influenced by effectiveness of ISA on speed

(0.37), equity on ISA equipment for different groups (0.31). Usefulness (0.13) and equity of ISA depending on level of penetration (0.11): drivers who find ISA effective and useful will accept ISA more. Also the lower the penetration level is before installing ISA, and if more intervening types of ISA are chosen for the different groups, the higher is the acceptability. Remarkable is the finding that the willingness to pay has a very small direct effect (0.02) on the acceptability. Drivers who like higher speed limits and speeding will accept ISA less (-0.09 in high speed zones; -0.08 in low speed zones). Respondents who would rather choose social aims (0.04) in driving and drivers who use the car as main transport mode to work (0.07) are more willing to accept ISA. Drivers between 25 and 45 years old (-0.04) prefer ISA less.

Total Effects
Finding *ISA effective in reducing speeding* (0.62) will have a very high influence on the *acceptability of ISA*. This was also expected. Also being convinced that other *ITS systems are effective* (0.21) will highly influence acceptability. In this way we can assume that drivers who are convinced that technology can help to support their driving behaviour will accept ISA better. Also being convinced that ISA is *beneficial for most of the groups of certain type of drivers (equity)* (0.32) will increase the acceptability. The lower the *ISA penetration level* has to be, the higher (0.12) the acceptability can become. Believing that ISA can be *useful* and *satisfying* will increase the level of acceptability. These two items were already proven as relatively good predictors of ITS and ISA acceptance (Varhelyi et al. 2004, Vlassenroot et al. 2007). *Satisfaction* (0.68) will highly influence *usefulness*. Drivers who *like to speed in high-speed zones* (-0.14) (as part of the factor problem awareness) will accept ISA less. Rating ISA as *efficient* (0.12) related to other speed reducing measures will also increase the acceptability. Drivers between the age of *25 and 45 years* (-0.14) will accept ISA less. A higher value for *social aims* (0.23) will increase the acceptability. While in many trials *willingness to pay* has been stated to be a good predictor for acceptance, this was not found in our model. Also the *secondary effects of ISA* will not have a high influence on the level of acceptability.

Drivers who are not influenced by the *equity level of penetration of ISA* are more *satisfied* (0.19) and will rate ISA more as *useful* (0.19). Also these drivers are highly *willing to pay* for ISA (0.51). *Effectiveness of ISA* (between 0.22 and 0.59) on speed and speeding seems to be a good predictor for all of the system-related indicators except for usefulness and satisfaction. *Efficiency* (between 0.07 and 0.17) will also influence all the other system-related indicators, except usefulness and *satisfaction*. The same can be found for the total effects on *effectiveness of ITS*.

A high valuation of the *responsibility* of the different actors to counter speed will influence the *efficiency* of ISA (0.17) related to other measures. Being aware of responsibility can also lead to finding ITS and ISA more *effective* (0.11 and 0.13) and a higher willingness to pay (0.13). People who *like to speed* will *accept* ISA less (-0.14 in high speed zones and -0.08 in low speed zones) and will find it less effective (-0.06 and -0.13). Being convinced that certain driving behaviour and contextual issues (items from the *attitudes on safety*) can cause accidents could lead to a higher *responsibility awareness* (0.22), higher valuation on the *effectiveness of ITS* (0.18) and finding ISA *beneficial for different groups of drivers* (0.12) (as part of the factor *equity*). Personal and social aims will have a high influence (higher than 0.10) on many of the variables (except on usefulness and knowledge about ISA). *Social norms* will mostly influence personal and *social aims* (0.19).

Going by *car to work* can also increase the *acceptability* of ISA (0.11). *Mileage* will decrease the use of a car as *transport to work* (-0.11 and -0.19): people who drive less than 25,000 km on a yearly basis will use the car less as a transport mode to work. *Having children* would mainly influence the *efficiency of ISA* (0.09) but would slightly lead to *speeding in low speed zones* (-0.05).

Two age groups were kept in the model as the only groups that have significant influence on the other variables. *Drivers between 25 and 45 years* will be less likely to *accept ISA* (-0.14). This is also the group with the most children younger than 12 years old (0.47). Social norms (0.13) and personal and social aims (0.17) will be highly affected by this age group of drivers. Drivers aged between 25 and 45 will have mainly a negative effect on most of the 'device-specific indicators' (between -0.08 and -0.15). *Younger drivers* (<25 years) are less convinced that certain behaviour or accidents could cause accidents (*attitudes on safety*: -0.12); these drivers will also evaluate *responsibility awareness* (-0.13) and *efficiency* (-0.13) lower. *Female* drivers will speed less in *high-speed zones* (-0.15) and are less *informed about ISA* (-0.15).

Conclusion

The lack of a theory and definitions of acceptability has resulted in a large number of different attempts to capture or measure ITS acceptability, often with quite different results. In our research we have tried to make a clear distinction between acceptance and acceptability. Some existing theories, like TPB and TAM, were developed in a certain timeframe and place and for specific audiences. Although these models are frequently used, rarely has anyone questioned whether they are good enough to be used to study the problem of speeding.

One of our main ambitions was to derive a model to define acceptability with respect to ITS. However, taking into account such a large variety of indicators resulted in a model that is still rather complex. This suggests that defining acceptability is rather complex. We are also aware that some of the selected topics to define the indicators could be improved. However, this research has resulted in improved insight into the opinions and attitudes that can influence acceptability of ISA.

Many different items influence acceptability, directly or indirectly. It is important to understand these in order to develop implementation strategies. Increasing the support of ISA has to be established at different levels.

Our model shows that the willingness of drivers to adopt ISA increases if they are convinced that ISA does what it is designed to do. The issue of 'equity' has rarely been investigated in other ITS or ISA studies. However, in other types of traffic and transport studies (e.g., tolling), equity has been investigated. Often when a new driver support technology is introduced – especially when it could restrict certain freedom in driving – a majority of the population is reluctant to 'buy or use' the system. In the Ghent ISA trial, it was noted that most of the drivers were convinced of the effectiveness and were highly in favour of the supportive system but they stated that they would only use ISA further when more or certain groups of drivers would (also/be forced to) use the system (equity on level of penetration). In the development of implementation strategies this is a very important issue. Therefore, policymakers should be aware that, if they want to introduce certain types of ITS, the penetration level should be sufficient from the start to convince others to adopt these devices (Brookhuis and De Waard 2007). Promoting ITS by implementing it in certain groups of vehicles, for instance, those driven by professionals (bus-, taxi-, van-, truck-drivers) or younger drivers, may be helpful to introduce certain systems (equity related to the equipment of certain groups). It is assumed that implementing ITS in the fleet of professional vehicles would be very effective in increasing acceptability rates. Our model showed that willingness to pay was not a major indicator influencing acceptability of ISA. However, others have reported that price-policy, subsidies and so on could be good instruments to increase the level of acceptability for a policy measure.

Our study aimed at an understanding of the indicators associated with acceptability that may support decision-makers in developing an appropriate implementation strategy. Through the construction of a feasibility framework, we are able now to provide decision-makers with methods and procedures that are easy to use and understand, based on well-accepted socio-psychological models.

References

Adell, E. 2008. *The concept of acceptance*. ICTCT-workshop, Riga, Latvia.

Agerholm, N., Waagepetersen, R., Tradisauskas, N., Harms, L. and Lahrmann, H. 2008. Preliminary results from the Danish intelligent speed adaptation project pay as you speed. *IET Intelligent Transport Systems*, 2(2): 143–53.

Ajzen, I. 2002. *Attitudes, personality and behaviour*. Buckingham, UK: Open University Press.

Ausserer, K. and Risser, R. 2005. *Intelligent transport systems and services – chances and risks*. ICTCT-workshop, Helsinki, Finland.

Biding, T. and Lind, G. 2002. *Intelligent Speed Adaptation (ISA), Results of Large-scale Trials in Borlange, Lidkoping, Lund and Umea during the periode 1999–2002.* Vägverket, Borlange, Sweden.

Broeckx, S., Vlassenroot, S., De Mol, J. and Int Panis, L. 2006. *The European PROSPER-project: Final results of the trial on Intelligent Speed Adaptation (ISA) in Belgium.* ITS World Conference, London.

Brookhuis, K.A. and De Waard, D. 2007. Intelligent Transport Systems for Vehicle Drivers. In *Threats from car traffic.* Edited by T. Gärling and E.M. Steg. Amsterdam: Elsevier.

Chen, C., Fan, Y. and Farn, C. 2007. Prediction electronic toll collection service adoption: an integration of the technology acceptance model and the theory of planned behaviour. *Transportation Research Part C*, 15(5): 300–311.

Corbett, C. 2001. Explanations for "understating' in self-reported speeding behaviour. *Transportation Research Part F: Traffic Psychology and Behaviour*, 4(2): 133–50.

Davis, F., Bagozzi, R. and Warshaw, P. 1989. User acceptance of computer technology: a comparison of two theoretical models. *Management Science*, 35: 982–1003.

De Mol, J., Broeckaert, M., Van Hoorebeeck, B., Toebat, W. and Pelckmans, J. 2001. *Naar een draagvlak voor een voertuigtechnische snelheidsbeheersing binnen een intrinsiek veilige verkeersomgeving.* Centre for Sustainable Development/Ghent University—BIVV, Ghent, Belgium (in Dutch).

Driscoll, R., Page, Y., Lassarre, S. and Ehrlich, J. 2007. LAVIA – An evaluation of the potential safety benefits of the French Intelligent Speed Adaptation project. *Annual Proceedings—Association for the Advancement of Automotive Medicine*, 51: 485–505.

Fischbein, M. and Ajzen, I. 1975. *Belief, attitude, intention, and behavior: an introduction to theory and research.* Reading, MA: Addison-Wesley.

Garvill, J., Marell, A. and Westin, K. 2003. Factors influencing drivers' decision to install an electronic speed checker in the car. *Transportation Research Part F: Traffic Psychology and Behaviour*, 6(1): 37–43.

Goldenbeld, C. 2002. *Publiek draagvlak voor verkeersveiligheid en veiligheidsmaatregelen. Overzicht van bevindingen en mogelijkheden voor onderzoek.* SWOV, Leidschendam, the Netherlands (in Dutch).

Golob, T. 2003. Structural equation modeling for travel behavior research. *Transportation Research Part B: Methodological*, 37(1): 1–25.

Harms, L., Klarborg, B., Lahrmann, H., Agerholm, N., Jensen, E. and Tradisauskas, N. 2007. *Effects of ISA on the driving speed of young volunteers: a controlled study of the impact information and incentives on speed.* Sixth European Congress on Intelligent Transport Systems and Services, Aalborg, Denmark.

Hedge, J.W. and Teachout, M.S. 2000. Exploring the concept of acceptability as a criterion for evaluating performance measures. *Group and Organisation Management*, 25(1): 22–44.

Hjalmdahl, M. and Várhelyi. A. 2004. Speed regulation by in-car active accelerator pedal: effects on driver behaviour. *Transportation Research Part F: Traffic Psychology and Behaviour*, 7(2): 77–94.

Lahrmann, H., Agerholm, N., Tradisauskas, N., Juhl, J. and Harms, L. 2007. *Spar paa farten: an intelligent speed adaptation project in Denmark based on pay as you drive principles*. ITS Europe Conference, Aalborg, Denmark.

Molin, E.J.E. and Brookhuis, K.A. 2007. Modelling acceptability of the intelligent speed adapter. *Transportation Research Part F: Traffic Psychology and Behaviour*, 10(2): 99–108.

Nelissen, W. and Bartels, G. 1998. De transactionele overheid. In *De transactionele overheid. Communicatie als instrument: zes thema's in de overheidsvoorlichting*. Edited by G. Bartels, W. Nelissen and H. Ruelle. Utrecht: Kluwer BedrijfsInformatie (in Dutch).

Parker, D. and Stradling, S. 2001. *Influencing driver attitudes and behaviour*. DETR, London.

Regan, M.A., Young, K.L., Healy, D., Tierney, P. and Connelly, K. 2002. *Evaluating in-vehicle Intelligent Transport Systems: a case study*. Road Safety Research, Policing and Education Conference, Adelaide, Australia.

Regan, M.A., Young, K.L., Triggs, T.J., Tomasevic, N., Mitsopoulos, E., Tierney, P., Healey, D., Tingvall, C. and Stephan, K. 2006. Impact on driving performance of intelligent speed adaptation, following distance warning and seatbelt reminder systems: key findings from the TAC SafeCar project. *IEE Proceedings: Intelligent Transport Systems*, 153(1): 51–62.

Rogers, E.M. 2003. *Diffusion of innovations*. New York: Free Press.

Schade, J. and Baum, M. 2007. Reactance or acceptance? Reactions towards the introduction of road pricing. *Transportation Research Part A: Policy and Practice*, 41(1): 41–8.

Schade, J. and Schlag, B. 2003. Acceptability of urban transport pricing strategies. *Transportation Research Part F: Traffic Psychology and Behaviour*, 6(1): 45–61.

Schlag, B. and Teubel, U. 1997. Public acceptability of transport pricing. *IATSS Research*, 21: 134–42.

Schuitema, G. and Steg, L. 2008. The role of revenue use in the acceptability of transport pricing policies. *Transportation Research Part F: Traffic Psychology and Behaviour*, 11(3): 221–31.

Shinar, D., Schechtman, E. and Compton, R. 2001. Self-reports of safe driving behaviours in relationship to sex, age, education and income in the US driving population. *Accident Analysis and Prevention*, 33(1): 111–16.

SpeedAlert. 2005. *Evolution of SpeedAlert concepts, deployment recommendations and requiremenuts for standardisation, version 2.0*. ERTICO, Brussels, Belgium.

Steg, L., Dreijerink, L. and Abrahamse, W. 2005. Factors influencing the acceptability of energy policies: a test of VBN theory. *Journal of Environmental Psychology*, 25(4): 415–25.

Steg, L., Vlek C. and Rooijers T. 1995. Private car mobility: problem awareness, willingness to change, and policy evaluation, a national interview study among Dutch car users. *Studies in Environmental Science*, 65: 1173–6.

Stern, P. 2000. Toward a coherent theory of environmentally significant behaviour. *Journal of Social Issues*, 56: 407–24.

Stradling, S., Campbell, M., Allan, I., Gorell, R., Hill, J. and Winter, M. 2003. *The speeding driver: who, how and why?* Scottish Executive Social Research, Edinburgh.

Ullman, J.B. 2007. Structural Equations Modelling. In *Using Multiuvariate Statistics*. Edited by B.G. Tabachnick and L.S. Fidell. Boston: Pearson.

Van der Laan, J. D., Heino, A. and De Waard, D. 1997. A simple procedure for the assessment of acceptance of advanced transport telematics. *Transportation Research Part C: Emerging Technologies*, 5(1): 1–10.

Várhelyi, A., Hjalmdahl, M., Hyden, C. and Draskoczy, M. 2004. Effects of an active accelerator pedal on driver behaviour and traffic safety after long-term use in urban areas. *Accident Analysis and Prevention*, 36: 729–37.

Venkatesh, V., Morris, M., Davis, G.B. and Davis, F.D. 2003. User acceptance of information technology: Toward a unified view. *MIS Quarterly*, 27(3): 425–78.

Vlassenroot, S., Broeckx, S., Mol, J.D., Panis, L. I., Brijs. T. and Wets, G. 2007. Driving with intelligent speed adaptation: final results of the Belgian ISA-trial. *Transportation Research Part A: Policy and Practice*, 41(3): 267–79.

Vlassenroot, S., Brookhuis, K., Marchau, V. and Witlox, F. 2010. Towards defining a unified concept for the acceptability of Intelligent Transport Systems (ITS): a conceptual analysis based on the case of Intelligent Speed Adaptation (ISA). *Transportation Research Part F: Traffic Psychology and Behaviour*, 13(3): 164–78.

Vlassenroot, S., De Mol, J., Dedene, N., Witlox, F. and Marchau, V. 2008. *Developments on speed limit databases in Flanders: a first prospective.* ITS World Conference, New York.

Vlassenroot, S., Molin, E., Kavadias, D., Marchau, V., Brookhuis, K. and Witlox, F. 2011. What drives the acceptability of intelligent speed assistance (ISA)? *European Journal on Transport and Infrastructure*, 11(2): 256–73.

Wijnen, K., Janssens, W., De Pelsmacker, P. and Van Kenhove, P. 2002. *Martonderzoek met SPSS: statistische verwerking en interpretatie.* Antwerp: Garant.

Young, K.L., Regan, M.A., Mitsopoulos, E. and Haworth, N. 2003. *Acceptability of in-vehicle intelligent transport systems to young novice drivers in New South Wales.* Monash University Accident Research Centre, Report no. 199. Monash University, Victoria, Australia.

Chapter 5

Modelling Driver Acceptance: From Feedback to Monitoring and Mentoring Systems

Mahtab Ghazizadeh and John D. Lee
Department of Industrial and Systems Engineering
University of Wisconsin-Madison, USA

Abstract

This chapter discusses dimensions of driver support systems, ranging from feedback from technology to monitoring and mentoring by coaches (e.g., parents or safety supervisors), and draws on previous work in information technology, organisational behaviour and driving safety domains to propose a framework for evaluating drivers' acceptance of such driver support systems. The proposed model, comprised of a trust-augmented version of the Technology Acceptance Model (TAM) by Davis, Bagozzi and Warshaw (1989) and the model of organisational trust by Mayer, Davis and Schoorman (1995), views acceptance of the technological and coaching components of the system as determinants of acceptance of the support system as a whole. Device characteristics, driver characteristics, driving behaviour, context and culture and coaching characteristics are introduced and discussed in the context of driver support system acceptance. These factors capture the multidimensional nature of acceptance and can guide the development of effective support systems. System effectiveness often depends on the degree to which these factors combine to create a mentoring, rather than a monitoring, system.

Driver Feedback, Monitoring and Mentoring Systems

According to Donmez, Boyle and Lee (2009: 519), feedback in the driving context is defined as 'the information provided to the driver regarding the state of the driver-vehicle system'. Driving performance in itself provides feedback (e.g., lane position); however, this feedback can be augmented with additional feedback by in-vehicle devices, in the form of warnings or alerts and/or driving performance reports. Although both types of feedback can enhance driving safety, their aims are different: warnings help drivers avoid looming imminent hazards, whereas cumulative feedback (e.g., weekly reports) goes beyond guiding a particular single action and aims to shape longer-term attitudes, habits and behaviours (Donmez et al. 2009). As such, the combination of the two can be most promising (Donmez, Boyle and Lee 2008, Lee 2009), especially because the effect of poor driving performance on safety may not always be obvious, as a dangerous driver can avoid crashing for many years. In this chapter, both warnings and cumulative feedback, provided by technology and people, are considered and discussed in the context of driver acceptance.

The benefits of driving feedback have been established (Donmez et al. 2008, McGehee et al. 2007b); however, for these benefits to be realised, the feedback system needs to be accepted by drivers. As such, an understanding of the factors that determine acceptance and ways to incorporate them in the design are imperative to success. An important consideration pertains to encouraging a mentoring (as opposed to a monitoring) relationship between the driver and the coach. Data collected by a particular in-vehicle device can be used to form a meaningful mentor-protégé relationship with the driver or it can solely facilitate a monitoring protocol that records violations. Likewise, different device implementations might enforce one role or the other. This chapter provides background on feedback, monitoring and mentoring systems and proposes a model to assess the acceptance of a driver support system.

Feedback from Technology

Feedback via in-vehicle devices can be presented using different modalities, on different timescales and with different levels of information. Auditory, visual and tactile alerts are examples of feedback modalities. Major classes of feedback in terms of timing are concurrent (milliseconds), delayed (seconds), retrospective (minutes, hours) and cumulative (days, weeks, months) (Donmez et al. 2009). Furthermore, feedback can be as simple as a blinking LED (Hickman and Hanowski 2011, McGehee et al. 2007b), or it can provide details like alerts or warnings that specify the nature and type of the risky behaviour, for example, a lane deviation. Feedback devices and protocols can be designed for the general driver population or target specific populations like teenagers (Farmer, Kirley and McCartt 2010, McGehee et al. 2007a, McGehee et al. 2007b, Carney et al. 2010), older drivers (Lavallière et al. 2012, Marottoli et al. 2007) or commercial vehicle drivers (Hickman and Hanowski 2011, Lehmer et al. 2007, Orban et al. 2006).

Donmez, Boyle and Lee (2003) developed a taxonomy of distraction mitigation strategies composed of 12 categories, along two dimensions: level of automation and whether the strategy is driving-related or not. Each of the driving-related and non-driving-related strategies was further divided into system-initiated and driver-initiated. Although this taxonomy was intended to classify distraction mitigations systems, the feedback strategies in the driver-related, system-initiated group (i.e., intervening, warning and informing) can be used to describe different types of feedback systems in general. At the lowest level of automation, *informing* involves providing drivers with necessary information that they would miss themselves. At the middle level, *warning* systems alert the driver to take action, but do not intervene. Finally, at the highest level, there is *intervening* that refers to taking control of the vehicle in hazardous situations, when the driver seems unable to manage the situation safely. An alternative perspective takes the way in which feedback is communicated to the driver into account, distinguishing between monitoring and mentoring roles. The mentor-monitor dimension adds a dimension to Donmez et al.'s (2003) taxonomy that may be particularly important for understanding driver acceptance.

Monitoring

Monitoring operator performance via video and other electronic recordings is not a new approach, but one that has found application recently in driving safety research, as well as the insurance industry for both driver training and for premium adjustment. Electronic performance monitoring (EPM) is defined as a system in which electronic technology is used to collect, store, analyse and report the actions or performance of people while working on the job (Nebeker and Tatum 1993). In the United States, 66 per cent of employers monitor employee Internet connections, 43 per cent monitor email, 45 per cent track time spent on the telephone and numbers called, 48 per cent use video monitoring to counter theft, violence and sabotage and 7 per cent use video surveillance to track employee's on-the-job performance (American Management Association [AMA] and the ePolicy Institute 2007). This type of monitoring mainly aims at enhancing productivity by evaluating performance and controlling operator behaviour.

Although known to benefit performance (at least for simple and familiar tasks) (Stanton and Barnes-Farrell 1996, Aiello and Kolb 1995), EPM systems have shown to increase worker stress, both directly and through their effect on job design (Carayon 1993, 1994, Smith et al. 1992). Workers' perceptions of fairness of work standards, fairness of the measurement processes and fairness in applying measurements to worker evaluation are among the main determinants of worker stress, and consequently workers' satisfaction with and acceptance of the EPM system (Westin 1992). Management style, organisational structure and work environment determine the physical and mental workload, which in turn influence the level of worker stress (Carayon 1993, 1994). Additionally, a technology that respects workers' privacy is better trusted (Muir 1987, Jones and Mitchell 1995) and, therefore, accepted by the workers (Stanton 2000, Tabak and Smith 2005, Alder, Noel and Ambrose 2006).

The goal of in-vehicle EPM systems is often related to improving safety and mitigating risky behaviours. A wide variety of parameters can be monitored and performance data can be collected on vehicle speed, location, acceleration, braking patterns, fuel consumption and so on. Some monitoring systems collect video data from inside and outside of the driver's cab, providing data on behaviours such as failure to use seatbelts, inattention and distraction and fatigue (Horrey et al. 2011).

In-vehicle monitoring systems also differ in that some log data continuously, while others only record data surrounding safety-critical events. An important factor that differentiates in-vehicle monitoring systems from each other is the channel for communicating feedback to the driver; the feedback can flow directly to the driver or it can be delivered through a third party. This factor and others, such as the quality of the feedback, can influence the effectiveness and acceptance of the system. Even the mere presence of the device in the vehicle may have positive or negative effects on drivers' behaviour (Hickman and Hanowski 2011, Horrey et al. 2011).

Recently, the insurance industry in the United States has started voluntary installation of monitoring devices in customers' vehicles in order to tie insurance premiums to each driver's level of safe driving and also to enhance driver safety. The Progressive Corporation, for example, offers a system called Snapshot® that tracks parameters of travel, for example, speed, time, mileage and distance, as well as the frequency of abrupt braking events, and transmits these data back to the company wirelessly (Progressive Corp. 2012). The incentive for safe driving can be up to a 30 per cent reduction in premiums while there is no penalty for unsafe driving. A similar program is the Teen Safe Driver Program by American Family Insurance, which provides free DriveCam video feedback to families with insured teenage drivers, with the goal of helping teenagers reduce their risky driving habits (American Family Insurance 2007). It should be noted that different monitoring systems may pursue varying goals; while Snapshot® is more geared towards identifying safe and unsafe drivers, the goal of using DriveCam by the Teen Safe Driver Program is mainly to enhance teen driver safety and facilitate mentorship by parents. The DriveCam system is used by several commercial and government fleets, in areas such as construction, distribution and energy (DriveCam). Although the technological component of these systems might be similar, the relationship that is established between the driver and the system governs its acceptance and effectiveness. Systems that are viewed as monitoring are likely to be much less effective and accepted than those that are viewed as supporting mentoring.

Mentoring

While the central focus of monitoring is on *evaluating* performance, the emphasis of mentoring is on *enhancing* performance by establishing a nurturing, insightful, supportive and protective relationship with the protégé (Buchanan, Gordon and Schuck 2008, Anderson and Shannon 1988). In the context of driving, the same video recording methods that can be used to monitor performance can also mentor drivers to achieve higher degrees of knowledge and proficiency in driving. The idea of using video to provide feedback has been pursued in education. For example, in performance courses of the communication discipline, reviewing one's recorded performance is found beneficial when accompanied by instructors' or peers' constructive feedback (Quigley and Nyquist 1992). A very similar approach can be undertaken in driving: recordings of safety-relevant events can be reviewed by the driver and a safety supervisor or parent with the goal of helping the driver identify his/her areas of vulnerability and work on them to achieve a higher level of safe driving.

A system that collects data from an operator's performance and shares it with a third party can be perceived as playing a monitoring role, a mentoring role or a role between these two poles. The monitoring role is generally characterised by control and reactive advice, whereas mentoring is characterised by care and proactive advice (Barry 2000). The goals of implementing an EPM system guide decisions about the technical features of the system (e.g., type of data collected) and the supervisory approaches undertaken to incorporate EPM into the work system. In the driving domain, an EPM system can be used as a tool for collecting quantitative performance data to facilitate the enforcement of safety standards or, alternatively, as a means of providing feedback that aims to enhance drivers' safety and well-being. Quantitative measures of performance that are compared against strict work standards are often suggestive of a monitoring role. However, the type of feedback provided is also important: does the feedback merely include numerical comparisons against standards (e.g., maximum number of abrupt braking events per week), requiring those who performed poorly to exercise more effort (negative feedback; 'monitoring'), or does it additionally provide support by facilitating safer practices (constructive feedback; 'mentoring')? Moreover, feedback that is tied to each driver's unique characteristics can motivate a mentoring relationship, just like feedback that compares the driver's performance to others can strengthen the perception of being monitored.

It is also important to consider the interface between the EPM device and the driver: Does the driver know when the system is recording? Is the driver able to control when and what is recorded? These relate to the concept of privacy that requires that employees be notified when they are monitored (Stanton and Barnes-Farrell 1996). Such transparency can be achieved by efforts on several levels: from training that takes place before/during the driver-system interaction, to real-time indicators that signal monitoring (e.g., a signal light). A transparent system is better perceived and trusted as a co-operator (Muir 1987, Jones and Mitchell 1995) and can support a mentoring relationship, whereas a system that is poorly understood will enforce the perception of monitoring.

In commercial driving applications, the way the EPM system is incorporated into job design is crucially important in the perceptions of its mentoring or monitoring role. The elements of job design can be grouped into three categories: job demands, job control and social support (Carayon 1993). Some factors relevant to these categories have been discussed earlier in this chapter; for example, performance standards (job demands) and type of feedback and supervision (social support). Job-demand factors influence the perceptions of fairness – whether the standards and measurement processes are reasonable – and the climate of employee trust in management (Westin 1992). If the management is perceived as enforcing unfair and unrealistic standards by observing drivers' performance in unexpected ways, then a monitoring role is more likely to be realised. Conversely, if the EPM data collection is well understood and the measurements are used in a way perceived as fair by the drivers, they are more likely to see themselves in a protégé-mentor relationship with the safety supervisors. The job control element can also shape the perceptions of the EPM role: degree of autonomy is a function of the type of input received – whether certain practices are mandated or workers are given latitude in how to use the feedback.

The influence of the EPM system's technical characteristics on drivers' attitudes can go beyond perceptions of transparency, privacy and fairness to determine the degree of trust in the system. In cases where the system is designed to provide performance feedback (e.g., performance-related warnings and performance synopses), false information delivered by the system can lead to system's credibility loss and workers' mistrust (Breznitz 1984), degrading the EPM system to the level of an obtrusive and incompetent monitoring device.

Culture and context influence the EPM system implementation in several ways: they determine management's attitude towards system specifications and implementation procedures, as well as drivers' reaction to the system. In a supportive atmosphere, trust between workers and supervisors leads to positive perceptions of the EPM (Alder 2001) and the belief that the system is designed to enhance their well-being (mentoring role). These positive perceptions are further reinforced by a system's fairness, respect for privacy and information accuracy.

Figure 5.1 summarises the mentoring role versus monitoring role discussion above and integrates the factors that encourage one role or the other for an EPM system. These factors range from the aim of the system (control of the driver versus care for the driver) to respect for privacy and system transparency. Perceptions of mentoring or monitoring can move along a spectrum, with perceptions anywhere between the two extremes possible.

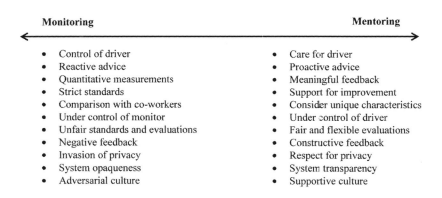

Monitoring   **Mentoring**

← ——————————————————————————————→

- Control of driver
- Reactive advice
- Quantitative measurements
- Strict standards
- Comparison with co-workers
- Under control of monitor
- Unfair standards and evaluations
- Negative feedback
- Invasion of privacy
- System opaqueness
- Adversarial culture

- Care for driver
- Proactive advice
- Meaningful feedback
- Support for improvement
- Consider unique characteristics
- Under control of driver
- Fair and flexible evaluations
- Constructive feedback
- Respect for privacy
- System transparency
- Supportive culture

Figure 5.1   Factors suggesting a mentoring versus a monitoring role

Assessing the Acceptance of Feedback, Monitoring and Mentoring Systems

In this section, a theoretical model for assessing drivers' acceptance of a driver support system is proposed that brings together drivers' acceptance of in-vehicle feedback and monitoring devices and their acceptance of monitoring or mentoring exercised by a coach (e.g., parent, transportation company manager). The Technology Acceptance Model (TAM) framework (described below) will be used for assessing attitudes towards the feedback technology. The TAM is augmented by trust to better account for drivers' perceptions of the in-vehicle system (Lee and See 2004) and is further extended by constructs from Mayer et al.'s (1995) model of organisational trust to capture perceptions of the coaching protocols. Variables that can influence acceptance are added to the model as external variables. The modelling framework is developed based on the different approaches to providing feedback as described earlier in this chapter. This section describes the TAM framework and some of its related applications, lists the categories of external variables relevant to driver support system acceptance and proposes a model for assessing driver support system acceptance.

The Technology Acceptance Model

Several frameworks and methodologies exist that describe people's acceptance of technology (see also earlier chapters in this volume). Within the driving domain, a simple method that assesses system usefulness and satisfaction has been particularly dominant (Van der Laan, Heino and De Waard 1997). In other domains, the Technology Acceptance Model (TAM) has successfully predicted technology use, and as such has been broadly used since its introduction more than two decades ago. The TAM (Davis et al. 1989), built upon the Theory of Reasoned Action (TRA) of Fishbein and Ajzen (1975), posits that perceived usefulness and perceived ease of use are the main determinants of attitude towards a technology, which in turn predicts behavioural intention to use and ultimately, actual system use. Since attitude is found to only partially mediate the effect of perceived usefulness on intention to use, a parsimonious TAM is suggested that excludes attitude, as shown in Figure 5.2 (Davis and Venkatesh 1996, Venkatesh and Davis 2000). The TAM constructs have been found to be highly reliable, valid and robust to measurement instrument design (Davis and Venkatesh 1996).

The TAM has recently been applied in studies assessing the acceptance of driving assistance systems. Xu et al. (2010) used the TAM to assess acceptance of advanced traveller information systems, incorporating four domain-specific constructs (i.e., information attributes, trust in travel information, socio-demographics and cognition of alternate routes). Chen and Chen (2011) used the TAM for evaluating acceptance of GPS devices, adding perceived enjoyment and personal innovativeness constructs to the model. A few other studies have also used the TAM constructs in their analysis of driving assistance systems (Adell 2010, Meschtscherjakov et al. 2009), finding that perceived system disturbance

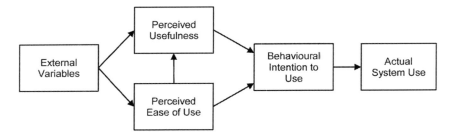

Figure 5.2   Technology Acceptance Model. Adapted with permission from Davis et al. © 1989, the Institute for Operations Research and the Management Sciences, 7240 Parkway Drive, Suite 300, Hanover, Maryland 21076

and perceived risk, as well as social factors, strongly influence the behavioural intention to use a system. These studies demonstrate the aptness of the TAM for driving assistance systems assessment and provide background and structure for future driving technology evaluations. Longitudinal studies that trace perceptions over time and as a function of system use would be most helpful in identifying elements of the technology and the human-technology interaction that shape dynamics of acceptance and eventually, long-term adoption decisions (Ghazizadeh, Lee and Boyle 2012a, Kim and Malhotra 2005, Bajaj and Nidumolu 1998).

Factors Influencing Driver Support System Acceptance

Several factors can influence a driver's perception of a feedback, monitoring and mentoring system. In this subsection, these factors will be grouped into five major categories that define important constructs influencing acceptance. This list of constructs spans from the device and the driver (driver characteristics and driving behaviour), to the context and culture in which the driving task takes place and finally, to the characteristics of the coaching system, describing a driver's work system. Figure 5.3 provides a schematic representation of factors surrounding a driver, with feedback from the driver support system shown as dashed arrows. The variables within each category will enter the driver support system acceptance model as external variables, influencing acceptance indirectly through their impact on perceptions of the system (Davis et al. 1989).

Device Characteristics
The properties of the feedback device can largely influence perceptions of the system and eventually, decisions to accept or reject feedback. Drivers' perceptions of the system can be formed along several dimensions, ranging from evaluations of system effectiveness to the degree of annoyance induced by the system. Various factors influence these perceptions: feedback content, style,

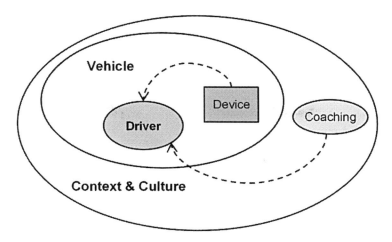

Figure 5.3   **A schematic representation of the driving system with dashed arrows representing feedback from the driver support system**

timing, frequency, precision, style (positive or negative) and method of delivery (Arroyo, Sullivan and Selker 2006, McLaughlin, Rogers and Fisk 2006a and 2006b, Huang et al. 2005, Huang et al. 2008). Purely informative systems are likely to be more accepted than systems that force changes (Van der Laan et al. 1997). A driver support system may involve a device that merely provides feedback (e.g., real-time or delayed) or may also collect data that is shared with a safety coach (e.g., parents, safety supervisors at a trucking company) for follow up with the driver. Based on the specific goal of a system, other factors can be added to the list. For example, in the case of auditory alerts, parameters like format, sound type, pulse duration and inter-pulse interval can affect annoyance (Marshall, Lee and Austria 2007).

When systems have a monitoring component, invasion of privacy can become an issue, to the point that hinders the willingness to install the device. For example, parents have shown concern about installing monitoring systems in their teens' vehicles because they perceived monitoring to be an invasion of privacy (McCartt, Hellinga and Haire 2007). Invasion of privacy has long been a concern with EPM systems and merits careful attention (Zweig and Webster 2002, Alder 2001). The type of data collected (private behaviours and driving outcomes) and whether another person views the data (as opposed to data being shared only with the driver) can influence perceptions of privacy invasion. While important for most people, privacy can become a major concern for specific groups of drivers: older drivers may not want to share their driving data out of fear of losing their right to drive. Commercial drivers may also find it disturbing to be monitored and with no way to avoid it, might even resort to sabotaging the system (Hickman and Hanowski 2011).

Driver Characteristics

The characteristics of drivers, such as age, gender and driving purpose, can influence his/her adoption decisions. Age and gender are important factors in that they can influence the relative importance of determinants of acceptance. Venkatesh and colleagues (Morris and Venkatesh 2000, Venkatesh, Morris and Ackerman 2000) conducted a series of studies to evaluate the role of age and gender in the relative importance of attitude, subjective norm (i.e., perceived social pressure to perform a behaviour) and perceived behavioural control (predictors of system use based on the Theory of Planned Behavior, Ajzen 1991). Their findings showed that younger workers' technology-use decisions are more strongly influenced by their attitudes towards technology, whereas in older workers, subjective norm and behavioural control are the main determinants of technology adoption (Morris and Venkatesh 2000). A similar pattern was observed in the comparison between men and women: men's decisions are more strongly influenced by their attitude, whereas women's decisions are primarily driven by subjective norm and perceived behavioural control (Venkatesh et al. 2000). In the case of driver support systems, these findings suggest that, when encountered with the same system, different driver groups based their assessments on various criteria and that those criteria might be weighed differently from driver to driver. Furthermore, assessments of the system along each criterion can be different based on a driver's characteristics. In the assessment of distraction mitigation systems, older drivers accepted the system more than middle-aged drivers – a pattern that was attributed to older drivers' diminished driving performance and their low self-confidence, which in turn led to higher trust in support systems (Donmez, Boyle and Lee 2006). Another study showed that, while younger drivers were somewhat dissatisfied with a driving tutoring system, older drivers held a positive attitude towards it (De Waard, Van der Hulst and Brookhuis 1999).

The purpose of driving – that is, driving for personal reasons or as part of one's job – is another factor to consider. Those who only drive for personal purposes and those who are employed to drive (e.g., commercial truck drivers) may have different perceptions of a particular technology. In commercial driving situations, the decision as to whether or not to use a driver support system is typically made by management. This is different from when drivers voluntarily decide to use a driver support system to lower their insurance premium or increase their safety – a distinction between mandatory and voluntary use. With voluntary adoption, the driver has freedom to decide to use the system, whereas with mandatory adoption the user is forced to use the system (Rawstorne, Jayasuriya and Caputi 1998). When evaluating acceptance of a mandatory system, the effect of perceived ease of use on intention to use was found to be larger than the effect of perceived usefulness on intention to use (Brown et al. 2002), contrary to the finding in many voluntary use cases that perceived usefulness is the primary determinant of intention to use (e.g., Venkatesh and Davis 2000, Gefen, Karahanna and Straub 2003, Karahanna, Agarwal and Angst 2006). One explanation for this pattern is that, because people know that they have to use the system, their focus shifts from usefulness to how easy or difficult the system is to work with (Rawstorne 2005). These observations, both from within and outside the driving safety community, underscore the importance of considering driver characteristics in predicting acceptance of driver support systems.

Driving Behaviour

Drivers differ in their driving skills, their degree of commitment to safety and their compliance with traffic laws. These differences can translate to their acceptance of support systems and the effectiveness of these systems in encouraging behavioural changes. Novice (usually younger) drivers are still in the process of developing driving skills, whereas experienced drivers have already developed such skills. For younger drivers, feedback and training aims at shaping safe driving behaviours and habits; however, for experienced drivers, the goal of support systems is to modify their risky habits. While both efforts can lead to promising results, helping younger drivers can be more readily accepted and effective. This asymmetry was noted by Hickman and Hanowski (2011) in explaining the differences between their study on commercial vehicle drivers and the McGehee et al.'s (2007b) study: the drivers in the McGehee et al. study were novices, whereas the drivers in the Hickman and Hanowski study were experienced professional drivers.

The degree to which a driver support system is compatible with drivers' perceptions of appropriate driving behaviour can be a determining factor in the driver's acceptance of the system. According to the Theory of Diffusion of Innovations, one of the major factors that determine adoption of an innovation is compatibility; that is, the degree of consistency between a technology and users' values, past experience and needs (Rogers 1995). As such, any target population of drivers might evaluate a particular system differently than the others. Furthermore, different types of drivers within the same population (e.g., safe and risky drivers) might have varying perceptions of the same system. An example of such a difference was observed in a recent preliminary analysis of commercial truck drivers' perceptions of an on-board monitoring system: the majority of drivers with moving violations in their history had moderately positive perceptions towards the feedback system, while many with clean records had negative perceptions (Peng et al. 2012). This association might be due to the identification of the need for behavioural improvement by those with a history of violations.

Context and Culture

Drivers are not isolated operators: driving happens in a social context and is influenced by culture. Drivers experience the driving environment through some context: their car is their immediate environment, which is in turn (especially with the recent proliferation of infotainment devices) connected to other people and places. Driver support systems, especially those with a monitoring or mentoring component, can change the role of context in one's driving by making the influence of others, such as parents or safety supervisors, more tangible. Depending on the characteristics of the context, as well as social and organisational culture and management style, the reaction of the driver to such change can be different.

One example of the role of context and culture on the attitudes of drivers towards support systems was demonstrated by studies of truck drivers in the United States and China. The comparison between results demonstrated that Chinese truck drivers had a more positive attitude towards feedback received from technology compared to US truck drivers who strongly preferred feedback from a human to feedback from technology (Huang et al. 2008, Huang et al. 2005). This example, while limited in scope, clearly demonstrates the inadequacy of assessments made without considering context and culture.

Coaching Characteristics
System characteristics, both those related to the device and those related to the human coach, can influence perceptions of mentoring and monitoring. For example, a device that is transparent and indicates when video recordings are being made or when a report is being transmitted to a supervisor/parent is likely to be perceived more positively and better trusted (Muir 1987, Lee and See 2004, Jones and Mitchell 1995). A coach's approach to providing feedback and the style in which feedback is delivered (negative or positive) is also critically important in defining a monitoring or mentoring role for the coach and the acceptance of the support system.

A Model for Assessing Acceptance of Driver Support Systems

Figure 5.4 integrates the broad range of factors affecting acceptance of driver support systems using the TAM framework. This model can guide measurement of the various factors that influence drivers' acceptance of an existing system or can be used as a predictive model that can guide design a new driver support system. Variables along the five categories described above (i.e., device characteristics, driver characteristics, driving behaviour, context and culture, and coaching characteristics) are considered as external variables. As the TAM is mainly concerned with the technological component of the system, the model is augmented by constructs from the model of organisational trust by Mayer et al. (1995). Both TAM and the model of organisational trust have been extensively cited and validated in the literature (e.g., see Pavlou 2003, Mayer and Davis 1999, Szajna 1996, Hu et al. 1999).

The model consists of two parts: Technology Acceptance and Coaching Acceptance. The Technology Acceptance part is based on the TAM, with the trust in device construct added as a predictor of technology acceptance. Trust is a major determinant of reliance on and acceptance of automation, standing between people's beliefs towards automation and acceptance of it (Lee and See 2004, Lee and Moray 1992 Lee and Moray 1994, Parasuraman, Sheridan and Wickens 2008, Gefen et al. 2003, Pavlou 2003, Carter and Bélanger 2005). Because previous studies suggest that trust does not fully mediate the effect of beliefs on behavioural intentions (Lee

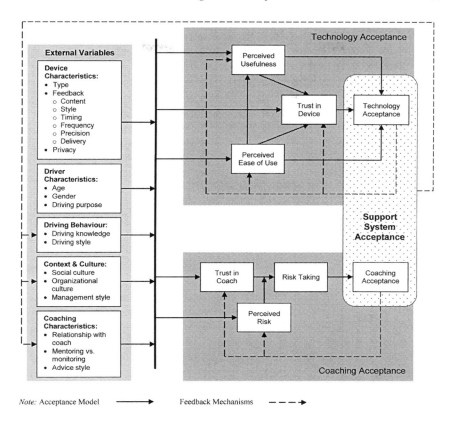

Figure 5.4   Driver support system acceptance model

and See 2004), the direct effects of perceived usefulness and perceived ease of use on acceptance are retained (Ghazizadeh et al. 2012b). External variables related to device characteristics, driver characteristics, driving behaviour, context and culture and coaching characteristics can influence acceptance through their effect on perceived usefulness, perceived ease of use and trust in device, as indicated in Figure 5.4. Feedback mechanisms (dashed arrows) emphasise the dynamic nature of acceptance decisions – just like perceived usefulness, perceived ease of use, and trust in device influence Technology Acceptance, acceptance and use influence perceptions (Kim and Malhotra 2005, Ghazizadeh et al. 2012a).

The Coaching Acceptance part of the model captures the effect of device characteristics, driver characteristics, driving behaviour, context and culture and coaching characteristics on acceptance of coaching. It is proposed that these variables influence drivers' trust in a coach, which, together with the perceived risk of the situation (itself influenced by external variables), determines willingness to take risks of being coached and acceptance of coaching (Mayer et al. 1995).

Coaching Acceptance can in turn influence the trust drivers place in their coach and the perceived risk of being coached. The dashed arrows show feedback mechanisms.

If the support system includes both technology and coaching components, then these components are often interwoven into a single entity. The acceptance of the support system as a whole will then be determined by drivers' perceptions of the technology-coaching hybrid, shown as the Support System Acceptance construct in Figure 5.4. The acceptance of the support system would in turn influence drivers' behaviour, the context they find themselves in and also the relationship formed between the driver and the coach. Dashed arrows from Support System Acceptance to these categories of variables highlight the dynamic nature of the system adoption. There is no one direction for the influences – external variables influence perceptions, trust and finally use, while using the system also influences these factors.

Conclusion

The attributes of a driver support system design are not the sole determinants of acceptance – decisions pertaining to acceptance and use are made in a context that encompasses the driver, the feedback and/or monitoring device and the coaching techniques employed. Acceptance and effectiveness of these systems often depends on the degree to which these factors combine to create a mentoring, rather than a monitoring system. This chapter proposed a model that views acceptance as a function of perceptions of both the technological component and the coaching component of a driver support system. Not every driver support system is comprised of both components: some are merely a feedback device and some follow only a monitoring or mentoring protocol, often based on video recordings. Nonetheless, the acceptance assessment model proposed here can be used to provide a framework to conceptualise acceptance, and thus effectiveness, of a broad range of driver support interventions. Interestingly, a mapping is evident between the categories of variables in the proposed acceptance model and the five components of the work system model by Smith and Carayon-Sainfort (1989); that is, person, tasks, technology and tools, environment and organisation. Just as the work system model emphasises the interplay between these factors in affecting workers and outcomes, we believe that all these factor categories play a role in shaping a driver's perception of the support system, while also influencing the other factors. A more elaborate account of these interactions can shed light on the dependencies of the elements – an important consideration when designing a support system.

References

Adell, E. 2010. Acceptance of driver support systems. European Conference on Human Centred Design for Intelligent Transport Systems, Berlin, Germany, 2010.

Aiello, J.R. and Kolb, K.J. 1995. Electronic performance monitoring and social context: Impact on productivity and stress. *Journal of Applied Psychology* 80 (3): 339.

Ajzen, I. 1991. The theory of planned behavior. *Organisational Behavior and Human Decision Processes* 50 (2): 179–211.

Alder, G.S. 2001. Employee reactions to electronic performance monitoring: A consequence of organisational culture. *Journal of High Technology Management Research* 12 (2): 323–42.

Alder, G.S., Noel, T.W. and Ambrose, M.L. 2006. Clarifying the effects of Internet monitoring on job attitudes: The mediating role of employee trust. *Information and Management* 43 (7): 894–903.

American Family Insurance. 2007. Teen safe driver program: Guidance for teens after they become drivers. Available at: http://www.teensafedriver.com/ [accessed 11 June 2012].

American Management Association (AMA) and the ePolicy Institute. 2007. Electronic monitoring and surveillance survey. Available at: http://www.epolicyinstitute.com/survey2007Summary.pdf. [accessed 11 June 2012].

Anderson, E.M. and Shannon, A.L. 1988. Toward a conceptualisation of mentoring. *Journal of Teacher Education* 39 (1): 38–42.

Arroyo, E., Sullivan, S. and Selker, T. 2006. CarCoach: A polite and effective driving coach. In Proceedings of the CHI '06 Extended Abstracts on Human Factors in Computing Systems, 357–62.

Bajaj, A. and Nidumolu, S.R. 1998. A feedback model to understand information system usage. *Information and Management* 33 (4): 213–24.

Barry, M. 2000. The mentor/monitor debate in criminal justice: 'What works' for offenders. *British Journal of Social Work* 30 (5): 575.

Breznitz, S. 1984. *Cry Wolf: The Psychology of False Alarms*, 9–16. Hillsdale, NJ: Lawrence Erlbaum Associates.

Brown, S., Massey, A., Montoya-Weiss, M. and Burkman, J. 2002. Do I really have to? User acceptance of mandated technology. *European Journal of Information Systems* 11 (4): 283–95.

Buchanan, J., Gordon, S. and Schuck, S. 2008. From mentoring to monitoring: The impact of changing work environments on academics in Australian universities. *Journal of Further and Higher Education* 32 (3): 241–50.

Carayon, P. 1993. Effect of electronic performance monitoring on job design and worker stress: Review of the literature and conceptual model. *Human Factors: The Journal of the Human Factors and Ergonomics Society* 35 (3): 385–95.

Carayon, P. 1994. Effects of electronic performance monitoring on job design and worker stress: Results of two studies. *International Journal of Human-Computer Interaction* 6 (2): 177–90.

Carney, C., McGehee, D.V., Lee, J.D., Reyes, M.L. and Raby, M. 2010. Using an event-triggered video intervention system to expand the supervised learning of newly licensed adolescent drivers. *American Journal of Public Health* 100 (6): 1101.

Carter, L. and Bélanger, F. 2005. The utilisation of e government services: Citizen trust, innovation and acceptance factors. *Information Systems Journal* 15 (1): 5–25.

Chen, C.F. and Chen, P.C. 2011. Applying the TAM to travelers' usage intentions of GPS devices. *Expert Systems with Applications* 38: 6217–21.

Davis, F.D., Bagozzi, R.P. and Warshaw, P.R. 1989. User acceptance of computer technology: A comparison of two theoretical models. *Management Science* 35 (8): 982–1003.

Davis, F.D. and Venkatesh, V. 1996. A critical assessment of potential measurement biases in the technology acceptance model: Three experiments. *International Journal of Human-Computer Studies* 45 (1): 19–45.

De Waard, D., Van der Hulst, M. and Brookhuis, K.A. 1999. Elderly and young drivers' reaction to an in-car enforcement and tutoring system. *Applied Ergonomics* 30 (2): 147–57.

Donmez, B., Boyle, L.N. and Lee, J.D. 2003. Taxonomy of mitigation strategies for driver distraction. In Proceedings of the Human Factors and Ergonomics Society 47th Annual Meeting, 1865–9.

———. 2006. The impact of distraction mitigation strategies on driving performance. *Human Factors: The Journal of the Human Factors and Ergonomics Society* 48 (4): 785–804.

———. 2008. Mitigating driver distraction with retrospective and concurrent feedback. *Accident Analysis and Prevention* 40 (2): 776–86.

———. 2009. Designing feedback to mitigate distraction. In *Driver Distraction: Theory, Effects, and Mitigation*, 519–31. Edited by Regan, M., Lee, J. and Young, K. Boca Raton, FL: CRC Press.

DriveCam. Available at: http://www.drivecam.com. [accessed 19 December 2012].

Farmer, C.M., Kirley, B.B. and McCartt, A.T. 2010. Effects of in-vehicle monitoring on the driving behavior of teenagers. *Journal of Safety Research* 41 (1): 39–45.

Fishbein, M. and Ajzen, I. 1975. *Belief, Attitude, Intention, and Behavior: An Introduction to Theory and Research*. Reading, MA: Addison-Wesley.

Gefen, D., Karahanna, E. and Straub, D.W. 2003. Trust and TAM in online shopping: An integrated model. *MIS Quarterly* 27 (1): 51–90.

Ghazizadeh, M., Lee, J.D. and Boyle, L.N. 2012a. Extending the Technology Acceptance Model to assess automation. *Cognition, Technology and Work* 14 (1): 39–49.

Ghazizadeh, M., Peng, Y., Lee, J.D., Boyle, L.N. 2012b. Augmenting the technology acceptance model with trust: Commercial drivers' attitudes towards monitoring and feedback In Proceedings of the Human Factors and Ergonomics Society 56th Annual Meeting, 2286–90.

Hickman, J.S. and Hanowski, R.J. 2011 Use of a video monitoring approach to reduce at-risk driving behaviors in commercial vehicle operations. *Transportation Research Part F: Traffic Psychology and Behaviour* 14 (3): 189–98.

Horrey, W.J., Lesch, M.F., Dainoff, M.J., Robertson, M.M. and Noy, Y.I. 2011. On-board safety monitoring systems for driving: Review, knowledge gaps, and framework. *Journal of Safety Research* 43: 49–58.

Hu, P.J., Chau, P.Y.K., Sheng, O.R.L. and Tam, K.Y. 1999. Examining the technology acceptance model using physician acceptance of telemedicine technology. *Journal of Management Information Systems*: 91–112.

Huang, Y.H., Rau, P.L.P., Zhang, B. and Roetting, M. 2008. Chinese truck drivers' attitudes toward feedback by technology: A quantitative approach. *Accident Analysis and Prevention* 40 (4): 1553–62.

Huang, Y.H., Roetting, M., McDevitt, J.R., Melton, D. and Smith, G.S. 2005. Feedback by technology: Attitudes and opinions of truck drivers. *Transportation Research Part F: Traffic Psychology and Behaviour* 8 (4–5): 277–97.

Jones, P.M. and Mitchell, C.M. 1995. Human-computer cooperative problem solving: Theory, design, and evaluation of an intelligent associate system. *Systems, Man and Cybernetics, IEEE Transactions* 25 (7): 1039–53.

Karahanna, E., Agarwal, R. and Angst, C.M. 2006. Reconceptualising compatibility beliefs in technology acceptance research. *MIS Quarterly* 30 (4): 781–804.

Kim, S.S. and Malhotra, N.K. 2005. A longitudinal model of continued IS use: An integrative view of four mechanisms underlying postadoption phenomena. *Management Science* 51 (5): 741–55.

Lavallière, M., Simoneau, M., Tremblay, M., Laurendeau, D. and Teasdale, N. 2012. Active training and driving-specific feedback improve older drivers' visual search prior to lane changes. *BMC Geriatrics* 12 (5).

Lee, J.D. 2009. Can technology get your eyes back on the road? *Science* 324 (5925): 344–6.

Lee, J.D. and Moray, N. 1992. Trust, control strategies and allocation of function in human-machine systems. *Ergonomics* 35 (10): 1243–70.

———. 1994. Trust, self-confidence, and operators' adaptation to automation. *International Journal of Human Computer Studies* 40 (1): 153.

Lee, J.D. and See, K.A. 2004. Trust in automation: Designing for appropriate reliance. *Human Factors: The Journal of the Human Factors and Ergonomics Society* 46 (1): 50.

Lehmer, M., Miller, R., Rini, N., Orban, J., McMillan, N., Stark, G., Brown, V., Carnell, R. and Christiaen, A. 2007. Volvo trucks field operational test: Evaluation of advanced safety systems for heavy trucks. US Department of Transportation, National Highway Traffic Safety Administration.

Marottoli, R.A., Van Ness, P.H., Araujo, K.L.B., Iannone, L.P., Acampora, D., Charpentier, P. and Peduzzi, P. 2007. A randomised trial of an education program to enhance older driver performance. *Journals of Gerontology Series A: Biological Sciences and Medical Sciences* 62 (10): 1113–19.

Marshall, D.C., Lee, J.D. and Austria, P.A. 2007. Alerts for in-vehicle information systems: Annoyance, urgency, and appropriateness. *Human Factors: The Journal of the Human Factors and Ergonomics Society* 49 (1): 145–57.

Mayer, R.C. and Davis, J.H. 1999. The effect of the performance appraisal system on trust for management: A field quasi-experiment. *Journal of Applied Psychology* 84 (1): 123.

Mayer, R.C., Davis, J.H. and Schoorman, F.D. 1995. An integrative model of organisational trust. *Academy of Management Review*: 709–34.

McCartt, A.T., Hellinga, L.A. and Haire, E.R. 2007. Age of licensure and monitoring teenagers' driving: Survey of parents of novice teenage drivers. *Journal of Safety Research* 38 (6): 697–706.

McGehee, D.V., Carney, C., Raby, M., Lee, J.D. and Reyes, M.L. 2007a. The impact of an event-triggered video intervention on rural teenage driving. In Proceedings of the Fourth International Driving Symposium on Human Factors in Driver Assessment, Training and Vehicle Design, 565–71.

———. 2007b. Extending parental mentoring using an event-triggered video intervention in rural teen drivers. *Journal of Safety Research* 38 (2): 215–27.

McLaughlin, A.C., Rogers, W.A. and Fisk, A.D. 2006a. How effective feedback for training depends on learner resources and task demands. In Proceedings of the Human Factors and Ergonomics Society 50th Annual Meeting, 2624–8.

———. 2006b. *Importance and Interaction of Feedback Variables: A Model for Effective, Dynamic Feedback*. Atlanta: Georgia Institute of Technology.

Meschtscherjakov, A., Wilfinger, D., Scherndl, T. and Tscheligi, M. 2009. Acceptance of future persuasive in-car interfaces towards a more economic driving behaviour. In First International Conference on Automotive User Interfaces and Interactive Vehicular Applications (AutomotiveUI 2009), Essen, Germany, 81–8.

Morris, M.G. and Venkatesh, V. 2000. Age differences in technology adoption decisions: Implications for a changing work force. *Personnel Psychology* 53 (2): 375–403.

Muir, B.M. 1987. Trust between humans and machines, and the design of decision aids. *International Journal of Man-Machine Studies* 27 (5–6): 527–39.

Nebeker, D.M. and Tatum, B.C. 1993. The effects of computer monitoring, standards, and rewards on work performance, job satisfaction, and stress. *Journal of Applied Social Psychology* 23 (7): 508–36.

Orban, J., Hadden, J., Stark, G. and Brown, V. 2006. Evaluation of the Mack Intelligent Vehicle Initiative Field Operational Test: Final report. Washington, DC: Department of Transportation, Federal Motor Carrier Safety Administration.

Parasuraman, R., Sheridan, T.B. and Wickens, C.D. 2008. Situation awareness, mental workload, and trust in automation: Viable, empirically supported cognitive engineering constructs. *Journal of Cognitive Engineering and Decision Making* 2 (2): 140–60.

Pavlou, P.A. 2003. Consumer acceptance of electronic commerce: Integrating trust and risk with the technology acceptance model. *International Journal of Electronic Commerce* 7 (3): 101–34.

Peng, Y., Ghazizadeh, M., Boyle, L.N. and Lee, J.D. 2012. Commercial drivers' initial attitudes toward an on-board monitoring system. In Proceedings of the Human Factors and Ergonomics Society 56th Annual Meeting, 2281–5.

Progressive Corp. 2012. How Snapshot® works. Available at: http://www.progressive.com/auto/snapshot-how-it-works.aspx [accessed 11 June 2012].

Quigley, B.L. and Nyquist, J.D. 1992. Using video technology to provide feedback to students in performance courses. *Communication Education* 41 (3): 324–34.

Rawstorne, P. 2005. A systematic analysis of the theory of reasoned action, the theory of planned behaviour and the technology acceptance model when applied to the prediction and explanation of information systems use in mandatory usage contexts. University of Wollongong, Australia.

Rawstorne, P., Jayasuriya, R. and Caputi, P. 1998. An integrative model of information systems use in mandatory environments. In Proceedings of the Nineteenth International Conference on Information Systems, 325–30.

Rogers, E.M. 1995. *Diffusion of Innovations (4th Edition)*. New York: Free Press.

Smith, M.J. and Carayon-Sainfort, P. 1989. A balance theory of job design for stress reduction. *International Journal of Industrial Ergonomics* 4 (1): 67–79.

Smith, M.J., Carayon, P., Sanders, K.J., Lim, S.Y. and LeGrande, D. 1992. Employee stress and health complaints in jobs with and without electronic performance monitoring. *Applied Ergonomics* 23 (1): 17–27.

Stanton, J.M. 2000. Reactions to employee performance monitoring: Framework, review, and research directions. *Human Performance* 13 (1): 85–113.

Stanton, J.M. and Barnes-Farrell, J.L. 1996. Effects of electronic performance monitoring on personal control, task satisfaction, and task performance. *Journal of Applied Psychology* 81 (6): 738.

Szajna, B 1996. Empirical evaluation of the revised technology acceptance model. *Management Science* 42 (1): 85–92.

Tabak, F. and Smith, W.P. 2005. Privacy and electronic monitoring in the workplace: A model of managerial cognition and relational trust development. *Employee Responsibilities and Rights Journal* 17 (3): 173–89.

Van der Laan, J.D., Heino, A. and De Waard, D. 1997. A simple procedure for the assessment of acceptance of advanced transport telematics. *Transportation Research Part C: Emerging Technologies* 5 (1): 1–10.

Venkatesh, V. and Davis, F.D. 2000. A theoretical extension of the technology acceptance model: Four longitudinal field studies. *Management Science* 46 (2): 186–204.

Venkatesh, V., Morris, M.G. and Ackerman, P.L. 2000. A longitudinal field investigation of gender differences in individual technology adoption decision-making processes. *Organisational Behavior and Human Decision Processes* 83 (1): 33–60.

Westin, A.F. 1992. Two key factors that belong in a macroergonomic analysis of electronic monitoring: Employee perceptions of fairness and the climate of organisational trust or distrust. *Applied Ergonomics* 23 (1): 35–42.

Xu, C., Wang, W., Chen, J., Wang, W., Yang, C. and Li, Z. 2010. Analysing travelers' intention to accept travel information: Structural equation modeling. *Transportation Research Record: Journal of the Transportation Research Board* 2156: 93–110.

Zweig, D. and Webster, J. 2002. Where is the line between benign and invasive? An examination of psychological barriers to the acceptance of awareness monitoring systems. *Journal of Organizational Behavior* 23 (5): 605–33.

PART III
Measurement of Driver Acceptance

Chapter 6

How Is Acceptance Measured? Overview of Measurement Issues, Methods and Tools

Emeli Adell
Trivector Traffic, Sweden

Lena Nilsson
Swedish National Road and Transport Research Institute (VTI), Sweden

András Várhelyi
Lund University, Sweden

Abstract

This chapter describes how acceptance has been measured and identifies various measurement categories. The relationship between these measurement methods and the different definitions of acceptance appearing in the literature is described and the lack of correspondence between definition and measurement is highlighted. The chapter illustrates the different outcomes of acceptance measurements depending on choice of assessment method and gives some guidance that could be used, depending on the purpose of the assessment.

Introduction

As seen throughout this book, substantial efforts have been put into the research and development of various driver assistance systems, and such systems are now being introduced in vehicles at a faster and faster rate. In this connection, it is important to remember that the expected impact of a driver assistance system will be realised only if the system is used. As Van der Laan, Heino and De Waard (1997: 1) put it, 'It is unproductive to invest effort in designing and building an intelligent co-driver if the system is never switched on, or even disabled'.

The effect of a system intended to assist drivers will be influenced by drivers' experiences and acceptance of it. Therefore, it is important to include assessment of acceptance in the process of system development and deployment, preferably in the early concept and design phases and again in iterative assessments along the development chain. Knowledge about user acceptance is valuable for understanding humans in complex environments (like the transport system) and for establishing contextual possibilities and limitations. In relation to driver assistance systems, understanding the human includes understanding his/her views and values, actions and behaviour, and human-system performance, as well as the outcomes and consequences of these. The establishment of contextual possibilities and limitations is an important prerequisite for enabling predictions and estimations (e.g., to forecast possible system benefits, user intentions and interests) and for adoption of measures. Possibilities and limitations are also vital ingredients in impact analyses (using data on system use and take up) and in generating testable recommendations for improved system design. Thus, measures of acceptance represent pieces of information of great value for guiding system development towards successful and used solutions.

In the field of developing new in-vehicle systems, major research projects undertaken previously have recognised the need for acceptance measurement as well as access to relevant acceptance measurement methods and tools. In Europe, such projects include ADVISORS, FESTA and euroFOT.

The ADVISORS project aimed to develop an integrated methodology and relevant criteria for the assessment of road network efficiency, traffic safety and environmental impact, as well as usability and user acceptance of ADAS (Advanced Driver Assistance Systems) (Brook-Carter et al. 2001). In that project, the literature was reviewed for instruments to assess acceptance of ADAS. However, the result was disappointing and led to the project team pointing out what was an obvious lack of standardised and reliable procedures and tools for assessing acceptance. Although ADVISORS did not develop a reliable and valid instrument for the assessment of acceptance, it was recommended that acceptance be measured by questionnaire, based on a component model integrating three dimensions considered to constitute acceptance: usability, driver comfort and safety benefits. For the measurement of acceptance-related features, the following measures were suggested and applied by ADVISORS (SWOV 2003): the usefulness/satisfaction scale (Van der Laan et al. 1997), the usability questionnaire (Brooke 1996), the driving quality scale (Brookhuis 1993) and a willingness to pay questionnaire (Brookhuis, Uneken and Nilsson 2001).

In the FESTA project, a handbook of good practice for the evaluation of ADAS using field operational tests (FOTs) was developed (FESTA consortium 2008, FOT-NET consortium 2011). The need to measure driver acceptance of investigated system(s), including willingness to purchase, was put forward as general advice in the part of the handbook dealing with aims, research questions and hypotheses definitions (see Annex B of the Handbook). In the work undertaken, however, FESTA refers to acceptability rather than acceptance; what acceptability is, as well as tools for measuring acceptability of technology have been considered (Kircher et al. 2008). Kircher et al. (2008: 36) state that 'regarding the field of "technology acceptance", the term acceptability indicates the degree of approval of a technology by the users, which can be measured by the frequency of use'.

Also, the context where a specific technology is (or is supposed to be) used is pointed out as a factor of great importance when evaluating acceptability.

As the driving context is constantly changing, understanding if drivers are willing and capable of accepting assistance systems has to be examined in a variety of contexts. FESTA concludes that acceptability of technology is comprised of different (described) dimensions and that no unique model or theory of it exists. To measure these dimensions, tools like standardised questionnaires, focus groups, individual interviews and self-reporting methods are suggested. No specific instrument for acceptance (or acceptability) measurement was, however, proposed as part of the evaluation methodology described in the FESTA handbook.

The euroFOT project (euroFOT 2012) was aimed at evaluating the impact of active safety systems by applying the common European approach described in the FESTA handbook. Systems already on the market or sufficiently mature to represent commercial applications were examined – for example, adaptive cruise control (ACC), lane departure warning (LDW) and forward collision warning (FCW). Impacts at the traffic system level in terms of safety, efficiency and environmental friendliness were investigated as well as effects at the individual level in terms of driver behaviour, system interaction and user acceptance. Driver behaviour and acceptance were analysed to assess the impact of the involved driver assistance systems based on real data and to improve awareness about the potential of the systems. A common core acceptance measurement questionnaire was developed enabling addition, but not deletion, of items by each research team. The questionnaire was sequentially distributed during the consecutive phases of the one-year-long tests of system exposure: prior to the test, after baseline/before system exposure, at several occasions during system exposure and at the end of the system exposure (test period).

There have also been other joint European initiatives to develop methodologies and guidelines for the assessment of driver assistance systems where, surprisingly, work on acceptance measurement was not included at all. One example is the HASTE project, which aimed to develop methodologies and guidelines for the assessment of IVIS (In-Vehicle Information Systems). In HASTE, behavioural, psycho-physiological and self-report measures were studied with a focus on driving performance, while work on acceptance measures was omitted (Carsten et al. 2005).

Despite the recognised importance of acceptance, there is no established definition of acceptance, and there are almost as many ways to measure acceptance as there are researchers trying to do so. Considering the many different ways of defining acceptance (see Chapter 2 and other relevant chapters in this book) it is hardly surprising that no consistent way of measuring it exists. Besides, the definition, and in turn the understanding and meaning of 'acceptance', are usually taken for granted in research dealing with driver assistance systems, and researchers mostly measure acceptance without defining it. The large differences in defining and measuring acceptance point to a large discrepancy in understanding the acceptance concept and make comparisons between systems, designs, settings and studies almost impossible.

Measuring Acceptance

A thorough review of studies assessing acceptance was reported by Adell (2009). The review shows that a number of different ways to measure acceptance have been employed previously. However, even when results concerning acceptance are presented, how it has been measured and how the results have been obtained are not always described, and the reliability and validity of the measures are seldom explored.

The numerous ways of assessing acceptance found in the literature were categorised by Adell (2009) into eight different groups, with 25 sub-groups (see Table 6.1). Most researchers use more than one measure to assess acceptance, either from the same category or from different categories. The measurements used are most frequently derived from questionnaires (questions and/or rating scales), but there are measurements derived from interviews, focus groups, system use and driving performance.

Table 6.1   Measures used to assess acceptance, based on the literature review; main categories with subcategories (adapted from Adell 2009). For source references see Adell (2009). Includes simulated driving as well as actual on road driving

Category	Subjective measures	Observed behaviour*	Physiological measures
Using the word 'accept'/'acceptable'	X		
Considered acceptability of the system			
Satisfying needs and requirements/ Sum of attitudes	X		
Usefulness/satisfaction scale (Van der Laan et al. 1997)			
Perceived usefulness/perceived comfort			
Usefulness			
Satisfaction			
Willingness to use	X		
Willingness to use			
Willingness to pay			
Willingness to buy			
Willingness to accept			
Willingness to have			
Willingness to keep			
Willingness to install in own car			

Wish to shut down the system		
Actual use	X	X
Voluntary use		
Frequency of use		
Action to shut down the system		
General assessment	X	
Judgment of the concept/idea		
Being in favour of the system		
Ranking by popularity		
Recommend others to use		
Supporting implementation		
Importance of the system	X	
Expressed necessity		
Ranking by importance		
Reliability	X	
Credibility of the system		
HMI assessments	X	
Opinions about the HMI		

Acceptance Measures and Their Relation to Acceptance Definitions

The different categories of acceptance measurements presented in Table 6.1 are discussed below. First the categories that match any of the definition categories presented in Chapter 2 are presented. Thereafter, those categories that do not relate to any definition of acceptance are presented.

Using the Word 'Accept/Acceptable'
Some researchers define acceptance by using the term 'accept'. Consequently, the term 'accept' or 'acceptable' is used when measuring acceptance. This is relatively common and usually measurements in this category use questions and rating scales with phrases like 'would you accept … ?' or 'how acceptable is … ?' (e.g., Menzel 2004, Parker et al. 2003, Stradling, Meadows and Beatty 2004). Another way of measuring acceptance assigned to this group is the usage of questions or ratings of the 'willingness to accept' something (e.g., a driver assistance system). Using the word 'accept' clearly relates to the acceptance definition using the word 'accept', but does not provide any further information or explanation about the concept and/ or meaning of acceptance.

Satisfying Needs and Requirements and/or Sum of Attitudes

The definition of acceptance as 'satisfying needs and requirements' implies that the assessment of acceptance should focus on whether the system satisfies the needs and requirements of the user, whereas defining acceptance as the 'sum of attitudes' demands an aggregation of attitudes in some way.

The most used instrument for measuring acceptance, the usefulness/satisfaction scale developed and proposed by Van der Laan et al. (1997), is one example of this aggregation. The tool is a standardised instrument used to estimate the 'usefulness of' and the 'satisfaction with' a driver assistance system. The 'acceptance' of the system in question is estimated by rating nine bipolar items (useful–useless, pleasant–unpleasant, bad–good, nice–annoying, effective–superfluous, irritating–likeable, assisting–worthless, undesirable–desirable and raising alertness–sleep-inducing) on five-point rating scales. The ratings on the bipolar scales are then combined into one usefulness score and one satisfaction score for the system. When launching the usefulness/satisfaction scale, the developers presented information about the reliability of the instrument as well as information on how to instruct the participants doing the estimations.

Törnros et al. (2002) used the Van der Laan scale (Van der Laan et al. 1997) to measure drivers' acceptance of an ACC system (Adapted Cruise Control) in motorway and rural road driving in a moving base simulator. The Van der Laan scale (Van der Laan et al. 1997) was mandatorily applied also in the pilot studies carried out in the ADVISORS project. It was concluded that acceptance was high and consistent over conditions for ACC, low for urban ACC with S&G (Adapted Cruise Control with Stop & Go), especially for young drivers, and that getting a drowsiness warning when driving with a DMS (Driver/Drowsiness Monitoring System) increased acceptance (Nilsson et al. 2003). Other examples where the Van der Laan scale has been used are studies reported by Adell, Várhelyi and Hjälmdahl (2008), Duivenvoorden (2008), Van Driel (2007), Broekx et al. (2006), Van Winsum, Martens and Herland (1999) and Várhelyi, Comte and Mäkinen (1998).

There exist also a variety of measurements designed to assess whether systems 'satisfy the needs and requirements' of users. In some studies, 'satisfaction' is measured in another way (than the Van der Laan et al. 1997 scale) to assess acceptance. Examples are questions about general assessment of satisfaction with the system, opinions on whether having the system is an advantage or disadvantage, the attractiveness/unattractiveness of the system, whether the system is disturbing or annoying, and whether it is supportive or constructive; see, for example, Ervin et al. (2005), Menzel (2004), Nilsson, Alm and Janssen (1992) and Turrentine, Sperling and Hungerford (1991).

Many studies also assess 'usefulness' to obtain data on acceptance (with methods other than the Van der Laan et al. 1997 scale). Examples are questions about whether and how much the system facilitates the driving task, and affects one's own driving performance and/or the driving performance of others; see, for example, Najm et al. (2006), Stanley (2006), Collins et al. (1999), Kuiken and

Groeger (1993) and Turrentine et al. (1991). Asking for users' opinions about the effectiveness of the system as well as what kinds of instructions/corrections they want from the system belong to this measurement category; see, for example, Young et al. (2007) and GM and Delphi-Delco Electronic Systems (2002).

Concerning the relationship with acceptance definitions, the Van der Laan usefulness/satisfaction scale (Van der Laan et al. 1997) reflects the definition category dealing with 'the sum of attitudes' (the third definition category in Chapter 2 of this book). Usefulness and satisfaction measurements may be associated also with the acceptance definition dealing with 'needs and requirements' (the second definition category in Chapter 2 of this book).

Willingness to Use – or to Submit to Something
The definition of acceptance as the 'willingness to use' a system implies the straightforward assessment of willingness to use and some studies apply this way of measuring acceptance; see, for example, Cherri, Nodari and Toffetti (2004) and Chalmers (2001). However, quite a few studies measure willingness to pay, either by posing an open-ended question or a closed one with different price intervals; for example, Adell and Várhelyi (2008), Najm et al. (2006), Piao et al. (2005), Comte, Wardman and Whelan (2000) and Carsten and Fowkes (1998). Further, willingness to buy, accept, have, keep and install a system, as well as the wish to shut down a system, have been proposed as indicators to assess acceptance; see, for example, Adell and Várhelyi (2008), Van Driel (2007), Broekx et al. (2006), Marchau et al. (2005) and Nilsson and Nåbo (1996).

The willingness to pay for different driver assistance systems, both for installing and for initial purchase when buying a new car, was investigated in the ADVISORS project by letting the test participants choose between given price intervals. The results showed that the test drivers were willing to pay a moderate price for the ACC and the LSS (Lateral Support System), a low price for the DMS (Driver/Drowsiness Monitoring System) and that about one-third of the drivers did not want to pay anything at all for the urban ACC with S&G (Adapted Cruise Control with Stop & Go) (Nilsson et al. 2003).

Actual Use
The last approach to defining acceptance (mentioned in Chapters 1 and 2 of this book) is by reference to actual use of the system. Measurements of voluntary use of the system, frequency of system use and acts of shutting down the system are examples of this. These measures are most frequently derived from questions and rating scales but are sometimes derived also by observing or recording drivers' driving behaviour. Drivers' statements on, for example, how often they override the system may be seen as an indirect measure of system use; it depends on whether one sees overriding the system as not using the system or as utilising a certain feature of the system. Examples of studies that have used this way of measuring acceptance are Vlassenroot et al. (2007), Broekx et al. (2006),

Ervin et al. (2005), Philipps and Schmitz (2001), Nilsson and Nåbo (1996) and Kuiken and Groeger (1993).

Apart from the methods of measurements that build on a definition or just happen to be in line with someone else's definition, there are also a number of acceptance measurement methods that do not have support in any definition.

General System Assessment

Some researchers use a general approach to measure acceptance. Examples of this way of measuring are judgements of the concept/idea of a system generally, of the perceived popularity of the system, whether the driver is in favour of the system and whether the user would recommend to loved ones that they use the system or appreciate it if they did; see, for example, Adell and Várhelyi (2008), Najm et al. (2006), Piao et al. (2005), Chalmers (2001), Marell and Westin (1998), Várhelyi et al. (1998). Such general statements do not relate clearly to any acceptance definition found in the literature (see Chapter 2 in this book). However, the measurements indicate, to some degree, the attitude towards the system ('sum of attitudes') and the usefulness of the system ('satisfying needs and requirements').

Importance of the System

The importance of a system is seen by some researchers as reflecting acceptance. The importance of the system is measured, for example, by ranking it compared to other systems (or measurements) or by judging its necessity. Measures of whether an implementation of the system in question is supported are also included in this measurement category; see, for example, Molin and Brookhuis (2007), Cherri et al. (2004), Matsuzawa, Kaneko and Kajiya (2001). The importance of a system does not clearly relate to any of the acceptance definitions. However, also for this measurement category, the measures may, to some degree, mirror the attitude towards the system ('sum of attitude') and the usefulness of the system ('satisfying needs and requirements').

Reliability of the System

System reliability in terms of drivers' level of trust in the system or the credibility of the system has been used to measure acceptance in a few studies; see, for example, Stanley (2006) and Philipps and Schmitz (2001). Again, the reliability of a system does not clearly relate to any acceptance definition but may be associated with the 'sum of attitudes' and 'satisfying needs and requirements'.

HMI Assessments

The HMI (Human–Machine Interaction) of a new driver assistance system is the system's 'face' toward the driver; hence, it is important for the intended use of it. Thus, in some studies, acceptance is measured by assessing the HMI of the system. The measures used cover mainly driver experiences of various HMI design issues like the timing of presented information and interventions, the intensity of feedback given by the system, if the reasons for presented alerts are understood, if information and interventions are startling and so on; see, for example, Najm et al. (2006), Stanley (2006) and Collins et al. (1999), Nilsson and Nåbo (1996) and Nilsson, Alm and Janssen (1992).

Summary
The many different ways of measuring acceptance may cause confusion and therefore lead to incorrect conclusions and interpretations. One illustration of this problem was found in the European PROSPER project where two Intelligent Speed Adaptation (ISA) systems – BEEP (auditory warning when exceeding the speed limit) and AAP ('active accelerator pedal,' with upward pressure when exceeding the speed limit) – were evaluated in field trials in Hungary and Spain (Adell et al. 2008). Most of the drivers had a positive attitude to the concept of the tested ISA systems (*general system assessment*). Both systems were considered 'good', 'effective', 'useful', 'assisting' and 'raising alertness', all of which are items relating to 'usefulness' in the Van der Laan scale (Van der Laan et al. 1997). The BEEP system was considered 'annoying' and 'irritating' (both 'satisfaction' items in the Van der Laan scale), but also 'raising alertness' more than the AAP (*sum of attitudes*). In spite of this, the drivers were more positive about having the BEEP system in their own cars as compared to the AAP. When choosing between the systems, more drivers selected the BEEP over the AAP and more drivers wanted to keep the BEEP system than the AAP (*willingness to have*). However, drivers willing to pay to keep the system were generally willing to pay 40 per cent (Hungary) and 75 per cent (Spain) more for the AAP than for the BEEP (*willingness to pay*). This clearly illustrates the great influence the choice of measurement might have. Two of the satisfaction items in the Van der Laan scale (Van der Laan et al. 1997) were rated negatively for the BEEP, which would imply a higher acceptance of the AAP. Nonetheless, the drivers' willingness to have the system in their cars and the choice between the systems show a higher acceptance of the BEEP system.

One interpretation of the different results is that the measurements used do not measure the same kind of acceptance. The concept of the system and the usefulness/satisfaction scale relate to the 'sum of all attitudes', while the willingness to keep relates to the 'willingness to use'. However, even if more measurements assessing the same kind of acceptance were used, there would be no guarantee that the results would concur, since validations of the measurements used are virtually non-existent. Most researchers define acceptance implicitly by the measurement tools they use to assess it, making validation impossible.

The present situation is troublesome. If acceptance has not been defined, then we cannot be sure that the tool we use to measure it will give valid results. The inconsistency of acceptance definitions (implicitly defined or not) and of measurements, and thereby the diversity of results even though collected in the same experiment, presents a breeding ground for misinterpretations and misuse of the results. What is more, it makes comparisons between systems, studies and settings almost impossible.

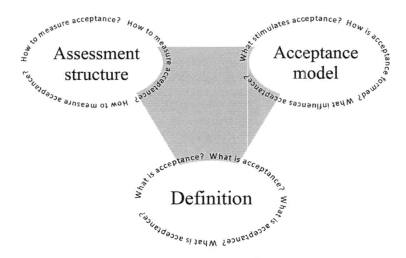

Figure 6.1   The three pillars of the acceptance concept

Framework for Measuring Acceptance

It is argued here (and also in Chapters 2 and 3 of this book) that acceptance measurement is one of the three closely related pillars of the acceptance concept (Figure 6.1). It rests on a definition of acceptance and has its theoretical foundation in an acceptance model with its constituting items and constructs as well as relationships between these constructs. Measuring acceptance requires well-defined measurement methods and tools based on knowledge of what acceptance is (definition) as well as its delimitation in terms of the context or field of application of the acceptance measurement (e.g., transport system, driver assistance system). Both the acceptance definition and the acceptance model are necessary for enabling establishment of a valid and reliable method for acceptance measurement.

One starting-point for measuring acceptance has been usability engineering, which is the foundation for Nielsen's (1993) framework of acceptability, where the focus is '*Can* an individual use the system?' According to Nielsen (1993), the general acceptability of an interactive system depends on whether the system can satisfy the needs and expectations of its users. The relationships between the concepts included in Nielsen's framework (1993) are described as 'System acceptability' branched out to 'Practical acceptability' (in terms of cost, compatibility, reliability etc.) and then to 'Usefulness' and 'Usability' (in terms of ease to learn, efficiency to use, ease to remember, few errors, subjectively pleasing). Thus, several of the features being components in this framework appear as acceptance measures in various studies evaluating driver assistance systems.

Another point of departure for measuring acceptance is system adoption patterns, modelled, for example, by Rogers (1995). The focus here is 'Who *will use* the system?' The process of different user groups successively adopting a new technology/driver assistance system is described from the first users (innovators), over early adopters, early majority, late majority and finally incorporating the laggards, until a 'saturated' level of usage is reached. From this framework, acceptance measures reflecting actual use and willingness to use can be identified.

Attempts have also been made to more strictly separate acceptance from acceptability and system uptake (euroFOT 2012) for enabling better guidance on how to measure acceptance. It has been suggested that acceptance should mean 'how much a system is/would be used', in line with the definition proposed by Adell (2009), while acceptability should mean 'how much a system is liked' and uptake should mean 'how likely it is that someone would buy a system' (Jamson 2010).

Conclusions

In this chapter we have described various ways of measuring acceptance. How to measure acceptance in a valid way depends on how acceptance is defined. It is not surprising, therefore, that the weak common ground regarding an acceptance definition has resulted in a large number of different attempts to measure acceptance. The large differences in the measures used indicate quite a large discrepancy in the understanding of acceptance, as well as in what are believed to be important and valid indicators of acceptance.

The many different ways of measuring acceptance may cause confusion and lead to incorrect conclusions or interpretations. This is clearly illustrated in the PROSPER project (Adell et al. 2008), where several measurements of acceptance of a driver assistance system were used, in parallel, with different results.

Today the field of acceptance measurement is surprisingly immature. No methods and tools that are widely agreed and accepted by the scientific community exist. The Van der Laan scale (Van der Laan et al. 1997) is the only instrument for obtaining knowledge and data on drivers' acceptance of assistance systems that was developed from scientific work and also validated to a certain extent. The availability of this scale and the scientific presentation/publication of it has made it frequently used, but it could still be discussed whether the final usefulness and satisfaction scores truly reflect acceptance. The acceptance questionnaire used for acceptance measurement in the euroFOT project was developed and used by several European research teams together and has therefore been relatively widespread. The measurement approach distinguishes between liking, using and adopting a system. Other instruments applied are more ad hoc and designed more personally and for a specific study. More joint scientific activities are needed to develop reliable and validated methods and tools for the measurement of acceptance. To be more successful in this work, a couple of issues, in particular, have to be solved. One is to come to an agreement about whether manifestation in use of a system (willingness and/or actual) should be required for acceptance to occur and thus be the focus for measurement. Another issue is to more concretely define the difference between acceptability and acceptance, and also separate acceptance clearly from usability. Many of the tools applied to measure acceptance in terms of various variables/dimensions probably work well. The problem is whether these variables reflect acceptance.

With our current state of knowledge, it is not presently possible to give any concrete advice on how to measure acceptance, since there is no universally accepted definition of the term. However, there is some guidance that could be used depending on the purpose of the assessment:

1. If the main goal is to investigate the acceptance of a driver assistance system, make sure that you define what you mean by acceptance and stick to that definition when choosing/constructing the measurement tool;
2. If the main goal is to compare your system to another, already investigated system, use the same method to assess acceptance;
3. Instead of developing new tools, use and adapt tools that are already frequently used (e.g. the Van der Laan et al. 1997 scale); and
4. If possible, use more than one way of measuring acceptance.

It is desirable to invest in fundamental research in the area of acceptance to help build a common definition, model and measurement tool. This would facilitate and improve the quality of applied research and development of new technologies – and this is of fundamental importance.

Meanwhile however, all studies of acceptance can contribute to the furthering of knowledge in the area by clearly defining what they mean by 'acceptance' and by consequently using that definition when measuring the concept. In this way we will start shaping more detailed knowledge of what acceptance is and how to measure it.

Acknowledgements

This chapter draws on the dissertation 'Driver experience and acceptance of driver assistance systems – A case of speed adaptation' (Adell 2009) as well as a number of methodological efforts within the European research framework programs.

References

Adell, E. 2009. Driver experience and acceptance of driver support systems – A case of speed support. Bulletin 251. PhD thesis, Lund University, Sweden.
Adell, E. and Várhelyi, A. 2008. Driver comprehension and acceptance of the active accelerator pedal after long-term use. *Transportation Research Part F*, 11(1): 37–51.

Adell, E., Várhelyi, A. and Hjälmdahl, M. 2008. Auditory and haptic systems for in-car speed management – A comparative real life study. *Transportation Research Part F*, 11(6): 445–58.

Broekx, S., Vlassenroot, S., de Mol, J., and Int Panis, L. 2006. The European PROSPER-project: Final results of the trial on Intelligent Speed Adaptation (ISA) in Belgium. Proceedings of the 13th World Congress and Exhibition on Intelligent Transport Systems, London.

Brook-Carter, N., Parkes, A., Ernst, A.C., Jaspers, I., Nilsson, L., Brookhuis, K., de Waard, D., Damiani, S., Tango, F., Boverie, S., Danglemaier, M., Knoche, A., Yannis, G. and Golias, J. 2001. Development of multiparameter criteria and a common impact assessment methodology. Deliverable D4.1, EU project ADVISORS (Advanced Driver Assistance Systems), European Commission. TRL, Transport Research Laboratory, Crowthorne, England.

Brooke, J. 1996. SUS – A quick and dirty usability scale. In *Usability Evaluation in Industry*, 189–94. Edited by Jordan, P.W. et al. London: Taylor and Francis.

Brookhuis, K.A. 1993. Geïntegreerde informatiesystemen en taakbelasting (Integrated information systems and work load). Traffic Research Centre, University of Groningen: Haren. VK-93–10.

Brookhuis, K.A., Uneken, E. and Nilsson, L. 2001. *ADVISORS Common Measures*. Internal Deliverable RUG ID5_1_1, ADVISORS project, EC Programme Competitive and Sustainable Growth.

Carsten, O.M.J. and Fowkes, M. 1998. External vehicle speed control, Phase I results, Executive summary. Institute for Transportation Studies, University of Leeds, UK.

Carsten, O.M.J., Merat, N., Janssen, W.H., Johansson, E., Fowkes, M. and Brookhuis, K.A. 2005. HASTE final report. EU project HASTE (Human machine interaction And the Safety of Traffic in Europe), European Commission.

Chalmers, I.J. 2001. User attitudes to automated highway systems. Advanced Driver Assistance Systems International Conference (IEE Conf. Publ. No. 483), 6–10.

Cherri, C., Nodari, E. and Toffetti, A. 2004. Review of existing Tools and Methods. D2.1.1 AIDE, Information Society Technologies (IST) Programme.

Collins, D.J., Biever, W.J., Dingus, T.A. and Neale, V.L. 1999. Development of human factors guidelines for Advanced Traveler Information Systems (STIS) and Commercial Vehicle Operations (CVO): An examination of driver performance under reduced visibility conditions when using an In-Vehicle Signing and Information System (ISIS). Virginia Polytechnic Institute and State University, Centre for Transportation Research, Blacksburg, USA; Federal Highway Administration, Report Number: FHWA-RD-99–130.

Comte, S., Wardman, M. and Whelan, G. 2000. Drivers' acceptance of automatic speed limiters: Implications for policy and implementation. *Transport Policy*, 7: 259–67.

Duivenvoorden, K. 2008. Roadside versus in-car speed support for a green wave: A driving simulator study. Master's thesis, TNO Defence, Security and Safety, Soesterberg, the Netherlands.

Ervin, R., Sayer, J., LeBlanc, D., Bogard, S., Mefford, M., Hagan, M., Bareket, Z. and Winkler, C. 2005. Automotive collision avoidance system field operational test report: Methodology and results. University of Michigan Transportation Research Institute, General Motors Corporation, Research and Development Centre, US Department of Transportation, National Highway Traffic Safety Administration, DOT HS 809 900.

euroFOT. 2012. Final results: User acceptance and user-related aspects Deliverable 6.3.

FESTA Consortium. 2008. FESTA Handbook, version 2. Deliverable 6.4, EU project FESTA (Field opErational teSt support Action), European Commission.

FOT-NET consortium. 2011. FESTA Handbook, version 4. EU project FOT-NET (Field Operational Tests Networking and Methodology Promotion), European Commission, DG Information Society and Media, Seventh Framework Programme.

GM and Delphi-Delco Electronic Systems. 2002. Automotive collision avoidance system field operational test: Warning cue implementation, summary report. US Department of Transportation, National Highway Traffic Safety Administration, DOT HS 809 462.

Jamson, S. 2010. Acceptability data – What should or could it predict? International seminar on acceptance, Paris, 25 November.

Kircher, K., Gelau, C., Vierkötter, M., Rakic, M., Bärgman, J., Kotiranta, R., Victor, T., Regan, M., Saad, F., Engel, R., Page, Y., Rognin, L., Brouwer, R., Feenstra, P., Hogema, J., Malone, K., Minett, C., Schrijver, J., Barnard, Y., Jamson, H., Jamson, S., Hammarström, U., Hjälmdahl, M., Kircher, A., Tapani, A., Luoma, J., Peltola, H. and Rämä, P. 2008. Comprehensive framework of performance indicators and their interaction. Deliverable 2.1, EU project FESTA (Field opErational teSt supporT Action), European Commission.

Kuiken, M. and Groeger, J. 1993. Effects of feedback on driving performance at crossroads and on curves. Traffic Research Centre, University of Groningen, the Netherlands.

Marchau, V., van der Heijden, R. and Molin, E. 2005. Desirability of advanced driver assistance from road safety perspective: The case of ISA. *Safety Science*, 43(1): 11–27.

Marell, A. and Westin. K. 1998. Bilisters inställning till dynamisk hastighetsanpassning – ett försök med hastighesvarnare i Umeå. TRUM-report 1998:01, Umeå University, Sweden (in Swedish).

Matsuzawa, M., Kaneko, M. and Kajiya, Y. 2001. User acceptability of the safe-driving support system for winter roads. Proceedings of the 8th World Congress on Intelligent Transport Systems, Sydney, Australia.

Menzel, C. 2004. Basic conditions for the implementation of speed adaptation technologies in Germany. PhD thesis, Technische Universität Kaiserslautern, Germany.

Molin, E. and Brookhuis, K. 2007. Modelling acceptability of the intelligent speed adapter. *Transportation Research Part F*, 10(2): 99–108.

Najm, W.G., Stearns, M.D., Howarth, H., Koopmann, J. and Hitz, J. 2006. Evaluation of an automotive rear-end collision avoidance system (Chapter 5). US Department of Transportation, National Highway Traffic Safety Administration, DOT HS 810 569, Research and Innovative Technology Administration, Volpe National Transportation Systems Centre, Cambridge, MA, USA.

Nielsen, J. 1993. *Usability Engineering.* San Diego, CA: Academic Press.

Nilsson, L., Alm, H. and Janssen, W. 1992. Collision Avoidance Systems – Effects of different levels of task allocation on driver behaviour. Reprint from DRIVE Project V1041 (GIDS), September 1991, VTI Reprint No. 182, VTI (The Swedish National Road and Transport Research Institute), Sweden.

Nilsson, L. and Nåbo, A. 1996. Evaluation of application 3: Intelligent cruise control simulator experiment, effects on different levels of automation on driver behaviour, workload and attitudes. Reprint of Chapter 5 in Evaluation of Results, Deliverable No. 10, DRIVE II Project V2006 (EMMIS).

Nilsson, L., Törnros, J., Parkes, A., Brook-Carter, N., Dangelmaier, M., Brookhuis, K., Roskam, A-J., De Waard, D., Bauer, A., Gelau, C., Tango, F., Damiani, S., Jaspers, I., Ernst, A. and Wiethoff, M. 2003. An integrated methodology and pilot evaluation results. Deliverable D4/5.2 part III, EU project ADVISORS (Advanced Driver Assistance Systems), European Commission. VTI, Swedish National Road and Transport Research Institute, Linköping, Sweden.

Parker, D., McDonald, L., Rabbitt, P. and Sutcliffe, P. 2003. Older drivers and road safety: The acceptability of a range of intervention measures. *Accident Analysis and Prevention*, 35(5): 805–10.

Philipps, P. and Schmitz, P. 2001. From information/service platforms to services in the transport and tourism sector – The European capitals approach. Proceedings of the 8th World Congress on Intelligent Transport Systems, Sydney, Australia.

Piao, J., McDonald, M., Henry, A., Vaa, T. and Tveit, O. 2005. An assessment of user acceptance of intelligent speed adaptation systems. Proceeding of the IEEE conference on Intelligent Transportation Systems, Vienna, Austria: 1045–9.

Rogers, E. 1995. *Diffusion of Innovations.* New York: Free Press.

Stanley, L.M. 2006. Haptic and auditory interfaces as a collision avoidance technique during roadway departures and driver perception of these modalities. PhD thesis, publication number 3207055, Montana State University, US.

Stradling, S., Meadows, M. and Beatty, S. 2004. Attitudes to telemetric driving constraints. In *Traffic and Transport Psychology, Theory and Application, Proceedings of the ICTTP*, 333–8. Edited by Rothengatter, T. and Huguenin, R.D. London: Elsevier.

SWOV. 2003. Action for advanced driver assistance and vehicle control systems implementation, standardization, optimum use of the road network and safety. EU project ADVISORS (European Community, GRD1 2000 10047). Final publishable report. SWOV Institute for Road Safety Research, Leidschendam, the Netherlands.

Törnros, T., Nilsson, L., Östlund, J. and Kircher, A. 2002. Effects of ACC on driver behaviour, workload and acceptance in relation to minimum time headway. Ninth World Congress on Intelligent Transport Systems, Chicago, 14–17 October.

Turrentine, T., Sperling, D. and Hungerford, D. 1991. Consumer acceptance of adaptive cruise control and collision avoidance systems. *Transportation Research Record: Journal of the Transportation Research Board*, 1318: 118–21.

Van der Laan, J.D., Heino, A. and De Waard, D. 1997. A simple procedure for the assessment of acceptance of advanced transport telemetics. *Transportation Research Part C*, 5(1): 1–10.

Van Driel, C. 2007. Driver support in congestion – An assessment of user needs and impacts on driver and traffic flow. PhD thesis, TRAIL Thesis Series, T2007/10. Netherlands: TRAIL Research School.

Van Winsum, W., Martens, M. and Herland, L. 1999. The effects of speech versus tactile driver support messages on workload, driver behaviour and user acceptance. TM-99-C043, TNO Human Factors, the Netherlands.

Várhelyi, A., Comte, S. and Mäkinen, T. 1998. Evaluation of in-car speed limiters: Final report. Deliverable D11, MASTER-project.

Vlassenroot, S., Broekx, S., de Mol, J., Int Panis, L., Brijs, T. and Wets, G. 2007. Driving with intelligent speed adaptation: Final results of the Belgian ISA-trial. *Transportation Research Part A*, 41(3): 267–79.

Young, K.L., Regan, M.A., Triggs, T.J., Tomasevic, N., Stephan, K. and Mitsopoulos, E. 2007. Impact on car driving performance of a following distance warning system: Findings from the Australian Transport Accident Commission Safe Car Project. *Journal of Intelligent Transportation Systems, Technology, Planning and Operations*, 11(3): 121–31.

Chapter 7

Measuring Acceptability through Questionnaires and Focus Groups

Eve Mitsopoulos-Rubens
Monash University Accident Research Centre, Monash University, Australia

Michael A. Regan
Transport and Road Safety Research, University of New South Wales, Australia

Abstract

This chapter explores how focus groups and questionnaires can be used to measure acceptability of new vehicle technologies. It is intended to serve as a practical guide to assist researchers and system developers in choosing the most appropriate method for assessing acceptability given their needs and expertise. We argue that the evaluation aims, the precise stage of system development and definition of acceptability being adopted are at the core of this choice. Guidance in each of these three areas is offered, as is more specific guidance on focus group design and conduct, and questionnaire design and administration. Case studies are also presented to exemplify the implementation of each approach in the measurement of acceptability.

Introduction

As seen in other chapters of this book, the extent to which users find a system 'acceptable' plays an important role in the ultimate effectiveness of that system. Thus, the ability to measure the acceptability of a system at critical points during the system's development is paramount.

This chapter explores how focus groups and questionnaires can be used to measure driver acceptability of new vehicle technologies. Guidance is offered on which method might be the more appropriate to use given the circumstances. While the tendency in this chapter is to focus on acceptability *prior* to a system's implementation, we contend that the issues raised and guidance offered here are also applicable to the evaluation of a system's acceptance *post* its implementation. Further, while we concentrate on focus groups and questionnaires, it is important to note that there are other methods available for measuring acceptability. These include one-on-one interviews and direct observations of system use. The current emphasis is motivated, at least partly, by the popularity of focus groups and questionnaires in acceptability evaluations.

Defining Acceptability in an Operational Sense

Having an operational definition of acceptability to guide the evaluation is paramount. As seen elsewhere in this book, the term 'acceptability' is generally used to refer to what is quite an abstract construct and whose meaning is likely to differ from one person to the next, and from one point in time to another as one's experience with a new technology increases. For example, issues relating to interface look and feel may figure prominently in one's assessment of acceptability when a user first starts to interact with a given technology but less so later on once the user has had the chance to adapt or compensate for any deficiencies in the system's look and feel. At this point, 'deeper' issues such as whether or not the system is actually meeting a need and/or working reliably, will be assigned a higher weighting in a user's overall assessment of a new system's acceptability (Chau 1996, Neilsen 1993).

While a number of formal definitions of acceptability exist, common to most is that acceptability constitutes a multidimensional construct. Examples of common dimensions include usefulness, satisfaction and ease of use (e.g., Davis 1989, Regan et al. 2006, Van der Laan, Heino and De Waard 1997, Vlassenroot et al. 2010). These dimensions too, should be operationally defined. Having at the outset clear, unambiguous, operational definitions of exactly what it is that one is trying to assess will ensure that the design of the data collection tool and its composition remain focused and that the tool includes only those questions that are considered central to the assessment of the acceptability of the particular system under study. It will also help to ensure that questions are unambiguous and do not require any guesswork to answer; and that responses, once collated, can be analysed and interpreted directly.

What Are Questionnaires?

In broad terms, a questionnaire is a data collection method that comprises a series of questions presented in a written format, which could be on paper or computer-based, including online. Questionnaires provide a systematic means through which to collect information about individuals' knowledge, beliefs, attitudes and behaviour (Boynton and Greenhalgh 2004).

Questionnaires are conducive to the collection of both quantitative and qualitative information; although a distinct advantage of questionnaires over other subjective assessment methods is in the provision of numerical data which can be analysed using quantitative techniques where appropriate. Questionnaires are intended for completion by individual respondents, although administration of a questionnaire to individual respondents simultaneously (i.e., in a group setting) may sometimes be appropriate and more cost-efficient.

What Are Focus Groups?

As succinctly defined by Stewart and Shamdasani (1998: 505), 'a focus group involves a group discussion of a topic that is the "focus" of the conversation'. In essence, focus groups involve several individuals brought together to discuss a particular topic under the guidance of a skilled moderator.

Focus groups are best suited to the collection of qualitative data. That is, the data collected are generally not suited to further scrutiny using quantitative techniques for collation and analysis. In part, this is because focus group samples are not a representative sample of the population and tend to be quite small (Morgan and Krueger 1993). The real value of focus groups is in their ability to generate in depth information on the topic of interest – information which may be enriched through the group dynamic. Focus groups should not be used if the goal of the investigation is to resolve conflicts, build consensus or to change attitudes but are appropriate if the goal, in general, is to bring out the various points-of-view of the individuals taking part.

Should Questionnaires or Focus Groups Be Used?

Based on our experience, we believe that there are three overarching, non-independent issues that one must consider when deciding whether to use questionnaires or focus groups to explore the acceptability of a new technology: (1) what are the evaluation questions, (2) what is the stage of system development, and (3) which aspects of acceptability are to be explored?

What Are the Evaluation Questions?

The *primary* issue to consider when contemplating the use of questionnaires or focus groups is to ask, what are the evaluation questions? That is, what are the aims of the data collection exercise? In the case of technology acceptability, the overarching aim will be invariably to gauge the acceptability of a system. However, more specific aims must also be defined.

Specific aims could include the identification of particular barriers to acceptability, as opposed, or in addition, to the provision of a more global estimate of acceptability, which may be used for comparison and/or benchmarking purposes. For example, questionnaires would be more appropriate when a goal of the evaluation is to obtain a global measure or numerical estimate of acceptability for comparison over time. Acceptability is generally considered to be a dynamic construct (e.g., Várhelyi 2002). As such, the ability to capture changes in users' acceptability of a new technology as a function of their increasing exposure to, experience with, and proficiency in using, the new technology, may be a worthwhile goal of the research.

While both focus groups and questionnaires lend themselves to the collection of information on barriers to acceptability, focus groups, being qualitative in nature, may be more appropriate when the identification of barriers and, in particular, brainstorming and discussing potential ways to overcome or address these barriers, is the main goal.

What is the Stage of System Development?

In deciding whether questionnaires or focus groups are more appropriate, the evaluation aims must be considered in the context of the stage of system development. The use of questionnaires for gauging acceptability may be more appropriate when there is a partial or fully functional prototype of the new technology with which users can interact directly. In contrast, focus groups may be better suited to gauging the acceptability of systems that are earlier in the design process – for example, at the concept stage, where the system may exist simply as an idea or as an early, two-dimensional or low fidelity, prototype.

Which Aspects of Acceptability Are You Interested in Measuring?

Some aspects of acceptability may be better suited to examination through either questionnaires or focus groups. For example, more concrete, specific issues regarding interface usability may be adequately addressed through a questionnaire, which is administered once the participant has had the opportunity to experience the system. On the other hand, more conceptually abstract areas such as perceived effectiveness and usefulness may be better served through focus groups, as the context provides participants with greater opportunity to justify and expand upon their views.

The Approach Has Been Decided. What Next?

Once it has been decided which approach would be most appropriate for measuring acceptability, attention can turn to making decisions regarding the composition of the participant sample and the design of the data collection instrument itself. Specific guidance in each of these areas is given below for focus groups and questionnaires. A case study is presented to exemplify the use of each approach. It is beyond the scope of this chapter to provide specific guidance on the analysis, interpretation and reporting of questionnaires and focus groups. For guidance in these areas, refer to Boynton (2004) for questionnaires and Morgan (1997) for focus groups.

Focus Group Design and Conduct

Participants and Group Composition

A key to the success of focus groups is the composition of the groups, the number of participants in each group and the number of groups. It is generally recommended that each focus group involve 6 to 10 participants. If there are too few participants, there is the risk that the discussion will stagnate and that too few perspectives will be canvassed. If there are too many participants, there is the potential for participants to break off into smaller groups that can disrupt the flow of the core group, and make it difficult to get back on course.

At the centre of the issue of group composition is an acknowledgement by the investigators that the quality of the discussion is dependent on both the individuals who make up the group *and* the dynamics of the group as a whole (Morgan 1997). Individuals who are unwilling to express their views in a group setting are perhaps not appropriate as focus group participants. In a similar vein, conducting groups composed of individuals who, given the research topic, are not homogenous may make some individuals less likely to voice their views. Attempting to run a group composed of young novice drivers and older/middle-aged experienced drivers may prove counterproductive as the immediate needs from the technology and experiences of the two groups would be expected to differ quite markedly.

The more heterogeneous the desired participant sample with respect to the research area, the greater the number of groups that may need to be conducted in order to ensure that, within each group, participants are as homogenous on the critical variables as is practicable. Beyond this consideration the decision of the number of focus groups to conduct will typically have a pragmatic basis relating to the availability of resources – namely, money, time, staffing and, critically, the size of the participant pool (overall and within each sub-group) from which individuals can be recruited to take part in the focus groups. While willingness to participate will influence the size of the participant pool, participant availability will also be a factor. Scheduling focus groups for a day of week and time of day (e.g., weekends, evenings) when participants are most likely to be available will help maximise likelihood of attendance. Giving participants more focus group timing options will also help in this regard.

Discussion Guide and the Role of the Moderator

The ideal focus group discussion should be free-flowing and should never take the form of a simple question-and-answer session. Achieving these goals requires a well-constructed discussion guide and a skilled moderator.

In formulating the discussion guide, it is important to be mindful of the overall time allotted to the focus group – that is, usually one and a half to two hours. This is not a lot of time in practice and so it is important that an attempt is made not to cover more topics than is needed to address the evaluation questions adequately.

A good discussion guide comprises a list of general open-ended questions, or loosely phrased questions, about the topics of interest. Examples of probes for further information may also be included for use if needed. Questions that are too specific are not ideal as these may give rise to one word or sentence answers, which are difficult, and perhaps even pointless, to explore in greater depth. Moreover, questions that are too specific have the potential to stagnate the discussion and disengage the participants. The discussion should flow and progress naturally and logically and, as such, the ordering of questions is important.

An effective way to start a focus group discussion is with a very general question about the area of interest. In the case of technology acceptability research, the focus group session would usually begin with a demonstration of the system. Nielsen (1997) recommends that participants be presented with the most concrete examples of the technology being discussed as is possible. The group can be asked about their first impressions of the technology. This can lead to more targeted questions framed around the constructs of acceptability of interest, for example. A useful way to end a focus group is to ask participants to provide a 'take-home message' – that is, their views on the one or two most important issues raised during the discussion. In the acceptability context, this could involve asking participants to comment on the aspects of the technology that they liked most, the aspects that they liked least, and what feature or features of the technology they would most like to see changed and in what ways.

The role of the moderator, in effect, is to get useful information from the participants. The moderator needs to be well-prepared, with sufficient domain expertise, and, during the discussion, be attentive (Morgan and Krueger 1993). The moderator must keep the discussion on path without inhibiting the flow of ideas and comments (Nielsen 1997). The moderator must know when to probe further and when not to do so. Thus, the moderator does much more than simply keep time and deliver the questions – the moderator is integral to data quality.

It is often impractical for the moderator to take detailed notes while also facilitating the discussion. To aid accurate data collection, it is generally recommended that a note-taker be enlisted and/or an audio recording of each focus group be taken for later review and to facilitate extraction of key themes. When reporting the results, preferably, these key themes should be organised according to the evaluation aims, and considered in the context of the group composition, the definition of acceptability which was adopted and the stage of system development.

Case Study 1

To exemplify the use of focus groups, we present as a case study a project which we completed for the Royal Automobile Club of Victoria in Australia (Regan et al. 2002). The purpose of the project was to assess the acceptability to Victorian car drivers of certain in-vehicle intelligent transport systems which were judged at the time to have high safety potential.

There were two key research phases. Phase 1 involved determining on which technologies the research should focus and to determine the composition of the groups. Phase 2 involved gauging the acceptability of the selected technologies through focus groups involving members of the driver sub-groups identified in Phase 1. Of specific interest was the identification of any barriers to the use of the technologies in the manner intended by system developers.

Acceptability was defined as comprising five constructs: usefulness, effectiveness, usability, affordability and social acceptability. To be useful, the user must perceive the system to serve a purpose. To be effective, the user must believe that the system does what it is designed to do. To be usable, the user must perceive the system to be easy to use. Affordability concerns whether users can afford to purchase and maintain the system, while social acceptability is concerned with the broader social issues (e.g., privacy) that may be taken into account by users.

Seven technologies were selected for study: Forward Collision Warning, Intelligent Speed Adaptation (ISA), Emergency Notification, Electronic Licence, Alcohol Interlock, Fatigue Monitoring and Lane Departure Warning. Analyses of the most recently available Victorian road-crash data were conducted to identify the driver sub-groups that are over-represented and those that are involved most in the crash types for each of the seven selected technologies. The outcomes of these analyses served as the primary basis for selecting the driver sub-group composition of the eight focus groups. Other considerations were that there be no more than two technologies for discussion in any one focus group to ensure that there was sufficient opportunity to discuss each technology, and that the age range be homogenous within each focus group (e.g., 18 to 24 years) to ensure that participants did not feel inhibited from freely expressing their opinions. Where this was not feasible, the range of ages spanned no more than two consecutive age groups (e.g., 18 to 24 and 25 to 39 years). Further, it was felt that, provided the age range was homogenous, a group comprised of males and females was not inappropriate. The composition of each of the eight focus groups is shown in Table 7.1, along with the technologies discussed in each group.

A total of 52 drivers took part, with most focus groups each involving six or seven participants. All participants were naive users of the technologies under study. A list of open-ended questions was developed to guide the focus group discussions. An extract of the discussion guide is given in Table 7.2, along with examples of probing questions. These questions covered the five constructs of acceptability as defined in the current research (see Table 7.2). Brief video clips demonstrating each of the technologies were also developed to provide participants with information prior to the discussion regarding the look and functionality of the technologies and of the type of warnings that the technologies issue. All systems shown in the videos were prototype versions, although some systems were a little more developed than others. As participants were not being given the opportunity to interact directly with the technologies, issues relating to usability were generally given less emphasis and allotted less time in the discussion than issues relating to perceived usefulness, effectiveness, affordability and social acceptability.

Table 7.1   Focus group composition and technologies for discussion

Focus Group	Technology	Driver sub-groups	
		Males	**Females**
1	Intelligent speed adaptation	18 to 24 years	18 to 24 years
2	Intelligent speed adaptation	25 to 39 years	25 to 39 years
3	Forward collision warning	25 to 39 years	25 to 39 years
4	Forward collision warning	40 to 64 years	40 to 64 years
5	Alcohol interlock	18 to 24 years; 25 to 39 years	NA
6	Lane departure warning; fatigue monitoring	18 to 24 years	18 to 24 years
7	Emergency notification; electronic licence	25 to 39 years; 40 to 64 years	NA
8	Lane departure warning	65 years and over	65 years and over

Source: Table adapted from Mitsopoulos, Regan and Haworth 2002.

Table 7.2   Extract of focus group discussion guide from Regan et al. (2002) also showing link between question and acceptability dimension

Question	Acceptability dimension
What are your first impressions of the technology?	–
Do you think that driving a car with this system will make you drive any differently? In what ways? Under what conditions?	Effectiveness
How useful would you find the technology? Would it serve a purpose for you? In what ways? Under what conditions?	Usefulness
Can you think of any potential problems or concerns that you might have in using the system? Source of distraction? Potential for over-reliance? Reliability issues? Issues with the look and feel of the warnings?	Effectiveness; usability
If you were buying a car and the system were not a standard feature, would you buy the system? How much would you be willing to pay? What would encourage you to buy the system? What would stop you from buying the system?	Affordability

How would you feel if it were compulsory for you to fit this technology to your vehicle? Any concerns?	Social acceptability
How do you think the system could be better designed to be more appealing to you?	–

It is beyond the scope of the current chapter to present the findings of the research. The reader is directed to Regan et al. (2002) or Mitsopoulos et al. (2002).

Questionnaire Design and Administration

Existing Questionnaires

Having made the decision that a questionnaire is an appropriate method to pursue, a consideration is whether there already exists a questionnaire which will adequately meet the needs of the evaluation were this questionnaire to be used. If suitable, there are several advantages in taking such an approach. These include savings in cost and time, which might otherwise have been spent on design and development activities, and the ability to make inter-study comparisons (Boynton and Greenhalgh 2004).

In the vehicle technology domain, as seen elsewhere in this book, an example of an existing questionnaire for assessing acceptability is that developed by Van der Laan et al. (1997). This questionnaire gives a score for *usefulness* and also a score for *satisfaction*, facilitating comparisons across studies, systems and time (with increasing system experience; e.g., before and after use of the technology). Despite its advantages, this particular questionnaire would not be appropriate were a broader definition of acceptability being adopted and/or if the elicitation of information on specific barriers to acceptability were a goal of the evaluation. In this case, the use of additional, or alternative, methods would be required.

Participant Sample

Individuals from the intended user group(s) should form the participant sample in technology acceptability research. This is as true for questionnaires as it is for focus groups. Having a clear understanding of the user group(s) is an essential early step in the questionnaire design process. Not only will it help target participant recruitment efforts, but such knowledge will help guide the look and feel of the questionnaire itself.

The target number of participants will depend largely on the study design, which, naturally, will have been determined by the evaluation objectives. Time and money available will also play a role. For example, the purpose of the study may be solely to gauge the acceptability of a range of in-vehicle technologies, which are still in the concept or idea stage of development. A further purpose may be to explore the extent to which certain demographic, behavioural and/or attitudinal factors might, in principle, influence the acceptability of each of those technologies. In this example, in order to achieve the goals of the investigation, a relatively large number of participants overall might need to be asked to complete the questionnaire.

As a further example, a questionnaire or series of questionnaires for gauging acceptability might be administered to participants who are taking part in a study, the primary aim of which is to explore objectively the effect of a given new technology or technologies on certain measures of driving performance. In this context, and relative to the previous example, a smaller number of participants overall may be asked to complete the questionnaires. Here the goal may be to explore changes in acceptability as a function of increasing experience with using the technology in a single, relatively homogenous group of individuals. The implication is that the homogeneity of the participant sample is an important determinant of the desired sample size: the more heterogeneous the sample, the larger the recommended sample size. Further, as it is often desirable to subject the numerical data deriving from questionnaires to statistical analysis, it is worthwhile noting here the relationship which exists between sample size and effect size. That is, the smaller the effect that one is interested in detecting, the larger the sample needed in order to detect that effect if it indeed exists.

Question Types

Broadly, questions can be categorised as either 'closed' or 'open-ended'. Both categories have a place in technology acceptability studies. There are several types of closed question, but common to all is that the method in which participants should articulate their response is provided. Moreover, the data provided are typically numerical, or can be coded numerically. These factors together facilitate later data collation and analysis. Riffenburgh (2012) distinguishes between the following types of closed question: *dichotomous* (e.g., 'yes' or 'no'), *multiple-choice* (i.e., participants are required to select one or several options from a set of options which are not necessarily orderable), *ranked* (i.e., participants are required to place a set of possible factors in rank order), *continuous* (i.e., participants are requested to provide a number or place a mark on a visual analogue scale) and *rated* (i.e., participants select a category from among an ordered set of categories; for example, numbered one to five, where one means 'never' and five means 'always'). Open-ended questions allow for a free-text, narrative response. A typical treatment of such responses is to scrutinise them for key themes.

With one of the main advantages of questionnaires being that they provide a mechanism through which numerical data can be collected, it is recommended that open-ended questions be used only if necessary and sparingly. If the questionnaire ends up consisting mainly of open-ended questions, this may raise the issue of whether an alternative data collection approach, such as the focus group, may be more appropriate.

Question Wording and Presentation

Critical to questionnaire reliability is that each of the questions is understood by those completing the questionnaire as is intended by the investigators. Questions need to be simply worded, with the use of technical jargon best avoided. Questions should not be difficult or impossible to answer. Also, it is best to avoid questions with ambiguous wording, double-barreled questions and leading or loaded questions (Marshall 2005, Riffenburgh 2012). Where the response to a question is dependent on questionnaire timing, setting the time frame for the question is crucial (Riffenburgh 2012). That is, in determining their response to a given question, should participants be thinking about the present, the last week or the last month, for example? Marshall (2005) also cautions against asking individuals to recount their experiences from more than six months ago – as responses to such questions tend to be less accurate than those which ask participants to recall their more recent experiences.

Clarity of expression is also important for any accompanying instructions to participants. Further, the sequencing of the questions should be logical, and filtering questions should be used when appropriate to ensure that, for a given respondent, the questionnaire does not take more time than is necessary to complete. Questionnaires which take in excess of 20 minutes to complete are best avoided. Such measures help to ensure that participants remain engaged and cooperative and increase the likelihood that participants will complete the questionnaire – in other words, they help to increase the response rate. A professional layout, with sufficient spacing for responses, and the use of an appropriately sized and styled typeface, will also contribute in this regard (Boynton and Greenhalgh 2004, Marshall 2005, Riffenburgh 2012).

Administration Mode

Common modes of questionnaire administration are 'paper-and-pencil' and 'Web-based'. With more people having access to the Internet and with the availability of 'easy-to-use' software tools for questionnaire implementation, Web-based administration has increased in popularity. Among other considerations, Web-based administration allows for more efficient data collection and coding of the data in preparation for analysis. Nonetheless, paper-and-pencil may still be the preferred mode for some potential user groups – for example, the elderly. Thus, selecting the most appropriate mode of administration given the particular needs and preferences of the intended participants is an important determinant of response rates, and as such ought to be factored into administration planning discussions.

Piloting and Preparing for Data Collection

A sometimes underrated yet crucial step in questionnaire development is the process of piloting. Piloting the questionnaire prior to its administration proper will provide an indication of the reliability and validity of the questionnaire. In this regard, a main purpose of the piloting exercise would be to identify any questions or instructions in need of rewording and refinement, and any redundant, superfluous or inappropriate questions and/or response categories for potential exclusion. A further purpose of piloting would be to ensure that the data are being recorded accurately and as intended, particularly in the case of a Web-based administration mode, and that the data are in a form suitable for analysis and subsequent interpretation in the context of the evaluation aims and definition of acceptability that is being adopted.

Case Study 2

As a case study in the use of questionnaires to measure the acceptability of new in-vehicle technologies, we present the 'Transport Accident Commission (TAC) SafeCar Project' (Regan et al. 2006). The primary aim of the project was to evaluate the potential safety benefits of a suite of in-vehicle intelligent transport systems. A further aim was to explore the degree to which drivers found the technologies acceptable, and whether this level of acceptability varied as a function of experience with a given technology. Barriers to system acceptability were also of interest. In presenting this case study, the focus is on the timing of questionnaire administration given the study aims, study design and the adopted definition of acceptability. For further detail on the questions themselves and on the results obtained, refer to Regan et al. (2006).

Each of 15 test vehicles was equipped with four technologies: Intelligent Speed Adaptation (ISA), Following Distance Warning (FDW), Seat-Belt Reminder (SBR) and Reverse Collision Warning (RCW). Each of 23 drivers drove one of the vehicles for approximately 16,500 kilometres. Drivers were volunteers from participating organisations, which had agreed to lease at least one of the test vehicles for dedicated use by their employees. Participants belonged to either the treatment or control group.

The study comprised 'Before' (2), 'During' (3) and 'After' (3) periods. No technologies were active during Before 1. In Before 2, SBR and RCW were enabled and remained enabled for the rest of the study. In a given After period, the systems (i.e., ISA by itself, FDW by itself and ISA plus FDW) that were enabled in the preceding During period were no longer active.

The definition of acceptability was the same as that in Case Study 1: usefulness, effectiveness, usability, affordability and social acceptability. An overview of the questionnaires administered is given in Table 7.3. An example question for ISA and for each acceptability dimension is provided in Table 7.4.

The first questionnaire was administered to all participants at the beginning of the study and before actual use of the systems. This questionnaire provided a baseline measure of acceptability and comprised questions for assessing the acceptability of all four technologies under study. With the exception of usability, all acceptability dimensions were assessed as part of the baseline questionnaire. Usability assessments of the technologies occurred once only for each system and early in participants' first period of exposure to a given system. Questionnaires to assess usefulness, effectiveness, affordability and social acceptability (subsequent to the baseline) were administered during each of the three After periods. The same questions were used in all three questionnaires and the baseline to enable examination of the effects of system exposure on usefulness, effectiveness, affordability and social acceptability.

Table 7.3   Questionnaires administered in the TAC SafeCar on-road study to assess acceptability

Administration timing	Acceptability dimension	Group	Systems being assessed
Before 1	Usefulness; effectiveness; affordability; social acceptability	Treatment and control	ISA; FDW; SBR; RCW
Before 2	Usability	Treatment and control	SBR; RCW
During 1	Usability	Treatment	ISA or FDW or (ISA and FDW)
After 1	Usefulness; effectiveness; affordability; social acceptability	Treatment	ISA; FDW; SBR; RCW
		Control	SBR; RCW
During 2	Usability	Treatment – only if did not experience *both* ISA and FDW in During 1	ISA or FDW
After 2	Usefulness; effectiveness; affordability; social acceptability	Treatment	ISA; FDW; SBR; RCW
		Control	SBR; RCW
After 3	Usefulness; effectiveness; affordability; social acceptability	Treatment and control	ISA; FDW; SBR; RCW

Source: Table adapted from Mitsopoulos et al. (2003).

Table 7.4   Extract of questionnaires for ISA from Regan et al. (2006) also showing link between question and acceptability dimension

Question	Acceptability dimension
To what extent do you feel that the Speed Warning System will be of use to you? A score of 0 means that the system will be of no use to you while a score of 5 means that the system will always be of use to you. 0 1 2 3 4 5 If you gave a score of '0', why do you think the Speed Warning System will be of no use to you (select only one response)? • I never exceed the speed limit. • It is my choice whether I speed or not. • I never get caught for exceeding the speed limit. • The speed limits are too low. • It is out of my control if cars around me carry me over the speed limit. • It would take away the enjoyment of driving. • I am a good driver anyway and I know when it is safe to exceed the speed limit. • Exceeding the speed limit does not make any difference to my safety. • Other, please specify: _____	Usefulness

What effect on travel speed will the Speed Warning System have on drivers who exceed the speed limit for the following reasons?
 	
(a) Speed inadvertently
    Increase speed   No change   Decrease speed
(b) Believe they can control their car safely at any speed
    Increase speed   No change   Decrease speed
(c) In a hurry
    Increase speed   No change   Decrease speed
	Effectiveness

How much would you be willing to pay for the Speed Warning System if it was able to be retrofitted to an existing car? Purchase $____ Installation $____ Maintenance/service $____ (assume yearly)	Affordability

To what extent do you agree or disagree with each of the following statements? The options range from strongly disagree on the left to strongly agree on the right.
 	
(a) The Speed Warning System takes too much control away from the driver.
    Strongly disagree   Disagree   Neither agree nor disagree
    Agree   Strongly agree
(b) The Speed Warning system should be compulsory for all drivers.
    Strongly disagree   Disagree   Neither agree nor disagree
    Agree   Strongly agree
	Social acceptability

Did you have any difficulty seeing the Speed Warning System flashing visual icon (i.e., miniature speed limit sign with flashing red circle) on the visual warning display? Yes No If you responded 'Yes', was it because (you can select more than one response) • There was too much glare on the screen. • There was too much reflection on the screen • The screen was too bright • The screen was not bright enough • The flashing visual icon rarely attracted my attention • The display is too far over to the left, making the screen difficult to view • The visual icon was too small • The visual icon was too blurry • The red circle was flashing too quickly • The red circle was flashing too slowly • Other, please specify:_____	Usability

Concluding Remarks

Questionnaires and focus groups can be powerful tools for gauging the acceptability of new vehicle technologies. However, they are not without their drawbacks. The best defence here is knowledge of what the methods can and cannot do and then selecting the method that best meets the needs of the evaluation in terms of the aims, the stage of system development and the aspects of acceptability of interest.

In this chapter, beyond the provision of general guidance, we also provide specific guidance on what constitutes good practice in the design and conduct/administration of focus groups and questionnaires. In the case of focus groups, elements of good practice include a well-constructed discussion guide and the use of a skilled moderator. In the case of questionnaires, question wording and the process of piloting are paramount. Finally, we exemplify the application of each method through case studies.

References

Boynton, P.M. 2004. Administering, analysing, and reporting your questionnaire. *British Medical Journal*, 328: 1372–5.

Boynton, P.M. and Greenhalgh, T. 2004. Selecting, designing, and developing your questionnaire. *British Medical Journal*, 328: 1312–15.

Chau, P.Y.K. 1996. An empirical assessment of a modified technology acceptance model. *Journal of Management Information Systems*, 13: 185–204.

Davis, F.D. 1989. Perceived usefulness, perceived ease of use, and user acceptance of information technology. *MIS Quarterly*, 13: 185–204.

Marshall, G. 2005. The purpose, design and administration of a questionnaire for data collection. *Radiography*, 11: 131–6.

Mitsopoulos, E., Regan, M.A. and Haworth, N. 2002. Acceptability of in-vehicle intelligent transport systems to Victorian car drivers. *Proceedings of the 2002 Road Safety Research, Policing and Education Conference*. Adelaide, Australia.

Mitsopoulos, E., Regan, M. A., Triggs, T. and Tierney, P. 2003. Evaluating multiple in-vehicle intelligent transport systems: The measurement of driver acceptability, workload, and attitudes in the TAC SafeCar on-road study. *Proceedings of the 2003 Road Safety Research, Policing and Education Conference*. Sydney, Australia.

Morgan, D.L. 1997. *Focus groups as qualitative research*. Thousand Oaks, CA: Sage.

Morgan, D.L. and Krueger, R.A. 1993. When to use focus groups and why. In *Successful Focus Groups: Advancing the State of the Art*, 3–19. Edited by D.L. Morgan. Newbury Park, CA: Sage.

Neilsen, J. 1993. *Usability Engineering*. San Diego, CA: Academic Press.

———. 1997. The use and misuse of focus groups. Available at http://www.useit.com/papers/focusgroups.html [accessed: 17 September 2012].

Regan, M.A., Mitsopoulos, E., Haworth, N. and Young, K. 2002. *Acceptability of in-vehicle intelligent transport systems to Victorian car drivers* (Report No. 02/02). Melbourne, Australia: Royal Automobile Club of Victoria Ltd.

Regan, M. A., Triggs, T. J., Young, K. L., Tomasevic, N., Mitsopoulos, E., Stephan, K. and Tingvall, C. 2006. *On-road evaluation of Intelligent Speed Adaptation, Following Distance Warning and Seatbelt Reminder Systems: Final results of the Australian TAC SafeCar project* (Report No. 253). Clayton, Australia: Monash University Accident Research Centre.

Riffenburgh, R.H. 2012. *Statistics in Medicine* (3rd edition). Amsterdam: Elsevier.

Stewart, D.W. and Shamdasani, P.N. 1998. Focus group research: Exploration and discovery. In *Handbook of Applied Social Research Methods*, 505–26. Edited by L. Bickman and D.J. Rog. Thousand Oaks, CA: Sage.

Van der Laan, J.D., Heino, A. and De Waard, D. 1997. A simple procedure for the assessment of acceptance of advanced transport telematics. *Transportation Research Part C*, 5: 1–10.

Várhelyi, A. 2002. Speed management via in-car devices: Effects, implications, perspectives. *Transportation*, 29: 237–52.

Vlassenroot, S., Brookhuis, K., Marchau, V. and Witlox, F. 2010. Towards defining a unified concept for the acceptability of Intelligent Transport Systems (ITS): A conceptual analysis based on the case of Intelligent Speed Adaptation (ISA). *Transportation Research Part F*, 13: 164–78.

Chapter 8

The Profile of Emotional Designs: A Tool for the Measurement of Affective and Cognitive Responses to In-Vehicle Innovations

Robert Edmunds and Lisa Dorn
Cranfield University, UK

Lee Skrypchuk
Jaguar Land Rover, UK

Abstract

Driver acceptance of in-vehicle technology design is seen by Original Equipment Manufacturers as dependent upon its emotional, cognitive and experiential effect on the consumer. This chapter describes the development and validation of an instrument aimed at measuring consumer's affective responses to innovative in-vehicle technologies and referred to as the 'Profile of Emotive Designs' (PED). A literature review generated the items and existing scales for the construction of the PED, and Principal Components Analysis of responses from 674 participants for three different in-vehicle technologies, revealed four scales (Technology Acceptance, Moderating Factors, Affective Appraisal and Emotional Valence). The results from Study 1 showed that these scales discriminated between three in-vehicle designs and were predictive of intentions to purchase the vehicle. Study 2 found that there was no difference in the PED scales for level of information provided about the in-vehicle technology design and scores were also very similar for pre and post in-vehicle experience. The differences that did emerge were concerned with ease of use and the anticipated help required using the technology. This is intuitive, as only in-vehicle experience will give rich information of a design's usability. The findings are discussed with reference to driver acceptance.

Introduction

The automotive industry has a number of challenges with respect to gaining driver acceptance of the use of technology within a vehicle and differentiating their products in the marketplace. When developing technology for the modern-day vehicle, not only does the interface need to be easily usable and not distracting, but it has to engage the user and give them a sense of enjoyment and satisfaction. This means that in-vehicle interfaces have developed a sense of 'form as well as function' which complements the previous mantra of purpose to achieve a function. Therefore the quicker and more effective a design is at capturing the imagination of the user, the more likely the innovation will gain driver acceptance. Achieving customer satisfaction can be done in many ways. Some individuals will get excited by the look of a car, some by the technology and some by the brand name or the way the vehicle sounds. However, the common measurable component across all these features is the driver's emotional responses and their attachment to the overall product.

When developing technology, the Original Equipment Manufacturer (OEM) must understand the impact of these innovations and ensure in-vehicle design solutions invoke a positive response from the customer. This is especially true for premium cars in which consumer expectation for market differentiation is high. Therefore, the development of an understanding of the fundamental relationship between drivers' affective and cognitive response to technology and how this affects their perception of a product or service is critical and one which can help make decisions on how and what is delivered in the next generation of in-vehicle technologies. Driver acceptance of in-vehicle technology design is seen by OEMs as dependent upon its emotional, cognitive and experiential effect.

In recent times, the proliferation of technology into vehicles has created an explosion of new features that the user can interact with. This is a double-edged sword in terms of customer satisfaction as there is a balance between the amount and complexity of the features available and the effect of the technology in terms of driver distraction. There are many challenges associated with this and it is becoming increasingly more difficult as drivers appear to want and expect to execute secondary tasks whilst driving. The automotive industry frequently introduces innovative designs and car models to produce a better product and also create consumer demand. In such a competitive sector it is of the utmost importance to produce a design that meets consumer needs; otherwise, the competitive edge will be lost. It is therefore vital that OEMs understand which technologies will offer both significant customer satisfaction and help prioritise which features will attract the next generation of customers that achieves the 'wow' factor whilst being fun and safe to use.

While a great deal of market research is conducted to position a brand and also to gauge consumer reaction to the finished product, there seems to be little systematic research to understand customer responses to new technologies in the early design stages. Rather, this is guided by the intuition of the designer using past experience and product history (Jordan 2000). Therefore, the efficacy of measuring the affective impact of a product during early design stages was also investigated.

Development of the Profile of Emotive Designs (PED)

This chapter describes the development of the PED instrument designed to measure consumer's affective responses (defined as any positive or negative feelings and emotions) to new in-vehicle technologies. A three-year research program undertaken by Cranfield University was commissioned by Jaguar Land Rover (JLR) to develop the PED as a measure for the assessment of in-vehicle innovations for all stages of product development and for use across all types of in-vehicle technologies, such as a new satellite navigation system, a novel interior car control or the automation of some function. JLR required an instrument that would measure in-vehicle innovations reliably across different modes of presentation; for example, a storyboard, a video, a prototype or the in-vehicle realisation of the design.

Consumer affect on interacting with a product is likely to engage complex aspects of cognition and emotion, so a multidimensional scale was required to capture the important facets of human responses generated. The instrument should be capable of discriminating between a generally liked and disliked innovation and also be sensitive enough to differentiate between quite similar technologies. The instrument should be short to complete and capable of identifying individual differences. While to be developed initially for the Jaguar XF model, it should also work across contexts; that is, for other models of car. Finally, the scores obtained by the instrument should correlate with some other meaningful variable for validation purposes, especially intention to purchase the car.

The research undertaken addressed three main questions. Firstly, could an instrument be designed that measures affective responses to in-vehicle innovations that has an element of user interaction? Secondly, could this instrument provide comparable and discriminating scores across different types of innovation? Thirdly, could the instrument measure responses at different stages of the design process in a meaningful way that would give an indication of how favourably the design would be evaluated when in production and installed in the target vehicle? To address these questions, a wide-ranging literature review was conducted covering areas of psychology, marketing and ergonomics, in order to design a framework to theoretically underpin the development of the PED and inform its construction.

Summary Findings of the Literature Review

Given space constraints, only a summary of the key findings of the literature review is presented here. In-vehicle technologies usually require some level of user interaction; thus, understanding the impact of relevant product function on the individual is important. Within the cognitive domain, there is extensive literature covering technology acceptance and usability. As seen in other chapters of this book, perhaps the most widely cited is the Technology Acceptance Model (TAM; Davis 1989). That model is particularly appropriate here as it concerns individuals' *perception* of technology, rather than just an analysis of product functionality, measuring two factors: 'perceived ease of use' and 'perceived usefulness'. However, the TAM does not encompass many possible drivers of technology acceptance. Venkatesh and Davis (2000) attempted a comprehensive revision to the TAM to include four main determinants of an intention to use technology; these were performance expectancy, effort expectancy, social influence and facilitating conditions (Venkatesh et al. 2003), indicating that individuals' perception of product technology could be multidimensional. In particular, the hedonic aspects of consumer interaction or perceived interaction with a product can have a significant effect on the satisfaction with the product at a level beyond that captured by just its utilitarian aspects. Yi et al. (2006) provide a further example of an integrative approach outlining a predisposed tendency towards adopting an innovation, and Bruner and Kumar (2005) also extended the TAM by incorporating hedonic aspects of technology use. Similarly Lu et al. (2009) found that perceived 'enjoyment' significantly influenced attitude towards using a technology.

Jordan (2000; see also Chapter 18 in this volume) proposed that, once a product has become functional and easy to use, the consumer searches for a product that is pleasurable to use. However, the experiential outcome of interaction with a product is likely to be context dependent. How the formal and experiential properties of a product link to each other is a vital step in understanding affective product design. *Kansei engineering* developed by Misuto Nagamatchi considers consumers' feelings and image using statistical techniques and so captures the idea of affective design (Nagamatchi 1995, Schutte et al. 2004). While this technique has proved successful, there are a number of problems in its use. Firstly the Kansei words for analyses come from divergent sources and the choice of words is subjective. Secondly, a cluster analysis requires a large number of examples of the product properties and this may not be feasible. Finally, and most importantly for the present research, Kansei knowledge will be specific to the product and may not generalise to other aspects of the overall design or even similar products.

For the present research aims it was also important to consider what types of product attributes are likely to generate positive emotions. Kano (1984) proposes three distinct types of needs plotted on two axes, one for subjective satisfaction level (satisfied to dissatisfied) and the second an objective measure on how well each need has been executed ('very well' to 'not at all') (see Figure 8.1). The three types of needs plotted on these axes are Basic needs, Performance needs and Excitement needs (Fuller and Matzler 2007).

The model suggests two routes to increase customer satisfaction: either by increasing some scalar quality such as performance/economy, or alternatively, surprise the customer with some innovation that meets a latent demand. Basic product qualities are not a route to increase satisfaction, but may lead to dissatisfaction if not executed properly. While this model provides an interesting framework to understand the different dimensions of product quality, it does not provide any detailed prediction of what attributes may lead to positive affect.

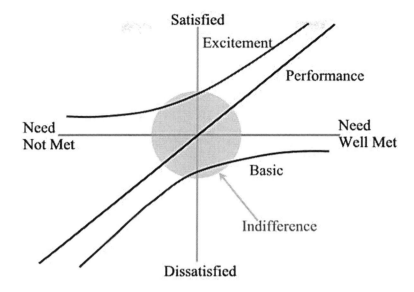

Figure 8.1   Kano Model Dimensions

Evans and Burns (2007) termed extreme customer satisfaction as 'delight' defined as a positive emotional state resulting from having one's expectations exceeded to a surprising degree (Burns 2003, Rust and Oliver 2000). In their study, 34 attribute-based delight reactions that could not be ascribed to either of Kano's two routes to satisfaction could be found. For the Evans and Burns (2007) study, none of the attributes could be ascribed as either unexpected or high performance; yet they 'delighted' the customers. Some of the products' attributes were commonplace in cars, but delighted customers because of the novel way they were delivered. Others delighted not because they had very high performance, but because the level of performance was 'just right'. A third category for delight was also revealed; it seems that some attributes were satisfying as they were part of the holistic appeal of the car. This enhances our understanding of the possible routes to positive customer affect or delight, but 'distinctive delivery', 'just the right performance' and 'holistic appeal' still need to be further unpacked.

In designing an instrument to measure customer affect or delight when interacting with a product, it is clear that mood and emotion must be considered. There is extensive psychological literature focusing on mood and emotion, in particular, the idea that emotions are object driven and evaluative (Scherer 2005) is particularly relevant for the present research; in determining the affective impact of a novel in-vehicle design the change in affect should be about the design after some evaluation has taken place. In Scherer's terms, individuals will be required to undertake a cognitive appraisal of how they are feeling at that particular moment when evaluating the design which may differ from the fast and most likely automatic appraisal that takes place during the onset of an emotional event, but one will influence the other and lead to 'response synchronisation'. Cognitive and physiological components are mobilised together so that an emotional event will elicit a number of changes in the individual, each of which can potentially be measured following an interaction with in-vehicle innovations and lead to a 'behavioural impact' (Scherer 2005). This suggests that a both persistent and measurable change may occur on presentation of the target innovation.

Mehrabian and Russell's (1974) theory, the Pleasure, Arousal and Dominance paradigm of affect (PAD), is particularly useful here and asserts that three dimensions are needed to assess the individual's feelings, and these factors in turn also influence behaviour (Kulviwat et al. 2007). The first dimension is pleasure and refers to an enjoyable reaction to the object; the second dimension relates to arousal and excitement. Dominance refers to the level with which an individual feels they are in control or controlled by a stimulus. Dominance is relevant when considering consumer reaction to new interactive innovations such as a complex design which is not intuitive to use. This may be important for novel design solutions since new technology may have an aversive effect if the customer feels submissive and unable to master a new innovation, and this may impact on the affective state of self-efficacy.

Based on the literature reviewed to inform the design and development of the PED, there are four core areas of relevance for its theoretical underpinning:

- Incorporating a measure of technology acceptance;
- Moderating effects of social and attitudinal factors;
- Affective appraisal dimensions, such as delight and surprise;
- Valence (mood/emotion).

The exact composition of the PED scales will not be described here as it was developed for JLR and their use; however an extensive set of items was required to cover these core areas and existing scales were included in the pilot version of the PED where relevant; for example, the Pleasure, Arousal and Dominance scales taken from the PAD.

Study 1: A Comparison of Three Design Innovations

Aim

The aim for Study 1 was to examine the factor structure of the PED using a large sample and to determine whether the instrument was sensitive to different types of in-vehicle technologies. That is, does it meet the main goal of producing lower scores for less well-liked designs compared to designs that are known to be well received by consumers?

Stimuli

Three technologies were chosen for this study. The first was the Jaguar gear selection control (JaguarDrive™), which utilises a rotary control that rises from the centre console on start-up and allows selection of the automatic drive. Second, the Jaguar cabin light (SenseLights™), which turns on when it is touched, eliminating the need to find a switch. Both these technologies are in production and known to be delivering customer satisfaction. The third technology chosen is not in production and has had a less favourable, more mixed, reaction: the so-called SenseWindows™, which opens the window depending on what position is touched on the window pillar. It was hypothesised that the two production technologies would be rated as more favourable than the window-opening design.

Participants and Method

Six hundred and seventy-four JLR employees based at Coventry, UK, completed the PED: 258 for the JaguarDrive survey, 206 for the SenseLights and 210 for the SenseWindows. A survey was conducted for each technology. Participants could view a picture and a description of one of the designs over the intranet including a short video of the design being used. Following this they were asked to complete an online pilot version of the PED.

Principal Components Analysis

The responses across the three innovations were subjected to principal components analysis to identify the constructs underlying each of the four scales (Technology Acceptance, Moderating Factors, Affective Appraisal and Emotional Valence). The number of factors to extract was determined by considering the parallel analysis of 1,000 random correlation matrices using the program written by O'Connor (2000), scree plot and Eigen one rule (Factors with an Eigenvalue ≥ 1 are accepted as salient). Principal axis analysis was used to extract the relevant number of factors, and these were submitted to oblique rotation using a quartimin procedure (Direct Oblmin) to achieve simple structure. Item loadings greater than 0.30 were regarded as important for interpreting the factors so as to retain as many items as possible at the early stage of the PED development. The final instrument will accept higher loadings of 0.40 or 0.50, so as to reduce the number of items for each of the factors.

The items yielding salient loadings of the magnitude of at least 0.30 on each factor were taken to define a sub-scale, and each participant was assigned scores on each sub-scale by calculating the mean of their responses to its constituent items. The reliability of each sub-scale was estimated using Cronbach's coefficient alpha.

For Technology Acceptance, two factors emerged: Usefulness and Ease of Use. Each sub-scale was found to yield values of coefficient alpha of 0.918 and 0.917, respectively, regarded as satisfactory according to conventional criteria.

Moderating Factors load onto three factors that captured ideas about 'Attitude to the Technology', the amount of 'Available Help' required to use the innovation and issues over 'Anxiety' about the innovation. After reversing the scale score for the two negatively correlated items on the Anxiety scale, each sub-scale was found to yield values of coefficient alpha of 0.894, 0.869 and 0.762, respectively. Following parallel analysis it was determined that two factors should be extracted for the Affective Appraisal sub-scale; these two factors were concerned with the concepts of delight and novelty, yielding values of coefficient alpha of 0.956 for Delight and 0.825 for Novelty. The final scale measured valence using the PAD model (the Pleasure, Arousal and Dominance paradigm of affect described earlier). Analysis found that the three sub-scales were replicated in the current context of car design, with factors of Pleasure, Arousal and Dominance, each yielding respective values of coefficient alpha of 0.929 for Pleasure, 0.842 for Arousal and 0.807 for Dominance. Overall, the analyses indicate the internal reliability of the PED is adequate.

Results

Calculating mean response scores for each sub-scale revealed an encouraging pattern of results according to the three different technologies. SenseWindows™ was scored least positively compared with the JaguarDrive™ and SenseLights™. SenseWindows™ was seen as less easy to use and useful, produced a less positive attitude and lower levels of delight and pleasure. Respondents reported that SenseWindows™ was less easy to control, but created higher anxiety and was thought to require more help to use. In contrast the SenseLights™ scored highest for most sub-scales, requiring little help to use, but was more useful and easy to use than the other two technologies. Somewhat surprisingly though, the SenseLights™ scored lowest in novelty.

When the scores were summed across all the scales using a simple algorithm that reversed some scores for negative scales, the results shown in Table 8.1 were found.

Table 8.1   Mean score* technology collapsed across scales

Questionnaire ID	Mean	Std. Deviation
JaguarDrive™	3.47	0.56
SenseLights™	3.41	0.53
SenseWindows™	3.07	0.61

* Higher mean scores indicate greater levels of acceptance

ANOVA revealed that there was a significant difference between the three technologies; F(2,636) = 29.54, p < 0.001 *Partial eta squared* 0.085. Cronbach's alpha = 0.835. Post hoc Tukey tests found a significant difference in scoring only between the SenseWindows™ technology and JaguarDrive™ and SenseLights™ t (p < 0.05), suggesting that the SenseWindows™ rated as least appealing across all the scales compared with the JaguarDrive™ and SenseLights™; these latter two designs were found to have no significant difference in responses between them.

Participants were also asked if the inclusion of the technology would encourage them to purchase the car; responding via a five-point scale from 'not at all' to 'very much so'. This item was used as a measure of intention to purchase to provide some support for the validity of the instrument and the results are presented in Table 8.2.

Table 8.2   Regression results 'intention to purchase'

Model		Unstandardised coefficients		Standardised coefficients		
		B	**Std. error**	**Beta**	**t**	**Sig.**
1	(Constant)	-1.920	0.243		-7.915	0.000
	All scales with PAD and 2 factors of moderating factors reversed	1.292	0.072	0.585	17.999	0.000

Note. Dependent Variable: E3: Do you feel the presence of the technology would play a part in encouraging you to purchase the car?

Table 8.2 shows the results from a linear regression, with the algorithm score as the predictor and the intention to purchase score as the dependent variable. The analysis reveals the standardised coefficient of Beta = 0.585, t = 17.999, p < 0.001. The t-test on B suggests that the findings are not close to zero and t is significant suggesting that the model is meaningful. This indicates that scale scores predict participants' responses about intention to buy the product based on the innovation.

In summary, the findings suggest that the PED successfully discriminates between the technologies and, as predicted, the SenseWindows™ design was least liked. When the scores were summed across the scales, they not only significantly discriminate between the three technologies, JaguarDrive™, SenseLights™ and SenseWindows™, but the algorithm score is also a significant predictor of intention to purchase.

It would seem from this study that the PED meets the objectives as set out and highlights important indicators of consumer affect and cognition about in-vehicle innovations, discriminating between favourable and less favourable designs. It should be noted at this point that the PED scores only make sense when they are used to compare different designs; it is not possible to specify an absolute criterion value that indicates whether it is a good or bad design at this stage. Whilst technologies with different functions were tested in Study 1, the instrument is required to measure the emotional impact of the design with functional components being part of this evaluation. The importance of the functional component of innovative designs is explored further in Study 2. However, we assessed system mock-ups in Study 1 rather than actual systems. Study 2 uses an actual system and compares PED responses across different modes of presentation.

Study 2: Modes of Design Presentation

Aims

The primary aim of the second investigation was to determine how different modes of design presentation would affect PED scores. That is, if the innovation is presented in a degraded form such as a picture or video, will the results be indicative of the scores obtained when seated in the production car and interacting with the design innovation? This is an important aspect as it is hoped that the instrument will be useful at early stages of the design process as well as at the physical prototype or production stage. A secondary aim of Study 2 was to investigate another design innovation that is both functional, in-vehicle and may have emotive appeal, while extending the research beyond Jaguar cars. A design that meets these criteria was the BMW iDrive.

Stimuli

While aesthetics may be possible to judge from a graphic representation, what the design function is like to use is more difficult to determine using this mode of presentation. For this reason a description of how the design functions was presented alongside any picture or video of the product. Similarly, the context in which the design is set may have an effect, and so pictures of the target car and cabin were necessary to include for rating the innovation. Generally, it is known that the more information that is provided about a design, the greater the individual's engagement with the product and the more positive their impression of that product becomes (Meyers-Levy 1989, Nagaraj 2007, Castle and Chattopadhyay 2010). To address the second aim of Study 2, the BMW iDrive was chosen as, like JaguarDrive™, it is multifunctional, controlling many aspects of the technology inside the car and intended to neaten the interior of the car, reducing the number of switches necessary, while looking good and impressing the consumer. Finally, it has some similarities in appearance to the JaguarDrive™, but has a completely different function.

Participants and Procedure

Twenty-five participants took part in Study 2 drawn from employees and students based at Warwick University. JLR provided a top of the range 6 series BMW with second generation iDrive (Figure 8.2). Participants were invited to take part by email and respondents were randomly assigned to one of two conditions and sent either the Text/Picture information about the iDrive or sent this information in conjunction with a short video clip of the iDrive being used. They were instructed to view the information and then complete the PED online, save their responses and send it back to the researcher as an email attachment.

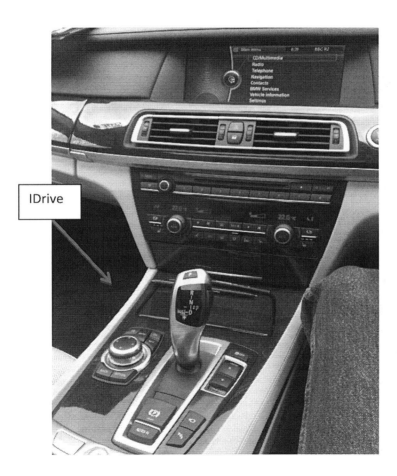

Figure 8.2   Interior and iDrive in target vehicle

After completing the PED, participants were asked to attend a designated location a few days later to sit in the car and experience the iDrive first hand. Each participant was asked to sit in the drivers' seat while the researcher sat in the front passenger seat. A brief description of the purpose of the study was given, indicating that their evaluation of the iDrive was important and that they would be asked to complete a paper version of the PED once more at the end of the study. Usability of the design solution whilst driving was not assessed during this study; however, future research will employ driving simulator-based methodology to investigate how different driving experiences might impact on the cognitive and affective aspects of different design solutions.

The researcher then instructed participants to use the iDrive and its various functions, such as radio, satellite navigation, heating and vehicle information. Some of these functions required the participant to navigate through a number of sub-menus and scroll through a number of options displayed on the centre screen. When all the tasks had been attempted, participants were asked to complete the PED for a final time, which took an average of three minutes.

Results

The study employed a 2x2 factorial design with one within-subjects' factor of Experience with two levels (Information and In-vehicle) and one between-subjects' factor of Information Type with two levels (Text/Picture and Text/Picture + Video). The effect of sitting in the car and using the iDrive on PED ratings was also investigated across groups.

No statistically significant differences were found for the PED scales between the two different levels on information (Information and In-vehicle) or pre and post experience of the actual design in-vehicle. Multivariate analysis found the only statistically significant difference between conditions was for the 'Anxiety' and 'Available Help' scales. Before sitting in the car, respondents were more anxious about the design after only viewing the picture and description information compared to those that also viewed the video in addition to this information. This difference disappeared after sitting in the car. However, when scoring the design at this point, respondents felt that more help was needed to use the device compared to when they had just viewed the information supplied.

Another way to analyse the effect of information types was to determine whether the scores from the video and text material predict the scores that were later obtained in the car. In other words, do the ratings individuals give from impoverished information still produce results comparable to that found when the finished product is presented in the vehicle. Results of a regression analysis indicated that individuals' scores before sitting in the car were predictive of the pattern of scoring found after sitting in the vehicle, except for Ease of Use where the regression was not significant (see Table 8.3).

Table 8.3   Regression results pre and post

Predictor	Dependent	Adjusted R2	Beta	t	Sig.
Usefulness pre	Usefulness post	0.240	0.522	2.932	$p < 0.01$
Ease of use pre	Ease of use post	0.007	0.221	1.086	$p = 0.289$
Attitude to tech pre	Attitude to tech post	0.727	0.859	8.048	$p < 0.001$
Help required pre	Help required post	0.426	0.671	4.339	$p < 0.001$
Anxiety pre	Anxiety post	0.232	0.482	2.635	$p = 0.015$
Delight pre	Delight post	0.360	0.622	3.807	$p < 0.01$
Novelty pre	Novelty post	0.286	0.535	3.038	$p < 0.01$
Pleasure pre	Pleasure post	0.415	0.645	4.043	$p < 0.01$
Arousal pre	Arousal post	0.314	0.560	3.241	$p < 0.01$
Dominance pre	Dominance post	0.377	0.614	3.734	$p < 0.01$

In summary, there was no significant difference between the two conditions for level of information; Text/Picture or Text/Picture + Video. Scores were also very similar for pre and post in-vehicle experience. Also, Pre in-vehicle scores seemed predictive of scores after participants sat in the car and used the iDrive. The differences that did emerge seemed centred around ease of use and the anticipated help required using the system. This is intuitive, as only in-vehicle experience will give rich information of its usability.

This suggests that the PED may provide an indication of emotive reactions to innovations at various stages of the design process. Printed descriptions and photographic information alone seem to provide some cues for individuals to gauge their feeling about the design, except those usability dimensions that may vary across presentation types. It would still be sensible to suggest though, that when directly comparing design innovations, they should be presented in the same modalities for a valid comparison of the scores, since the regression models, while statistically significant, explain only a portion of the variance. This study provides some early evidence that individuals can meaningfully score different innovations from different types of presentation material and these scores are to some extent predictive of the scores they would give when sitting in the vehicle interacting with the design innovation itself.

Conclusion

The findings indicate that driver acceptance of in-vehicle technology design may be a multidimensional concept dependent upon the emotional, cognitive and experiential effect on the consumer. The complexity required to measure the individual's response to an automotive innovation is an important message for this chapter. Other attempts have been made to measure the driver's acceptance of vehicle technologies, such as the scale suggested by Van der Laan, Heino and De Waard (1997), which has a usefulness (similar to the TAM) and satisfaction component. While scales like these are short and simple to use, they do not consider the potential moderating emotional and cognitive factors in such depth as the PED. The PED was constructed to capture the main drivers of affect for novel in-car designs as fully as possible and its advantage lies in this and its demonstrated usefulness at various stages of the design process.

The PED scale and sub-scales do seem to measure important aspects of consumer affect and cognitions to do with design innovations. Focusing here on the Jaguar drive selector, cabin lights and window opening, the scales discriminated between these three in-vehicle design technologies in the expected direction. Considering the different modes of presentation for the BMW iDrive, since responses to graphical representations of a design correlated with responses to the in-vehicle experience, this suggests that the instrument may give a reasonable indication of reactions to the final product from early renderings of that design.

While the PED captured responses that differentiated the technologies tested, measurement in general may be more complex and suggests further areas of development for the instrument. For instance, the factors which are important to measure acceptance in prestige models, often purchased for aesthetics and performance reasons, may differ when similar technologies are in more mundane vehicles. Also the change in appraisal of the innovation over time needs to be considered; certain designs may be frustrating to begin with, but become more satisfactory as its functions are learned. However, it is also possible that a design that is acceptable to begin with, may become just a novelty over time and retain less affective appeal.

A final thread of future investigation for instruments such as the PED, could be to investigate whether such scales are sensitive enough to discriminate across different levels of acceptance between quite similar technologies, such as subtly different types of drive selector. It is reasonable to conclude that driver acceptance as a concept needs to incorporate a broad range of factors in order to understand its impact on behaviour. As in-car innovations are increasingly incorporated into vehicle design this is an important area to pursue.

References

Bruner II, G.C. and Kumar, A. 2005. Applying TAM to consumer usage of handheld Internet devices. *Journal of Business Research*, 58: 553–8.

Burns, A.D. 2003. Phenomenology of customer delight: A case of product evaluation. PhD thesis, Cranfield University (Thesis-C).

Castle, C. and Chattopadhyay, S.P. 2010. *Text messages as advertisements: An empirical investigation of mobile phones as media.* International Conference on Business and Economic Research, Kuching Sarawak, Malaysia.

Davis, F.D. 1989. 'Perceived usefulness, perceived ease of use, and user acceptance of information technology'. *MIS Quarterly,* 13(3): 319–39.

Evans, S. and Burns, A.D. 2007. An investigation of customer delight during product evaluation: Implications for the development of desirable products. *Journal of Engineering Manufacture*, 221(B): 1625–40.

Fuller, J. and Matzler, K. 2007. Customer delight and market segmentation: An application of the three-factor theory of customer satisfaction on life style groups. *Tourism Management*, 29: 116–216.

Jordan, P.W. 2000. *Designing pleasurable products: An introduction to the new human factors.* London and New York: Taylor and Francis.

Kano, N. 1984. Attractive quality and must-be quality. *Journal of the Japanese Society for Quality Control*: 39–48.

Kulviwat, S., Bruner II, G.C., Kumar, A., Nasco, S.A. and Clark, T. 2007. Toward a unified theory of consumer acceptance technology. *Psychology and Marketing*, 24(12): 1059–84.

Lu, Y., Zhou, T. and Wang, B. 2009. Exploring Chinese users' acceptance of instant messaging using the theory of planned behavior, the technology acceptance model, and the flow theory. *Computers in Human Behavior*, 25: 29–39.

Mehrabian, A. and Russell, J.A. 1974. *An approach to environmental psychology.* Cambridge, MA: MIT Press.

Meyers-Levy, J. 1989. Priming effects on product judgments: A hemispheric interpretation. *Journal of Consumer Research*, 16: 76–86.

Nagamatchi, M. 1995. Kansei engineering: A new ergonomic consumer-oriented technology for product development. *International Journal of Industrial Ergonomics*, 15: 3–11.

Nagaraj, S. 2007. *The impact of consumer knowledge, information mode and presentation form on advertising effects.* PhD thesis, KTH Royal Institute of Technology, Stockholm, Sweden.

O'Connor, B.P. 2000. SPSS and SAS programs for determining the number of components using parallel analysis and Velicer's MAP test. *Behavior Research Methods, Instrumentation, and Computers,* 32: 396–402.

Rust, R.T. and Oliver, R.L. 2000. Should we delight the customer? *Journal of the Academy of Marketing Science*, 28(1): 86–94.

Scherer, K.R. 2005. What are emotions? And how can they be measured? *Social Science Information,* 44: 695–729.

Schutte, S.T.W., Eklund, J., Axelsson, J.R.C. and Nagamatchi, M. 2004. Concepts, methods and tools in Kansei engineering. *Theoretical Issues in Ergonomics Science.* 5(3): 214–31.

Van der Laan, J.D., Heino, A. and De Waard, D. 1997. A simple procedure for the assessment of acceptance of advanced transport telematics. *Transportation Research*, 5(1): 1–10.

Venkatesh, V. and Davis, F.D. 2000. A theoretical extension of the Technology Acceptance Model: Four longitudinal field studies. *Management Science,* 46(2): 186–204.

Venkatesh, V., Morris, M.G., Davis, G.B. and Davis, F.B. 2003. User acceptance of information technology: Toward a unified view. *MIS Quarterly,* 27(3): 425–78.

Yi, M.Y., Jackson, J.D., Park, J.S. and Probst, J.C. 2006. Understanding information technology acceptance by individual professionals: Toward an integrative view. *Information and Management,* 43: 350–63.

Chapter 9

An Empirical Method for Quantifying Drivers' Level of Acceptance of Alerts Issued by Automotive Active Safety Systems

Jan-Erik Källhammer
Autoliv Development, AB, Sweden

Kip Smith
Naval Postgraduate School, USA

Erik Hollnagel
University of Southern Denmark, Denmark

Abstract

This chapter addresses three issues related to false alarms and the development of automotive active safety systems. First, it is prudent for system developers to acknowledge that false alarms are inevitable considering the rarity of accidents and to focus on achieving driver acceptance for alerts that are false alarms. Second, system developers who consider false alarms to be an integral part of the design of active safety systems that address potential accident situations can take advantage of the drivers' subjective perception of those situations to guide the specification of the system's alerting criteria. Third, this approach to the development of active safety systems is likely to produce systems that achieve relatively higher levels of driver acceptance.

Introduction

This paper presents a review of issues raised by the prevalence of false alarms by automotive active safety systems. It has three sections. The first defines the terms 'driver acceptance' and 'false alarms', discusses driver acceptance of false alarms and reviews false alarms within the context of the development of automotive active safety systems. The second section urges that system developers acknowledge that false alarms are not only pragmatically unavoidable, but have genuine utility when developing active safety systems. The third section argues that high levels of driver acceptance of issued alerts should become one of the main targets of system development. It presents an empirical methodology for system development based on drivers' acceptance of alerts in situations where false alarms are unavoidable.

The discussion in this review is restricted to driver assistance systems that issue alerts but that do not intervene to initiate a vehicle response. Much of the discussion can be extended to include systems that do take control in some form. Issues related to alarm reliability (Bliss and Gilson 1998), the ability of a human observer (driver) to detect and act on an issued alert, and the influence of alert modality and intensity on driver acceptance are beyond the scope of this review.

Active Safety Systems

An active safety system may be defined as any automotive safety system with functionality that is activated before the collision. They are designed to assist drivers in the detection of potential accident situations. Such systems are also referred to as primary safety systems. In contrast, passive safety systems – or secondary safety systems – are designed to mitigate the consequences after a collision has occurred.

There are two broad classes of active safety systems, those that issue alerts and those that autonomously intervene to initiate a vehicle response. The effectiveness of systems that issue alerts will depend on timely and appropriate action by the driver. Systems that autonomously take steps to avoid an accident will not depend on either the driver's reaction or level of acceptance of the system or of its response. However, any system activation – either alert, intervention or both – will likely modify the driver's attitude towards the system. Accordingly, the benefits of any active safety system that issues alerts will depend, in part, on driver acceptance and the adequacy of the technology.

Driver Acceptance and False Alarms

In line with Abe and Richardson (2005), Breznitz (1983) and Vlassenroot et al. (2010), we define 'driver acceptance' as the driver's attitude towards an installed system where the degree of acceptance is influenced by the rate and nature of its misses and false alarms. This definition ties driver acceptance to well-known constructs of Signal Detection Theory (Green and Swets 1966), a familiar and powerful tool for quantifying the accuracy and bias of decision outcomes. Signal Detection Theory stands as the basis of our discussion of false alarms, although we acknowledge that other definitions have been proposed, for example, Xiao and Seagull (1999).

We use the word 'alert' as a general description of any response issued by an active safety system, independent of its modality or objective or subjective validity. We use the term 'false alarm' exclusively for alerts that are false according to a strict definition: alerts to predicted events that do not occur. We use the term 'miss' for events that do occur but are not predicted by the system.

As the consequences of a missed detection can be catastrophic, it often is more important to reduce misses than to eliminate false alarms when designing systems that can reduce the risk of fatalities (Rice and Trafimow 2010, Zabyshny and Ragland 2003). Reducing the rate of alerts that are false alarms by setting the decision criteria more conservatively often leads to a delayed activation of the alert. Alerts that come too late are often mistrusted (Abe and Richardson 2005). However, conventional wisdom states that false alarms will reduce the trust in system reliability and, as a consequence, compliance with issued alerts. This distrust is often referred to as the 'cry wolf' syndrome (Breznitz 1983). It can lead the driver to neglect the system or to find creative ways of bypassing it. The erosion of confidence in the system may lead to underuse and even to disuse of the system (Farber and Paley 1993, Lerner et al. 1996, Parasuraman and Riley 1997).

A false alarm is always a post hoc categorisation. To determine whether the alert was correct or a false alarm, it must be known whether the event occurred or not. In the context of vehicle collisions, an alert that is a false alarm is an alert to any set of conditions that could be associated with a collision but which do not lead to one (non-collision event). Strictly speaking, this definition implies that even an alert to a situation where the driver avoided the collision is a false alarm – as the event (collision) did not take place. The alternative, classifying the event as a true alarm with a successful outcome, is likely to be difficult to justify, as the influence of any alert or response by the system may be hard to demonstrate.

Our use of the term 'alert' does not imply any level of correctness or immediacy. It adheres to the condition in Signal Detection Theory where the null hypothesis has been rejected and the issued alert may be either a hit or a false alarm. We do not follow Bliss and Gilson (1998) to differentiate a 'warning' from an 'alarm'. We believe our terminology is more in line with those more traditionally used in discussions of automotive safety.

A complication inherent to any discussion of false alarms in the context of automotive safety is confusion about what actually constitutes a false alarm. One line of thinking holds that false alarms are system failures of some kind and should be avoided. For example, Lerner et al. (1996) stated that false alarms imply some type of hardware failure or a situation where an algorithm has incorrectly identified a non-threatening situation as a hazard. We suggest that this line of thinking is not very useful. It is often difficult to establish whether or not a system has failed or evasive actions were initiated by either party.

False Alarms and Nuisance Alerts

Researchers in Europe, Japan and the United States have conducted many studies on the impacts of alerts that are false alarms on driver acceptance of active safety systems. Many of these studies distinguish between useful alerts that are false alarms and those that are a nuisance. This distinction may originally have been drawn in studies of Forward Collision Warning (FCW) systems sponsored by the US National Highway Traffic Safety Administration (NHTSA 2005). FCW systems are designed to advise drivers to impending rear-end collisions. Kiefer et al. (1999) used the two terms 'nuisance alarms' and 'nuisance alerts' interchangeably and defined them as alerts issued by the FCW system that the driver believes are not justified by the situation. They described the ideal system as one that acts like an always attentive passenger, providing a crash alert only when he or she becomes alarmed. Kiefer et al. acknowledged that the identification of an alert as either a nuisance or as welcome is necessarily subjective. Harrington et al. (2008) discuss the subjective nature of nuisance alerts and suggest that the criteria that make an alert a nuisance are driver- and context-dependent.

Kiefer et al. (1999) distinguished three cases of nuisance alerts: those caused by noise or interference when there is no object present, out-of-path nuisance alerts and in-path nuisance alerts. Out-of-path nuisance alerts are caused by objects that are not in the path of the subject vehicle. Some in-path nuisance alerts are caused by vehicles that are in the path of the subject vehicle but are at a distance or moving at a speed that drivers do not perceive as alarming. Other in-path nuisance alerts are issued in situations where the driver can avoid a collision by his or her normal braking behaviour and intensity.

As Kiefer et al. (1999) point out, alerts that should trigger a driver response have to be issued early enough to allow an inattentive driver to take appropriate action. As a result, an alert that some drivers consider a nuisance may be accepted by others as both valid and helpful. This has a clear implication for system design: the rate of acceptance is likely to be increased if drivers can adjust the level of threshold. However, such adjustments should never frustrate system functionality.

Najm et al. (2006) defined false alarms in the context of rear-end collision avoidance systems as alerts issued in situations where the host vehicle is not on a rear-end crash course with an in-path obstacle. They also defined the terms 'conflicts' and 'near-crashes' along a continuum of situations that require driver (re)action at the last second. A situation that is a conflict requires driver braking or steering at normal response levels, whereas a near-crash requires hard deceleration or steering manoeuvres at the last second. Implicit in these definitions is the contention that only alerts to events that would lead to a collision would be considered a true alert. However, Najm et al. pragmatically acknowledged the fuzziness of false alarms and driver acceptance. On the one hand, they pointed out that alerts may be considered nuisances by some drivers if they are deemed to be unnecessary or come too early. On the other hand, they suggested that out-of-path alerts (which they defined as false alarms) may be helpful in getting a distracted driver to refocus attention back to the road. Thus, Najm et al. considered alerts that are false alarms to be helpful if they get a positive result.

Lees and Lee (2007) distinguished between false alarms and unnecessary alarms. They defined false alarms as non-useful, unintended alerts that are either inconsistent with the design of the system or characterised by unpredictable activation.

This line of thinking implies that alerts are false alarms if they do not match a threat or are not understood by the driver. In contrast, they defined 'unnecessary alarms' as alerts that are predictable and understood by the driver, but that are not considered useful. Thus, an unnecessary alarm is a nuisance that is fully consistent with the design of the system.

Our reading of Kiefer et al. (1999), Najm et al. (2006), Lees and Lee (2007) and others understands them to argue that the distinction of whether or not alerts are false should be less relevant than whether or not they are useful. A correct alert to conditions that often precede a collision where the collision is avoided (due to driver action) is a false alarm, according to our strict definition, and may be considered a nuisance by many drivers but not by all (Smith and Zhang 2004). Indeed, many drivers are likely to find a false alarm caused by an object or event in a situation normally associated with the risk of an accident to be understandable, acceptable and useful.

In summary, many (but not all) automotive researchers have concluded that the distinctions between a false alarm that is a nuisance and a false alarm that is helpful and acceptable depends on both context and driver judgment. Many alerts that are false alarms are in fact useful.

False Alarms Are Unavoidable and Have Utility

The base rate for crashes is low. Actuarial statistics based on US data indicate that fatal motor vehicle crashes can be expected approximately once every 5,000 driver years, a crash resulting in an injury can be expected approximately once every 100 driver years and property damage crashes approximately once every 50 driver years (NHTSA 2005). Data from Sweden, the UK and the Netherlands – the countries of the European Union with the best traffic safety records – show that fatal crashes can be expected approximately once every 8,000 driver years, severe injuries approximately every 900 years and slight injuries approximately every 200 years (Koornstra et al. 2002). The implication of these actuarial data is that the activation of a specific active safety system in response to a situation that actually would lead to bodily injury (if the active safety system were not present) will occur, on average, less than once in the lifetime of every driver.

Because accidents are rare, the base rate of true alerts – alerts that are *not* false alarms – is necessarily low. This fact has two undesirable consequences if designers strive to eliminate all false alarms. First, the few true alerts would be so rare as to be utterly unfamiliar and the driver's reaction unpredictable, even if the alert had succeeded in directing the driver's attention appropriately. Second, the frequency of alerts would be insufficient to enable drivers to calibrate trust in the system.

Driver awareness of system functionality is influenced by the frequency of alerts. It is an irony of automation that efficient recall of how to react depends on the frequency of use (Bainbridge 1983). Accordingly, the rarity of alerts that are not false alarms reduces the likelihood that drivers would be able to respond appropriately and in a timely fashion (Lee et al. 2002, Parasuraman, Hancock and Olofinboba 1997). An alert only in situations leading to a crash would be so rare that it might aggravate an already critical situation.

Any system designed to provoke the driver to take action requires driver awareness of the meaning and utility of its alerts. This awareness enables the driver to develop trust in the system (Riley 1996). The development of trust may rely on hearsay or reputation but is more generally based on experiencing many alerts and forming an opinion about the system's reliability and predictability (Lee and See 2004).

Due to the low base rates of traffic collisions, we believe that the only type of system activation frequent enough to provide this experience is the false alarm. Accordingly, we encourage designers of active safety systems to accept that the system will issue alerts that are false alarms and to work to ensure that those alerts are sufficiently common and predictable that drivers accept them and can calibrate their trust in the system.

In summary, alerts that are false alarms will likely be accepted if they provide useful, trustworthy information to the driver. It also implies that false alarms are the only source of alerts frequent enough to allow the driver to develop trust in the system. These considerations support our proposal that system designers need to recognise that false alarms are not necessarily undesirable and, in fact, can support drivers' development of trust in the system and their acceptance of the system.

False Alarms Are Not Necessarily Bad

The negative consequences of alerts that are false alarms can be largely mitigated if they are predictable (Lee and See 2004). Large, but predictable, errors may affect driver trust in the system less than minor unpredictable faults (Muir and Moray 1996). The effectiveness of the system depends on whether the driver will become aware of the risky situation earlier with the system than without it (McLaughlin, Hankey and Dingus 2008). To be effective, alerts need to be rare, predictable and informative. Drivers can benefit from alerts that are false alarms if they can adopt a strategy that can benefit from the imperfect information they provide.

Even somewhat unreliable systems can aid distracted users (Dixon and Wickens 2006, Maltz and Shinar 2007). There is a body of research that indicates that drivers may accept alerting systems that produce nuisance alerts systematically and relatively frequently. For example, vehicles in the field experiments reported by Lerner et al. (1996) issued nuisance alerts routinely once or twice a week. Drivers reported minimal annoyance levels. Learner et al. concluded that intrusive alarms are acceptable at modest rates. LeBlanc et al. (2008) suggest that 15 alerts per 100 miles (160 km) driven may be an acceptable level of nuisance alerts. Precisely how they reached that number is, however, unclear.

The goal of minimising misses and false alarms is important, but the overriding goal should be to prompt appropriate driver behaviour in response to all alerts, especially those that are 'true alarms'. We argue that this can be achieved if and only if the driver regularly accepts most of the alerts issued by the system.

Our view of false alarms takes into account the utility of the information and the driver's perception of the situation. For example, an alert by a pedestrian detection system to a child in the street will likely be considered useful information even though there is no immediate risk of a collision. Many drivers will likely accept an alert for this type of situation. False alarms are useful when they match the driver's expectation for an alert given the situations.

Rethinking False Alarms

Our discussion of these issues supports the claim that drivers' perceptions of issued alerts can provide useful information in the development of alerting strategies that better match the drivers' expectations.

Sullivan, Tsimhoni and Bogard (2008) noted that the subjective assessment of the reliability of a system may be the most important influence on the driver's response to its alerts. As this reliability needs to be measured, they suggested an approach that asks the driver to directly assess the system. The method we propose – driver acceptance ratings to calibrate system design – follows their suggestion.

We propose that drivers' acceptance should shape the system's activation requirements (alerting criteria). To achieve this goal, designers need to focus on the driver's expectations and to set the target of making driver acceptance as high as possible, rather than focusing on reducing the false alarm rate. We must therefore rethink the concept of false alarms. False alarms – alerts to situations that may develop into a crash but do not – provide useful information to the driver and, when issued in situations where the risk of an accident is self-evident, are likely to be received with relatively high levels of driver acceptance. Just as Aven (2009: 929) defined safety 'by reference to acceptable risk', the performance of an active safety system may be defined by reference to acceptable false alarm rates.

A major implication of our argument to reconsider the utility of false alarms is that researchers and designers should seek to understand the factors that influence driver acceptance of alerts that are false alarms. Understanding these factors makes it possible to define objective criteria that are suitable for implementation in the decision algorithms used by active safety systems.

Designing for Acceptance of False Alarms

This re-evaluation of the utility of false alarms leads us to propose an empirical methodology for the design of active safety systems. If we can define the factors that predict when most drivers will accept alerts that are false alarms, it becomes reasonable to focus the design of an active safety system on driver expectations for when the system should issue an alert. By conforming to driver expectations, the system should be able to achieve a relatively high level of user acceptance and become an effective partner in the driver-vehicle system.

Our method uses video recordings of a large number of actual traffic situations recorded from the driver's point of view. The recordings are captured by an active safety system that has a reasonable (e.g., prototype) level of performance in the field, on the road. A single frame from an event recorded using a Night Vision system is shown in Figure 9.1.

Figure 9.1   A typical alert issued by a Night Vision system with pedestrian alert

The system couples a far infrared (FIR) sensor, pedestrian recognition algorithms, alerting logic and a console display. Prior to presentation in the laboratory, the video clips are reviewed to eliminate both incorrectly detected cars, pedestrians or other objects and traffic situations where the reason for an alert might be ambiguous or unrelated to the purpose of the study. Presenting these recordings in a lab environment provides experimental control of stimuli while retaining much of the high ecological validity of actual traffic events.

Participants in the lab are experienced drivers. In a self-paced task, they view a recording and rate the acceptability of an alert to its traffic situation. The ratings can be elicited using either a slider bar like that shown in Figure 9.2 or a set of radio buttons that create a Likert-type scale. The scale bar is anchored at one end by 'Reject' and at the other by 'Accept'. When asked, participants have indicated that they understood the scale to represent a continuum from 'completely acceptable' to

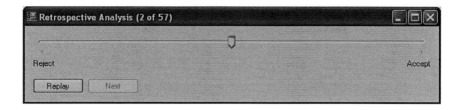

Figure 9.2   Continuous scale to rate the level of acceptance of an assumed alert

'completely unacceptable'. The task is complete when the participant has viewed and rated the alerts to the full set of video recordings.

Our method builds upon two established procedures. The first is the self-report rating tool discussed by Van der Laan, Heino and De Waard (1997) that may be the most widely used technique for assessing driver acceptance of new automotive technology. This tool measures driver acceptance by asking participants to rate nine different attributes of the evaluated system using an anchored five-point scale. It has been used to compare driver responses to a variety of systems.

The second foundation for our method is the hazard perception test that is part of UK driving tests (Jackson, Chapman and Crundall 2009). In this test, participants are required to watch a series of video clips recorded from the driver's point of view and to push a button when he or she detects a hazard. The response time from the appearance of the hazard is the dependent measure. Example hazards are pedestrians, parked cars, cyclists and other vehicles, regardless of traffic lane or direction.

Like the hazard perception test, our approach quantifies the relative level with which drivers are likely to accept an alert from an active safety system and how that level varies across situations. It follows Van der Laan et al. (1997) to collect subjective ratings rather than the response times. Instead of the nine scales advocated by Van der Laan et al., our approach uses a single measure of alert acceptability to simplify and clarify the participants' task. The ratings are ranked to control for individual differences in scale use and the ranks thereafter used in within-subjects statistical analyses.

Application

A measure of agreement between the field and laboratory can be gained by comparing the post hoc, laboratory rating by the drivers who experienced the recorded events to the responses of the participants who did not. The correlations between the drivers' ratings and the means of the others' ratings has always exceeded 80 per cent. These results suggests that the ratings are robust and generalisable to the population sampled.

In one study, drivers had a two-button response unit (Accept/Reject) that they used to indicate at the time of pedestrian encounters in the field whether or not they found the alerts issued by the system to be acceptable. Our review of the in-field button pushes and the ratings provided in the laboratory suggests that the laboratory results are highly consistent with responses elicited in the field.

This method for extracting useful information from alerts that are false alarms can be used to assess driver acceptance at the various stages of system development or during FOT. For example, Källhammer et al. (2007) applied the method to elicit drivers' assessments of a variety of naturalistic traffic situations. It has also been used by Smith and Källhammer (2010) to assess the risk posed by intersection encroachments.

Källhammer and Smith (2012) used the method to assess the acceptability of 57 pedestrian alerts that were all false alarms and to identify factors that influence drivers' ratings. A regression analysis identified two factors, pedestrian location and velocity. A follow-up study generated a best-fit regression model using pedestrian location and velocity as predictor variables and mean ratings as the response variable (Smith and Källhammer 2012). The model explained more than 60 per cent of the variability in driver ratings.

Discussion

The method can be used to obtain immediate feedback following replays of actual false alarm events or to obtain retrospective laboratory analysis by the drivers themselves or by others. The method is proving to be a cost-effective tool for bridging the gap between field experiments with their high level of ecological validity and lab-based experiments with their high level of experimental control.

Eliciting driver ratings of the acceptance of alerts that are false alarms can be used to test alternate criteria for issuing an alert. By varying the criteria that generate the alerts for the collected incidents, it is possible to test hypotheses about their impact on driver acceptance of the system. The method can also be applied to video data created in a simulated or animated environment, benefiting from the additional experimental possibilities simulators provide. Driver acceptance ratings of animations based on reconstructed accidents could provide valuable additional data when actual crashes are not available for the driver to assess.

Disclaimers

We do not claim that the proposed method will work for all automotive active safety applications. The driver's ability to judge system performance will depend on his or her ability to judge the condition determining an alert (Sullivan et al. 2008). A forward collision warning is one example where the method may have limited utility if the alerting system acts on information that the driver has difficulty perceiving or understanding or both. As we have yet to extend the method to investigate behavioural responses to alerts that are false alarms, we cannot comment on the linkage between false alarms and driver behaviour.

Chen and Terrence (2009) showed that the response to systems that are prone to issuing false alarms and/or misses varies across individuals and their scores on a test of attentional control. Participants with relatively high scores found false alarms to be more disadvantageous than did participants with relatively low scores. Conversely, miss-prone systems were deemed more disadvantageous by participants with low scores on the test.

Conclusion

The involvement of drivers in the function of active safety systems implies that the systems cannot be seen as technical systems alone. Knowledge of how the driver and the vehicle function together as a joint cognitive system (Hollnagel and Woods 2005) is critical to achieving a successful design. Driver acceptance has to be an important design goal for any active safety system. To achieve this goal, designers need to focus on the driver's expectations and to set the target of making driver acceptance as high as possible. Finding the proper balance between false alarms that are perceived as useful and as nuisances will require extensive empirical work both in the laboratory and the field. We have developed and are using a method that elicits from drivers assessments of the utility of alerts that are false alarms. These assessments help to tune the design of active safety systems and promote their acceptance.

References

Abe, G. and Richardson, J. 2005. The influence of alarm timing on braking in low-speed driving. *Safety Science* 43: 639–54.

Aven, T. 2009. Safety is the antonym of risk for some perspectives of risk. *Safety Science* 47: 925–30.

Bainbridge, L. 1983. Ironies of automation. *Automatica* 19: 775–9.

Bliss, J.P. and Gilson, R.D. 1998. Emergency signal failure: Implications and recommendations. *Ergonomics* 41: 57–72.

Breznitz, S. 1983. *Cry-wolf: The psychology of false alarms*. Hillsdale, NJ: Erlbaum.

Chen, J.Y.C. and Terrence, P.I. 2009. Effects of imperfect automation and individual differences on concurrent performance of military and robotics tasks in a simulated multitasking environment. *Ergonomics* 52: 907–20.

Dixon, S.R. and Wickens, C.D. 2006. Automation reliability in unmanned aerial vehicle control: A reliance-compliance model of automation dependence in high workload. *Human Factors* 48: 474–86.

Farber, E. and Paley, M. 1993. *Using freeway traffic data to estimate the effectiveness of rear-end collision countermeasures.* Proceedings of the Third Annual Meeting of Intelligent Vehicle Highway System (IVHS) America, Washington, DC.

Green, D.M. and Swets, J.A. 1966. *Signal detection theory and psychophysics.* New York: Wiley.

Harrington, R., Lam, R., Nodine, E., Ference, J.J. and Najm, W.G. 2008. *Integrated vehicle-based safety systems light-vehicle on-road test report.* Online: National Highway Traffic Safety Administration Report DOT HS 811020, Washington, DC. Available at: http://deepblue.lib.umich.edu/bitstream/ 2027.42/58195/1/100880.pdf.

Hollnagel, E. and Woods, D.D. 2005. *Joint cognitive systems: Foundations of cognitive systems engineering.* Boca Raton, FL: Taylor and Francis.

Jackson, L., Chapman, P. and Crundall, D. 2009. What happens next? Predicting other road users' behaviour as a function of driving experience and processing time. *Ergonomics* 52: 154–64.

Källhammer, J.-E. and Smith, K. 2012. Assessing contextual factors that influence acceptance of pedestrian alerts by a night vision system. *Human Factors* 54: 654–62.

Källhammer, J.-E., Smith, K., Karlsson, J. and Hollnagel, E. 2007. *Shouldn't the car react as the driver expects?* Proceedings of the 4th International Driving Symposium on Human Factors in Driver Assessment, Training, and Vehicle Design (Stevenson, WA, US).

Kiefer, R., LeBlanc, D., Palmer, M., Salinger, J., Deering, R. and Shulman, M. 1999. *Development and validation of functional definitions and evaluation procedures.* Online: National Highway Traffic Safety Administration Report DOT HS 808 964, Washington, DC.] Available at: http://www.itsdocs.fhwa. dot.gov/jpodocs/repts_te/87L01!.pdf.

Koornstra, M., Lynam, D., Nilsson, G., Noordzij, P., Pettersson, H.E., Wegman, F. and Wouters, P. 2002. *Sunflower: A comparative study of the development of road safety in Sweden, the United Kingdom and the Netherlands.* Online: Leidschendam: SWOV Institute for Road Safety Research. Available at: http:// www.swov.nl/rapport/Sunflower/Sunflower.pdf.

LeBlanc, D., Bezzina, D., Tiernan, T., Freeman, K., Gabel, M. and Pomerleau, D. 2008. *System performance guidelines for a prototype integrated vehicle-based safety system (IVBSS) – Light vehicle platform.* Online: University of Michigan Transportation Research Institute Report UMTRI-2008-20, Ann Arbor, MI, USA.] Available at: http://www.umtri.umich.edu/content/DOT_ HS_811_020.pdf.

Lee, J.D., McGehee, D.V., Brown, T.L. and Reyes, M.L. 2002. Collision warning timing, driver distraction, and driver response to imminent rear-end collisions in a high-fidelity driving simulator. *Human Factors* 44: 314–34.

Lee, J.D. and See, K.A. 2004. Trust in automation: Designing for appropriate reliance. *Human Factors* 46: 50–80.

Lees, M.N. and Lee, J.D. 2007. The influence of distraction and driving context on driver response to imperfect collision warning systems. *Ergonomics* 50: 1264–86.

Lerner, N.D., Dekker, D.K., Steinberg, G.V. and Huey, R.W. 1996. *Inappropriate alarm rates and driver annoyance.* Online: National Highway Traffic Safety Administration Report DOT HS 808 533, Washington, DC. Available at: http://www.itsdocs.fhwa.dot.gov/JPODOCS/REPTS_TE/3757.pdf.

Maltz, M. and Shinar, D. 2007. Imperfect in-vehicle collision avoidance warning systems can aid distracted drivers. *Transportation Research Part F* 10: 345–57.

McLaughlin, S.B., Hankey, J.M. and Dingus, T.A. 2008. A method for evaluating collision avoidance systems using naturalistic driving data. *Accident Analysis and Prevention* 40: 8–16.

Muir, B.M. and Moray, N. 1996. Trust in automation: 2. Experimental studies of trust and human intervention in a process control simulation. *Ergonomics* 39: 429–60.

Najm, W.G., Stearns, M. D., Howarth, H., Koopmann, J. and Hitz, J. 2006. *Evaluation of an automotive rear-end collision avoidance system.* Online: National Highway Traffic Safety Administration Report DOT HS 810 569, Washington, DC. Available at: http://www.nhtsa.dot.gov/portal/nhtsa_static_file_downloader.jsp?file=/staticfiles/DOT/NHTSA/NRD/Multimedia/PDFs/Crash Avoidance/2006/HS910569.pdf.

National Highway Traffic Safety Administration (NHTSA). 2005. *Traffic safety facts 2004.* Online: NHTSA Report DOT HS 809 919, Washington, DC. Available at: http://www-nrd.nhtsa.dot.gov/Pubs/TSF2004.PDF.

Parasuraman, R. and Riley, V. 1997. Humans and automation: Use, misuse, disuse, abuse. *Human Factors* 39: 230–53.

Parasuraman, R., Hancock, P.A. and Olofinboba, O. 1997. Alarm effectiveness in driver centered collision-warning systems. *Ergonomics* 40: 390–99.

Rice, S. and Trafimow, D. 2010. How many people have to die over a type II error? *Theoretical Issues in Ergonomics Science* 11: 387–401.

Riley, V. 1996. Operator reliance on automation: Theory and data. In *Automation and Human Performance: Theory and Applications*, 19–35. Edited by R. Parasuraman and M. Mouloua. Mahwah, NJ: Erlbaum.

Smith, K. and Källhammer, J.-E. 2010. Driver acceptance of false alarms to simulated encroachment. *Human Factors* 52: 446–76.

Smith, K. and Källhammer, J.-E. 2012. *Experimental evidence for the field of safe travel.* Proceedings of the 56th Annual Meeting of the Human Factors and Ergonomic Society, 2246–50. Boston.

Smith, M. and Zhang, H. 2004. *SAfety VEhicles using adaptive interface technology (task 8): A literature review of intent inference.* Washington, DC: Research and Innovative Technology Administration. Available at: http://www.volpe.dot.gov/coi/hfrsa/work/roadway/saveit/docs/dec04/litrev_8a.pdf.

Sullivan, J., Tsimhoni, O. and Bogard, S. 2008. Warning reliability and driver performance in naturalistic driving. *Human Factors* 50: 845–52.

Van der Laan, J.D., Heino, A. and De Waard, D. 1997. A simple procedure for the assessment of acceptance of advanced transport telematics. *Transportation Research Part C* 5: 1–10.

Vlassenroot, S., Brookhuis, K., Marchau, V. and Witlox, F. 2010. Transport Systems (ITS): A conceptual analysis based on the case of Intelligent Speed Adaptation (ISA). *Transportation Research Part F* 13: 164–78.

Xiao, Y. and Seagull, F.J. 1999. *An analysis of problems with auditory alarms: Defining the roles of alarms in process monitoring tasks.* Proceedings of the Human Factors and Ergonomics Society 43rd Annual Meeting. Santa Monica, CA: HFES.

Zabyshny, A.A. and Ragland, D.R. 2003. False alarms and human-machine warning systems. Online: UC Berkeley Traffic Safety Center UCB-TSC-RR-2003-07, Berkeley, CA. Available at: http://repositories.cdlib.org/its/tsc/UCB-TSC-RR-2003-07.

PART IV
Data on Driver Acceptance:
Case Studies

Chapter 10

Driver Acceptance of In-Vehicle Information, Assistance and Automated Systems: An Overview

Gary Burnett

Human Factors Research Group, Faculty of Engineering, University of Nottingham, Nottingham, UK

Cyriel Diels

Coventry School of Art and Design, Department of Industrial Design, Coventry University, Coventry, UK

Abstract

This chapter provides an overview of Human Factors issues relevant to the acceptance by drivers of technology-based systems within vehicles. A distinction is made between issues relevant to systems providing information to support driving-related tasks (e.g., navigation), systems that provide some degree of control-based assistance (e.g., Adaptive Cruise Control) and those systems which automate the driving task (e.g., platooning). It is recognised that a range of Human Factors issues will have a direct influence on the acceptance of these systems, including those related to distraction, trust and reliability. Moreover, it is apparent that acceptance itself will impact on system usage, primarily raising issues of reliance. The chapter concludes by highlighting some topics which have received relatively little consideration, but will be critical for the ultimate acceptance of in-vehicle systems.

Introduction

It is widely acknowledged that vehicles are experiencing a revolution in design as increasing amounts of computing and communications technologies are being introduced within everyday driving situations. Many technologies are utilised, but from the driver's perspective, systems can be classified into three broad categories:

1. Systems that provide information or warnings of relevance to the driving task (e.g., navigation, traffic and travel, driver status monitoring, lane departure warnings);
2. Systems that aim to assist the driver in fundamental and specific vehicle control tasks (e.g., adaptive cruise control, collision avoidance, intelligent speed adaptation); and
3. Systems that replace the driver in a range of vehicle control tasks, ultimately automating driving (e.g., platooning, driverless cars).

In addition, it is important to note that a range of systems provide information and services related to other salient goals: for instance, for comfort/entertainment purposes or to enhance working productivity (e.g., email/Internet access). These systems are important in terms of the impact they can have on primary driving tasks (distraction, behavioural adaptation etc.).

The acceptance of such technology by end users (predominantly drivers and their passengers) is important for several reasons. Firstly, and perhaps most importantly, systems must be accepted if they are then to be used (a utility argument), such that the fundamental design goals for a system (safety, driving efficiency and so on) have the potential to be met. Secondly, an understanding of acceptance is required when considering the closely related issues of usability and satisfaction (see also the chapters in this book by Green and Jordan, Chapter 18; and Stevens and Burnett, Chapter 17). As noted by Faulkner (2000), there is no universal view on how these various 'soft' terms should be defined, but it is clear that they impinge on each other. Finally, acceptance is highly relevant to key issues of trust and reliance for in-vehicle technology (see also the chapter in this book by Ghazizadeh and Lee, Chapter 5). When new systems are wholly accepted, trust levels may be overly high and there may be a mismatch between objective and subjective levels of reliability for a system. Consequently, complacency effects may arise (e.g., following instructions from a navigation system when it is inappropriate to do so). Conversely, a system considered unacceptable to users may be deemed untrustworthy and may be used in an inappropriate fashion (misuse effects). Such behavioural adaptation is a common result of new technological interventions within an overall system perspective (Wickens et al. 2004).

Considerable data have been collected by the research community relating to the acceptance of in-vehicle information, assistance and automating systems. This chapter sets the scene for subsequent chapters in this section by highlighting the breadth of studies that have been conducted. In particular, we will discuss acceptance issues for three distinct example systems: vehicle navigation (information), adaptive cruise control (assistance) and platoon driving (automating). In these three types of systems, there are considerable differences in the maturity and adoption of the technology. Moreover, the level of automation associated with the technology rises with each subsequent example, leading to differences in the fundamental Human Factors issues of interest. As a consequence of such variation, the nature of research and conclusions that can be drawn can be expected to be significantly different.

Acceptance Issues for Specific Systems

Vehicle Navigation Systems

Vehicle navigation systems are an example of a ubiquitous information technology where there has been considerable Human Factors research both before and after widespread implementation (e.g., Ross et al. 1995, Forbes and Burnett 2007). These systems aim to support drivers in the strategic and tactical components (planning and following routes, respectively) of the driving and navigating task. They have become increasingly popular in recent years, across many countries, as costs have reduced and the technology has matured. Three broad types of system now exist, each with their own distinct advantages and disadvantages:

1. Original equipment manufacturer (OEM) systems (integrated within a vehicle);
2. Personal navigation devices (PNDs – nomadic devices, designed specifically to support navigation); and
3. Smartphones (multifunctional, small screen devices which possess navigation functionality).

There has been an evolution in design of these systems since the 1990s, reflecting technological advances as well as acceptance issues for users. Originally, navigation systems used by drivers were wholly OEM solutions. More recently, PNDs have been dominant, reflecting consumers' desires for affordable, dedicated and portable devices. Nevertheless, this situation is changing as increasingly people exploit the convenience of navigation functionality within a smartphone device. Indeed, there is now some linking between system types as vehicle manufacturers have developed 'human–machine interfaces' (HMIs) which can adapt an in-vehicle system to account for the presence of a smartphone (e.g., by utilising the preferences stored on a smartphone but presenting information within the vehicle according to an OEM solution – larger display, in-vehicle speakers, steering wheel controls, etc.). For current trends in navigation system design, see http://www.autoevolution.com/newstag/satnav/.

Considerable literature has focused on the human-centred design issues for vehicle navigation systems (see reviews by Srinivisan 1999 and Burnett 2009). It is evident that the majority of research in this area assumes that issues of distraction and excessive workload are the most likely concerns that navigation system designers must be aware of. The fundamental supposition is that acceptance will not arise unless basic safety concerns are accounted for. Previous authors have noted that driving is already a complex, largely visual task and navigation system

HMI designs will be associated with divided attention and additional information processing (Srinivisan 1999, Moriarty and Honnery 2003). Consequently, there is potential for drivers to make fundamental errors whilst engaging with a navigation system, such as failing to observe in time a lead vehicle slowing down or wavering out of lane (Green 2007).

There is considerable literature (especially from the 1980s and 1990s) focusing on the distraction effects of navigation systems (visual, auditory, cognitive and biomechanical) and several influential design guidelines handbooks have been produced, informed by research studies with this focus (e.g., Ross et al. 1995, Green et al. 1997; Campbell, Carney and Kantowitz 1998). These handbooks provide a wide range of guidance for designers concerning issues as diverse as the choice of modality for interfaces, the content and timing of voice messages, display position, colour combinations, font types/sizes, orientation of map displays and so on. Clearly, such handbooks can be important source documents for Human Factors professionals in an industry wishing to argue a case for a specific HMI.

Based on the authors' understanding of the content of these handbooks, it is tempting to speculate that many of the current vehicle navigation HMIs have been influenced by the available guidance. In particular, many vehicle navigation systems are clearly designed to make the workload associated with the navigation task low. This is often achieved using simple turn-by-turn instructions given in the auditory modality, combined with predominantly arrow-based graphics and distance-to-turn information. In some respects, this could be argued as a success for Human Factors research. Studies were conducted (often on public roads, but occasionally within simulators) to provide the 'believable' empirical data for guidelines; which accordingly have informed best practice (e.g., Burnett and Joyner 1996, Dingus et al. 1989). Unfortunately, however, as a result of the recent mass uptake, additional issues have come to light that impinge on safety/comfort, routing efficiency and ultimately acceptance, but may be of larger concern to drivers than distraction. In particular, two key issues relating to the automation effects of navigation systems have been found to be significant, which can be considered broadly under the headings of reliability and reliance.

Reliability

Surveys, in conjunction with considerable anecdotal evidence, have demonstrated the problems associated with unreliable guidance information from vehicle navigation systems. The resulting problems have obvious safety implications (e.g., when a driver turns the wrong way down a one-way street) and can have a considerable impact on the efficiency of the overall transport system (e.g., when a lorry gets stuck under a bridge).

Forbes (2009) (also reported in Forbes and Burnett 2007) conducted a survey of 872 navigation system owners, which established that 85 per cent had received inaccurate guidance. When asked about guidance that was considered dangerous/illegal, 23 per cent of respondents admitted to obeying the instructions on at least one occasion. Importantly, there was a clear relationship with age, such that older drivers were more likely to follow the unreliable guidance than their younger counterparts.

From an acceptance perspective, it is most interesting to consider here (a) why certain individuals are prone to following such instructions and (b) which characteristics of the HMI can contribute to the problem. This is an area around which there has been very little research to date. With respect to the former question, Forbes (2009) employed detailed follow-up diary studies with 30 navigation system users and used the data to hypothesise that, for certain drivers in specific situations, a *trust* explanation could be given. Specifically, there was evidence for over-trust (or complacency); that is, drivers saw the relevant road sign/cue, but chose to ignore it and favour the navigation instruction. In other contexts, there was evidence for an *attention*-based explanation, since drivers did not believe they saw or processed the relevant road sign/cue. In these cases, it is possible that characteristics of the system user-interface disrupted drivers' normal allocation of attention. More recent work conducted by Large and Burnett (2013, in press) considered these issues in a driving simulator context using eye tracking and confirmed objectively that two distinctive mechanisms are involved in this problem. For system acceptance, each of the mechanisms is likely to affect where drivers place the blame for their routing errors (agency) – either with themselves, the system or the surrounding road infrastructure.

Reliance

A further issue concerns drivers' long-term dependency on navigation systems, an outcome explicitly linked to overly high levels of system acceptance (Burnett 2009). Specifically, it has been noted that current technology automates core aspects of the navigation task, including trip planning (where the user's role is essentially to confirm computer-generated routes) and route following (where users respond to computer-generated filtered instructions) (Adler 2001, Burnett and Lee 2005, Reagan and Baldwin 2006). As a result, drivers are largely passive in the navigation task and consequently, fail to develop a strong mental representation of the space in which they are travelling, commonly referred to as a cognitive map. Several empirical studies have demonstrated this effect for drivers (Jackson 1998, Burnett and Lee 2005).

Several authors have provided convincing arguments as to why this issue is of concern (Burnett and Lee 2005, Jackson 1998). Specifically, it is noted that the following advantages exist for individuals who possess a well-formed cognitive map of an environment:

- Enhanced navigational ability – such people are able to accomplish navigation tasks with few cognitive demands based on their own internal knowledge. Indeed, it should be possible in certain environments (e.g., one's home town) to navigate using automatic processing, that is, with no conscious attention.
- Increased flexibility in navigation behaviour – informed individuals have the capacity to choose and then navigate numerous alternative routes to suit particular preferences (e.g., for a scenic versus efficient route) or in response to unanticipated situations (e.g., heavy traffic, poor weather, system failure or absence).
- Social responsibility – a well-formed cognitive map provides a wider transport efficiency and social function, since it empowers a person to navigate for others, for example, by providing verbal directions as a passenger, pedestrian or over the phone, sketching maps to send in the post and so on (Hill 1987).

This is essentially a complex trade-off problem. Notably, there is a conflict between the need to design navigation HMIs which enable an individual to acquire spatial knowledge (*active* navigation) and those which minimise the demands (or workload) of navigating (*passive* navigation). In this respect, authors have noted the potential for active, learning-oriented, HMIs for vehicle navigation systems, as an alternative to the current passive styles (Burnett and Lee 2005). Such interfaces would aim to provide navigation information in a form that ensures that the demands of the navigation task in wholly unfamiliar areas are at an acceptable, low level, whilst aiming to support drivers in the cognitive mapping process. In essence, these interfaces would aspire to move people onwards through the various stages of cognitive map development, ultimately to a level in which they are able to navigate effectively for themselves and others, independent of any external information. Some initial progress was made on this topic in a simulator study conducted by Oliver and Burnett (2008).

Adaptive Cruise Control (ACC)

Adaptive Cruise Control (ACC) is an example of a driver support (assistance) system which has been in production for several years. To date however, ACC is only offered as an optional feature in the luxury vehicle segment and the penetration rate is low as a consequence. Functionally, ACC will maintain a set speed as per conventional Cruise Control systems, when there is no traffic immediately ahead of the driver. In situations where traffic is ahead in the driver's lane, ACC uses radar to maintain a constant time headway to the vehicle ahead. This headway is kept constant by the system adjusting the speed of the vehicle to prevent exceeding a pre-defined time gap. First-generation ACCs require a minimum driving speed of typically 30 kph, below which the system is deactivated, requiring the driver to take over control below this speed. Similarly, manual control is regained when the driver deactivates the system by pressing the brake pedal. Second-generation ACCs have been developed that extend the utility of ACCs – by not only expanding the speed range to velocities below 30 kph but also bringing the vehicle to a complete stop and accelerating again if the preceding vehicle does so; a so-called Stop and Go function. Notwithstanding these significant system enhancements, ACCs have a limited deceleration level. Hence, under critical driving conditions, such as emergency braking situations, the driver is still required to regain control of the vehicle. It is for this primary reason that ACCs are marketed as comfort systems rather than safety systems.

ACC is predicted to have a number of positive effects. From the driver's perspective, it has already been shown that ACC can reduce workload and increase perceptions of comfort (e.g., Stanton, Young and McCaulder 1997). Furthermore, deployment of ACC is expected to lead to improved traffic safety, roadway capacity and environmental traffic impact (Vahidi and Eskandarian 2003). That is, shorter time headways as well as smoother acceleration and deceleration profiles help to increase road capacity and traffic flow whereas the minimum time headway adopted by ACC systems eradicate short, unsafe following distances. However, the extent to which these potential advantages materialise will be largely dependent on penetration rates which, at least for Europe, are predicted to be low in the foreseeable future – around 10 per cent in 2020 (Wilmink et al. 2008). A major factor in future deployment will be drivers' acceptance and willingness to engage with ACC systems.

User acceptance of ACC has been studied using a wide range of methods including interviews, questionnaire surveys, simulator experiments and field operational tests (FOTs). As part of the PROMETHEUS project, one of the earliest ACC acceptance studies was conducted by Becker et al. (1994) in which participants drove around in real traffic with prototype equipped vehicles. Results showed that ACC was perceived as a comfort-oriented and safety-enhancing driver assistance system. Overall, ACC was well received by participants and considered acceptable, comfortable, safe and relaxing. Similar results were obtained in a driving simulator study by Nilsson (1995) in which ACC was felt to add comfort and convenience to the driving experience. Fancher et al. (1995) conducted a field trial which showed that in comparison to conventional cruise control, ACC was perceived as more comfortable as it required fewer interventions. In dense traffic conditions, however, users tended to turn off the ACC as the system-defined headways were perceived to be too large resulting in other traffic cutting in.

Although these early studies suggest a high level of system acceptance, it is worth noting that acceptance may not be uniform across all users and may also depend on users' needs and motivations. For example, Hoedemaker and Brookhuis (1998) investigated ACC user acceptance as a function of users' driving style and found that, whereas ACC was perceived positively in terms of workload, comfort and usefulness, participants who liked to drive fast, as assessed using a driving style questionnaire, were less positive about it.

In 2005, the National Highway Traffic Safety Administration (NHTSA) in the United States reported the results of the Automotive Collision Avoidance System field operational test (ACAS FOT) program (NHTSA 2005). The FOT involved a 12-month period in which 11 cars equipped with ACC and Forward Collision Warning (FCW) were driven under natural conditions by a total of 96 participants. Each participant drove an equipped car for several weeks after which system acceptance was assessed using a combination of questionnaires, interviews and focus groups. Again, system acceptance was high with 75 per cent of participants intending to purchase an ACC system if they were buying a new car. When considering individual driver characteristics including age, gender, education and income, it was found that age was the best predictive factor, with older drivers reporting highest system acceptance. Notwithstanding the high acceptance levels, a number of ACC design characteristics were thought to benefit from future improvements. In particular, the maximum ACC speed and shortest available gap setting were considered too low. Some participants also mentioned the need to manually interfere due to the slowness for the system to decelerate and conversely, pick up speed in overtaking manoeuvres.

Similar to the ACAS FOT, Alkim, Bootsma and Looman (2007) reported the results of a Dutch field study which investigated drivers' use and acceptance of ACC, Lane Departure Warning (LDW), Headway Monitoring and Warning (HMW) and Lane Keeping Assistance (LKA) systems. Again, ACC enjoyed a high acceptance level. It further showed that the active assistance or intervention systems (i.e., ACC and LKA) enjoyed a higher level of acceptance than the warning systems (i.e., LDW and HMW). As pointed out by the authors, this was an unexpected finding given that drivers usually indicate a preference for an informative system rather than a system that takes over parts of the driving task. This difference in acceptance may be ascribed to the fact that the benefits of warning systems were not only perceived to be less apparent, but there was also a lack of system trust due to the high number of false alerts the systems produced. Furthermore, Alkim et al.'s (2007) findings that users are more positive after actual experience with such systems compared to a priori expectations clearly illustrates the point that the manner in which acceptance is evaluated (i.e., interview, on-road studies) affect users' perceptions and attitudes.

Most recently, Larsson (2012) conducted a questionnaire survey amongst 130 ACC owners regarding their daily use and experience. The study was limited in that it included only a specific ACC system. The results are nevertheless of interest and point towards some acceptance issues that have been consistently reported within the literature. These can be categorised as being either system limitations or communication errors. Regarding the latter, nearly a quarter of respondents indicated that they had forgotten at some point whether ACC was engaged and were subsequently surprised to find the vehicle braking or accelerating. This 'mode error', whereby the driver believes the system to be in one mode when it is actually in another, may not only negatively affect system acceptance but may also compromise road safety, as the unexpected vehicle behaviour may result in inappropriate or ill-timed driver responses. As pointed out by Larsson (2012), the occurrence of mode errors suggests that future systems would benefit from improved interface designs. The main factors affecting ACC acceptance, however, appeared to be related to system limitations. In particular, conditions in which radar contact was either lost (e.g., steep hills, sharp curves, roundabouts, exiting a motorway) or in which latches onto traffic in adjacent lanes led the vehicle to change its speed inappropriately were found to frustrate users. In addition, a recurrent complaint was the slow system response in picking up speed when changing lanes to overtake a vehicle.

Whereas the above-mentioned system limitations are clearly detrimental with regard to user acceptance, it can be concluded that overall, ACC enjoys a high level of user acceptance and is considered to significantly enhance drivers' comfort. Future system enhancements such as the incorporation of GPS, vehicle-to-vehicle, and vehicle-to-infrastructure data, as well as information on the driver's intentions (e.g., use of indicator) can be expected to eliminate or minimise current systems' limitations (e.g., Sol, van Arem and Hagemeier 2008), improving user acceptance even further.

Platoon Driving

Platoon driving is an example of a future technology where the Human Factors acceptance work is more speculative and usually simulator-based. Platoon driving, also referred to as platooning or road-training, can be viewed as a logical next step in road transport automation. It refers to the grouping of vehicles maintaining a short time headway achieved by using a combination of wireless communications, lateral and longitudinal control units, and sensor technology. Although different future platoon implementations can be envisaged (e.g., Martens et al. 2007), current concepts assume a system whereby the platoon is led by a trained, professional driver whilst the following vehicles are driven fully automatically by the system (Robinson, Chan and Coelingh 2010, Lank, Haberstroh and Wille 2011). Compared with ACC, platoon driving is extending the automation of the driving task considerably by adding lateral vehicle control which, in essence, leaves the driver free to relax or engage in non-driving tasks.

Platoon driving is predicted to provide a range of advantages (see Robinson et al. 2010 Lank et al. 2011). First, the small headways maintained in platoons result not only in a reduction in drag and associated energy efficiency, but also an increase in road network capacity due to the mere fact that less road space is required. A knock-on effect is that overtaking manoeuvres by other road users can be performed more quickly resulting in a more homogeneous traffic flow. Safety benefits are also expected: unlike drivers, the automated system does not suffer from distraction; and, secondly, the automated reaction times of the system are only a fraction of human responses times. Finally, the fact that the driving task is entirely taken over by the system is expected to result in enhanced driver comfort.

Regardless of the already proven technical feasibility as well as anticipated benefits (Lank et al. 2011), the success of platooning will depend ultimately on road users' as well as societal acceptance of the system. Compared to ACC, platooning creates a considerably more complex situation where we not only have to take into consideration the driver within the platoon but also road users' driving in their vicinity (Gouy et al. 2012). With regards to the former, there are some significant Human Factors challenges (see Larburu, Sanchez and Rodriguez 2010; Robinson et al. 2010, Martens et al. 2007). Taking the driver out of the loop raises questions about the effects on drivers' situational awareness, or their knowledge of the surrounding traffic and prevailing conditions. In particular, this may become a safety issue when the driver is required to switch from autonomous driving to normal driving or when responding to unexpected events due to system breakdowns. With the driver effectively becoming a passive monitor, the design of the human–machine interface will become a critical aspect for the success of such systems. In addition to these safety issues, user acceptance of platooning will depend on the extent to which the system is perceived to be accurate and reliable and its use to be considered both safer and more comfortable compared to normal driving.

As mentioned, the presence of platoons on normal motorways also creates an entirely new set of driving conditions for non-platoon road users. Although the exact consequences will be dependent on the specific design of platoons (e.g., what is the maximum number of vehicles; are vehicles allowed to leave or join a platoon from the side; see Robinson et al. 2010), for platooning to be acceptable it is important that the presence of platoons on normal motorways does not lead to actual or perceived negative consequences. Platoons may interfere with other road users in a number of ways. For example, entering and exiting a motorway and overtaking may be perceived to be less safe and more demanding. Also, what are the effects of the shorter time headways adopted in platoons? Could this result in behavioural adaptation whereby non-platoon drivers consciously or unconsciously also adopt shorter time headways possibly compromising road safety? Initial simulator studies on this topic have shown evidence for such effects (Gouy et al. 2012). These are only some of the type of questions that need to be answered to better understand the effects of platooning in the context of acceptability.

The first few studies have now been undertaken to start to better understand some of the above issues. The German national project KONVOI set out to conduct simulator and on-road testing of platoons consisting of coupled trucks (Lank et al. 2011). The platooning concept required the first driver of a platoon to manually control the truck with the other trucks following the lead truck fully automatically. To join a platoon, the driver was required to send a request via a touch screen when within 50 metres of the platoon. Following acceptance by the platoon lead vehicle, automation would set in and gradually close the gap to a distance of 10 metres from the truck at the end of the platoon. Similarly, to de-couple and leave the platoon, the driver would be sent a request and, following acknowledgement by the lead vehicle, the time headway would increase again to 50 metres followed by a visual-auditory countdown signal to indicate the end of the automated drive.

Within the KONVOI project, user acceptance was evaluated in three phases, starting out with focus groups, followed by simulator studies and on-road studies (Lank et al. 2011). This allowed for a clear demonstration of the effect that experience with a new technology might have on user acceptance. Before any actual experience with platoon driving, the initial focus groups revealed 80 per cent of the truck drivers to have a negative attitude towards the concept of platooning. However, following actual experience of platooning, a considerable shift was observed with an ultimate approval rate of 54 per cent.

System acceptance of non-platoon drivers was evaluated in a subsequent driving simulator study. Although some concerns were raised that platoon driving might lead to additional driving demands for some, the vast majority of drivers (80 per cent) showed a positive attitude towards platooning and thought of it as a sensible development. Platoons were regarded as reducing driver workload, in part due to the reduction in the number of overtaking manoeuvres required, and drivers expressed a preference for overtaking a platoon as opposed to individual trucks. On the other hand, concerns were raised regarding the additional demand and responsibility put on the driver of the lead vehicle. System over-reliance and subsequent inattention was feared to result in 'illusionary safety' and possibly increased accident risk. Although drivers reported little difficulties entering and exiting the motorway in the presence of platoons, the additional complexity of the traffic conditions was mentioned as a possible reason for lower acceptance levels by other road users. Respondents also mentioned the need for international standardisation regarding the legal length of platoons and the need for full development and testing before market deployment.

Most recently, user acceptance of platoon driving has been evaluated as part of the European project, Safe Road Trains for the Environment (SARTRE). Larburu et al. (2010) conducted a simulator study to assess drivers' responses to platoon driving. Again, acceptance was not only assessed from the perspective of the platoon driver but also from drivers encountering a platoon. The study evaluated various platoon configurations that varied in length and headways, and also included a prototype HMI which was incorporated to inform the driver during transition stages from manual to automatic driving, and vice versa. From a user acceptance perspective, the study showed some consistent gender effects with female drivers reporting to be less tolerant of shorter time headways when driving within the platoons than men. It was also found that the inter-vehicle distance at which participants reported to feel uncomfortable (16 metres) was well above the distance at which platoons are considered to become energy efficient and safe (Larburu et al. 2010). Acknowledging the inherent limitations of this kind of study (i.e., the lack of participants' experience and familiarity with platoons), these results illustrate the need for future studies to better understand the acceptability of short time headways. Regarding the HMI design, the provision of information during transition changes was considered imperative with a vast majority of participants referring to the need to include a driver acknowledgement step before starting a coupling or de-coupling manoeuvre.

When asked about their experience driving in proximity to a platoon, platoon length was one of the key parameters that affected users' acceptance. Whereas driving next to a five-vehicle-long platoon was perceived to be similar to normal driving – safe and not to cause any difficulties performing manoeuvres (e.g., exiting motorways). This was no longer the case with a platoon length of 15 vehicles. Platoons of vehicles longer than 25 were deemed unacceptable by 90 per cent of participants, suggesting this to be a maximum acceptable platoon length.

In summary, the results of the studies conducted so far indicate that platoon driving may become a near future reality. From a technical perspective, there are no barriers that would prevent such systems being implemented. However, these same studies also highlight several Human Factors and acceptance issues that require a better understanding before widespread introduction is feasible. Beyond obvious liability issues, system acceptance will be dependent on platoon configurations, protocols for transferring control between driver and vehicle, HMI design, system failure management procedures, as well as non-platoon drivers' interaction with and response to platoons. These fundamental questions require significant research efforts to provide the necessary empirical support before road authorities will be sufficiently confident to allow for platoon driving.

Overall Discussion and Conclusions

This chapter has raised a wide range of acceptance issues for in-vehicle technology by considering example systems according to their impact on the driving task, as well as their current level of maturity. It is clear that a broad range of automation-related effects are closely aligned with acceptance issues, whether dealing with information or assistance systems. For instance, issues concerning reliability, reliance and trust will be rich areas for future research. Whilst it is likely that the capabilities of these systems will increase with customer demand, it is unlikely that they will ever be 100 per cent reliable. Importantly, research from other application domains (e.g., process control) indicates that people find it particularly difficult to calibrate objective with subjective reliability when systems are close to perfect (Wickens et al. 2004). So, for example, if a system is objectively 99 per cent reliable, users are prone to treat it as 100 per cent reliable and may well adopt complacency-related behaviours accordingly.

Moreover, it is worth emphasising that different in-vehicle technologies will not be used independently, but in combination with each other. Research studies generally neglect this fact and consider the impact of drivers interacting with single systems. In reality, for many real-world driving situations there will be considerable interaction effects. For instance, a vehicle equipped with a system that automates longitudinal and lateral control of the vehicle (such as platooning) is likely to have a significant effect on the tasks that drivers are willing to undertake with other systems, for example, those providing entertainment or productivity services. How drivers will trade-off the various tasks that occur in future cars will be critical questions for research. In particular, the acceptance or otherwise of the different systems will have a profound effect on how they might be used as an integrated whole.

As a final point, it should be noted that the vehicle often incorporates a social environment when passengers are present, or even when communications are conducted with people remote/external to the vehicle (e.g., via a phone link). Previous research concerning acceptance issues has focused largely on the driver solely as an operator of the vehicle. In reality, the social context will also have a considerable impact on users' attitudes, behaviour and performance with new technology in many highly dynamic and complex driving situations. As an example, a recent study by Large and Burnett (2013) noted how the presence of passengers affected a driver's interactions with a navigation system, particularly related to the acceptance of voice instructions.

References

Adler, J.L. 2001. Investigating the learning effects of route guidance and traffic advisories on route choice behaviour. *Transportation Research Part C*, 9: 1–14.

Alkim, T., Bootsma, G. and Looman, P. 2007. The assisted driver, Report by the Dutch Ministry, "Rijkswaterstaat", Roads to the Future, Delft, April 2007.

Becker, S., Bork, M., Dorisen, H. T., Geduld, G., Hofmann, O., Naab, K. and NoÈcker, G. 1994. *Summary of experiences with autonomous Intelligent Cruise Control (AICC). Part 2: Results and Conclusions.* Proceedings First World Congress on Intelligent Vehicle Systems, 1836–143. Brussels: Ertico.

Burnett, G.E. 2009. On-the-move and in your car: An overview of HCI issues for in-car computing. *International Journal of Mobile Human-Computer Interaction*, 1(1): 60–78.

Burnett, G.E. and Joyner, S.M. 1996. Route guidance systems – Getting it right from the driver's perspective. *Journal of Navigation*, 49(2): 169–77.

Burnett, G.E. and Lee, K. 2005. The effect of vehicle navigation systems on the formation of cognitive maps. In *Traffic and Transport Psychology: Theory and Application*, 407–18 Edited by G. Underwood. Amsterdam; London: Elsevier.

Campbell, J.L., Carney, C. and Kantowitz, B.H. 1998. Human Factors design guidelines for advanced traveller information systems (ATIS) and commercial vehicle operations (CVO). Report no. FHWA-RD-98-057, Battelle Human Factors Transportation Center, Seattle.

Dingus, T., Hulse, M.C., Antin, J.F. and Wierwille, W. 1989. Attentional demand requirements of an automobile moving-map navigation system. *Transportation Research Part A: General*, 23(4): 301–15.

Fancher P.S., Baraket Z., Johnson G. and Sayer J. 1995. *Evaluation of human factors and safety performance, in the longitudinal control of headway.* Proceedings of the Second World Congress of Intelligent Transport Systems, Tokyo, 1732–8.

Faulkner, X. 2000. *Usability Engineering*. New York: Palgrave.

Forbes, N. 2009. Behavioural adaptation to in-vehicle navigation systems. Unpublished PhD thesis, School of Computer Science, University of Nottingham, June.

Forbes, N.L. and Burnett, G.E. 2007. Investigating the contexts in which in-vehicle navigation system users have received and followed inaccurate route guidance instructions, Third International Conference in Driver Behaviour and Training, Dublin, November.

Gouy, M., Diels, C., Reed, N., Stevens, A. and Burnett, G. 2012. The effects of short time headways within automated vehicle platoons on other drivers. In *Advances in Human Aspects of Road and Rail Transportation*. Edited by N. Stanton. Boca Raton, FL: CRC Press.

Green, P. 2008. Motor vehicle driver interfaces. In *The Human Computer Interaction Handbook*. Edited by J.A. Jacko and A. Sears. Lawrence-Erlbaum Associates; Mahwah, NJ, 701–19.

Green, P, Levison, W, Paelke, G. and Serafin, C. 1997. Preliminary human factors design guidelines for driver information systems. Report no. FHWA-RD-94-087; UMTRI, US.

Hill, M.R. 1987. 'Asking directions' and pedestrian wayfinding. *Man-Environment Systems*, 17(4): 113–20.

Hoedemaker, M. and Brookhuis, K.A. 1998. Behavioral adaptation to driving with an adaptive cruise control (ACC). *Transportation Research Part F*, 1: 95–106.

Jackson, P. 1998. In search of better route guidance instructions. *Ergonomics*, 41(7): 1000–1013.

KONVOI. Development and examination of the application of electronically coupled truck convoys on highways. Available at: http://www.ika.rwth-aachen.de/pdf_eb/gb6-24e_konvoi.pdf [accessed 5 October 2012].

Lank, C., Haberstroh, M. and Wille, M. 2011. Interaction of human, machine, and environment in automated driving systems. *Transportation Research Record: Journal of the Transportation Research Board*, 2243: 138–45.

Larburu, M., Sanchez, J. and Rodriguez, D.G. 2010. Safe road trains for environment: Human factors' aspects in dual mode transport systems. Paper presented at the ITS World Congress, Busan.

Large, D. and Burnett, G.E. 2013. Drivers' preferences and emotional responses to satellite navigation voices. *International Journal of Vehicle Noise and Vibration*, 9(1): 28–46.

Larsson, A.F. 2012. Driver usage and understanding of adaptive cruise control. *Applied Ergonomics*, 43(3): 501–6.

Martens, M., Pauwelussen, J., Schieben, A., Flemish, F., Merat, N., Jamson, S. and Caci, P. 2007. Human factors' aspects in automated and semi-automated transport systems: State of the art. CityMobil Deliverable 3.2.1, EU DG Research.

Moriarty, P. and Honnery, D. 2003. Safety impacts of vehicular information technology. *International Journal of Vehicle Design*, 31(2): 176–86.

National Highway Traffic Safety Administration (NHTSA). 2005. Automotive collision avoidance system field operational test. Report no. DOT HS 809 886. Washington, DC: NHTSA.

Nilsson, L. 1995. *Safety effects of adaptive cruise control in critical traffic situations.* Proceedings of the Second World Congress on Intelligent Transport Systems: Steps Forward, Vol. III, 1254–9. Yokohama: VERTIS.

Oliver, K.J. and Burnett, G.E. 2008. *Learning-oriented vehicle navigation systems: A preliminary investigation in a driving simulator.* Proceedings of the Tenth International Conference on Human-Computer Interaction with Mobile Devices and Services, 119–26. New York: ACM.

Reagan, I. and Baldwin, C.L. 2006. Facilitating route memory with auditory route guidance systems. *Journal of Environmental Psychology,* 26: 146–55.

Robinson, T., Chan, E. and Coelingh, E. 2010. *Operating platoons on public motorways: An introduction to the SARTRE platooning programme.* Seventeenth World Congress on Intelligent Transport Systems, 25–29 October, Busan, Korea.

Ross, T., Vaughan, G., Engert, A., Peters, H., Burnett, G.E. and May, A.J. 1995. *Human factors guidelines for information presentation by route guidance and navigation systems.* DRIVE II V2008 HARDIE, Deliverable 19. Loughborough, UK: HUSAT Research Institute.

Sol, E.-J., van Arem, B. and Hagemeier, F. 2008. *A 5 generation reference model for intelligent cars in the twenty-first century.* Proceedings of the Fifteenth World Congress on Intelligent Transport Systems, 16–20 November, New York.

Srinivisan, R. 1999. *Overview of some human factors design issues for in-vehicle navigation and route guidance systems.* Transportation Research Record 1694, paper no. 99-0884. Washington, DC: National Academy Press.

Stanton, N.A., Young, M.S. and McCaulder, B. 1997. Drive-by-wire: The case of driver workload and reclaiming control with adaptive cruise control. *Safety Science*, 27 (2/3): 149–59.

Vahidi, A. and Eskandarian, A. 2003. Research advances in intelligent collision avoidance and adaptive cruise control. *IEEE Transactions on Intelligent Transportation Systems.*, 4: 143–52.

Wickens, C.D., Lee, J.D., Liu, Y. and Becker, S.E.G. 2004. *An Introduction to Human Factors Engineering.* Upper Saddle River, NJ: Pearson.

Wilmink, I., Janssen, W., Jonkers, E., Malone, K., van Noort, M., Klunder, G., Rämä, P., Sihvola, N., Kulmala, R., Schirokoff, A., Lind, G., Benz, T., Peters, H. and Schönebeck, S. 2008. Impact assessment of intelligent vehicle safety systems. eImpact Deliverable D4, Version 1.0, Contract 027421, April.

Chapter 11

Driver Acceptance of Electric Vehicles: Findings from the French MINI E Study

Elodie Labeye and Corinne Brusque

*Institut Français des Sciences et Technologies des Transport,
de l'aménagement et des Réseaux (IFSTTAR), Bron, France*

Michael A. Regan

*Transport and Road Safety Research,
University of New South Wales, Australia*

Abstract

The electric vehicle (EV) has great potential to reduce the impact of transport on the environment and is being rolled out in increasing numbers by the automotive industry. Driver acceptance of this new type of vehicle is uncertain, however, due to the relatively limited range and higher price of the EV compared to conventional vehicles.

To assess driver acceptability of EVs, the MINI E France project was undertaken by IFSTTAR, in France, in cooperation with the vehicle manufacturer BMW Germany. Fifty private users from Paris responded to a set of questionnaires, focus group questions and travel diary items before and after six months of daily use of an electric MINI E. The results showed that the performance and the ease of use with respect to the EV are generally well judged by the participants. However, the analyses of purchase intention demonstrate that the barriers to EV acceptance are still present, even after a long period of use of the vehicle.

Introduction

The struggle against global warming is one of the major political issues of the twenty-first century. A front-line activity in this struggle is the reduction of greenhouse gas emissions. Given that the transport sector is a major CO_2 emitter (OECD/ITF 2010), various constraints are being imposed by governments on manufacturers so that they invest in research and development of new technologies, thus allowing for the construction of less polluting vehicles.

Among the range of vehicles proposed by manufacturers, the electric vehicle (EV) has reappeared, more efficient than ever, positioning itself as a new and potentially viable eco-friendly mode of transport, when the electricity that fuels it is generated in an eco-friendly manner. This vehicle, which does not emit CO_2 locally, seems to be able to meet the mobility needs of individuals with greater efficiency than those proposed in the 1990s, while reducing the harmful environmental impact of transportation.

Many governments stimulate deployment of this technology by encouraging the commercialisation of EVs; in France, for example, government aid for the purchase of an EV is in the order of €5,000 and a bulk order of 50,000 vehicles was made by the French authorities in 2010 (Negre 2009). The media coverage which has accompanied these policy directives, running in parallel with the rise in oil prices, has gradually transformed public opinion vis-à-vis mobility: the ecological necessity of a profound change in our patterns of mobility and in our energy sources has become a reality for citizens.

The reintroduction of the EV in the market has taken place in an unprecedented technological context, in which electromobility is made possible by superior performance of the lithium battery, regenerative braking and the deployment of public charging stations (and in parallel, the introduction of mobile applications which can support the location and reservation of charging stations). Nevertheless, even if the technological and ecological circumstances appear more favourable than 20 years ago, and despite the obvious investments of government and industry, the possibility of widespread introduction of EVs remains uncertain.

Recent studies of discrete choice analysis, stated preference surveys of the intention of purchasing hybrid vehicles and forecasting models of EV take-up, show that features limiting the adoption of these vehicles are mainly the price and performance of vehicles in terms of range, charging time and acceleration (Eggers and Eggers 2011, Lieven et al. 2011, Potoglou and Kanaroglou 2007), and that these limiting features have persisted through time as they were already identified in the 1990s (Cheron and Zins 1997, Golob and Gould 1998, Kurani and Turrentine 1996). For Kurani and Turrentine (1996), these different limitations suggest that the EV would be particularly attractive as a second car in the multicar household; however, for other authors, 'unless the limited driving range for electric vehicles is increased substantially this technology will not be fully competitive in the automobile' (Dagsvik et al. 2002: 383).

Finally, it is interesting to note that the importance of these limiting features may, however, be modified with actual use of the EV: some research shows that when individuals have the opportunity to experience the technology directly, and to see its impact on their mobility, their opinions on certain features seem to be positively influenced (Buehler et al. 2011, Gould and Golob 1998, Woodjack et al. 2012). Other studies, however, show that use of the EV does not alter initial impressions of some features. Regarding limited range, for example, individuals continue to perceive this as insufficient despite the observation after use that the EV meets overall their daily needs for mobility (Golob and Gould 1998).

There is currently considerable interest in whether long-term exposure to EVs can change the representation that individuals have of these limiting factors and thus influence behavioural intent vis-à-vis use of the concerned technology.

Objectives

The aim of the research reported in this chapter was to evaluate, for a French sample of potential buyers of EVs, their behavioural intention to use electric vehicles and, more generally, to examine driver acceptance of EV technologies. For this, we rely on the acceptance theory, presented in the next section, which suggests that when users are presented with a new technology, a number of different cognitive and social factors influence their decision about how and when they will use it.

In addition, the study reported here explores the impact of real and daily use of the EV on the main acceptance factors. We are interested in the evolution of participants' opinions concerning performance expectancy, ease of use expectancy and purchase intentions when using an electric vehicle during a six-month period. Here, the objective is to compare the main factors of acceptance at T0 month and T6 months, and their potential evolution, and to highlight features which can have a significant impact on the adoption of the EV in our mobility choices.

Finally, we conclude with a discussion of the main barriers related to EV uptake identified by individuals of the sample at the end of the study, which exposes the principal issues to address in stimulating future uptake of EVs.

Acceptance

Models of acceptance have identified several factors closely linked to whether or not an individual adopts new technologies. In general, two main factors are highlighted.

As noted elsewhere in this book, the first one corresponds to *performance expectancy* proposed by the Unified Theory of Acceptance and Use of Technology – UTAUT – which refers to 'the degree to which a person believes that using a particular system would enhance his or her job performance' (Vankatesh et al. 2003: 447; see also Davis 1989).

Performance expectancy of the EV, therefore, relates to the mobility needs of drivers, the ability of EV to suit the road transport environment and also to its ability to bring new opportunities in terms of mobility: ecological advantages (absence of CO_2 locally, possibility of travelling in a more eco-friendly way) or perceptual advantages (lack of vehicle noise at low speed).

The second main factor identified is *effort expectancy,* defined as 'the *degree of ease* associated with the use of the system' (Venkatesh et al. 2003: 450). In the case of the EV, the issue is how people find the usability of this new type of vehicle: is it easy to learn to use, to take into account the distance and the charging time and so on.

Some other factors are also highlighted by extant models of acceptance to account for the adoption of new technologies. Among these is the social influence factor that represents 'the degree to which an individual perceives that important others evaluate and use the system, and believe he or she should use this new system' (Fishbein and Ajzen 1975: 216; see also Ajzen 1991, Venkatesh and Davis 2000).

Finally, some models add a *facilitating conditions* factor underlying acceptance (Thompson, Higgins and Howell 1991, Venkatesh et al. 2003), which makes reference to the organisational and technical infrastructure that supports use of the new system. For the EV, for example, this is the infrastructure for charging the EV, or subsidies, which may drive societal acceptance of electric vehicles.

The Unified Theory of Acceptance and Use of Technology states that each of the factors mentioned above also appear to be moderated by factors including the gender, age and experience of individuals with respect to the new technology, and that together these things influence overall acceptance of technology (Moore and Benbasat 1991, Venkatesh et al. 2003).

Testing these different factors of acceptance will therefore help to identify the behavioural intention of individuals related to use of the EV. However, it is interesting to note that actual use of EV may have an impact on some of these factors and change the way they are perceived and evaluated. Like many new products, their advantages and defects can be overvalued or undervalued by lack of real user feedback. Our study aimed, therefore, to identify possible changes in certain factors of acceptance following use of EV.

Study Context

To examine driver acceptance of EVs in France, and the changes in behaviour and attitudes that occur over time with the use of this type of vehicle, an experiment, already conducted in Germany (Cocron et al. 2011a, Cocron et al. 2011b), the United States (Woodjack et al. 2012) and England (Everett et al. 2010) – for the manufacturer BMW – was replicated in Paris, France.

The originality of the French study rested on the fact that drivers responded to a set of questionnaires (and also focus groups, travel and charge diaries) at the beginning of the study, in the middle and after six months of EV use. Each driver utilised a MINI E electric car for their daily trips during a six-month period. Two waves of 25 drivers were tested – from December 2010 to June 2011 for the first wave, and from July to December 2011 for the second wave. In this chapter we will focus on the data that specifically concerns acceptance of the EV and its evolution after daily use over a period of several months, and on the negative features identified by individuals of the sample at the end of the study which serve to explain the acceptance highlighted.

Methodology

Electric Vehicles

Twenty-five MINI E prototypes were deployed. The MINI E is similar in external appearance to the MINI Cooper, but with only two seats and equipped with a lithium-ion battery. The average range of the MINI E is 160 km and the car has regenerative braking that slows the vehicle (while at the same time regenerating energy) from the moment the driver releases the accelerator pedal.

To charge the vehicle, each participant had a wall box of 12 amps installed in his or her home by the French electricity provider Électricité de France (EDF). Drivers could also charge their vehicles from Parisian public charging stations. A full charge took about nine hours to complete.

Data Collection

Data were collected from a set of questionnaires, focus groups and travel and charge diaries. These research tools were designed originally by the German research team which worked on the first MINI E study (Neumann et al. 2010). Data were compared across three time intervals: T0, T3 months and finally at T6 months after the start of the study.

The procedure was as follows. At T0, two questionnaires were completed – one face-to-face and the other online – measured on a Likert scale of six points, ranging from one 'strongly disagree' to six 'strongly agree' or a one 'very unimportant' to six 'very important'. Some of the questions were open questions. Several issues were addressed: the prospective views and expectations of future users about the electric vehicle, their considerations of the ecological aspects and techniques of EVs, and their driving habits in traditional cars.

In parallel, the travel diary was administered. It related to drivers' use of their own (private) car during a typical week. For seven days, participants used it to register all their trips, detailing the trip distance, means of transport taken, purpose of the trip and so on.

After three months of using the EV participants were asked to complete two further questionnaires – one face-to-face and another online – containing items that were either already presented at T0 or were new. These items concerned the experience and appreciation of participants of the use of the MINI E on a daily basis. Participants were also required to complete again a travel diary, relating this time to use of the MINI E. Users were also required to complete a charge diary detailing all charges made during a week. Users reported place of charge, charge status at the beginning and the end of the charging process, and the reasons for the charge.

Finally, at six months, participants completed a questionnaire which was administered face-to-face. The majority of items were identical to items from previous questionnaires. Finally, participants were asked to complete a travel diary and a charge diary similar to the previous ones.

In this chapter, we focus only on data for T0 and T6 months; given that the objective is to see if daily use of an EV influences and modifies purchase intent, it is interesting to consider the maximum duration of use, namely those drivers evaluated after six months (for more details, see Labeye et al. in press, and Labeye et al. 2012).

Participants

More than 900 people applied online (via the MINI.fr site) to participate in the study. A first selection was made based on the following criteria: being a resident of the Paris area, having a garage or a dedicated place to park the MINI E, being able to provide payment for leasing the vehicle (€475 per month; insurance included) and having access to a suitable electrical power supply.

Fifty subjects were chosen based on the number of kilometres they were driving each day, and the selection was aimed at maximising the number of women in the sample and to have a majority of drivers who had no experience with electric or hybrid vehicles.

Data for only 40 subjects were analysed because of the difficulty caused by the six-month duration of the experiment: two subjects dropped out and eight subjects did not respond to the final questionnaire.

The profile of the final sample was as follows: 7 women and 33 men, with an average age of 43.9 years (SD = 8.029). The number of people per household was on average 3.53; 78 per cent of selected participants had a university level qualification; 25 per cent were driving more than 70 km per day; 30 per cent had already had a MINI; 23 per cent had experienced an electric vehicle, and 20 per cent a hybrid vehicle; and, finally, 25 per cent of participants did not have more than one vehicle at home.

It should be noted that those who participated in the MINI E study presented a particular profile that was not typical of the French population. They lived in Paris and its suburbs, they had a high level of income, and they drove on average for 60 km per day (SD = 30.108). Moreover, the sample was especially representative of those that might eventually buy electric vehicles. Indeed, individuals selected were potential early adopters of EV (selected on the MINI E website), which explains their high experience of electric vehicles (nearly a quarter), and their great interest in innovative technology and the environmental benefit of EVs. Indeed, participants' motivation to take part in this study was mainly due to two factors: the attractiveness of the EV being innovative and the attractiveness of the environmental benefits that the EV was considered to induce (and to a lesser extent, attachment to the brand and reduced energy costs).

Results

Acceptance at T0 Month

The questionnaires of the MINI E study addressed several different research issues. Only items related directly to, and exemplary of, the acceptance factors (defined in the introduction) are discussed here: items which refer to performance expectancy and ease of use expectancy in connection with the electric vehicle, items related to the subjective norms which can influence expectations of individuals, and finally, participants' overall purchase intentions with respect to this new type of vehicle.

It should be noted that, because of conditions related to the constitution of the sample (which, as noted, was not balanced in terms of gender and age), we do not in this chapter study the moderating effects of gender, age and experience of the individuals.

Performance Expectancy

Concerning this acceptance factor, 12 items are studied regarding how the MINI E would meet the mobility needs of responders, regarding their global satisfaction and the general added value they expect from the MINI E; and specifically on the ecologically-related added value related to its use.

Table 11.1 below presents the mean for each item and the standard deviation. For items presented in the questionnaire in a negative turn (for example, 'The limited range of the MINI E will not permit me to do all of my normal driving'), both positive and negative means are indicated in Table 11.1. The calculation of the global performance expectancy factor mean average is formed from the set of positive values.

The performance expectancy factor for acceptance is generally high (M = 4.3, SD = 0.4). This suggests that participants consider highly that the MINI E is expected to satisfy their daily mobility needs (M = 5.03, SD = 1.00) and is suitable for everyday use (M = 5.10, SD = 0.74), even if they are aware of some difficulties caused by the limited range (M = 3.33, SD = 1.42). This indicates that participants expect that the EV would meet a large part their mobility needs.

Furthermore, it seems that participants consider the MINI E to be a safe vehicle (M = 4.80, SD = 0.82) and more satisfying to drive than a conventional car (M = 4.05, SD = 1.11). Finally, concerning the ecological added values, the means show how the participants associate these issues with the electric vehicle. They agree highly that the EV is a good solution to reduce noise and CO_2 pollution (M = 5.28, SD = 0.82; M = 5.02, SD = 0.73).

Table 11.1   Performance expectancy items means and standard deviations at T0 month

Performance expectancy factor	Items	Means	Std. Dev.
Mobility needs satisfaction	The MINI E will satisfy my daily mobility needs	5.03 +	1.00
	Electric vehicles are suitable for everyday use	5.10 +	0.74
	My driving patterns will be different when I drive a MINI E	3.60 — 3.40 +	0.84
	The limited range of the MINI E will not permit me to do all of my normal driving	3.33 — 3.67 +	1.42
	The MINI E will change my mobility behaviour essentially	3.20 +	1.24
Global satisfaction/ general added value	The MINI E will bring me reliably from one place to another	4.80 +	0.82
	The MINI E will be more satisfying to drive than a conventional car	4.05 +	1.11
	The MINI E will be an affordable transport option for me	3.88 +	0.88
	The MINI E is more useful than a conventional car in meeting my mobility needs	3.40 +	0.84
Ecologic added values	I think that electric vehicles are a good solution to reduce the noise in town	5.28 +	0.82
	I think that electric vehicles are a good solution for reducing the CO_2 emissions in France	5.02 +	0.73
	Electric vehicles are a key solution to reduce air pollution in France	4.77 +	0.92
Global performance expectancy factor mean average		**4.30**	**0.40**

Ease of Use Expectancy

The *ease of use expectancy* factor of EV acceptance was assessed with a six-item scale corresponding to the facility to learn expected by the subjects and the facility to use the EV.

Table 11.2 below presents the average mean for each item and the standard deviation. For items presented in a negative turn in the questionnaire, both positive and negative means are indicated in Table 11.2. The calculation of the global *ease of use* factor mean average is formed from the set of positive values.

Table 11.2   Ease of use expectancy items means and standard deviations at T0 month

Ease of use expectancy factor	Items	Means	Std. Dev.
Facility learning	It will be easy to learn how to handle the MINI E	4.93 +	0.94
	I need to learn a lot of things before I could get going with the MINI E	2.17 – 4.83 +	1.04
Global facility to use	The MINI E is easy to use	4.53 +	0.60
	Having to take into account my route length and charging times will make using the MINI E a big challenge	3.95 – 3.05 +	1.28
	The MINI E will be easier to use than a conventionally powered vehicle	3.68 +	0.89
	The mental workload required to drive the MINI E will be greater than that for a conventional car	2.93 – 4.08 +	0.89
	I am worried that the charging times will not suit my daily routine	2.85 – 4.15 +	1.12
Global ease of use expectancy factor mean average		**4.18**	**0.46**

The general mean ease of use expectancy factor is also high (M = 4.18, SD = 0.46). Globally, the results show that participants consider the learning to drive and use of the EV as easy (M = 4.93, SD = 0.94; M = 4.53, SD = 0.60), even though they think they should also take into account their route length and the charging times when they use the MINI E (M = 3.95, SD = 1.28).

Subjective Norms

To study the place of the subjective norms in the acceptance of the EV, five items were analysed. Table 11.3 below presents the means for each item and the standard deviation.

Table 11.3   Subjective norms items means and standard deviations at T0 month

Subjective norms	Items	Means	Std. Dev.
	People who are like me would probably like to drive an electric vehicle	4.45 +	0.60
	People who are important to me believe that electric vehicles are the future of transport	4.18 +	0.78
	People who are important to me would like to buy an electric vehicle	3.60 +	0.98
	Most of the people who are important to me think I should own an electric vehicle	3.35 +	0.98
	People expect me to buy an electric vehicle	2.72 +	1.01
Global subjective norms factor mean average		**3.66**	**0.63**

The analysis of the subjective norms shows this factor is less important than the previous (M = 3.66, SD = 0.63); and moreover, items have varying importance for participants – they think people like them would like to drive an EV (M = 4.45, SD = 0.60), but they are less likely to think people who are important to them would like to buy an EV (M = 3.60, SD = 0.98), and not many people expect them to buy an EV (M = 2.72, SD = 1.01).

Table 11.4   Use and purchase intention items means and standard deviations at T0 month

Use and purchase intentions	Items	Means	Std. Dev.
	I will use the MINI E very frequently	4.70 +	0.52
	For me, the MINI E could only be a second car	4.53 – 2.48 +	1.36
	I will seriously consider buying an electric vehicle after taking part in this study	4.18 +	1.04
	I think in _____ years it will be the right time to buy an electric vehicle	3.55 Years	2.92
Global use and purchase intentions factor mean average		**3.79**	**0.63**

Use and Purchase Intentions

Finally, four items related to the use and purchase intentions of the EV were studied. Table 11.4 presents the mean for each item and the standard deviation. For items presented in a negative turn in the questionnaire, both positive and negative means are indicated in Table 11.3. Moreover, the calculation of the global use and purchase intention factor mean average is formed from the set of positive values.

The mean of the use and purchase intention item is $M = 3.79$, $SD = 0.63$. In detail, the results show that the EV is not envisaged in the immediate term as a main car; the MINI E can only be a second car ($M = 4.53$, $SD = 1.36$), even if participants expect to use it very frequently ($M = 4.70$, $SD = 0.52$).

These considerations with respect to the EV demonstrate that some parameters of acceptance are supported but that, overall, they may not be sufficient to give the EV the first place in the household's vehicle fleet.

Finally, it can be argued that the size of the vehicle or the two seats would play a role in viewing the EV as a second rather than primary vehicle. Indeed, it is possible that the lack of a seating place could have modified mobility patterns of the participants by limiting the possible trips and adding to the constraints of range.

Acceptance after Six Months of Use

To analyse the impact of the use of the EV on the acceptance of this new mode of transport, we compared using a t-test the means obtained at T0 and T6 months on the principal factors of the acceptance. The factors' subjective norms and ecological factors were not analysed, as they are relatively independent of the real use of the car in everyday life.

No difference was found between the global means for the performance expectancy factor at T0 and T6 months, and neither was one found for the ease of use expectancy factor. However, results show some significant differences for specific items.

Performance Expectancy

Concerning the performance expectancy factor, means for the item 'The limited range of the MINI E will not permit me to do all of my normal driving' are significantly different between the beginning and the end of the experiment, t (39) = 2.06; $p < 0.05$. The means increase from $M = 3.33$ ($SD = 0.88$) at T0 month, to $M = 3.83$ ($SD = 1.15$) at T6 months, meaning the impact of the limited range of the EV was initially underestimated by the participants.

The analysis of open questions and travel dairies at T6 months shows that participants mention having to get used to the handling of range by planning their trips according to distance, and they used their own private car for long trips

(M = 4.29, SD = 1.1). However, we have to keep in mind that they highly estimated the MINI E as being satisfying (mean performed on the six months because the values do not differ significantly: M = 4.90, SD = 0.81) and suitable (mean calculated on the six months: M = 5.20, SD = 0.55) for their daily mobility needs, throughout the experiment. Moreover, at the end of the experiment, participants felt that the EV had satisfied their needs for daily mobility, M = 4.8 (SD = 1.19). Globally, these results are not necessarily contradictory. Even if there remain long trips for which participants need to use another mode of transport, the EV is generally meeting their needs for mobility.

Another item, 'The MINI E will be more satisfying to drive than a conventional car', also shows a significant difference between T0 and T6 months means, $t(39) = 3.90; p < 0.05$. The means increase from M = 4.05 (SD = 1.11) at T0 month, to M = 4.58 (SD = 1.11) at T6 months. The results indicate the importance of real use of the electric car in order to appreciate its features and thus show that some aspects of acceptance can be positively modified.

Finally, the last item of the performance expectancy factor presents significantly different means: $t(39) = -2.52; p < 0.05$. For 'The MINI E will be an affordable transport option for me', means decreased from M = 3.88 (SD = 0.88) at T0 month, to M = 3.05 (SD = 1.06) at T6 months. It seems that, over time, the financial asset is not as important as expected. Nevertheless, we must qualify this analysis since the participants rented the car for each month of the experiment, and hence the payments over time can reduce the financial benefits at the beginning.

Ease of Use Expectancy

Concerning the ease of use expectancy factor, several items were assessed differently at the beginning and the end of the experiment. Globally, all of them show that use of the EV is simpler than was expected initially by participants, even if at the beginning of the experiment users' expected difficulties were already rated as low.

For the item 'I needed to learn a lot of things before I could get going with the MINI E', the means decrease from M = 2.17 (SD = 1.04) at T0 month, to M = 1.48 (SD = 0.82) at T6 months: $t(39) = -3.62; p < 0.05$. Significant decreases were also observed for the item 'The mental workload required to drive the MINI E will be greater than that for a conventional car' (M = 2.93, SD = 0.89 at T0 month; M = 2.42, SD = 1.45 at T6 months: $t(39) = -2.11; p < 0.05$), and for the item 'Having to take into account my route length and charging times will make using the MINI E a big challenge', (M = 3.95, SD = 1.28 at T0 month; M = 3.35, SD = 1.25 at T6 months: $t(39) = -2.66; p < 0.05$). Overall, the results demonstrate that the ease of use expectancy factor is modulated by use of the EV, in relation to learning how to use the vehicle and in dealing with the range issues of the vehicle.

Finally, the item 'I am worried that the charging times will not suit my daily routine' is also rated differently at T0 month (M = 2.85, SD = 1.12) and at T6 months (M = 3.43, SD = 1.45): $t(39) = 2.23; p < 0.05$. Thus, the item concerns the particular issue of the charging time and the French infrastructure utilised during the experiment. Indeed, 42 per cent of users found that the charge process time – nine hours on average – didn't fit with their daily routine (participants reported that six hours would be more acceptable and three hours would be a good time). This was due to the 12 amps sockets that all users charged from. The use of 32 amps sockets could divide by two the time necessary to charge and thus make the process more suitable for everyday use.

Use and Purchase Intention

The analysis of the results corresponding to the use and purchase intention items was not significant between the means obtained at T0 and T6 months. Likewise at T0 month, the EV is not envisaged, even after six months of use, as a main car; the MINI E is still considered suitable as a second car only (means calculated on the six months: $M = 4.66$, $SD = 1.04$). However, it is interesting to note that the participants declared that they would use the EV very frequently ($M = 4.47$, $SD = 1.04$) and considered seriously buying an electric vehicle after the study (means calculated on the six months: $M = 4.21$, $SD = 0.91$).

These considerations with respect to the EV demonstrate that some factors of acceptance are validated but, overall, they may not be sufficient to give the EV first-car status in the household.

Main Barriers to Acceptance of the MINI E

We turn now to a final analysis of the negative features identified by participants at the beginning and end of the study, which can explain the mitigated acceptance of the EV. Participants were asked what important changes in different areas would be necessary for them to consider buying an EV in the future. The results were not significantly different before and after the six-month study period. Participants were concerned mainly about the potential range of travel, the durability of the batteries, the purchase price, the charging time and the construction of public charging infrastructure.

A t-test did show, however, a significant difference between the means related to the purchase price item at T0 month and at T6 months: $t (39) = -3.64$; $p < 0.05$ – the means increase from $M = 4.95$, $SD = 0.88$ to $M = 5.4$, $SD = 0.71$. This last result suggests that using the vehicle for six months makes the vehicle more concrete in the life of participants such that they can actually project themselves into the process of buying the car. They seem to have, henceforth, a more accurate idea of how much they are willing to pay to purchase an electric vehicle.

Finally, we can note that, among the main barriers to EV use, the charge infrastructure issue is of great importance because the actual charging points are not sufficient, and the charging time seems too long to participants. Thus, charging and price issues strongly influence EV acceptance and they represent the facilitating conditions factor proposed by Thompson et al. (1991). Thus, the technical infrastructure that supports use of the EV and the subsidies provided must increase in order to enhance acceptance and uptake of the electric vehicle.

Discussion

The study analysed how the different factors of acceptance related to the electric vehicle are appreciated by a potential early adopter population in Paris, France.

It appears that expectancy concerning performance and ease of use of the electric vehicle are generally well judged by the participants of the study. Indeed, they expect that the learning and use of the EV will be simple and that the EV will respond well to their daily mobility needs, while providing more ecological value. This new type of vehicle seems to yield many positive aspects and thus could pass for a very acceptable technology.

However, the analyses of intention to use and purchase intention demonstrate that the barriers to EV acceptance are still present. Even after a long period of use of the vehicle, the barriers are still the same and concern the purchase price, the charging and batteries issues and finally the limited range of the vehicle.

On this last factor, the inability of the car to support long trips prevents the EV from being perceived as a potential main use car in the household, despite the suitability of the EV to satisfy the rest of the mobility needs of participants. Concerning the charging time, the results showed that there was no change in participants' apprehension of this constraint. It is true that the charging time in France is very long (nine hours on average) compared with other countries (e.g., Germany, where the charging time is around four hours) and this does not facilitate use of the vehicle (Labeye et al. 2012).

The idea that everyday use of the EV modifies the perception of range and charge issues is not shown; however, we must keep in mind that the tested sample is not representative of the French population. They were people who already have a favourable opinion vis-à-vis electric technology and who were curious about the product (23 per cent had experienced an EV before the study). Use of a more heterogeneous population would probably have shown an effect of EV use on the acceptance of EVs. In this case, the daily use of the EV could be more influential and modify the representation they would have.

In the end, to see the EV feature more prominently in our mobility choices, EV technology and charging infrastructure have to be refined and accepted by potential owners. It is reasonable to assume that EV technology will continue to become increasingly innovative in order to overcome the problem of range; and finally, the generalisation of the infrastructure would potentially reduce purchase cost, which remains undeniably a major problem.

Acknowledgements

We are grateful to BMW Germany and BMW France for providing us with the opportunity to conduct this study. In particular, we thank Michaela Luehr, Roman Vilimek, Michael Hajesch, Maximillian Schwalm and Jean-Michel Cavret and his colleagues for their support, and for their input to the project. We thank also colleagues from CEESAR, for their important role in recruiting participants and collecting the data for this study; in particular, Julien Adrian, Annie Langlois and Reakka Krishnakumar. Finally, we thank Julien Delaitre, Magalie Pierre (EDF) and Julien Augerat (Veolia) for their support.

References

Ajzen, I. 1991. The theory of planned behavior. *Organizational Behavior and Human Decision Processes*, 50(2): 179–211.

Buehler, F., Neumann, I., Cocron, P., Franke, T. and Krems, J.F. 2011. *Usage patterns of electric vehicles: A reliable indicator of acceptance? Findings from a German field study.* Proceedings of the Ninetieth TRB Annual Meeting, Washington, DC, January.

Cheron E. and Zins, M. 1997. Electric vehicle purchasing intentions: The concern over battery charge duration. *Transportation Research Part A: Policy and Practice*, 31(3): 235–43.

Cocron, P., Bühler, F., Franke, T., Neumann, I. and Krems, J. 2011a. *The silence of vehicles – Blessing or curse?* Proceedings of 90th TRB Annual Meeting, Washington, DC, January.

Cocron, P., Bühler, F., Neumann, I., Franke, T., Krems, J.F., Schwalm, M. and Keinath A. 2011b. Methods of evaluating electric vehicles from a user's perspective – The MINI E field trial in Berlin. *IET Intelligent Transport Systems*, 5(2): 127–33.

Dagsvik, J.K., Wennemo, T., Wetterwald, D.G. and Aaberge, R. 2002. Potential demand for alternative fuel vehicles. *Transportation Research B*, 36: 361–84.

Davis, F.D. 1989. Perceived usefulness, perceived ease of use, and user acceptance of information technology. *MIS Quarterly*, 13(3): 319–39.

Eggers, F. and Eggers, F. 2011. Where have all the flowers gone? Forecasting green trends in the automobile industry with a choice-based conjoint adoption model. *Technological Forecasting & Social Change*, 78: 51–62.

Everett, A., Walsh, C., Smith, K., Burgess, M. and Harris, M. 2010. *Ultra low carbon vehicle: Demonstrator programme.* Proceedings of the Twenty-fifth World Battery, Hybrid and Fuel Cell Electric Vehicle Symposium and Exhibition, EVS 25, Shenzhen, China.

Fishbein, M. and Ajzen, I. 1975. *Belief, attitude, intention, and behavior: An introduction to theory and research.* Reading, MA: Addison-Wesley.

Golob, T.F. and Gould, J. 1998. Projecting use of electric vehicles from household vehicle trials. *Transportation Research A*, 32: 441–54.

Gould, J. and Golob, T.F. 1998. Clean air forever? A longitudinal analysis of opinions about air pollution and electric vehicles. *Transportation Research D: Transport and the Environment*, 3: 157–69.

Kurani, K.S. and Turrentine, D.S. 1996. Testing electric vehicle demand in hybrid households' using a reflexive survey. *Transportation Research D*, 1(2):131–50.

Labeye, E., Adrian, J., Hugot, M., Regan, M. and Brusque, C. In press). Daily use of an electric vehicle: Behavioural changes and potential support of ITS. IET Intelligent Transport Systems.

Labeye, E., Hugot, M., Regan, M. and Brusque, C. 2012. Electric vehicles: An eco-friendly mode of transport which induce changes in driving behaviour. In *Human Factors of Systems and Technology*. Edited by D. de Waard, N. Merat, H. Jamson, Y. Barnard and O. Carsten. Maastricht, the Netherlands: Shaker Publishing.

Lieven, T., Muhlmeier, S., Henkel, S. and Waller, J.F. 2011. Who will buy electric cars? An empirical study in Germany. *Transportation Research D*, 16: 236–43.

Moore, G.C. and Benbasat, I. 1991. Development of an instrument to measure the perceptions of adopting an information technology innovation. *Information Systems Research*, 2(3): 192–222.

Negre, L. 2009. Rapport du Senat. Structuration de la filière des véhicules décarbonés. Available at: http://www.ladocumentationfrancaise.fr/var/storage/rapports-publics/114000055/0000.pdf [accessed 1 February 2013].

Neumann, I., Cocron, P., Franke, T. and Krems, J.F. 2010. *Electric vehicles as a solution for green driving in the future? A field study examining the user acceptance of electric vehicles.* Proceedings of the European Conference on Human Interface Design for Intelligent Transport Systems, 29–30. Edited by J.F. Krems, T. Petzoldt and M. Henning (Hrsg). Berlin, Germany, April 2012 (445–53). Lyon: Humanist Publications.

OECD/ITF. 2010. *Reducing Transport Greenhouse Gas Emissions Trends and Data 2010.*

Potoglou, D. and Kanaroglou, P.S. 2007. Household demand and willingness to pay for clean vehicles. *Transportation Research Part D*, 12(4): 264–74.

Thompson, R.L., Higgins, C.A. and Howell, J.M. 1991. Personal computing: Toward a conceptual model of utilization. *MIS Quarterly*, 15(1): 124–43.

Venkatesh, V. and Davis, F.D. 2000. A theoretical extension of the technology acceptance model: Four longitudinal field studies. *Management Science*, 45(2): 186–204.

Venkatesh, V., Morris, M.G., Davis G.B. and Davis, F.D. 2003. User acceptance of information technology: Toward a unified view. *MIS Quarterly*, 27(3): 425–78.

Woodjack, J., Garas, D., Lentz, A., Turrentine, T., Tal, G. and Nicholas, M. 2012. *Consumers' perceptions and use of electric vehicle range changes over time through a lifestyle learning process.* Proceedings of Ninetieth TRB Annual Meeting, Washington, DC, January.

Chapter 12

User-Centred Design and Evaluation as a Prerequisite for the Success of Disruptive Innovations: An Electric Vehicle Case Study

Roman Vilimek and Andreas Keinath
BMW Group, Germany

Abstract

Introducing electromobility and transportation to customers can be a potentially disruptive innovation. To make this innovation successful and to design the BMW i3, the BMW Group's first purpose-built electric vehicle, we applied a customer-centred development and evaluation cycle to bridge the gap between early adopters and late adopters. In this paper, the results of fundamental field trials to define and verify basic customer requirements are reported where the BMW Group gathered a lot of information about usage of electric vehicles in a real-world setting. These results have been incorporated in development and have been continuously tested and evaluated by customers. We believe that the results of this customer-oriented development process will contribute significantly to make the potentially disruptive innovation of electromobility a success in terms of driver acceptance.

Customer-Centred Development as a Means to Ensure Driver Acceptance

In 1908 when the first Ford Model T vehicles were produced, the internal combustion engine vehicle could not generally be described as a disruptive innovation as it had been around well before this date. By definition, disruptive innovations define a new market, are revolutionary, but not evolutionary, and typically come up with new customer segments (Christensen 2012). However, what is described as a disruptive innovation was that the Ford Model T made the internal combustion engine vehicle affordable, and therefore completely changed individual transportation (Christensen and Raynor 2003). Even before the Ford Model T, there were electric vehicles; even electric cabs had been seen on the streets of New York and Boston and other cities around 1896 (Kirsch 1997). However, the principle behind the completely new approach of BMW i, the new sub-brand of the BMW Group, is a disruptive innovation with potential to change the whole market. BMW i will not only introduce new vehicle concepts like the all-electric purpose-designed BMW i3. It will also put a major focus on sustainable mobility by focussing not only on the vehicle itself but also by redesigning the entire value chain. Additionally BMW i will introduce innovative mobility services and develop technological innovations like fast charging to ensure the suitability of electric vehicles for the customer's everyday needs.

While there exists a whole body of literature on how to innovate and how to generate ideas for product innovations, there is much less information on how to accompany the development of a disruptive innovation to achieve maximum user acceptance and customer satisfaction. In other words, we will focus in this paper on how to carry out user-centred product evaluation even if it is about a radical technological innovation that will change a good deal of a customer's everyday behaviour. Standard theories and models of technology acceptance are only of limited use here, as they do not give a descriptive approach as to how to do user-centred development. Hence, one of the best-known models for technology acceptance by Davis (1993) only states that perceived usefulness and perceived ease of use are crucial for technology acceptance in terms of actual use of the product. However, the model does not describe how to achieve usefulness and ease of use from a developer's point of view, especially when it comes to such a complex system as transportation and use of electric vehicles (EVs). In his book, *The Invisible Computer*, Donald Norman (1998) combines insights from how disruptive innovations change whole markets with the theory about how early adopters lead market acceptance before late adopters join in. It is crucial for innovations to bridge the gap between early adopters and late adopters to become a success (Moore 1991). Norman states that this gap can only be bridged by products that are customer-driven or human-centred and provide good value with a good user experience. Figure 12.1, taken from Norman (1998), shows the change from technology-driven high technology that attracts early adopters to customer-driven user centred technology that attracts late adopters who are responsible for the success of a product innovation. In emphasising the importance of user experience and usability of a product to become a success, Norman comes close to the basic notions of Davis's (1993) acceptance model that states these same factors as being crucial for technology acceptance.

However, the combination of both models gives clear advice for anybody working on how to make a disruptive innovation successful by suggesting the application of a user-centred design and evaluation approach, even if the innovation at hand is something as big as electric vehicles in a mature transportation system. The special challenge with disruptive technologies is creating solutions that are actually optimised for everyday usage: because there is no 'everyday' context yet established during systems design, a process is needed that allows integration of user feedback in early phases of the design by deploying pilot use cases with target users. The user-centred design process proposes exactly this approach. In the following we will describe how we applied the basic user-centred design process to accompany the development of some aspects of the BMW Group's first series-built electric vehicle.

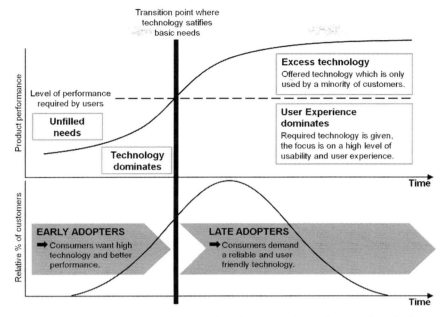

Figure 12.1 The transition from early adopters to late adopters in relation to technology development (reproduced from Donald A. Norman, *The Invisible Computer: Why Good Products Can Fail, the Personal Computer Is So Complex, and Information Appliances Are the Solution*, figure 2.4, © 1998 Massachusetts Institute of Technology, by permission of MIT Press)

Application of the User-Centred Design Framework within the BMW Group Development of Electric Vehicles

User-centred design requires above all a very good understanding of user requirements. In an early publication on this subject, Norman (1986) states that 'the needs of the user should dominate the design of the interface, and the need of the interface should dominate the design of the rest of the system'. The aspect 'rest of the system' cannot be overemphasised as in many cases user research focuses too much on the product itself and not enough on the ecosystem the product will exist in.

International Organization for Standardization (ISO) standard 9241-210 (2010), human-centred design for interactive systems, specifies general requirements for any user-centred design process. The process framework of the international standard can be regarded as the common foundation of most usability engineering and user experience process models. According to this standard, four main activities must take place during system development: (1) understand and specify the context of use, (2) specify the user and organisational requirements,

(3) produce design solutions that fulfil these requirements, and (4) evaluate designs against requirements from a user's perspective. Iteration loops may apply in any stage of the process.

On this basis, we applied the whole cycle of iterative design and evaluation several times with different levels of granularity: user context was analysed by literature research as well as doing interviews with experts from transportation. The human–machine interface (HMI) was evaluated with prototypes in driving simulators as well as using road trials. However, the most fundamental step was conducting international field trials with electric vehicles with customers to analyse what design decisions are necessary to meet the requirements of early adopters as well as late adopters and to make this innovation successful. We will focus in this chapter on the groundbreaking international field trials on electric vehicle usage that were undertaken to define user requirements for the BMW Group's new electric vehicle BMW i3.

The BMW Group launched two key learning projects in preparation for the BMW i3. The first step in this project was the MINI E which was specifically developed for field trials with customers and deployed in several test sites worldwide. The second step was the introduction of the BMW ActiveE that serves primarily to do research on EV-related technology, infrastructure, charging

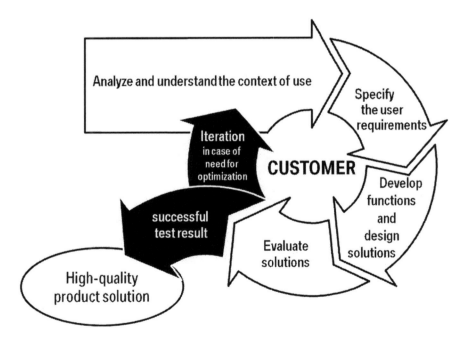

**Figure 12.2   The customer-centred development process as implemented
by the BMW Group's concept quality department**

solutions and service processes. It also played a significant role by providing the basis for evaluation iterations in user research on EV specific functions.

Analysing User Requirements: The MINI E Field Trials

The MINI E is a conversion of the MINI cooper and focuses on gaining customer experience and customers' requirements. The vehicle features a 204 hp electric motor, a torque of 220 Nm and a 35 kWh lithium-ion battery (28 kWh available). In real terms, and depending on the driving style, the MINI E's range is roughly 160 km (100 miles). Charging with a 32 Ampere wall box takes about 3.8 hours. With a conventional plug, using 12 Ampere and 240 volts, about 10.1 hours are needed for a full charge (from 0 to 100 per cent).

Understanding how EVs are used in real-world scenarios has substantially advanced during these field trials. The study was planned to provide answers to key topics, like the profile of customers currently interested in driving EVs as well as their expectations and motives. Objective long-term EV usage patterns on a day-to-day basis were analysed in combination with the users' subjective judgements on likes and dislikes about e-mobility[1] in general, and on specific vehicle characteristic of the MINI E in detail.

The MINI E trials, involving more than 600 vehicles since 2009, have been carefully planned and executed by the BMW Group in cooperation with public, private and university affiliates. The study was conducted in the United States, Germany, the United Kingdom, France (see also Chapter 11 in this volume), Japan and China. At the beginning of 2012, more than 15,000 people have applied worldwide to be test customers since the beginning of the trials. More than 16 million kilometres (10 million miles) driven by real customers have been logged. A subset of the MINI E customers, 430 private users, took part in in-depth interviews and research activities. Together with numerous users of 14 fleet companies, these private household customers were surveyed by a total of 15 research institutions in the participating six countries. The international MINI E field trials ended in early 2012. Until October 2013, the MINI E is still on the road in several local projects in Germany to address special research topics identified during the field trials and is helping to shed light on e-mobility in rural regions, to gain a first-hand insight into customer groups beyond early adopters, and to analyse in detail customers' driving and energy efficiency patterns when using EVs and combustion engine vehicles in direct comparison.

The compilation of data from the MINI E field trials has yielded arguably the most extensive results regarding everyday usage of electric vehicles worldwide. These results form the basis of early-phase user input on the context of use. In the following paragraphs, a short description of the methods used in the study will be

---

1   Electro mobility, or short e-mobility, refers to driving vehicles with electric powertrain technologies and on-board energy storage. While the general definition also includes hybrid vehicles, we will focus on full electric vehicles in this paper.

given, as well as selected results. The examples will depict how early customer feedback found its way into the development cycle.

Methods

The first steps in the field trial occurred in June 2009 in the United States and in Germany. Customers in New York and the New Jersey region as well as customers in Los Angeles rented the MINI E for at least one year. With 240 vehicles in fleet usage and 246 private customers, the US study formed the largest sample. The University of California at Davis carried out the scientific research with the household customers and in detail with 54 of those. Of the remaining private households, 72 customers took part in a survey that allowed for international comparisons with the other markets involved. Customers in the European and Asian field trials held their EV typically for six months. In Berlin, Germany, a total of 110 private customers took part in two separate test phases that were scientifically monitored by the Chemnitz University of Technology. An additional field trial in Munich involved 26 private customers. Moreover, 52 vehicles were deployed in fleets during the German trials. France (Paris) and the UK (Oxford, London) were further test sites in Europe with 50 private users/25 fleet vehicles and 40 private users/20 fleet vehicles, respectively. Research partners were the French Institute of Science and Technology for Transport, Development and Networks (IFSTTAR) and the Oxford Brookes University. The Asian trials started last in early 2011 in China (Beijing, Shenzhen) with 50 private users/25 fleet vehicles and 28 private customers/six fleet vehicles in Japan (Tokyo, Osaka). Research cooperation partners included the Chinese Automotive Technology and Research Center (CATARC), the market research company INS in Beijing, the Waseda University in Tokyo, Japan, and the Japanese market research company IID Inc. Following the explicit wish of customers, the study was continued after a short interruption following the Great East Japan Earthquake in March 2011.

The Berlin projects can be regarded as the blueprint for the MINI E trials in terms of methodology. The Institute of Cognitive and Engineering Psychology at the Chemnitz University of Technology has contributed their expertise in human-machine interaction and user research for in-vehicle systems. The Institute of Transportation Studies at the University of California at Davis has a long tradition in exploring alternative fuel vehicles and plug-in hybrid electric vehicles. Together with these partners a methods toolset was established that was used in similar form in all following MINI E projects. This basic set of methods is described in detail in Krems et al. (2010), Bühler et al. (2011) and Cocron et al. (2011).

Potential customers interested in participating in the trial applied via an online application form and had to supply information about relevant aspects of socio-demographic and psychographic background. Certain criteria were a prerequisite to be included (e.g., to be willing to actually use the car on a regular basis and to be willing to pay a monthly leasing fee). After being selected for the study, participants took part in telephone interviews asking about their motivation and attitudes. Interviews were held directly before the customers received the vehicle, after three months of usage and at the end of the leasing period. The interviews were face-to-face whenever possible; or if not possible, by telephone and supplemented with online questionnaires. Travel and charging diaries helped us to gain a deeper understanding about mobility needs and charging habits. Objective usage data was gathered with onboard data-loggers that recorded variables like trip length, speed, acceleration, frequency and duration of charging and battery status. Data on the vehicle's GPS position were not collected. In Germany, France, the United Kingdom and China a large proportion of the MINI Es were equipped with these data loggers. In the United States data on driving distances were read from AC Propulsion chips, which at least allow derivation of basic average values. The driving data were compared with a control group of privately owned, conventionally powered, vehicles which participated in a BMW Group research program.

In a field study of this magnitude and time frame it is not possible to keep all conditions constant, like in a controlled experiment. The field of e-mobility has undergone major changes during the last three years that made it necessary to add perpetually new questions to the interviews. It proved to be extremely difficult to translate all questions in an interculturally precise way into the different contexts. Therefore, it was not realistic to approach all countries with exactly the same set of interview items. Additionally, important events and major incidents during the field trials can be presumed to directly influence lives of the field trial participants like the earthquake and nuclear disaster in Japan, the public debates on fuel shortage, sustainable mobility, CO_2 and renewable energy as well as ongoing changes in transport and environmental policy. Although all care was exercised to take these effects into account in the design and interpretation of the study, a certain imprecision is inherent in the field trial method. Thus, numerical values (percentage agreement) should be interpreted as tendencies and, hence, only descriptive statistics are presented. Percentage values refer to answers on a Likert scale from one (do not agree at all) to six (fully agree). The top three values are grouped as 'agreement', the bottom three values as 'disagreement'.

User Motivations and Expectations

The high number of applicants (500–3,500 per country per phase) providing information about their background, allows a valid representation of the EV early adopter profile: typical applicants were male (approximately 80 per cent), around 40 years old (except for China: mean age 33), and well educated with above-average income and high self-reported affinity for new technology.

The most important motivation for participation was to experience a new clean and sustainable technology. Both factors were equally important. This tight combination of motivational influences is best reflected in the term 'sustainability meets technology'. Especially in the United States it was also important to gain independence from petroleum and to focus on the reduction of local emissions. Details on the US customers' motivations are described in Turrentine et al. (2011).

Before actually driving the MINI E, the majority of users expected to be constrained by the range and the missing cargo and back-seat passenger space. Between 19–60 per cent of users assumed that they would have to adapt their mobility behaviour. However, they were largely convinced that they were able to satisfy their daily mobility needs (between 88–100 per cent agreement rates).

Driving Experiences

Usage patterns in everyday mobility behaviour do not differ considerably between the MINI E and combustion engine vehicles in the same vehicle segment. Comparing the MINI E driving data to BMW 116i and MINI Cooper daily driving distances, the differences found are rather small with MINI E drivers in most locations using their car for even larger distances. The daily driving distance in the control group adds up to 43.4 km/27.0 miles (MINI Cooper) and 42.0 km/26.1 miles (BMW 116i). The MINI E drivers in France used their EV on average 44.2 km (27.5 miles) per day, in the UK 47.8 km (29.7 miles), in Germany 38.6 km (24.0 miles) and in China 49.0 km (30.4 miles). For these markets, details on the daily driving distance were available from the data logger. In the United States similar patterns arose with 50.9 km (31.6 miles) on West Coast test sites and 46.7 km (29.0 miles) on the East Coast. It is worthwhile comparing these data to mobility studies like the MiD study that analysed the mobility needs including all means of transportation in Germany. According to these results, the overall mobility need for an average citizen sums up to 39 km (24.2 miles) per day (Follmer et al. 2010). The US National Household Travel Survey reports that an average person travels 58.1 km (36.1 miles) per day (Santos et al. 2009). Both statistics are remarkably close to the usage pattern of the MINI E.

Detailed information about daily driving distance from data loggers are depicted in Figure 12.3. Clearly, the differences between the MINI Es and the conventionally powered vehicles from the same segment are very small. More importantly, the driving patterns of the 1 Series BMW and the MINI Cooper lie well within the range typical EVs can provide. Of course, there are also use cases that cannot be fulfilled using a pure battery electric vehicle. For instance, the driving distances of a 5 Series Diesel, which was also part of the study of BMW Group's data logger team, clearly cannot be covered with a range like that of the MINI E. Therefore, it can be expected that usage patterns for customers buying full EVs will resemble the use scenarios of the (quite voluminous) owner group of compact cars.

Based on the MINI E drivers' subjective estimations, and validated with travel diary data, the customers were able to undertake about 80 per cent of intended trips with the MINI E. The satisfaction of mobility needs ranges between a 77 per cent minimum (China) and an 84 per cent maximum (France), with an average value over all countries of 82 per cent. This can be increased on average to up to 91 per cent over all countries if the MINI E did not have the conversion vehicle

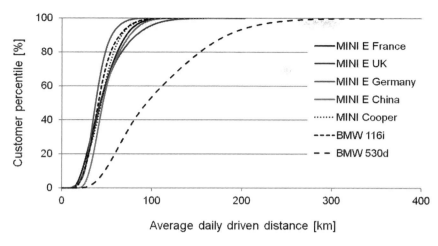

Figure 12.3   Accumulated MINI E and combustion engine vehicle daily driving distances

drawbacks of being only a two-seater without an adequate cargo or luggage compartment in the boot. On the one hand, these results imply that Original Equipment Manufacturers (OEMs) will need to offer their customers access to innovative mobility solutions if they want to use their EV as the only means of individual transport in their households. On the other hand, it clearly demonstrates that a purpose-built EV with the range of the MINI E will be sufficient for no less than 90 per cent of mobility needs, which is an extraordinary result considering the articulated scepticism about the range of EVs for everyday usage. It should be noted, however, that especially in winter the MINI E's range was not always available. If the batteries got too cold, range was significantly reduced or the vehicle was not able to operate as demanded in freezing temperatures during winter. The reason for this drawback, which the customers also criticised, was the thermal management of the MINI E's conversion vehicle battery system relying on air cooling only.

When asked for desired range options for future EVs, customers typically demand ranges like 200–250 km (125–155 miles). It is interesting to see that this is, of course, much less than the range of a combustion engine vehicle. Experienced EV customers take into account that higher ranges come with higher prices. This may have moderated their demand. Even more interesting is that, even with this additional range, they would not be able to cover the remaining 10 per cent of mobility needs. So the demand for more range seems to reflect in some ways the wish for a kind of safety buffer in range, as most customers recharge their battery evidently before the state of charge approaches very low values. Franke et al. (2011) add a very interesting perspective to this. They analysed how customers experience the limited range in an EV and how this is related to other variables, notably stress. They were able to show that users can indeed adapt to limited range, but that they utilise the available range sub-optimally. Certain personality traits and coping skills moderated the experience of comfortable range. Franke et al. (2011) conclude that it may be possible to change the personal feelings towards range by providing knowledge background, training and suitable driver information systems. In consequence, this may allow all EV drivers to use the available range to its full extent.

One lesson learnt in the MINI E field trials was also that customers need a simple and very direct way of extending the available range in unforeseen situations. Many drivers reported that they often tried to reduce energy consumption when battery state got low by driving more carefully and by switching off energy-consuming comfort functions like heating, ventilation or the radio. However, they did not know for sure which action in terms of switching off devices had which effect and how optimally they performed in driving efficiently or if they still drove too fast or maybe even unnecessarily slowly. This situation is already addressed in the BMW ActiveE. The ECO PRO mode assists drivers in reducing energy consumption. The acceleration behaviour is changed, energy supply for auxiliary systems is flattened and the driver is provided with hints if acceleration or velocity is too high. Resulting usage patterns and customer feedback will be subject to analysis in the BMW ActiveE user studies.

The available range of an EV, in real terms, in comparison with a combustion engine vehicle, is much more dependent on driving style and on the ability of the driver to use advanced efficiency features like regenerative braking. Regenerative braking refers to using the electric motor as a generator, thus recapturing energy otherwise lost during braking, coasting or downhill driving. The function is integrated in the accelerator pedal in the MINI E. Lifting the foot quickly from the pedal leads to quite strong deceleration of -2.25 m/s^2, which also triggers the braking lights to warn following traffic. This differs from most electric or hybrid vehicles and allows the car to be driven, basically, with only one pedal, activating the brakes only in cases of very strong or emergency braking events. For details on longitudinal dynamics refer to Eberl et al. (2012). Regenerative braking can be a powerful tool extending the available range, but it was largely unclear whether customers would accept the implementation in the accelerator pedal and if they would be willing to use it or try to avoid it. Therefore, the long-term evaluation of regenerative braking was one of the most important field study goals. The customer feedback was astonishing. Between 92 per cent (China, Japan) and nearly 100 per cent (Germany, United States, United Kingdom, France) of customers stated that they liked to be able to control the vehicle with just one pedal. They estimated using regenerative braking in 78 per cent to 92 per cent of all braking events, which mirrors almost exactly the objective data derived from data loggers. In detail, the customer feedback pointed out that the MINI E's high driving performance combined with regenerative braking provided a new form of sporty driving and allowed, at the same time, the ability to experience efficient driving and energy saving in a very immediate way. 'Single-pedal driving' became, for most customers, almost game-like. Trying to stop at every red traffic light without touching the brakes was a very common pattern. Using regenerative braking this way may even increase traffic safety as it is necessary to exercise anticipatory driving to be able to master single-pedal driving. Turrentine et al. (2011) discuss these behavioural patterns and potential factors to be adjusted in regenerative braking to enhance the experience.

From the users' point of view, the lack of engine noise is – besides regenerative braking – one of the most striking features in driving EVs. Drivers state unanimously that they liked the quiet operation of the MINI E (agreement in questionnaire items was between 95 per cent and 100 per cent). Especially customers in France (57 per cent) and China (65 per cent) rated the quiet interior to be a highly relevant comfort factor that would even justify a somewhat higher vehicle price if pushed to the maximum, while in the other countries the level of quietness already achieved seems to be satisficing (lower agreement rates for the demand of an even quieter interior range from 19 per cent to 32 per cent). However, when it comes to the low outside noise levels due to missing engine sounds, concerns about the potential danger of silent driving are often expressed in public discussion. Asked for their opinion, MINI E drivers in different countries report different experiences. Generally, lower concerns are expressed in Europe. While half of the German drivers (50 per cent) at the beginning of the field trials see a potential danger in not being heard at low speeds, there is a substantial reduction with growing experience, to 16 per cent at the end of the trial. In France, the figures are somewhat higher but there is still a reduction from 62 per cent to 50 per cent. The situation is different in Asia. Chinese customers start with a very low estimation of potential difficulties associated with silent driving (27 per cent) that increases over the trial to 69 per cent. In Japan, the estimation is already very high at the beginning, with 74 per cent, and decreases only marginally to the same level as in China, 69 per cent. It is quite likely that the traffic situations in the Asian mega-cities account for these risk estimations. Also in Europe, the traffic density in Paris is much higher than in Munich or Berlin.

Charging Experiences

Generally, there is a steep learning curve during the first one to two weeks of EV usage. While there is no problem at all with using the vehicle to get from A to B, users develop expertise in some areas of EV driving, like using regenerative braking. This expertise is most strongly reflected in charging behaviour. Most inexperienced EV drivers tend to charge their vehicle every time they get the chance to without reflecting their real range and charging needs. At the same time, most drivers of conventional vehicles do not think much about their daily mobility needs. When using an EV and dealing with the limited range, the MINI E drivers typically realised that they do not need the full range of the battery every day. Of course there are a lot of customers who make charging every night when they return home to their garage a habit. But on average, the data loggers in the MINI E show that most users switch from daily charging to only charging once every two or three days as can be seen in Figure 12.4 (mean charging events per week: Germany: 1.9, UK: 2.9, China: 2.5). This phenomenon is also found in other EV studies like the Ultra Low Carbon Vehicle Demonstrator Programme in the UK that also reported a charging frequency of less than once in every two days (Everett, Walsh, Smith, Burgess and Harris 2010). For technical reasons, the data logger results from France cannot be compared directly to the other markets reported below. As the MINI E customers in France did not charge at 32 amps like the other markets, their charging durations were much longer and they needed to rely on charging every night. As they thus did not have the chance to establish a different charging pattern, it is not surprising to find that Labeye et al. (2011) report an average charging frequency of 5.2 per week for French EV users (see also Chapter 11 in this volume).

Improving Usability: The BMW ActiveE Field Trials

Although the BMW ActiveE is also a conversion vehicle, based on a 1 Series Coupé and like the MINI E produced in small number for field trials, its architecture brings improvements for everyday usage with an unaltered passenger-compartment space giving access to four fully fledged seats and with a 200-litre luggage compartment. The car is able to accelerate from 0 to 100 km/h in nine

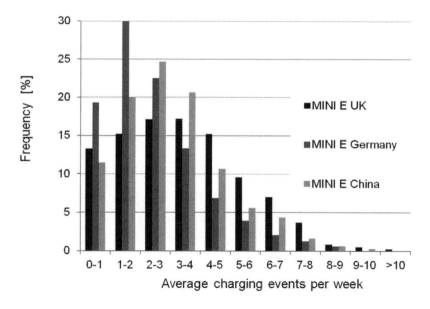

Figure 12.4   Charging frequency per week in markets with 32 amps wallbox

seconds with a power output of 170 hp and a torque of 250 Nm. Range and charging duration are comparable to the MINI E. However, as the newly conceived lithium-ion energy storage units, like the drive train a pre-production version of the BMW i3, have a cooling/heating system that tempers the liquid in the storage housing unit, the battery can be held in good operating temperatures in cold as well as in hot environments. This enables the BMW ActiveE battery system to compensate for low or high outside temperatures much better than the MINI E did.

A test fleet of more than 1,000 BMW ActiveE vehicles serves to deepen the knowledge gained in the MINI E study. Research focus in these field trials is shifted towards technological innovations and technical components. Of course, all standard development test procedures were performed before handing over the vehicle to customers, but valid data on long-term performance of EV components can only be gathered when analysing all possible everyday situations. The involvement of pilot customers ensures that usage scenarios are also covered that are currently unknown to engineers because of the still too small knowledge on EV customer behaviour. Furthermore, the BMW ActiveE can be seen as an iteration step in customer centred design of electric vehicles. Innovative functions specifically designed for EVs are implemented for the first time in a BMW. Whereas the MINI E studies aimed at learning about fundamental aspects of everyday life with EVs, user studies conducted with the BMW ActiveE will be oriented towards a usability testing approach of these new features. For instance, preconditioning allows customers to cool or heat the batteries and the vehicle interior before starting a trip, reducing energy consumption significantly while driving. Special menus in the Central Information Display allow control of advanced EV functions like programming a charge timer. Newly designed information displays are available that schematically represent vehicle energy flows in order to make the electric driving experience better perceptible and comprehensible. The already mentioned ECO PRO mode can be activated by a simple button press in the centre console and is intended to make energy saving easier if necessary: the accelerator pedal characteristic is changed, delivering less power, and systems like air conditioning are turned down. Several EV-related functions are also available remotely via smartphone apps like monitoring the charging progress or activating preconditioning. BMW ConnectedDrive functions do not only allow integration of smartphone apps and Internet-based functions like a charging station locator in the vehicle, it also serves as a feedback channel for data recorded during the field trials. Of course, only development-related data like the distance travelled or maximum vehicle range after charging are collected and anonymity of recorded data is guaranteed at all times.

Field trials with the BMW ActiveE started at the end of 2011 in Germany with 15 private and 15 corporate customers. Additional vehicles are part of governmentally funded research projects from 2013 onwards. At the end of the trials, about 190 vehicles will have contributed to projects in Germany. With the beginning of 2012, about 700 BMW ActiveE vehicles form the largest test fleet in the United States where they will not only provide insights in user experience but also in sales and handling processes and service infrastructure demands for higher numbers of electric vehicles. About 100 BMW ActiveE cars will begin their duty in 2013 in China. Additional vehicles are on the road in France, the United Kingdom, the Netherlands, Italy, Switzerland, South Korea and Japan.

Research projects with direct customer feedback are mainly carried out in the US, Germany and China. The early study in Berlin used a more qualitative approach to prepare surveys with a stronger quantitative component in the United States and in China. With data from Germany and the first preliminary data from the US, we will show through the exemplar of the ECO PRO mode how the results of these studies are transferred to series development.

BMW ActiveE Qualitative Pilot Study

The first BMW ActiveE user study with 15 private customers (14 male, 1 female, average age 46) was undertaken between December 2011 and March 2012. Ten of these customers were former MINI E drivers. Scientific cooperation partners were the Chemnitz University of Technology and the psychologically oriented market research agency Spiegel Institut Mannheim. Due to the novelty of several features of the BMW ActiveE, it was important to get very direct, unbiased feedback of the associated user experience. Therefore, the study design provided the customers with several opportunities to share their opinions. A three-phase research procedure began with open telephone interviews four weeks after having the vehicles handed over to the users. In this early stage of vehicle experience, customers freely stated their impressions, what excited them, what bothered them. Topics for focus group discussions as a second research step were defined partially following the customer feedback from the telephone interviews. The focus groups took place approximately after eight weeks of vehicle usage. The main topics were EV winter usage and preconditioning in one group and general system usability and efficient EV driving in the other group. At the end of the trial, users finally participated in an online survey which took the results from both the telephone interviews and the focus group discussions into account.

Concerning the case example ECO PRO mode, it was very interesting to see that it was not used in a uniform way at all. Customers already stated in the focus groups that their motives for using ECO PRO were different. Some of the customers used it almost always either because they needed to as they were driving longer distances regularly or simply because they were intrinsically motivated to save energy. Other customers stated that they only needed a kind of emergency option if they seem to be running out of charge while driving. Considering these results, it is not surprising to find that the ECO PRO usage varied from 5 per cent to 95 per cent share of total distance driven. Both extreme positions saw optimisation potential for the functions. One position was that the loss of comfort (e.g., the deactivation of seat heating) was rated to be too strong. The other position expected a stronger effect on driving dynamics and comfort functions when the usage intention was to maximally exploit a very low battery range. The layout of the ECO PRO mode in the BMW ActiveE was more strongly oriented towards the latter situation. If users prefer to drive with a permanently activated ECO PRO mode while at the same time the option for an 'emergency' range situation must be available, two different levels of this mode may better suit the range of use cases. Due to the small number of cases in the Berlin BMW ActiveE study, further evidence was gathered in the US field trial.

BMW ActiveE Quantitative Validation Study

Based on the final survey that was shaped during the BMW ActiveE field trial in Berlin and extended with insights from social media analysis conducted since the beginning of the field trials, an online survey was implemented at the beginning of 2013 addressing customers in the United States.

Focussing again on the example of ECO PRO mode, first data based on N = 79 participants replicated the results from the Berlin field trial. Figure 12.5 demonstrates again the existence of two extreme user groups providing further support for the idea to offer two different ECO PRO modes: enhanced energy efficiency versus maximum energy saving. The distinction between these two mode types into an ECO PRO and an ECO PRO + mode is part of the BMW i3 development plan, which was confirmed on the basis of these field trial results.

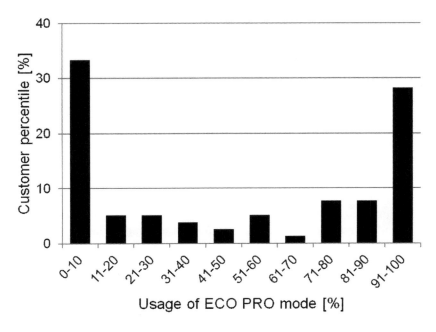

Figure 12.5 Usage of ECO PRO mode as percentage of daily driving

Conclusions

Disruptive innovations will only be a success if they are convenient and easy to use and meet the customer requirements of early adopters as well as those of late adopters. Research on how innovations gain market acceptance suggests that especially disruptive innovations should be designed according to a user-centred design and evaluation approach (Norman 1998). In this paper we described how this approach was applied to the development of the BMW i3 electric vehicle. The results from fundamental field trials of electric vehicle usage helped in defining customer requirements as well as in setting the benchmarks for later and more specific user testing (e.g., testing of HMI functionalities). In addition, specific EV related functions like the ECO PRO mode, preconditioning, remote charging control and so on were pioneered in the BMW ActiveE field trials in an approach similar to long-term usability testing, delivering additional insights into customer demands for future EVs. These requirements were not only fed into the development cycle as depicted in Figure 12.2, the analysis results were also translated into recommendations for further actions and distributed to relevant departments in marketing and sales, strategy and communication.

Several key deductions can be made from the results reported here. It became clear that the available range of the MINI E is an excellent basis for future EVs regarding the necessary trade-offs between required range for everyday mobility and costs. However, in order to be able to use an EV as the only vehicle in the household, the remaining 10 per cent gap needs to be closed. This will be done, for instance, by offering mobility services as well as enabling the BMW i3 for fast charging (DC charging) to increase flexibility. Details on mobility services, parking and charging solutions or assistance services are described on http://www.bmw-i.com.

The completely new feature 'single-pedal driving' as part of the regenerative braking concept was identified as being very important for EV customers. Although mainly rated as very unusual, customers quickly developed not only an acceptance in terms of driving efficiency, they strongly liked the associated efficiency experience and the exclusiveness of the emerging new driving style opportunities. Therefore, it has been decided to maintain single-pedal driving for the series vehicle.

Similarly surprising, especially for long-term combustion engine vehicle engineers, were the results on EV acoustics. The pilot customers did not miss the sound of a conventional engine at all. The new silence inside the MINI E was one of the most prominent positive experiences – although the MINI E as a conversion vehicle was not yet able to play out all advantages of a silent EV interior. Whereas most customers also liked that the vehicle did not emit any noise at all, some customers demanded that there should be a way of making an EV acoustically perceivable to other road users. In order to keep the advantages of silent driving and to follow the requests of the majority of users, customers will be able to control the active acoustic exterior vehicle sound if allowed by legislation of the country in which the EV is registered.

Additional research will be carried out to complement the findings of these field trials along the customer development cycle as depicted in Figure 12.2. We believe that with the information and knowledge gathered in this development cycle, we will be able to bridge the gap between early adopters and late adopters and make the disruptive innovation of electromobility a success.

Acknowledgements

The authors would like to acknowledge the *project i* team at the BMW Group (especially Julian Weber, Michael Hajesch, Søren Mohr and Jens Ramsbrock) for making these studies possible, the data-logger team around Tobias Karspeck and Katja Gabler, the MINI E user research team as well as Andreas Klein from Spiegel Institut Mannheim and Peter Dempster of the BMW Group Technology Office USA especially for their valuable contributions to the BMW ActiveE projects. We would like to thank our international research partners for bringing in their expertise and for conducting the research with extraordinary commitment and outstanding efforts. Parts of the research reported here were funded in Germany by the Federal Ministry for the Environment, Nature Conservation and Nuclear Safety and by the Federal Ministry of Transport, Building and Urban Development. The MINI E UK trial was funded by the Technology Strategy Board. Most of all, on behalf of everybody involved in the BMW Group's e-mobility projects, we would like to thank our pilot customers in the MINI E and BMW ActiveE field trials for their outstanding support of our research and their inspiring feedback.

References

Bühler, F., Neumann, I., Cocron, P., Franke, T. and Krems, J.F. 2011. *Usage Patterns of Electric Vehicles: A Reliable Indicator of Acceptance? Findings from a German Field Study*. Proceedings of the 90th Annual Meeting of the Transportation Research Board, TRB 90th Annual Meeting, Washington, DC, 23–27 January. Available at: http://amonline.trb.org/12jj41/1 [accessed: 10 January 2013].

Christensen, C.M. 2012. Disruptive Innovation. In *The Encyclopedia of Human Computer Interaction*. Edited by M. Soegaard and R. F. Dam. 2nd Edition. Aarhus, Denmark: Interaction Design Foundation. [Online]. Available at: http://www.interactiondesign.org/encyclopedia/disruptive_innovation.html [accessed: 9 January 2013].

Christensen, C.M. and Raynor, M.E. 2003. *The Innovator's Solution: Creating and Sustaining Successful Growth*. Boston: Harvard Business School Publishing.

Cocron, P., Bühler, F., Neumann, I., Franke, T., Krems, J.F., Schwalm, M. and Keinath, A. 2011. Methods of Evaluating Electric Vehicles from a User's Perspective: The MINI E Field Trial in Berlin. *IET Intelligent Transport Systems*, 5(2): 127–33.

Davis, F.D. 1993. User Acceptance of Information Technology: System Characteristics, User Perceptions and Behavioural Impacts. *International Journal of Man-Machine Studies*, 38: 475–87.

Eberl, T., Sharma, R., Stroph, R., Schumann, J. and Pruckner, A. 2012. Evaluation of Interaction Concepts for the Longitudinal Dynamics of Electric Vehicles. In *Advances in Human Aspects of Road Transportation*. Edited by N.A. Stanton. Boca Raton, FL: Taylor and Francis, 263–72.

Everett, A., Walsh, C., Smith, K., Burgess, M. and Harris, M. 2010. *Ultra Low Carbon Vehicle Demonstrator Programme*. Proceedings of the 25th World Battery, Hybrid and Fuel Cell Electric Vehicle Symposium and Exhibition (EVS 25). Shenzhen, China.

International Organization for Standardization (ISO). 2010. *Ergonomics of Human-System Interaction – Part 210: Human-Centred Design for Interactive Systems*. ISO 9241-210. Geneva: Beuth.

Follmer, R., Gruschwitz, D., Jesske, B., Quanst, S., Lenz, B., Nobis, C., Köhler, K. and Mehlin, M. 2010. Mobilität in Deutschland 2008. Bonn and Berlin: Institut für angewandte Sozialwissenschaft and Deutsches Zentrum für Luft- und Raumfahrt.

Franke, T., Neumann, I., Bühler, F., Cocron, P. and Krems, J.F. 2011. Experiencing Range in an Electric Vehicle: Understanding Psychological Barriers. *Applied Psychology*, 61(3): 368–91.

Kirsch, D.A. 1997. The Electric Car and the Burden of History: Studies in Automotive System Rivalry in America, 1890–1996. *Business and Economic History*, 26(2): 304–10.

Krems, J.F., Franke, T., Neumann, I. and Cocron, P. 2010. *Research Methods to Assess the Acceptance of EVs: Experiences from an EV User Study, in Smart Systems Integration*. Edited by T. Gessner. Proceedings of the Fourth European Conference and Exhibition on Integration Issues of Miniaturized Systems. Como, Italy: VDE.

Labeye, E., Hugot, M., Regan, M. and Brusque, C. 2012. Electric Vehicles: An Eco-Friendly Mode of Transport Which Induces Changes in Driving Behaviour. In *Human Factors of Systems and Technology*. Edited by D. de Waard et al. 2012. Maastricht, the Netherlands: Shaker Publishing.

Moore, G.A. 1991. *Crossing the Chasm: Marketing and Selling High-Tech Products to Mainstream Customers*. New York: Harper Business.

Norman, D.A. 1986. Cognitive Engineering. In *User Centered System Design: New Perspectives on Human-Computer Interaction*, 31–61. Edited by D.A. Norman and S.W. Draper. Hillsdale, NJ: Lawrence Erlbaum Associates.

———. 1998. *The Invisible Computer*. Cambridge, MA: MIT Press.

Santos, A., McGuckin, N., Nakamoto, H.Y., Gray, D. and Liss, S. 2009. *Summary of Travel Trends: 2009 National Household Travel Survey*. Report no. FHWA-PL-ll-022. Washington, DC: NHTSA.

Turrentine, T., Garas, D., Lentz, A. and Woodjack, J. 2011. *The UC Davis MINI E Consumer Study*. Davis: University of California Press.

Chapter 13

Motorcycle Riders' Acceptance
of Advanced Rider Assistance Systems

Véronique Huth

Institut Français des Sciences et Technologies des Transport,
de l'aménagement et des Réseaux (IFSTTAR), Bron, France

Abstract

Motorcycle riders have a pronounced vulnerability and crash risk with a high prevalence of human error as a contributory factor in crashes. So there appears to be substantial potential for Advanced Rider Assistance systems to improve their safety. However, the benefit of these systems for riding safety depends on riders' responses to them. This chapter presents research on the acceptability and acceptance of Advanced Rider Assistance Systems. Factors that influence riders' acceptance are identified and discussed. Indicators of the need for assistance technologies to enhance safety are contrasted with riders' views of different types of systems.

Introduction

Riding a motorcycle is challenging and it carries particular risk. Riders are clearly over-represented among crash victims all over the world (Peden et al. 2004). In Europe, the decrease of 30 per cent in the total number of traffic fatalities from 1999–2008 contrasts with a rise in motorcycle fatalities by 7 per cent (European Commission 2010). On the other hand, the popularity of motorcycle riding has increased in recent years, mirrored by an increasing number of registered motorcycles (Haworth 2012). Given the elevated crash risk and vulnerability of riders, riding safety is a relevant matter of concern. The prevalence of human error as a primary crash-contributing factor in motorcycle crashes of almost 90 per cent (Motorcycle Accidents In-Depth Study [MAIDS] 2004) suggests focusing measures especially on the riders and their interaction with other road users. Accordingly, it has been advised to investigate ways of effectively targeting the human factor in order to successfully improve riding safety (Elliott, Baughan and Sexton 2007). In addition to adequate rider licensing, education and training, Advanced Rider Assistance Systems (ARAS) could help prevent crashes that involve or are caused by human error. Although advanced assistance systems have been developed mainly to enhance car safety, they have considerable potential for motorcyclists as well (Ambak, Atiq and Ismail 2009), with crash reduction estimates reaching 40 per cent based on widespread use of ARAS (Rakotonirainy and Haworth 2006).

The following sections provide more detailed insights into the crash risk and vulnerability of motorcycle riders, the role of the human factor in motorcycle crashes and the ways in which advanced rider assistance systems could contribute to the enhancement of riding safety.

Crash Risk and Vulnerability of Riders

Riders are dramatically over-represented in road crashes (e.g., Peden et al. 2004, Haworth 2012) and the rising number of motorcycle fatalities in many European countries, and worldwide, contrasts with the overall reduction of traffic fatalities over the last decade (International Road Traffic and Accident Database [IRTAD] 2010). The typical crash scenarios that involve motorcycles result from the characteristics of the vehicle and the way the riders use it (e.g., Clarke et al. 2004). The most common one is the single-vehicle crash, where the riders lose control of their vehicle in a curve (Hurt, Ouellet and Thom 1981, MAIDS 2004, Traffic Accident Causation in Europe [TRACE] 2008). The relevance of this crash scenario is substantiated by its frequency and by the particular injury risk it carries. It holds a doubled fatality risk and a similarly increased probability of serious injuries compared with other motorcycle crashes (Clarke et al. 2004). Front-side crashes at intersections represent the second most prominent crash type for motorcyclists (Hurt et al. 1981, MAIDS 2004, TRACE 2008). Further typical accident types are rear-end crashes and side-side crashes, although these are not well represented in motorcycle crash databases (TRACE 2008).

Compared with cars, motorcycles are far less stable and riders are far less protected by their vehicle than car drivers. In a crash, the riders can easily be thrown off their motorcycle and their injury risk far exceeds the risk car drivers face in a collision (Elliot et al. 2007, Mayou and Bryant 2003, Pai 2011). Hence, the avoidance of any crash is both vital and challenging for motorcyclists. Given that riding is highly sensitive to unfavourable conditions, the necessary crash prevention measures should not interfere negatively with the control of the motorcycle.

Human Error in Motorcycle Accidents

Riding a motorcycle requires not only a high level of motor-skills, physical coordination and balance (Mannering and Grodsky 1995) but also involves a constant hazard-monitoring task (Haworth et al. 2005). Riders need to be able to anticipate and recognise risk-encumbered situations and to choose adequate crash avoidance behaviour (DEKRA 2010, Di Stasi et al. 2009). Such high demands make the riding task particularly susceptible to human error, which has been identified as the primary crash-contributing factor in 87.5 per cent of all crashes involving a motorcycle (MAIDS 2004). In 37.1 per cent of those crashes the riders committed errors, including inappropriate speed choice, short safety headways and failures when overtaking (DEKRA 2010). The remaining 50.4 per cent consisted of other road-users' errors, caused or amplified by the low sensory or cognitive conspicuity of the motorcycle (Brenac et al. 2006, Crundall et al. 2012). Although not causing the crash, riders might be contributing to these accidents with their riding style or they might not be able to carry out successful crash avoidance manoeuvres (Phan et al. 2010). This shortcoming has been detected in almost 30 per cent of all multi-vehicle crashes (MAIDS 2004) and may be influenced by overconfidence in the riders' own anticipatory capacities, and by speed (Phan et al. 2010).

The high prevalence of human failure in accidents involving a motorcycle creates a great potential for ARAS that may be developed to assist riders in monitoring the road situation for certain hazards and in preventing human error–related motorcycle crashes.

The Purpose of Advanced Rider Assistance Systems

Assistance systems can be classified into active and passive systems. The first category applies to technologies that influence crash risk by aiming to avoid crashes; the second category refers to systems that mitigate the consequences of a crash (i.e., primary and secondary safety systems, respectively). This chapter addresses only active assistance systems for riders, such as a Curve Warning system or Intelligent Speed Assistance.

A further distinction can be made regarding the way these systems interact with the rider. They can inform the rider about a risk-encumbered situation by transmitting a warning or they can intervene directly in the riding activity in such situations.

ARAS assist the riders in specific situations that may represent a threat and lead to a crash in the absence of an appropriate adjustment of riding behaviour. ARAS can reduce errors committed by the riders or increase the riders' alertness to possible errors of other road users as well as their preparedness to compensate for those errors. The purpose of ARAS is to increase the riders' safety margin by provoking cautious behaviour and by reducing reaction times in response to a potentially critical situation. These systems can either be factory-fitted by the manufacturer or available as after-market devices. Beyond the technical challenge of fitting such systems onto the motorcycle, the use of the ARAS by the riders is a crucial issue. Riders' acceptance of an ARAS will contribute to the safety potential of the system, since it will only take effect if riders acquire the system, install it on their vehicle and activate it during their rides. Furthermore, the riders must be willing to use it in the manner intended by the designers and follow the ARAS suggestions. As a consequence, successful implementation of ARAS, where technical benefits are effectively translated into safety benefits, requires investigation of the acceptability and acceptance by the riders and the integration of a user-centred evaluation of the support system into the system development process at an early stage (cf. Bayly, Hosking and Regan 2007). Furthermore, the identification of influencing factors on acceptance should help to improve ARAS and to create conditions that are favourable to widespread system use.

Rider Needs

The importance of knowledge about the acceptability and acceptance of assistance systems has been widely recognised and explored in the automotive context (e.g., Vlassenroot et al. 2010 and elsewhere in this book). However, the particular characteristics of motorcycle riding prevent it from being directly comparable to other modes of transport; road safety measures that might work for car drivers are not necessarily equally applicable to motorcycle riders. This may also hold true for the acceptance of technologies that relate to the riding activity. It therefore makes sense to consider the nature of riding and its possible implications for the acceptance of ARAS. Correspondingly, the Federation of European Motorcyclists' Associations (FEMA) adopted a positive attitude towards assistance systems for riders upon the condition that the development of such a system is driven by riders' needs and considers the particularities of their vehicle (FEMA 2011).

The Nature of Riding

The main psychological differences between riding and driving consist of the underlying motivations, the experience of the activity, the role of risk and social aspects. Riding a motorcycle has been described as a leisure activity that is driven by intrinsic motivations such as riding sensations rather than extrinsic motivations related to mobility needs (Broughton 2008). Emotions like thrill and feelings of freedom promote the enjoyment of the ride (Broughton and Stradling 2005, Broughton 2007, Haworth 2012) and intense sensations of dynamics and control can be achieved by the expressive riding style that the high manoeuvrability of the vehicle allows (e.g., Broughton 2005, Mannering and Grodsky 1995). Not surprisingly, passion for motorcycles and performance are common riding motivations (Christmas et al. 2009).

On the other hand, riding in an expressive manner and at high speed to increase riding sensation implies a noteworthy risk of losing control of the bike (Broughton et al. 2009, Moller and Gregerson 2008). Using the framework of the concepts of optimal task difficulty (Fuller 2005, Wilde 1982) or flow (Csikszentmihalyi 1997), risk can be understood as part of the riding performance and the riders may try to adapt the risk level of the riding activity to their own skill level. Few riders have been identified as active risk-seekers but almost 50 per cent have been identified as risk-acceptors who enjoy risk up to a certain threshold; that is, as far as it helps them to match their skills to the challenge of riding (Broughton and Stradling 2005).

Finally, riding a motorcycle serves as a mode of self-presentation and expression (Broughton 2007). Riding in groups is popular and may lead to the creation of strong relationships among the riders (Tunnicliff et al. 2011). As a consequence, specific group identities are often built and their norms of belief, expectation and behaviour may have a considerable influence on the members of the group (Tunnicliff et al. 2012). This influence can be reinforced in those contexts where relevant people are present (Parker et al. 1992).

These psychological characteristics of riding should be taken into account when developing ARAS. Their probable implications are presented in the following section.

Implications for Advanced Rider Assistance Systems

In view of the most common motivations for riding, and especially the significance of the riding experience, it seems essential that an assistance system does not alter the satisfaction of riding motives. The riders should not be annoyed by unnecessary or redundant warnings of the ARAS, but rather be alerted in specific situations that are relevant to individual riding safety. Regarding interference with riding sensations, the warning design must play a decisive role; the riders should feel assisted rather than disturbed. Besides, the strong emotional component of riding may influence the riders reasoning and decision-making. This has been shown for the riders' intention to speed (Elliott 2010) and may apply to the riders' intention to use an ARAS. That is why particular value should be attached to the assessment of riders' opinions in addition to objective indicators of system effectiveness.

Underestimations of the crash risk increase the chances of getting into a critical situation (Bellaby and Lawrenson 2001, Mannering and Grodsky 1995), especially if combined with an expressive use of the motorcycle. By adjusting a biased risk perception in specific situations, ARAS can avoid or rectify risky riding behaviours that would be carried out when riding without support. Yet, if the thresholds employed by the system differ excessively from the riders' accepted levels of risk, they may feel annoyed and disapprove of the ARAS. Hence, the acceptance of an ARAS might depend on the riders' awareness of the crash risk related to the situation the support function has been designed for. Nevertheless, riders might reject any ARAS if they feel the system is intervening too much in the riding activity, be it related to the sensation of ceding control to the system or to the possible risk of losing stability due to system intervention.

The identities of rider groups may give rise to a considerable social influence on the riders' choices and opinions. Such an effect of the reference group, 'fellow riders', has already been found in riders' intention to speed (Elliott 2010) and could be found in other types of safety-related behaviour, including the usage of novel ARAS.

These implications need to be taken into account when analysing riders' acceptance of ARAS, so as to determine how to design and implement systems that are compatible with rider needs. The following results on the acceptability and acceptance of ARAS represent first steps towards acquiring knowledge that will help to optimise the potential of ARAS and guide its successful implementation as a road safety measure.

Acceptability of Assistive Technologies

Before introducing or developing ARAS, riders' attitudes towards the potential assistance function can be gauged. As seen elsewhere in this book, this procedure can permit consideration of user needs at an early stage.

Schade and Schlag (2003) distinguish 'acceptability' from 'acceptance' by relating both concepts to the experience of the system or measure. While acceptance refers to the users' reaction after the introduction of the measure, acceptability denotes a prospective judgement of a measure that has not yet been experienced (Vlassenroot et al. 2010; see also Chapter 7 in this volume). Thus, acceptability is an attitudinal concept that excludes any behavioural or reactive aspect.

Since such measures of acceptability refer to a phase where the riders have not yet had the chance to interact with the system, opinions given by the riders are more hypothetical in character. Given the importance of acceptability and acceptance of a technology for its future implementation (Vlassenroot et al. 2010), it is nevertheless interesting to explore riders' acceptability of ARAS, and to compare it later with their acceptance measures.

Available Results on the Acceptability of ARAS

In this section, the results of the following four studies on the acceptability of ARAS are presented:

1. Simpkin et al. (2007) included rider ratings before system use in their test trials on Intelligent Speed Adaptation.
2. Cairney and Ritzinger (2008) analysed rider views on Intelligent Speed Adaptation through focus group interviews.
3. A focus group interview collected expert riders' opinions on several ARAS proposed by the European Commission's SAFERIDER project (Baldanzini 2008).
4. Within the European Commission's 2 BE SAFE project, focus group interviews and a questionnaire survey on a variety of assistance and information functions, including ARAS, were conducted with riders (Lenné et al. 2011, Oberlader et al. 2012).

The last two cover a range of systems, whereas the first two deal with a specific ARAS, that is, Intelligent Speed Adaptation (ISA). This system assists the riders in keeping the prevailing legal speed limit by indicating it on a display and emitting an alert (advisory) or applying a counterforce on the throttle (intervening) when the speed limit is exceeded.

The study on ISA (Simpkin et al. 2007) revealed that riders were hesitant in judging the potential usefulness of such systems before having tested them. Furthermore, they were more reluctant in doing so for an intervening version of ISA than for an advisory one. Cairney and Ritzinger (2008) also found that riders were rather sceptical regarding the effectiveness of ISA in improving riding safety. They were concerned with the reliability of the system and were in this respect far more reluctant to embrace active speed control by the ISA than an advisory version of the system. The riders showed a potentially higher interest for a system that would combine ISA with other support functions, such as navigation.

Concerns about the technical feasibility of reliable ARAS were also expressed in the SAFERIDER focus group (Baldanzini 2008). Although considerable benefits were expected especially from Frontal Collision Warning and Intersection Support, the experts wanted to test the device before passing their judgement. In general, they required adaptive systems, which can be personalised according to individual preferences and riding styles, and they asked for deactivation options and simple interfaces in order to avoid overloading the rider.

The 2 BE SAFE studies (Lenné et al. 2011, Oberlader et al. 2012) revealed that acceptability may be higher for those systems that are perceived as being more obviously useful, especially those that assist the riders in emergency situations. Acceptability was lower for systems that interfere with riding activity. Specifically, Adaptive Cruise Control, Intelligent Speed Adaptation and Lane Keeping Assistance received the lowest acceptability ratings. The riders expressed concerns about ceding the responsibility for a part of the vehicle control to a system. In addition, the riders were not convinced about the feasibility of fitting reliable ARAS on a motorcycle and they expected the cost of technically advanced systems would be too high. By comparison, acceptability was higher for systems that are already established and trusted as technologically mature by the rider population; for example, the Anti-lock Braking System (ABS). Moreover, the riders pointed out their doubts about the genuine interest of the industry in enhancing rider safety; rather, they felt that provision of such systems would be driven purely by commercial motivations.

Oberlader et al. (2012) concluded from their studies that the acceptability of rider assistance systems was rather low compared with systems that are available for passenger cars. Riders were sceptical about the safety potential of ARAS, since rider safety often depends on other road users' behaviour and their interaction with the riders. It seemed to be far-fetched for the riders to accept that ARAS could be beneficial in that context, and even in scenarios where riders need to avoid or resolve critical interaction situations, they tended to prefer not to rely on a system and expressed more interest in alternative measures, such as rider training. Beyond the concerns about reliability of the system, they felt the system could induce over-reliance and its usage could result in a deterioration of their own competences to safely manage a situation.

In summary, these studies of acceptability suggest that riders are generally reluctant to accept ARAS, an attitude that may be in contrast with expert opinion. They hesitate to put trust in their effectiveness and are more likely to reject systems that take over part of the riding task. The opinions expressed by the riders favoured rider training rather than assistance by ARAS.

Factors That Influence the Acceptability of ARAS

The 2 BE SAFE survey identified some factors that helped to distinguish between riders who expressed lower and higher acceptability of assistance systems. As was to be expected according to Schlag's (1997) assumption of problem awareness as a necessary condition for the acceptance of corresponding safety measures, riders who perceived risk as a downside of riding were more likely to belong to the higher acceptability group. Likewise, riders whose principal riding motive was fun showed less acceptability of assistance systems. Interestingly, a higher acceptability of assistance systems was found for riders who reported more risk-taking behaviour and attitudes. It seems that these riders recognise their higher need for assistance. However, self-reported data can easily be biased and riders who reject assistance could understate their risk behaviour in order to avoid cognitive dissonance that would result from admitting risky behaviour while rejecting safety-enhancing measures.

The survey revealed that direct experience with assistance systems was very rare among the riders. Exposure to the systems may well lead to changes in their attitude and it is necessary to assess acceptance once the riders have tested them.

Acceptance of Advanced Rider Assistance Systems

The approach to research on the acceptance of new technologies is underpinned by a range of theories and concepts, resulting in a variety of measurement procedures (Schade 2005, Adell 2009, and elsewhere in this book).

Schade and Schlag (2003) define acceptance as users' attitudes and behavioural reactions after the introduction of a measure. However, acceptance needs to be measured during the development process of a system, so as to benefit from the users' feedback for system improvement. Although usage behaviour is then not yet measurable, the riders can test the system and express their usage intention. The direct relationship between behavioural intention and actual behaviour has been postulated in the Theory of Planned Behaviour (Ajzen 1991) and confirmed by a number of studies with diverse backgrounds (e.g., Montada and Kals 2000). For novel ARAS, acceptance has thus been defined as the rider's intention to use the system if it was installed on the motorcycle (Huth and Gelau 2013) and is measured once the rider has experienced the ARAS. This is in line with the views of Ausserer and Risser (2005), who considered acceptance as the extent of the users' preparedness to use a system in the field of ITS, and with Adell (2009), who regarded acceptance as the driver's intention to use a system when driving a car. Further relevant concepts that allow for or hinder the actual usage of the ARAS by the rider are the willingness to acquire the system and to spend money on it. These two conditions for system usage are considered in the automotive domain (e.g., Arndt and Engeln 2008) and their measurement should also be included in the assessment of the acceptance of ARAS. Finally, acceptance can also be regarded as a positive behavioural response to the system. For instance, several studies on ISA measured acceptance in terms of behavioural changes when using the system as compared to driving without the system (cf. Vlassenroot et al. 2010). This also makes sense in the context of ARAS, given that the systems provide support to the riders in adapting their behaviour to the riding situation so as to stay within safety margins. This aspect of the riders' acceptance is crucial for the safety benefit that can be achieved by using the ARAS and can be seen as an indicator of acceptance, although it may not necessarily engender the usage intention that will lead to actual usage behaviour.

Available Results on the Acceptance of ARAS

This section reviews the results on the acceptance of three ARAS that have been tested with users:

1. Simpkin et al. (2007) rated riders' acceptance of an advisory and an assisting ISA and analysed the changes in the riding behaviour when using the system versions on a test track.
2. Within the SAFERIDER project, a Curve Warning system was tested in a simulator, comparing two rider interfaces – a force feedback throttle and a haptic glove (Huth et al. 2012a).
3. An Intersection Support system was evaluated in a further simulator study within SAFERIDER, again with the two rider interfaces (Huth et al. 2012b).

The system versions are referred to henceforth as 'advisory' and 'intervening'. Herein, the haptic glove is advisory, whereas the force feedback throttle and the assisting ISA are intervening. In all of the three studies, the output of the intervening system version was easily overruled by the riders, leaving them objectively in control of the vehicle.

Regarding acceptance as positive behavioural adaptation in response to the ARAS, the clearest results were achieved by the Curve Warning (CW) system. Riders responded to the warnings by adapting their behaviour earlier and better to the curve and increasing their safety margins. With the advisory CW, the riders even rode more cautiously through curves, provoking less warning situations. This finding gives hints to the possibility that some ARAS can have an educational effect. The riders might respond positively to the system by using it as a guidance to reset their personal threshold for safe riding in the target situation.

This effect could not be found, however, for the Intersection Support (IS). Rather, the riders relied on the system to detect a hazardous situation and then responded to the warning by reducing their approach speed to the intersection; and even then, the conducted simulator tests could only confirm such a behavioural response in right-of-way situations in higher speed environments, not in a typical urban setting. The riding behaviour with ISA did not change for cautious riders, since they already kept to the speed limits when riding without the system. Riders classified as aggressive based on their tendency to exceed the speed limit, however, adapted their behaviour to the assisting ISA by reducing speed violations and speed variations – an effect that was also detectable to a slighter extent for the advisory ISA.

Riders' subjective ratings revealed distinct preferences that are related to the system versions. While the intervening CW was rated negatively regarding its helpfulness to manage critical curve events, the advisory version received a positive evaluation in this respect. The appreciation of the CW as a whole and the ride with the system was also found to be very sensitive to the interface used. This finding was confirmed by the tests on the IS. Several subjective measures revealed that the positive ratings of the ARAS turn into neutral or negative ratings for the intervening version compared to the advisory version. Riders' comments indicated that this rejection was mainly attributable to the invasiveness of the warning strategy of the force feedback throttle. This result is in line with the outcome of the riders' evaluation of the ISA: The advisory version was assessed as more useful than the intervening one, which also received negative satisfaction ratings. Even so, it is noteworthy that, contrasting with the poor acceptability ratings of ISA, improved usefulness ratings were found after practical experience for the intervening ISA and for the advisory version as a tendency. Although the riders now acknowledged that the systems could enhance riding safety, they were concerned with a possible increase in rider irritations, feelings of being controlled and negative effects on riding pleasure – especially when using the intervening ISA.

Riders' interest in having ARAS installed on their motorcycle equally depended on the interface used. While high rates of willingness to have an ARAS have been found for the advisory versions, the same systems were mostly rejected if implemented with an intervening interface. Percentages of rejection and acceptance are compared in Table 13.1.

Regardless of the system version, riders showed a limited willingness to pay for a system in any of the three studies. About half of the participants were not willing to pay more than €100 for the CW or IS. For ISA, the great majority of the riders would not spend more than £100 (i.e., approximately €120). The top range any rider was willing to spend for the CW or the IS was limited to €500, and for the ISA to £200 (i.e., approximately €240).

Table 13.1   Acceptance of three types of ARAS: comparison of participants' interest to have advisory and intervening system versions

		Willingness to have the ARAS on their own motorcycle	
		No	Yes
ISA	advisory (n = 33)	12 (36%)	21 (64%)
	intervening (n = 33)	20 (61%)	13 (39%)
CW	advisory (n = 20)	5 (25%)	15 (75%)
	intervening (n = 20)	10 (50%)	10 (50%)
IS	advisory (n = 20)	6 (30%)	14 (70%)
	intervening (n = 20)	11 (55%)	9 (45%)
Total (systems)	advisory (n = 73)	23 (32%)	50 (68%)
	intervening (n = 73)	41 (56%)	32 (44%)

Finally, the usage intention found for the CW and IS was satisfactory. None of the participants expressed the intention not to activate the advisory CW at all if it was installed on the motorcycle, while 15 per cent of the riders stated this for the advisory IS. The rate of complete rejection regarding the usage intention was 30 per cent for both the intervening CW and the intervening IS. The majority of the participants chose the option to activate the advisory or intervening CW only in certain situations (characterised by constraints regarding the environmental conditions or the rider state), and almost 50 per cent of the riders intended to activate the advisory CW all the time (only 25 per cent for the intervening CW). A permanent activation of the advisory or intervening IS, in turn, was only intended by a few riders, with around 50 per cent of the riders preferring a selective activation depending on their familiarity with the environment and the visibility conditions. The tests on the ISA revealed a reluctance of the riders to use the system in most of the traffic situations. Overall, the usage intention was more positive for the advisory than for the intervening ISA.

Influencing Factors

One way of obtaining insight into possible reasons for limited acceptance and to determine relevant starting points for its improvement is to include the factors that might influence riders' acceptance of an ARAS into its evaluation. Predictors of the acceptance of ARAS have been identified and brought together in a model, which has been validated with data from user tests on four ARAS, including advisory and intervening system versions (Huth and Gelau 2013). Hereafter, the insights obtained in this study are combined with the results from the evaluation of the three ARAS presented above.

As a first predictor, the model for acceptance of ARAS includes the safety feeling when riding without assistance, which corresponds to the potential for experiencing benefits by using the system. In accordance with the relevance of problem awareness (Schlag 1997, Steg and Vleg 1997) and the perceived usefulness of technologies (e.g., Van der Laan, Heino and De Waard 1997, Venkatesh and Davis 2000) for acceptance, this predictor was expected to be especially important in view of possible self-efficacy conflicts between the conception of riding as a performance and being assisted by a system (cf. Bandura 1982). The predictor could not be confirmed by the empirical data from the SAFERIDER project, but since this result is likely to be because of the limitations of the study (due to a ceiling effect in the data; this factor might need to be measured more precisely) it is worth being further investigated.

Warning systems that require only very limited interaction with the rider may not be appraisable in terms of pleasantness as envisaged by some theories on acceptance of technologies (e.g., Van der Laan et al. 1997). Still, the possible interference of the system output with positive riding sensations or the rider's feeling of autonomy and control lends particular importance to interface design in the evaluation of ARAS functions. In line with the differences in the acceptance found between advisory and intervening versions of the three ARAS reported in this chapter, validation of the acceptance model confirmed the judgement that the interface is a powerful predictor of riders' acceptance of the ARAS (Huth and Gelau 2013). In particular, results on acceptance of the three ARAS reported here reveal the disadvantages of ARAS that are combined with an intrusive interface. On the one hand, riders deem it potentially dangerous since it could increase the instability of the motorcycle and lead to a conflict when rider intention differs from the intervening system behaviour (Huth et al. 2012b). By the same token, riders feel that their responsibility and control is taken away by the ARAS. In addition, the interference of the interface with riding operations may be related to a higher probability of distraction and increased irritation of the rider (e.g., Huth et al. 2012a). Although intervening versions of ARAS have shown to be at least as effective in improving the riding behaviour as the advisory versions, they tend to be rejected by riders due to the annoyance they provoke (Huth 2012, Simpkin et al. 2007). Regarding the modality of the warning transmission, vibration signals have proven to be effective and well accepted (Huth 2012), but they still need to be adjusted for real traffic environments. In contrast, visual messages are less recommendable owing to potential visual distraction and irritation of the rider (Simpkin et al. 2007).

Since motorcycle riding has a pronounced social aspect, the social norm should be considered as a predictor of behavioural intentions (Ajzen 1991). The social norm could be confirmed as a relevant predictor of road user behaviour more generally (Tunnicliff et al. 2011) and of riders' intention to speed in particular (Elliott 2010). In the validation of the acceptance model for ARAS (Huth and Gelau 2013), the expected opinion of fellow riders about the ARAS was a strong predictor of the riders' own acceptance of the system. This finding also corresponds with the influence of social norms on acceptance of ADAS in the automotive context (e.g., Adell 2010, and elsewhere in this book).

A comparison of riders' responses to the CW and IS allows initial conclusions to be drawn about the relevance of the type of support function. Curves are related to performance and positive riding sensations, whereas intersections are less related to riding pleasure and the rider is likely to be put at risk by others. However, the available test results are more positive for the CW than for the IS. This finding suggests that, rather than the target scenario, an appropriate choice of warning thresholds as well as compatibility of system output with riding intentions may be decisive factors for the acceptance of ARAS (Huth 2012).

In view of the existing diversity of riders (Haworth 2012), it is worth analysing the possible influence of rider characteristics on the acceptance of ARAS. The three studies on specific ARAS reported in this chapter did not allow for such an analysis due to their limited sample size, but the study on the ARAS acceptance model (Huth and Gelau in press) used the available dataset to detect the effects of age, annual mileage, riding frequency and riding motivation. However, none of these factors proved to have any influence on the intention to use an ARAS. Further studies are thus needed to capture possible impacts of different rider characteristics on acceptance. A promising starting point would be the comparison between novice and experienced riders. These groups may differ in their abilities to deal with hazards and their self-assessment of their own riding abilities (Liu, Hosking and Lenné 2009), which in turn may affect their acceptance of assistance.

Conclusions

As a result of concerns that are mainly related to the effectiveness of ARAS and surrender of control to the system, acceptability studies with ARAS show considerable signs of reluctance by the riders. By contrast, acceptance studies point towards good potential of ARAS. The riders had some reservations against the systems but it has to be considered that the studies were conducted with first prototypes that still need to be further improved. Adaptive assistance functions that warn riders whenever their behaviour deviates from a safe reference manoeuvre basically seem to be compatible with riding. Given that the systems aim at assisting the riders in situations where the they underestimate their risk, permanent activation should be targeted. The current results suggest that usage intention still lies below this target. Possibly, the usage intention would increase once the riders got more used to the system (Huth 2012) – an evolution found with car drivers' acceptance of ISA (Lai, Chorlton and Carsten 2007). The low willingness to pay for ARAS functions calls for the development of affordable technical solutions, especially those fitted by manufacturers.

Consistent with the relevance of curve crashes, the Curve Warning system has proven to be well accepted by riders. Riders' awareness of the risk in misjudging the approach speed in curves (Clarke et al. 2004) may contribute to this acceptance. On the other hand, the thresholds set by the ARAS may have been congruent with the riders' personal thresholds for enjoyment of risk. According to crash statistics, assistance at intersections has a similar relevance for riders. However, results of acceptance of the Intersection Support are less conclusive. This may be attributable to the fact that riders are not responsible for crashes in most of the cases or that the configuration of the ARAS has to be better adapted to the needs of riders in this setting. Intelligent Speed Adaptation obtained the most negative evaluation of the three ARAS tested for rider acceptance. This is in line with the objective need derived from crash statistics, since it is not speeding per se but riding too fast for the prevailing conditions that mainly contributes to motorcycle crashes (e.g., Clarke et al. 2004). Thus, ARAS should not take over the control from the riders, but assist them in correctly adapting their behaviour to riding situations where they deviate from safe reference manoeuvres.

Given that the behavioural response to a system can diverge from the attitude towards the ARAS or the intention to use it, it is important to conduct both objective and subjective evaluations of ARAS. In the presented studies, system versions that were objectively effective received considerably different subjective ratings depending on the interface used.

Intervening devices should be avoided, even though their low signal power may not destabilise the vehicle. Riders dislike any interference with vehicle control, considering it a disturbance to their feelings of autonomy, comfort and riding pleasure, and as a source of conflict with their riding intention.

Customisable systems have been promoted in the automotive context (e.g., Jiménez, Liang and Aparicio 2012) and have been requested for riders by FEMA (2011). The adaptability of warning thresholds and signals to individual needs could enhance the acceptance of ARAS and improve their effect on riding safety by higher rates of system activation (Huth et al. 2012b).

Apart from developing improved ARAS in order to enhance riders' acceptance, the issue of acceptability should not be neglected. During the first years of the introduction of an ARAS, many riders might not have the chance to test it. Taking into account the relevance of the social norm, it is important to create a favourable attitude towards ARAS among rider circles, as well as to avoid prejudices and misunderstandings of the system function. Finally, it could be beneficial to create awareness of riding risks and the corresponding safety benefits provided by ARAS, as prompted by the results of the acceptance of Curve Warning (Huth et al. 2012a) and on the acceptability of ARAS more in general (Oberlader et al. 2012).

References

Adell, E. 2009. Driver experience and acceptance of driver support systems – A case of speed adaptation. PhD thesis, Lund University.

————. 2010. *Acceptance of driver support systems*. Proceedings of the European Conference on Human Centred Design for Intelligent Transport Systems, 475–86. Berlin.

Ajzen, I. 1991. The theory of planned behaviour. *Organizational Behaviour and Human Decision Processes*, 50: 179–211.

Ambak, K., Atiq, R. and Ismail, R. 2009. Intelligent transport system for motorcycle safety and issues. *European Journal of Scientific Research*, 28(4): 600–611.

Arndt, S. and Engeln, A. 2008. Prädiktoren der Akzeptanz von Fahrerassistenzsystemen. In *Fortschritte der Verkehrspsychologie. 45. DGP-Kongress für Verkehrspsychologie*, 313–37. Edited by J. Schade and A. Engeln. Bonn: Deutscher Psychologen Verlag.

Ausserer, K. and Risser, R. 2005. Intelligent transport systems and services – Chances and risks. Paper presented at the Eighteenth ICTCT workshop. Helsinki, Finland, October.

Baldanzini, N. 2008. Initial project findings: Focus group. [Online]. SAFERIDER, First User Forum Meeting, Brussels, Belgium. Available at: http://www. saferider-eu.org/assets/docs/1st-user-forum/Annex_5_Saferider_User_Forum_Baldanzini.pdf.

Bandura, A. 1982. Self-efficacy mechanisms in human agency. *American Psychologist*, 37: 122–47.

Bayly, M., Hosking, S. and Regan, M. 2007. Intelligent transport systems and motorcycle safety. Paper No. 07-0301, 20th International Technical Conference on the Enhanced Safety of Vehicles (ESV), Lyon, France.

Bellaby, P. and Lawrenson, D. 2001. Approaches to the risk of riding motorcycles: Reflections on the problem of reconciling statistical risk assessment and motorcyclists' own reasons for riding. *Sociological Review*, 49(3): 368–89.

Brenac, T., Clabaux, N., Perrin, C. and Van Elslande, P. 2006. Motorcyclist conspicuity-related accidents in urban areas: A speed problem? *Advances in Transportation Studies*, 8: 23–9.

Broughton, P.S. 2005. Designing PTW training to match rider goals. In *Driver Behaviour and Training, Vol. 2*, 233–42. Edited by L. Dorn. Aldershot: Ashgate Publishing.

————. 2007. Risk and enjoyment in powered two wheeler use. PhD thesis, Transport Research Institute, Napier University, Edinburgh.

————. 2008. Flow, task capability and powered-two-wheeler (PTW) rider training. In *Driver Behaviour and Training, Vol. 3*, 415–24. Edited by L. Dorn. Aldershot: Ashgate.

Broughton, P.S. and Stradling, S. 2005. Why ride powered two wheelers? In *Behavioural Research in Road Safety*, 68–78. London: Department for Transport.

Broughton, P.S., Fuller, R., Stradling, S., Gormley, M., Kinnear, N., O'Dolan, C. and Hannigan, B. 2009. Conditions for speeding behaviour: A comparison of car drivers and powered two wheeled riders. *Transportation Research Part F: Traffic Psychology*, 12: 417–27.

Cairney, P. and Ritzinger, A. 2008. Industry and rider views of ITS for safe motorcycling. Paper presented at the 23rd ARRB Conference – Research Partnering with Practitioners. Adelaide, Australia.

Christmas, S., Young, D., Cookson, R. and Cuerden, R. 2009. Passion, performance, practicality: Motorcyclists' motivations and attitudes to safety – motorcycle safety research project. Published Project Report PPR 442. Crowthorne, UK: Transport Research Laboratory.

Clarke, D.D., Ward, P., Truman, W. and Bartle, C. 2004. An in-depth case study of motorcycle crashes using police road accident files. In *Behavioural Research in Road Safety*, 5–20. London: Department for Transport.

Crundall, D., Crundall, E., Clarke, D. and Shahar, A. 2012. Why do car drivers fail to give way to motorcycles at t-junctions? *Accident Analysis and Prevention*, 44(1): 88–96.

Csikszentmihalyi, M. 1997. *Finding Flow: The Psychology of Engagement with Everyday Life*. New York: Basic Books.

DEKRA Automotive GmbH. 2010. Motorcycle road safety report 2010 – Strategies for preventing accidents on the roads of Europe. DEKRA: Stuttgart, Germany.

Di Stasi, L.L., Álvarez-Valbuena, V., Cañas, J.J., Maldonado, A., Catena, A., Antolí, A. and Candido, A. 2009. Risk behaviour and mental workload: Multimodal assessment techniques applied to motorbike riding simulation. *Transportation Research Part F: Traffic Psychology*, 12: 361–70.

Elliott, M.A. 2010. Predicting motorcyclists' intentions to speed: Effects of selected cognitions from the theory of planned behaviour, self-identity and social identity. *Accident Analysis and Prevention*, 42(5): 718–25.

Elliott, M.A., Baughan, C.J. and Sexton, B.F. 2007. Errors and violations in relation to motorcyclists' crash risk. *Accident Analysis and Prevention*, 39: 491–9.

European Commission. 2010. EU energy and transport in figures: Statistical pocketbook 2010. Brussels: European Commission. Available at: http://ec.europa. eu/energy/publications/statistics/doc/2010 energy transport figures.pdf.

Federation of European Motorcyclists' Associations (FEMA). 2011. FEMA position paper on intelligent transport systems. Brussels: FEMA.

Fuller, R. 2005. Towards a general theory of driver behaviour. *Accident Analysis and Prevention*, 37(3): 461–72.

Haworth, N. 2012. Powered two wheelers in a changing world – Challenges and opportunities. *Accident Analysis and Prevention*, 44(1): 12–18.

Haworth, N., Mulvihill, C., Wallace, P., Symmons, M. and Regan, M. 2005. Hazard perception and responding by motorcyclists: Summary of background, literature review and training methods. Report no. 234 Monash University Accident Research Centre (MUARC), Melbourne.

Hurt, H.H., Ouellet, J.V. and Thom, D.R. 1981. Motorcycle accident cause factors and identification of countermeasures. Report no. DOT HS-5-01160. Traffic Safety Center, University of Southern California, Los Angeles.

Huth, V. 2012. *Riders' response to assistance functions – A comparison of curve and intersection warnings.* Proceedings of the European Conference on Human Centred Design for Intelligent Transport Systems, 177–84. June, Valencia, Spain.

Huth, V. and Gelau, C. 2013. Predicting the acceptance of advanced rider assistance systems. *Accident Analysis and Prevention*, 50: 51–8.

Huth, V., Biral, F., Martín, Ó. and Lot, R. 2012a. Comparison of two warning concepts of an intelligent Curve Warning system for motorcyclists in a simulator study. *Accident Analysis and Prevention*, 44(1): 118–25.

Huth, V., Lot, R., Biral, F. and Rota, S. 2012b. Intelligent Intersection Support for powered two-wheeled riders: A human factors perspective. *IET Intelligent Transport Systems*, 6(2): 107–14.

International Road Traffic and Accident Database (IRTAD). 2010. A record decade for road safety. Online: Press release, 9 September. Available at: http://www. internationaltransportforum.org/Press/PDFs/2010-09-15IRTAD.pdf.

Jiménez, F., Liang, Y. and Aparicio, F. 2012. Adapting ISA system warnings to enhance user acceptance. *Accident Analysis and Prevention*, 48: 37–48.

Lai, F., Chorlton, K. and Carsten, O. 2007. Overall field trial results: The ISA-UK project. Leeds: University of Leeds.

Lenné, M., Beanland, V., Fuessl, E., Oberlader, M., Joshi, S., Rössger, L., Bellet, T., Banet, A., Leden, L., Spyropulou, I., Roebroeck, H., Cavalhais, J. and Underwood, G. 2011. Relationships between rider profiles and acceptance of Advanced Rider Assistance Systems. D9 of Work Package 3.3, 2 Be Safe.

Liu, C.C., Hosking, S.G. and Lenné, M.G. 2009. Hazard perception abilities of experienced and novice motorcyclists: An interactive simulator experiment. *Transportation Research Part F: Traffic Psychology*, 12: 325–34.

Mannering, F.L. and Grodsky, L.L. 1995. Statistical analysis of motorcyclists' perceived accident risk. *Accident Analysis and Prevention*, 27(1): 21–31.

Mayou, R. and Bryant, B. 2003. Consequences of road traffic accidents for different types of road user. *Injury*, 34: 197–202.

Moller, M. and Gregersen, N.P. 2008. Psychosocial function of driving as predictor of risk-taking behaviour. *Accident Analysis and Prevention*, 40(1): 209–15.

Montada, L. and Kals, E. 2000. Political implications of psychological research on ecological justice and pro-environmental behaviour. *International Journal of Psychology*, 35: 168–76.

Motorcycle Accidents In-Depth Study (MAIDS). 2004. Final report 1.2: In-depth investigations of accidents involving powered two-wheelers. September 2004. Brussels: ACEM.

Oberlader, M., Füssl, E., Lenné, M., Beanland, V., Pereira, M., Simões, A., Turetschek, C., Kaufmann, C., Joshi, S., Rößger, L., Leden, L., Spyropoulou, I., Roebroeck, H., Carvalhais, J. and Underwood, J. 2012. *A group approach towards an understanding of riders' interaction with on-bike technologies: Riders' acceptance of advanced rider assistance systems.* Proceedings of the 3rd European Conference on Human Centred Design for Intelligent Transport Systems, 303–12. Valencia, Spain, June.

Pai, C.W. 2011. Motorcycle right-of-way accidents – A literature review. *Accident Analysis and Prevention*, 43(3): 971–82.

Parker, D., Manstead, A., Stradling, S.G., Reason, J.T. and Baxter, J. 1992. Intention to commit driving violations: An application of the theory of planned behaviour. *Journal of Applied Psychology*, 77: 94–101.

Peden, M., Scurfield, R., Sleet, D., Mohan, D., Hyder, A.A., Jarawan, E. and Mathers, C. 2004. *World Report on Road Traffic Injury Prevention.* Geneva: World Health Organization.

Phan, V., Regan, M., Leden, L., Mattson, M., Minton, M., Chattington, M., Basacik, D., Pittman, M., Baldanzini, N., Vlahogianni, E., Yannis, G. and Golias, J. 2010. Rider/driver behaviours and road safety for PTW. Online: Deliverable D1, 2 Be Safe Project. Available at: http://www.2besafe.eu/deliverables.

Rakotonirainy, A. and Haworth, N. 2006. Institutional challenges to ITS deployment and adoption. Paper presented at the Twenty-ninth Australasian Transport Research Forum. Gold Coast, Australia.

Schade, J. 2005. *Akzeptanz von Strassennutzungsgebühren: Entwicklung und Überprüfung eines Modells.* Lengerich, Germany: Pabst Science Publishers.

Schade, J. and Schlag, B. 2003. Acceptability of urban transport pricing strategies. *Transportation Research Part F: Traffic Psychology*, 6(1): 45–61.

Schlag, B. 1997. Road pricing-Massnahmen und ihr Akzeptanz. In *Fortschritte der Verkehrspsychologie, 36.BDP-Kongress für Verkehrspsychologie*, 217–24. Edited by B. Schlag. Bonn: Deutscher Psychologen Verlag.

Simpkin, F., Lai, B., Chorlton, K. and Fowkes, M. 2007. Intelligent Speed Adaptation (ISA) – UK: Results of motorcycle trial. Leeds: University of Leeds.

Steg, I. and Vleg, C. 1997. The role of problem awareness in willingness-to-change car-use and in evaluating relevant policy measures. In *Traffic and Transport Psychology*, 465–75. Edited by T. Rothengatter and E.C. Vaya. Amsterdam: Pergamon.

Traffic Accident Causation in Europe (TRACE). 2008. Deliverable 1.3: Road users and accident causation. Part 3: Summary report'. Project no. FP6-2004-IST-4 027763, June.

Tunnicliff, D.J., Watson, B.C., White, K.M., Hyde, M.K., Schonfeld, C.C. and Wishart, D.E. 2012. Understanding the factors influencing safe and unsafe motorcycle rider intentions. *Accident Analysis and Prevention*, 49: 133–41.

Tunnicliff, D.J., Watson, B.C., White, K.M., Lewis, I.M. and Wishart, D.E. 2011. The social context of motorcycle riding and the key determinants influencing rider behaviour: A qualitative investigation. *Traffic Injury Prevention*, 12(4): 363–76.

Van der Laan, J.D., Heino, A. and De Waard, D. 1997. A simple procedure for the assessment of acceptance of advanced transport telematics. *Transportation Research Part C: Emerging Technologies*, 5: 1–10.

Venkatesh, V. and Davis, F.D. 2000. A theoretical extension of the technology acceptance model: Four longitudinal field studies. *Management Science*, 46(2): 186–204.

Vlassenroot, S., Brookhuis, K., Marchau, V. and Witlox, F. 2010. Towards defining a unified concept for the acceptability of Intelligent Transport Systems (ITS): A conceptual analysis based on the case of Intelligent Speed Adaptation (ISA). *Transportation Research Part F*, 13(3): 164–78.

Wilde, G.J.S. 1982. The theory of risk homeostasis: Implications for safety and health. *Risk Analysis*, 2: 209–25.

Chapter 14

Driver Acceptance of Technologies Deployed Within the Road Infrastructure

Alan Stevens and Nick Reed

Transport Research Laboratory, UK[1]

Abstract

In this chapter, we review studies that have investigated aspects of drivers' acceptance of a range of road infrastructure-based technologies that have been proposed or introduced to support safe and efficient driving behaviour. We look at driving simulator studies of 'Active Traffic Management' and 'actively illuminated road studs' that have tried to gauge likely acceptance of the technology on the road by observing behaviour in the simulator and assessing drivers' confidence in the technology and understanding of how they are expected to respond to it. We also review real road experience of drivers' responses to roadworks delineated using flashing cones, and of speed monitoring and enforcement technology. In both cases we infer acceptance of the technology from drivers' behaviour.

Introduction

Road infrastructure is designed to provide an environment that supports efficient routing and engenders safe road user behaviour. It includes the following:

- Road surfaces, furniture and road markings such as barriers, speed bumps, lane delineations and road studs;
- Lighting and fixed signage such as speed limits, information and prohibition signs, high-visibility warnings and parking restrictions;
- Systems providing dynamic information or instruction such as traffic lights, and variable message signs; and
- Systems that monitor driving behaviour such as speed cameras and/or provide real-time and specific feedback.

As has been shown in many other chapters, there are challenges faced by designers of in-vehicle technologies in gaining drivers' acceptance of systems. In achieving

---

the objectives of supporting efficient routing and encouraging safe behaviour, the road infrastructure designer is faced with a similar set of challenges. As with in-vehicle systems, basic ergonomic principles must be adhered to but unlike in-vehicle systems, the information provided cannot be specifically directed towards an individual driver. The design of road infrastructure should provide information to road users that is

- conspicuous – it is easy for road users to detect the information presented to them in all environmental conditions;
- clear – there is no ambiguity or conflict in the information presented;
- intuitive – it is easy for road users to understand how to respond to the presented information;
- compatible with the driving task – attending to the presented information does not cause undue distraction to road users; and
- non-specific – which comprises features that are relevant for the broadest spectrum of road users, regardless of age, gender, experience or vehicle type.

Road infrastructure design is also likely to be constrained by regulatory and aesthetic considerations. The acceptance of road infrastructure is dependent upon how well these challenges have been addressed.

The use of lines and signs to provide information and guidance will be familiar to most road users. An example of the use of technology to enhance this guidance is dynamic road markings (DRM) for managing road layouts that have been tested in the Netherlands and Germany. Such schemes have the potential to increase road capacity by enabling dynamic changes in lane direction or availability to meet demand characteristics; often referred to as 'tidal' flow schemes. Figure 14.1 shows dynamic road markings whilst Figure 14.2 gives an outline of how the road markings change in a proposed tidal flow scheme based on DRM.

Whilst design of the road infrastructure can influence individual driver behaviour (positively or negatively with respect to safety), it cannot, ultimately, control; driving is a complex form of social interaction and individual behaviour depends on a wide range of factors that vary with time and circumstance. Compliance relies on drivers not only understanding but accepting the 'norms of behaviour' implied or required by the infrastructure such as signage (e.g., 'Lane Closed' or '50 mph') and on drivers accepting that they may be monitored, recorded and, where behaviour is deemed unacceptable, penalised.

In this chapter, we review studies that have investigated driver acceptance towards a range of road infrastructure technological interventions that have been proposed or introduced to support safe and efficient driving. We exclude, however, acceptance of real-time sources of information such as route guidance, congestion and location-specific traffic news that are transmitted *from* the infrastructure to be presented *within* the vehicle (issues around driver acceptance of such sources have been considered in previous chapters). Nevertheless, the availability of such

Figure 14.1   Dynamic road markings in the Netherlands (US Department of Transportation 2004)

additional sources is increasingly likely to affect drivers' acceptance of externally presented information or instructions. We also exclude a consideration of road pricing technology, chiefly because driver acceptance of such schemes is a complex subject and heavily determined by the acceptability of road pricing by society more generally. We examine drivers' acceptance of technology deployed within road infrastructure in two areas. Firstly, the use of a driving simulator to create an evidence base upon which to make decisions about infrastructure and,

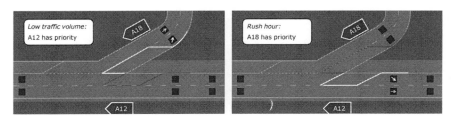

Figure 14.2   An example of a dynamic tidal flow scheme that has been proposed in the Netherlands (reproduced from Fafieanie and Sambell 2008, with permission)

secondly, drivers' acceptance of monitoring from the roadside, particularly of speed (vehicle-activated signs, fixed cameras and average speed cameras).

Using a Driving Simulator to Assess Behaviour in Response to Highway Technology

Driving is a complex, multifaceted, information processing task that humans undertake in routine situations with relative ease. However, mistakes, violations or misjudgements at inappropriate times can result in tragic consequences for the driver, passengers, other road users and/or bystanders.

The introduction of new technologies into the highway environment presents an opportunity to improve the quality of service experienced by road users. Testing such technologies to investigate expected and unexpected effects on driver behaviour prior to implementation is desirable, given the safety risk and potential costs should a new system be found to be ineffective, detrimental to driving practices or fail to reach standards of acceptability to the user.

Although analysis of driver behaviour through observation of performance in the real world produces data with the greatest validity, it is difficult to exert control over either the number or the types of vehicles or the demographics of the driving population involved in such a trial. When the context of a busy traffic environment is not required, tests can be conducted using a dedicated test track. This significantly increases experimental control and allows detailed behavioural assessment without placing drivers at risk of conflict with other vehicles. An example of such a study is described in Case Study 1.

When one is interested in drivers' responses to new road infrastructure features in the context of a busy traffic environment, test track studies become unfeasible as choreographing large numbers of other vehicles becomes unworkable and the safety of participants must be considered. Interactive driving simulation addresses these limitations. A simulator can provide detailed information about the behaviour of the driven vehicle, in relation to the test environment and to other vehicles whilst results can be supplemented by physiological (e.g., heart rate) and subjective (e.g., NASA-TLX workload questionnaire) measures. Scenarios are repeatable and can be tightly choreographed, both of which facilitate statistical analysis and efficiency, and participants involved are at no risk of real harm.

As a simulator presents a simulation of the real world, acceptance of on-road or vehicle technology in a simulator does not necessarily mean acceptance of that technology in the real world. The convergence of simulator acceptance and real-world acceptance obviously depends on the fidelity with which a simulator reproduces the real world but, formally, simulator studies that show acceptance of technologies can really only claim 'potential' for acceptance in the real world. Therefore, it is perhaps more correct to claim that positive driving simulator studies illustrate the acceptability of technologies (in principle) rather than (actual) acceptance as demonstrated by real-world usage. Thus, simulator

Figure 14.3   TRL Car Simulator during 'Red X' trial

studies claiming to measure acceptance of technologies depend on a participant sufficiently perceiving the relevant features of the test environment so that they produce behaviour that is representative of that which would be observed during real driving in the equivalent environment. Driving simulator validation studies (e.g., Törnros 1998, Diels, Robbins and Reed 2011) suggest that this is indeed the case.

Transport Research Laboratory (TRL) has successfully operated a driving simulator for more than 20 years. The latest (validated in Diels et al. 2011) uses a standard family hatchback, a limited motion platform and realistic graphics and sound (Figure 14.3). This system was used in the three case studies described below.

Case Study 1: Active Traffic Management

Congestion brings many vehicles into close proximity, raising the probability of collisions such as rear-end shunts or sideswipes (Webb 1995). As well as reducing congestion, there is continuing pressure to make better use of infrastructure and reduce vehicle emissions (Stern 2006). One such scheme, as part of 'Active Traffic Management' (ATM), was required to implement Variable Speed Limits (VSLs) under conditions of congestion and directing traffic to use the hard shoulder as an active traffic lane under conditions of heavy congestion (4-lane VSL). The acceptability of such additional control was hitherto untested, so it was important to assess potential driver acceptability (as demonstrated by behaviour including acceptance in a driving simulator) before implementation.

ATM involves gantries at 500 m intervals with Advanced Motorway Indicator (AMI) signs above each lane (including the hard shoulder), to provide lane-specific information and a Variable Message Sign (VMS) for the provision of general safety guidance as well as information about accidents, delays and weather conditions. One option was to use a blank AMI above the hard shoulder (whilst all other AMIs display the VSL), indicating to traffic that normal motorway rules apply to the hard shoulder; that is, it should be used for emergencies only. Alternatively, it had been proposed that a red X symbol should be used to give a definite signal to motorists that the hard shoulder is unavailable to traffic.

Prior to the implementation of hard shoulder running on the real motorway, it was possible to investigate the behaviour of drivers in response to these different signs using TRL's driving simulator (Thornton, Reed and Gordon 2005). Seventy-two participants were recruited and were assessed across experimental factors of Sign (Blank AMI vs. Red X AMI – to signal hard shoulder closure), Information (informed vs. uninformed about ATM) and Age (younger vs. older drivers).

During their drive, participants were instructed to hurry but then encountered clusters of simulated congestion. This was to encourage participants to make best progress along the route, using whatever road capacity they felt was open to them along the route. Analysis would then focus on the level of contravention and inappropriate use of the motorway. In one section of the route, there was no means by which a participant could overtake the congestion cluster unless they used the hard shoulder whilst it was closed to normal traffic. In another section, the hard shoulder was opened to traffic and the participant was thus able to overtake the simulated congestion traffic by travelling in the hard shoulder.

After completion of the trial, a questionnaire allowed assessment of the factors that were determinants in the decision by participants to use the hard shoulder, both at times when it was open and times when it was closed.

To maximise the benefits of ATM, drivers must accept the adoption of an unfamiliar driving practice (using the hard shoulder as a normal running lane). Participants who were aware of the operation of the ATM before taking part in the trial used the scheme more effectively than those who were uninformed. Results showed that informed participants had greater acceptance of the ATM measures as shown by them

- using the hard shoulder more often when it was appropriate to do so;
- using the hard shoulder sooner and for longer;
- being significantly more confident about using ATM in general; and
- being significantly more positive about the effect it will have on motorway travel and safety.

Once they had read the information leaflet post-trial, uninformed participants recognised how useful it would have been in raising their awareness of the operational regimes of ATM before entering the scheme. These results were used to highlight that the information strategy to publicise ATM must be comprehensive to raise driver acceptance of the scheme, thereby maximising correct use and ensuring that the potential benefits for ATM are fully achieved.

Since completion of the simulator study, the M42 ATM scheme has been successfully rolled out and has enjoyed remarkable success, delivering improved traffic flow and travel times, while having no detrimental effect on safety (Department for Transport 2008). Wider implementation of the ATM measures is now being planned and its success is owed, in part, to the simulator testing prior to commencement of the scheme.

Case Study 2: Actively Illuminated Road Studs

This study examined the potential improvement to road safety at night that may be achieved by illuminated road studs ('Active' studs, Figure 14.4) in place of standard

Figure 14.4   Active road studs

('Passive') retro reflective studs and to gain insight into drivers' acceptance of this technology (Reed 2006).

TRL's driving simulator was used to create a 37.1 km rural road within which a basic test section was repeated six times. This section was used to compare behaviour across the stud conditions, containing six critical corners in the basic section where the curve radius fell below 150 m and these were used for more detailed analyses.

Thirty-six participants were recruited from three age groups: Younger (17–25 years), Middle (26–54 years) and Older (55+ years), and each participant drove the trial route twice. In each drive, the participant experienced a simulated night-time environment and the road had sections with no studs and sections with studs. In one of their drives, the studded section had active studs; in the other drive it had passive studs. The studs were placed at varying intervals (based on the road characteristics) along the centreline of the road. Additional red studs (in both the active stud and passive stud versions) were placed on the nearside of the four sharpest bends in the repeat section used to create the trial route. The driven vehicle used dipped headlights throughout and no other traffic was present in the simulation.

Participants completed a post-drive questionnaire that asked them to indicate their feelings of safety and confidence in each driving condition. Picture cue cards were used to remind participants of the environments that they had seen. Positive responses in terms of safety and confidence were interpreted as representing acceptance in the simulator study and thus demonstrated (potential) acceptability in real road conditions.

Results from the simulator demonstrated that, in each age group, participants' average speed when driving was significantly higher (by around 5 kph) in both studded conditions, relative to the no stud condition (Reed 2006). However, there were no significant differences between the active and passive stud conditions across the age groups in terms of overall speed. Assessment of how participants controlled their lateral position revealed that older participants spent significantly less time with the right edge across the centreline of the road with active studs than they did with passive studs.

More detailed analysis of braking results in the critical corners suggests that participants were better informed about how they needed to control the vehicle in order to negotiate the bends when the active studs were present. Similarly, analysis of drivers' lateral position in the corners revealed a marked difference between the passive and active stud conditions in right turns and suggests that enhanced delineation of the offside road edge may promote improvements in drivers' lateral control of their vehicle.

Broughton and Buckle (2006) reported that loss of control was the only precipitating factor in the causation of accidents (of all severities) that had shown a significant increase since 1999. The results from this trial suggest that the active stud installation improved drivers' control, particularly in right turns and for older drivers. It is, therefore, possible that the introduction of active road studs may help to reverse this trend.

Participants reported that active studs encouraged them to drive faster than they would normally. However, this is contradicted by the simulator data, which showed that there were only very slight increases in speed with active studs. This discrepancy between drivers' opinion and observed behaviour highlights the benefit that simulation can bring in allowing schemes to be tested by real drivers in a naturalistic environment. Participants also reported that they believed active studs would be highly beneficial to road transport and road safety.

We interpreted this level of understanding and confidence as indicative of high acceptability of the technology and, overall, it was concluded that active studs are likely to be highly acceptable to drivers and offer a significant safety advantage over standard passive retro-reflective studs since they appear to improve lane guidance in right turns without causing drivers to proceed at higher speeds.

Lessons Learned Concerning the Role of Simulators in Assessing Acceptability

Based on the simulator work (and subsequent validation), a number of lessons can be drawn concerning aspects of road design for efficiency, safety and acceptability.

The study of 'Active Traffic Management' shows that systems designed to ease congestion can also have implications for safety. However, any potential safety problems can be mitigated by informing drivers and helping them to understand the new designs; for example, through appropriate signage. The information strategy needs to be comprehensive, to ensure that drivers both approach the scheme in the most positive frame of mind and, when using schemes, do so as safely and as comfortably as possible. Equally, the study of non-physical motorway segregation demonstrated that the anticipated benefits of a scheme may not be realised if drivers do not behave in the expected manner.

Studies of interventions designed for rural roads specifically to improve safety show that they may also have unanticipated consequences. For example, compared with standard passive retro-reflective studs, delineation of a road at night by 'actively illuminated road studs' offers a significant overall safety advantage that appears to outweigh the slight increase in drivers' speed choice.

These studies demonstrate that simulation can play a useful role in understanding changes in driver behaviour, anticipated safety outcomes and user acceptance in response to road infrastructure modifications. They allow testing under a wide range of conditions whilst ensuring participant safety and enable evidence-based decisions to be made before infrastructure is in place.

Assessing Behavioural Responses to Highway Technologies in the Real World

Once the design of a new technology scheme has been decided upon, it can be implemented in the real world. However, it is important to evaluate the system in situ to confirm that the expected benefits are achieved. Real-world evaluations can be difficult to conduct as it may be difficult to identify suitable control sites for comparison or have confidence that any changes observed in a before-after evaluation are a consequence of the installed intervention or some other factor. Furthermore, acceptance can only be inferred from behaviour since it is difficult and impractical to interrogate drivers who have recently experienced an intervention. In this section, we consider driver acceptance in two case studies of real-world evaluations of highway technology.

Case Study 1: Roadwork Delineation (Flashing Cones)

When motorway roadworks are in force, traffic is usually required to deviate from normal traffic lanes to accommodate the works site. This change from the usual situation can create risk, especially when traffic lanes are required to merge. Lane merging is typically achieved by applying standard configurations of traffic cones that (in conjunction with temporary signs) indicate to drivers the new geometry of the road ahead. A long row of cones guiding traffic out of an existing lane and into a new temporary lane or merging with an adjacent lane is termed a 'cone taper' (Figure 14.5). This represents a site of risk since it is where a driver must deviate from their routine behaviour in response to the presence of the cone taper. In 2001 and 2002, the UK Highways Agency found that around 50 per cent of near miss incidents in Area 4 of the UK's trunk road network (comprising a total route length of 451 km across the south and south-east of England) were cone taper strikes (Highway Agency Trials Team 2005) creating risk for the driver, other road users and road workers within the works region.

In 2002, the Highways Agency Trials Team began evaluating sequential flashing cone lamps (SFCLs) as an improved means by which to communicate upcoming road layout to drivers. In 2005, following off-road evaluation and confirmation that SFCLs were suitable for use on UK roads (for evaluation purposes), an on-road trial was conducted to assess drivers' use and acceptance of SFCLs.

A test site (with pre-existing temporary traffic management measures) was chosen on the M42 Active Traffic Management (ATM) section (prior to activation of any ATM features such as gantry signing) since this was equipped with inductive loops every 100m, enabling detailed interrogation of lane occupancy through the works region. The cone taper was demarcated during closure periods (22:00–03:00) by static cone lamps for three days and with SFCLs for three days, alternating daily between the two configurations over a period of six days.

Results indicated that, from 600 m upstream of the cone taper, significantly fewer vehicles were present in the closing/merging lane when demarcated by SFCLs than by static cone lamps. This indicates that drivers understood and responded to the message communicated to them by the presence of the SFCLs by choosing to leave the closing lane earlier than with the static cone lamps. Therefore, inferred from behaviour, the cone taper was judged as accepted by drivers in the simulator and, hence, likely to be acceptable to drivers in real road conditions.

Figure 14.5 An example of a cone taper used for temporary traffic management in the UK (closing the right lane of the carriageway in the left of the picture)

In addition, it was concluded that the use of SFCLs would, in the short term at least, increase the safety margin between traffic and the temporary traffic management site, thereby decreasing risk at temporary traffic management sites.

Case Study 2: Driver Acceptance of Speed Monitoring

Speed is a key determinant of both the number of road crashes and crash severity, so it is unsurprising that road authorities seek to influence drivers' speed. Traditionally this involves fixed signs showing the maximum permitted speed on a section or road but it is well known that limits are widely exceeded. In the Netherlands, for example, 20–40 per cent of vehicles exceed the posted limit on most road types (SWOV 2010a). Speed management has, more recently, employed a range of technology to measure speed of specific vehicles and this information can be used in a number of ways to influence drivers' speed choice.

Here we examine using immediate feedback to drivers, without enforcement or surveillance, using speed cameras that record vehicle identity.

Speed Indicating Devices

Speed indicating devices (SIDs) are temporary vehicle-activated signs that detect and display vehicle speeds providing direct real-time feedback to drivers (Figure 14.6). They are a relatively cheap method of speed management that aim to change drivers' speed behaviour and are increasingly being installed.

SIDs have been found to be effective at reducing vehicle speeds in urban areas when deployed for periods of a few weeks but that their effectiveness decreases over time (Walter and Knowles 2008). Also, their effectiveness has been reported to last only a short distance beyond the sign although this varies depending on site characteristics.

Whilst the effectiveness of SIDs in terms of speed management is limited, they tend to be regarded as acceptable (or harmless) because they are not covert and do not permanently record vehicle identity. Issues of acceptance and acceptability become much more acute for types of surveillance that record vehicle identity and location, as explored below.

Surveillance and Speed Enforcement

It is often claimed that we live in a surveillance society (Wood 2006). CCTV surveillance, however, is relatively acceptable in many countries as it is widely seen as protecting the public from the minority of criminals acting in a visibly deviant manner (Wells and Wills 2009).

Speed cameras are a specific form of surveillance and have been used extensively in Europe, particularly Great Britain, Germany and the Netherlands – and now increasingly in France; and in jurisdictions such as Victoria, Australia, and much less often in the United States and Canada.

Figure 14.6   Examples of a speed indicating device

Wherever used, there has been controversy – but resistance to speed cameras seems to be characterised not by complaints about civil liberties but in terms of unfairness or injustice.

It has been argued (Wells 2007) that whilst speed cameras were introduced as a method of risk reduction, the controversy around them can usefully be understood in risk terms where drivers view themselves as victims exposed to risk rather than protected from risk by the speed enforcement technology. With this view, the risk of punishment is more prominent than the risk of death or injury (Wells and Wills 2009). Part of the unacceptability seems to arise from the legal principle that the speeding offence does not distinguish between intentional and unintentional behaviour.

Drivers will often portray themselves as making intelligent judgements about the appropriate speed for a particular set of circumstances and contrast this with the fixed legal speed limits enforced by an 'oppressive state' through technology. To counteract this view, efforts are being made, in the Netherlands for example, to better match legal speed limits to road characteristics and perceptions in an attempt to make the speed limits more credible (SWOV 2010b).

A typology of drivers has been developed based on their response to speed cameras by Corbett (1995) and Corbett and Simon (1999):

- Conformers (those who report they never exceed limits);
- Deterred drivers (those put off speeding by the presence of cameras);
- Manipulators (those who slow only at camera locations); and
- Defiers (those who exceed limits regardless of cameras).

It could be inferred from their behaviour that Conformers and Deterred drivers accept speed cameras but Manipulators and Defiers do not.

Blincoe et al. (2006) interviewed a sample of road users who had been prosecuted for exceeding the speed limit in the rural county of Norfolk England and categorised them using the same groupings as Corbett and Simon. In a sample of 433, she found 31 per cent Conformers, 27 per cent Deterred drivers, 33 per cent Manipulators and 9 per cent Defiers. In her sample, speeding was perceived as widespread and normal, with many of the drivers resenting camera enforcement. For many, the prosecution experience resulted in distress, anger and anti-camera sentiments, predominantly because they expressed the belief that they were more skilled than other drivers. The Deterred drivers were most likely to express intentions to avoid further speeding and their speeding incident was found to be most likely to be accidental. Manipulators and Defiers tended to report that they had deliberately chosen to infringe the speed limits.

A strong body of research shows that speed cameras improve the behaviour of road users and reduces speeding and road crashes. For example, a four-year evaluation report (Gaines 2005) looked at 2,000 urban and rural sites in the UK where speed measurements were taken both before and after camera deployment. Analysis showed that once the cameras were operational there was

- a substantial improvement in compliance with speed limits;
- a particular reduction in extreme speeding; and
- a marked reduction in average speed at fixed sites.

In a comprehensive review (Allsop 2010), it was concluded that deployment of speed cameras leads to appreciable reductions in speed in the vicinity of the cameras and substantial reductions in collisions and casualties.

Nevertheless, opposition to speed cameras remains, as does some scepticism towards the scientific expertise drawn on by governments in setting their policy.

Speed cameras were first used for enforcement in Great Britain in 1992 and their rollout was accelerated between 2001 and 2005 in a national safety camera program. The Automobile Association (AA) has been monitoring the public acceptance of cameras in the UK for 10 years and the level of acceptance has been around 70 per cent. This latest poll (AA 2010) shows the highest levels of support ever.

In the Netherlands the annual perception study shows which types of speed enforcement Dutch drivers find more and less acceptable. They clearly find speed cameras at fixed positions more acceptable than methods that are not clearly visible (Table 14.1).

Table 14.1   Percentage of respondents who find certain types of speed enforcement (very) acceptable

Type of speed enforcement	2004	2005	2006	2007	2008	2010
Fixed position speed camera	67	67	66	72	80	73
Hidden police car	48	45	46	51	58	44
Stopping drivers (on-the-spot fine)	80	77	78	79	86	77
Laser gun	48	47	48	50	57	43
Video car	68	66	66	70	76	62
Average speed check	69	70	69	72	76	69

Source: Table reproduced from SWOV (2011), based on data from Poppeliers, Scheltes and In 't Veld 2009, Intomart Gfk 2010, with permission.

In Victoria, Australia, the speed camera program has been subject to persistent negative public perceptions and, in a 2010 survey, 69 per cent of respondents agreed that speed cameras are more about raising revenue than road safety (Pearson 2011). Public confidence in the reliability and accuracy of the technology has also been undermined by media reporting. Pearson lists (and refutes) a number of common misperceptions based on examination of media articles, public surveys and individual submissions:

- The purpose of the road safety camera program is to raise revenue;
- Low level speeding is safe;
- Road safety cameras don't reduce road trauma;
- Speed cameras should not be placed on freeways because freeways are safe; and
- The cameras are faulty.

Point-Based and Average Speed Enforcement Cameras

Drivers' understanding of the technology can have an effect on behaviour and this can be taken as one determinant of acceptance. Fixed single-point cameras have a zone of influence in the immediate vicinity but there is nothing to stop drivers slowing down and then speeding up after passing a camera. In one survey of 2,400 drivers (Swiftcover 2007), 53 per cent believed that fixed point cameras encourage people to drive more erratically and a further 56 per cent admitted to 'yo-yo driving' (speeding up between cameras) themselves.

Average speed cameras detect vehicles along a road and calculate average speed. Initially, acceleration/deceleration behaviour was also observed at average speed camera installations (Charlesworth 2008) but better signage and general awareness of how the system works has increased recognition and improved driver behaviour and compliance. In the survey cited above, 74 per cent of drivers said that they usually drive through entire 'average speed camera' areas, at the correct speed limit.

In Austria, positive effects of average speed enforcement were found in terms of speed reduction and in the number of crashes in tunnels (Stefan 2006).

According to Thornton (2010) whilst some drivers claim that driving through an average speed enforcement scheme is stressful, and that they cannot maintain a steady speed without constantly looking at their speedometer, the majority of drivers feel that the system makes the road less stressful, as other drivers are less prone to tailgating and 'bullying' and that there is less braking and lane-changing.

Summary Concerning Acceptability of Speed Cameras

There is a strong body of research showing that speed cameras improve the behaviour of road users and reduce speeding and road crashes. Nevertheless, a substantial minority of drivers still find such cameras unacceptable. This seems to be as a result of some misperceptions concerning the technology and its purpose, and also to a perceived unfairness around who 'gets caught'. Interestingly, there appear to be differences in acceptance rates between countries, suggesting that policy and public awareness can potentially be developed in a way to improve acceptance.

In terms of specific roadside technologies, vehicle-activated Speed Indicating Devices are relatively inexpensive but have limited effectiveness and are probably regarded as harmless and acceptable by a majority of drivers. Hidden speed cameras appear to be the least accepted form of speed enforcement. Cameras operating at fixed points are more acceptable but the most acceptable are average speed cameras. Their relative popularity seems to result from the robust and visible technology causing speed limits to be widely observed.

Conclusions

Interaction between drivers on the road depends on a wide range of factors that vary with time and circumstance, with safe interaction relying on drivers not only understanding but accepting the 'norms of behaviour' implied or required by the road layout and the information and other technology deployed within the infrastructure.

In this chapter we have illustrated how driver behaviour and acceptance can depend on the information provided, and that there can be unintended consequences from new system deployment. A driving simulator to investigate driver responses to new technology within the infrastructure can be very useful. We have shown that driver behaviour is affected by monitoring and feedback, although both the implementation approach and wider social factors influence the acceptability of that monitoring and the use of technology to implement it.

Acknowledgements

The content of this paper is the responsibility of the authors and should not be construed to reflect the opinions or policies of any organisation.

References

Automobile Association (AA). 2010. Speed cameras. [Online]. Available at: http://www.theaa.com/public_affairs/aa-populus-panel/aa-populus-increased-support-for-speed-cameras.html [accessed: 15 February 2012].

Allsop, R. 2010. The effectiveness of speed cameras – A review of evidence. [Online]. Available at: http://www.racfoundation.org/assets/rac_foundation/content/downloadables/speed%20camera%20effectiveness%20-%20allsop%20-%20report.pdf [accessed: 30 December 2011].

Blincoe, K.M., Jones, A.P., Sauerzapf, V. and Haynes, R. 2006. Speeding drivers' attitudes and perceptions of speed cameras in rural England. *Accident Analysis and Prevention* 38: 371–8.

Broughton, J. and Buckle, G. 2006. Monitoring progress towards the 2010 casualty reduction target – 2004 data. Report no. 653. Crowthorne: Transport Research Laboratory.

Charlesworth, K. 2008. The effect of average speed enforcement on driver behavior. Road Transport Information and Control. IET Publication.

Corbett, C. 1995. Road traffic offending and the introduction of speed cameras in England: The first self-report study. *Accident Anaysis & Prevention.* 27(3): 345–54.

Corbett, C. and Simon, F. 1999. The effects of speed cameras: How drivers respond. Department for Transport Local Government and the Regions (DTLR) Report no. 11. London: DTLR.

Department for Transport (DfT). 2008. Advanced motorway signalling and traffic management feasibility study – A report to the Secretary of State for Transport. London: Department for Transport (DfT) .

Diels, C., Robbins, R. and Reed, N. 2011. Behavioural validation of the TRL driving simulator DigiCar: Phase 1 – Speed choice. Presented at the International Conference of Driver Behaviour and Training, Paris, France, November.

Fafieanie, M. and Sambell, E. 2008. Dynamic tidal flow lane on provincial roads in the Netherlands. The Netherlands: University of Twente. [Online]. Available at: http://www.utwente.nl/ctw/aida/education/Final%20report%20ITS2%20Fafieanie%20and%20Sambell.pdf [accessed: 20 December 2011].

Gaines, A. 2005. The national safety camera programme: Four year evaluation report. London: PA Consulting Group and University College London (UCL). [Online]. Available at: http://eprints.ucl.ac.uk/1338/1/2004_31.pdf [accessed: 30 December 2011].

Highways Agency Trials Team. 2005. Safe temporary traffic management operations initiative – RoWSaF Trial report: Sequential flashing cone lamps. [Online]. Available at: http://www.dormanvaritext.com/pdf/sequentialflashingconestrialreport.pdf [accessed: 30 December 2011].

Intomart, GfK. 2010. Effectmeting Regioplannen 2010: Landelijke rapportage; Een internet-onderzoek in opdracht van het Landelijk Parket Team Verkeer van het Openbaar Ministerie. Intomart GfK, Hilversum.

Pearson, D.D.R. 2011. Road safety camera programme. Audit Report PP no. 3, Session 2011–12. Australian Victorian Government Printer, August. [Online]. Available at: http://www.audit.vic.gov.au/publications/20110831-Road-Safety-Cameras/20110831-Road-Safety-Cameras.pdf [accessed: 30 December 2011].

Poppeliers, R., Scheltes, W. and Veld, N. 2009. Effectmeting regioplannen (perceptieonderzoek). Landelijke rapportage 2008. Onderzoek in opdracht van het BVOM. NEA Transportonderzoek en -opleiding, Rijswijk.

Reed, N. 2006. Driver behaviour in response to actively illuminated road studs: A simulator study. Published Project Report PPR143. Crowthorne: Transport Research Laboratory.

Stefan, C. 2006. Section control – Automatic speed enforcement in the Kaisermühlen tunnel (Vienna, A22 Motorway). Vienna: Austrian Road Safety Board KvF.

Stern, N. 2006. Review on the economics of climate change. H.M. Treasury, UK, October. [Online]. Available at: http://www.sternreview.org.uk [accessed: 30 December 2011].

Swiftcover. 2007. Car Insurance: Accelerated roll-out of speed cameras costs motorists £144.6 million each year. Available at: http://www.swiftcover.com/about/press/speed_cameras_costs/.

Stichting Wetenschappelijk Onderzoek Verkeersveiligheid (SWOV). 2010a. Police enforcement and driving speed. SWOV fact sheet, June. [Online]. Available at: http://www.swov.nl/rapport/Factsheets/UK/FS_Surveillance.pdf [accessed: 30 December 2011].

SWOV. 2010b. Towards credible speed limits. SWOV factsheet, September. [Online]. Available at: http://www.swov.nl/rapport/Factsheets/UK/FS_Credible_limits.pdf [accessed: 30 December 2011].

SWOV. 2011. Speed cameras: How they work and what effect they have. SWOV fact sheet, October. 2011. [Online]. Available at: http://www.swov.nl/rapport/Factsheets/UK/FS_Speed_cameras.pdf [accessed: 30 December 2011].

Thornton, T. 2010. Reductions in fuel consumption and CO_2 emissions with specs average speed enforcement. Road Transport Information and Control Conference CD. Stevenage: Institution of Engineering and Technology.

Thornton, T., Reed, N. and Gordon, N. 2005. ATM – Driver behaviour during operational regimes. Presented at Smart Moving 2005, Birmingham, England.

Törnros, J. 1998. Driving behaviour in a real and a simulated road tunnel – A validation study. *Accident Analysis and Prevention*, 1998, 30(4): 497–503.

US Department of Transportation, Federal Highway Administration. 2004. Superior Materials, Advanced Test Methods, and Specifications in Europe, April 2004. Report no. FHWA-PL-04-007 [Online]. Available at: http://international.fhwa. dot.gov/superiormaterials/chap3.cfm [accessed: January 2012].

Walter, L.K. and Knowles, J. 2008. Effectiveness of speed indicator devices on reducing vehicle speeds in London. Report no. PPR314. Crowthorne: Transport Research Laboratory.

Webb, W.B. 1995. The cost of sleep-related accidents: A reanalysis (technical comments). *Sleep*, 18(4): 276–80.

Wells, H. 2007. Risk, respectability and responsibilisation: Unintended driver responses to speed limit enforcement. *Internet Journal of Criminology*. [Online]. Available at: http://www.internetjournalofcriminoogy.com [accessed: 30 December 2011].

Wells, H. and Wills, D. 2009. Individualism and identity: Resistance to speed cameras in the UK. *Surveillance and Society* 6(3): 259–74. [Online]. Available at: http://www.surveillance-and--society.org [accessed: 30 December 2011].

Wood, D. 2006. A report on the Surveillance Society for the Information Commissioner by the Surveillance Studies Network. [Online]. Available at: http://www. ico.gov.uk/upload/documents/library/data_protection/practical_application/ surveillance_society_full_report_2006.pdf [accessed: 30 December 2011].

Chapter 15

Operator Acceptance of New Technology for Industrial Mobile Equipment

Tim Horberry

Minerals Industry Safety and Health Centre, University of Queensland, Australia
Engineering Design Centre, University of Cambridge, UK

Tristan Cooke

Minerals Industry Safety and Health Centre, University of Queensland, Australia

Abstract

Using examples from mining and the wider minerals industry, this chapter focuses on operator acceptance of new technology for industrial mobile equipment. It initially takes a broad approach by introducing the mining/minerals industry and briefly describing the key elements in the mining system. Thereafter, it examines the development and deployment of new mining technologies, including the need for them and the types typically being introduced at mine sites. Operator acceptance of mining technologies, especially for mining vehicles, is then introduced: particularly considering how both design and deployment can assist in improving acceptance of such technology. A case study of recently undertaken research that considers operator acceptance of proximity warning systems for mining vehicles is then presented. Finally, the chapter concludes by stressing the importance of a user-centred design and deployment process for technology used in mining and elsewhere.

Introduction to Mining and the Minerals Industry

Given that the main focus of this chapter is not road transport, a short introduction to the domain being discussed here – mining and the minerals industry – is first given. The 'minerals industry' is an overall term for a group of activities related to mining (the extraction of minerals), ore/minerals processing and the transportation of minerals. Of course, such minerals include the more 'traditional' ones, such as coal, iron, gold and copper, plus those that have not been systematically exploited on a wide scale until quite recently (such as coal seam gas or oil sands).

The industry is both a significant worldwide employer and major revenue generator; for example, in recent years in Australia it was responsible for approximately half of the total exports of the country (Australian Bureau of Statistics 2010). It is present across virtually the whole globe, with some of the major mining areas being in Africa, Australia, North and South America, the former Soviet States, India and China. The worldwide injury and fatality rates for the industry vary greatly, ranging from usually single-figure deaths per annum in Australia and Canada, through to many hundreds being killed in developing countries (Simpson, Horberry and Joy 2009). Similarly, ill-health linked to high noise levels, hazardous manual tasks and respirable dust are still prevalent in some countries.

Elements in the Mining System

As with other complex socio-technical systems, such as road transport or aviation, there needs to be safe interactions between people, procedures, environments and equipment in the minerals industry. Focusing purely on mining, as noted by Horberry, Burgess-Limerick and Fuller (2013), the main elements include

- a varied group of people employed or contracted;
- a wide range of different mining jobs, tasks and roles;
- a specialised array of diverse mining equipment;
- many different mining equipment manufacturers, dealers and suppliers;
- a steady increase in the number of new technologies being designed and deployed;
- different worldwide mining companies;
- varying procedures, rules, practices and cultures at individual mine sites;
- a diverse range of national laws, regulations and guidelines;
- differences in the built environment and precise mining method used; and
- uncertainties in the natural environment being mined.

The focus in this chapter is upon several elements in the above list: namely, the interaction of the human element with the new technologies being deployed. Before specific operator acceptance issues are examined, some of the new technologies in the minerals industry are discussed.

Development and Deployment of New Mining Technologies and Automation

The Need for New Technologies in Mining

The ongoing imperative by mining companies and regulators for safe and healthy workplaces has been one of the major drivers for the introduction of new technologies into mining (Burgess-Limerick 2011). Theoretically, such new technologies can offer great potential to improve operator safety and health; for example, they could form another layer of protection to add to any physical and organisational measures already employed. In addition to potentially removing operators from hazardous mining situations (especially underground), they could also present new ways of addressing long-standing safety issues, such as lessening the severity post-injury by means of better incident detection and response systems (Horberry, Burgess-Limerick and Steiner 2010).

Safety and health concerns are, of course, not the only reason to introduce new technologies. As previously noted by Horberry et al. (2010), other factors include the following:

- Lower cost of mineral production. Examples include more ore transported, or more efficient process control operations.
- Enhanced precision. A simple example is automated blast-hole drilling in hard-rock mining – where not only is there a potential safety benefit by removing the operator, but the correct location of the blast holes can theoretically be more accurately achieved through automated systems.
- Less environmental impact. In theory, new technologies can minimise the need for land reclamation (e.g., by using keyhole mining methods, rather than more disruptive approaches) and require less energy to extract and process the commodity.
- Being able to mine areas previously inaccessible. This might include being able to mine in hard-to-reach locations that previously could not be mined economically.
- Reduced manning. Although it is a myth that automation fully removes the need for all human involvement in mining (Sanders and Peay 1998), in some cases it may reduce the need for humans, at least those on the front line (e.g., remotely controlled trucks not requiring an operator to drive them from within the vehicle's cabin).
- More data/information. The capacity to collect more data, often in real time, on the performance and state of mining equipment can be of considerable advantage for issues such as maintenance scheduling or appropriate responses in emergency situations.

The Scope of New Mining Technologies

Intensive research and development work in automation and new mining technologies is currently being undertaken by many major mining companies, equipment manufacturers and universities (Burgess-Limerick 2011). Some of the main mining equipment types being particularly investigated include haul trucks, blast-hole drills, rock crushers and ore trains. In future, it seems likely that automation or new technologies in some form will be further applied across virtually all mining equipment (e.g., shovels and excavators) and mining methods (Lynas and Horberry 2011).

Similar to many other industrial and transport domains that have not yet fully embraced the need to systematically integrate the human element, the rapid development and growth in new systems has often seen them being deployed as the prototype technology becomes available, without them being systematically designed, integrated into work environments and evaluated from a user-centred perspective. For example, new systems must meet the requirements of the job/ task, work in emergency/abnormal operational states, support operators and be acceptable to the eventual end-users. As such, devices in mining have generally been designed from a technology-centred perspective, rather than first seeing what are the needs and what safety or performance benefits they might bring (Li, Powell and McKeague 2012). Sadly, with the exception of the study reported later in this chapter, little research has been undertaken into operator acceptance of new technologies in the mining domain. If operator requirements and preferences are not well understood before new systems are introduced, the systems may be unacceptable when deployed. In mining, or indeed in other industrial domains where the use of the deployed technology is often mandatory, those technologies that are not accepted by operators are less likely to be used properly and are more likely to be sabotaged or misused; thus, any inherent potential for increasing safety or efficiency may not be fully achieved (Horberry and Lynas 2012).

Overview of Types of New Technologies Used at Mine Sites

A database of current and emerging technologies in mining, and the likely human element implications of such technologies, was recently produced (Horberry and Lynas 2012). Mining is already highly mechanised, but new technologies identified in the database ranged from full automation of complete mining processes (e.g., blasting in hard-rock mining before the ore is removed) to more piecemeal technologies such as proximity warning systems for mobile mining equipment. In the database, the specific technologies were grouped by 'degrees of automation', such as fully automated and partially automated systems; assistance devices such as proximity detection/warning systems; and other relevant technologies. The lower level automation entries where the operator is still directly in the system control loop (i.e., those technologies that are more within the focus of this book) include warning systems such as collision detection systems, and technologies that signal when maintenance of equipment is due. In this category, the operator remains in full control of the system, with the technology providing warnings, information or assistance. Roughly half of all the entries in the database were from this category: this high proportion might be explained partially by these systems being simpler to develop compared with large-scale fully automated systems (Horberry and Lynas 2012).

In terms of trends to be distilled from the database, in addition to the general growth of new technologies being introduced into this field, the authors noted that there was a lack of user-centred design, with approximately only one-third of the database entries mentioning explicitly how the technologies might impact upon the operator (Horberry and Lynas 2012). The implication here is that either human operators and maintainers are no longer important for safe and efficient mining, or they have simply been largely overlooked by the engineering-focused developers of these technologies.

Acceptance of Mining Vehicle Technologies

Given the general lack of an operator-centred focus in the development and deployment of new mining technologies, it is perhaps no surprise to note that there are few systematic and widely used measures to optimise operator acceptance of new technologies from mine sites. Despite this, two general approaches are sometimes undertaken (Horberry et al. 2010).

1. Safe Design

'Safe design' is slowly becoming more widespread in the minerals industry (Horberry et al. 2013). Also known as 'safety in design', 'safety by design' or 'prevention through design', the broad process aims to eliminate health and safety hazards, or minimise potential risks, by systematically involving end-users and decision makers in the full life cycle of the designed product or system. In the traditionally conservative mining and extractive industries, examples of its application include Cooke and Horberry (2011a) for mining equipment, Bersano et al. (2010) for extractive activity start up and management and Kovalchik et al. (2008) for preventing mining operator hearing loss.

A task-oriented safe design and risk assessment process that focuses on Human Factors risks related to mobile mining equipment design was recently created, validated and then applied to the design of several types of mobile mining equipment (Horberry et al. 2013). It is outside the scope of this chapter to present detailed safe design results, but recent trials by Cooke and Horberry (2011a) covered mining equipment issues such as equipment access and egress and safe design for maintenance tasks. At the heart of this safe design method is a 'participatory ergonomics' process whereby equipment operators and maintainers were actively engaged in both critiquing design problems with their existing equipment and helping to develop safer generations of new equipment. To date, the safe design method has largely been applied to 'traditional' mining equipment, but a recent trial with a new in-vehicle mining technology produced positive results (Cooke and Horberry 2011b). As will be noted in the case study later in this chapter, the assumption is that actively involving end-users (equipment operators and maintainers) in the technology design process would lead to them better accepting the technology when it is deployed.

2. Deployment Strategies and Operator Skills Requirements

As noted earlier in this chapter, the use of new technologies in mining is usually mandatory once they are introduced. So, whilst their use is generally officially compulsory, non-acceptance is often revealed by equipment being broken, sabotaged or otherwise neglected (Horberry et al. 2004). Similarly, poor operator acceptance of new technologies/automation after they are introduced might be evidenced by negative opinions of the new devices (e.g., not trusting the outputs of the new technology).

Although user-centred safer initial design is necessary for operator acceptance, it is not usually sufficient unless consideration is also given as to how the new technologies will actually be deployed at a mine site. User-centred deployment of the technology by means of operator consultation, understanding the exact requirements of the tasks and an ongoing feedback process to/from management can help reduce problems with new mining technology acceptance (Horberry et al. 2010).

Another key deployment aspect for operator acceptance is ensuring that operators have sufficient skills and training to effectively use the technology. Due to the ever-evolving nature of mining technology being developed, the exact skills and capability requirements cannot always be specified a priori; however, it is still of critical importance to have an ongoing process to identify and address skills gaps and training needs (Horberry et al. 2011). To give an example, an analysis by Dudley, McAree and Lever (2010) suggested the minerals industry would require a large number of new automation support staff if widespread automation was introduced. The exact skills and cognitive capabilities required by these automation technicians would depend on the tasks performed and the technologies worked with, but four general skills gaps identified by Dudley et al. (2010) were communication, problem solving, planning and organization, and technology. An interesting feature here is that of these four general skills gaps, the first three are largely 'non-technical' skills. The implication is that many of the skills and capabilities required to successfully use the new automated technology (and hence, indirectly, optimise acceptance of that technology) are not technical skills: instead, problem solving, planning and communication abilities need to be sufficient if the technology is to be successfully introduced and accepted.

Case Study: Acceptance of Collision Detection Systems in Mining Vehicles

Background

As noted above, a great deal of research and development effort is currently taking place with new mining technologies: this is particularly true for collision detection and proximity warning systems for mobile mining vehicles (Cooke and Horberry 2011b). In part this is because of the high percentage of mine site incidents that somehow involve collisions – especially between mobile mining equipment (such as large haul trucks and bulldozers, and light vehicles used for maintenance), or between mining vehicles and pedestrian workers (Horberry et al. 2010). This, in turn, is partly because in recent years there are more mobile mining vehicles, especially bigger equipment with more blind spots (Bell 2009).

Collision detection and proximity warning systems are also becoming increasingly important to regulators (particularly in North America and Australia);

in some locations their use is being strongly encouraged (even compelled) by the appropriate safety authorities. They argue that collision/proximity detection technologies are now a mature enough technology to become a valuable control, especially when used in conjunction with other measures such as traffic management, barriers and vehicle separation (Bell 2009).

Types of Collision Detection and Proximity Warning Systems in Mining

Like the term 'automation' more generally, collision detection and proximity warning systems cover a wide variety of technologies; they differ in where, when and how they can be used. In mining, no single type fits all areas (Horberry et al. 2010). Often they are low-level warnings of another vehicle (or pedestrian worker) nearby and only a few systems are specifically designed to take control (e.g., to intervene by applying the vehicle's brakes in response to a likely collision being detected).

A wide range of sensor technologies and associated tools are being used: radar, Wi-Fi, cameras, radio frequency identification (RFID), databases of static obstacles, global positioning systems (GPS), 3D mapping and ultrasonics. Some of these work better in specific environments (Cooke and Horberry 2011b). For example, many sensor types will not work underground, and the intrinsic safety and resulting certification requirements associated with underground coal mines in particular create additional challenges to the introduction of technology in general, and proximity detection in particular. Surface mining has an advantage over underground mining in that it can more easily build on previous work in other domains – most notably, collision detection technologies developed by the land transport/automotive domains (Horberry et al. 2010).

The mine used for the research described here was an underground gold mine in central Queensland, Australia. It had previously installed a radio frequency identification system to track vehicle movement. The system was primarily installed at the mine to improve the monitoring of gold production. However, subsequently the mine's management recognised that there was an opportunity to add a proximity warning system to, hopefully, reduce the risk of collisions between vehicles at the site. 'Tags' were mounted on all vehicles that would enter the mine. 'Readers' were mounted on heavy vehicles in the mine that had large blind spots, such as haul trucks and ore loaders (Cooke and Horberry 2011b).

In these heavy vehicles, a visual display was provided to the drivers via a touch screen tablet computer. This was mounted on the right side of the driver for both haul trucks and loaders. The system detected the presence of any vehicles in range, not just those that were determined to be dangerous or require action: as such, the driver needed to interpret and take a necessary course of action. An auditory warning (of alterable volume) occurred on detection, and a visual warning (a line on the vehicle's touch screen) flashed. Both warnings continued until the screen was physically touched as acknowledgement (Cooke and Horberry 2011b). Therefore, on some occasions when another vehicle was detected nearby, the design of the system was such that the operator was theoretically required to undertake two simultaneous tasks: take evasive action to prevent a collision, and physically touch the screen to cancel the warning. In practice, the second of these tasks was rarely undertaken due to it being of far less importance than preventing a collision.

Research Undertaken

Previous work by Cooke and Horberry (2011b) has used a variety of Human Factors methods to investigate these prototype proximity detection systems at the mine site. This past work included measuring detection distances at different points around the mine site and undertaking a usability audit of the prototype interface.

Building on this earlier work, research was conducted to understand the proximity detection device within the overall gold mining system, including other collision prevention controls present. The impetus for this subsequent research was that an accident involving a collision between a light and heavy vehicle had recently occurred at the mine. Following the accident, a number of changes were made to the proximity detection system interface, and testing driver/operator acceptance of these changes was therefore a key component of the work.

Following investigation of the above-mentioned accident, changes made to the proximity detection system included altering the auditory warning (to make it more salient) and modifying the visual information (e.g., adding newly detected vehicles to the top of the list on the computer screen, rather than to the bottom of the screen, where they were added previously). Full details of the changes made are given in Cooke and Horberry (2011c).

Method

To gauge the effects of the interface changes that were made, operators of the heavy vehicles employed at the mine were surveyed. The mine was a fairly small site with a total population of only 20 drivers operating heavy mobile equipment. Eighteen of these drivers completed the survey. It was conducted primarily to determine how accepting the drivers were of the initial system in comparison with the altered system. Drivers were also asked about the importance of other controls relative to the proximity detection systems.

Given the lack of technology acceptance work previously undertaken in mining, the research built on the long-established method for measuring driver acceptance developed by Van der Laan, Heino and De Waard (1997). This method was selected because it had been applied in several different studies of measuring acceptance of in-vehicle systems, as seen in other chapters in this book. Using the technique, a five-point rating scale was used for nine questions rating acceptance of the initial and altered interfaces of the proximity detection system. The heavy-vehicle drivers were required to select between five boxes placed between two opposing qualitative words (the position of the positive and negative words was sometimes reversed). The positive words are shown in italics below (adapted from Cooke and Horberry 2011c):

1. *Useful* ☐ ☐ ☐ ☐ ☐ Useless
2. *Pleasant* ☐ ☐ ☐ ☐ ☐ Unpleasant
3. Bad ☐ ☐ ☐ ☐ ☐ *Good*
4. *Nice* ☐ ☐ ☐ ☐ ☐ Annoying
5. *Effective* ☐ ☐ ☐ ☐ ☐ Superfluous
6. Irritating ☐ ☐ ☐ ☐ ☐ *Likeable*
7. *Assisting* ☐ ☐ ☐ ☐ ☐ Worthless
8. Undesirable ☐ ☐ ☐ ☐ ☐ *Desirable*
9. *Raising Alertness* ☐ ☐ ☐ ☐ ☐ Sleep Inducing

In the scoring system for the scales previously used by Van der Laan et al. (1997), the middle box represented a score of 0, the boxes either side represented -1 to +1 and the outer boxes +2 or -2. However, in our case, the scoring system was adapted to be positive numbers only (1–5) to allow shape plotting on a radar graph. By joining up each of the ratings, an irregular polygon was formed to visually communicate the overall change. The sum of all responses made up a score for acceptance. For Van der Laan et al. (1997), the nine questions assess system acceptance on two dimensions, a Usefulness scale (questions 1, 3, 5, 7 and 9) and a Satisfying scale (questions 2, 4, 6 and 8).

Results

Using the mean ratings from the participants, a graphical representation was plotted to reveal driver acceptance with the initial and revised proximity detection systems. This is shown in Figure 15.1: the acceptance of the initial system is shown in the lighter grey and the acceptance of the revised interface is shown in darker grey. A theoretical maximum acceptance score of five is shown by a solid outer line that joins Q1–Q9 in Figure 15.1: this has been included to show how far the ratings are from this ceiling. Also shown in Figure 15.1 with a dotted line is a 'positive-negative line', representing the midpoint score of three. Scores that are inside the 'positive-negative line' therefore represent a negative view of the interface on each of the nine questions. Full results can be found in Cooke and Horberry (2011c).

Discussion

The results show that, before the system changes, drivers, on average, were not accepting of the device: finding it neither particularly useful nor satisfying. After the system changes, all measures saw a more positive rating. On seven of the nine measures with the revised interface, drivers gave overall positive ratings for the system. Both the two negative measures (Q4 and marginally Q6) are in the

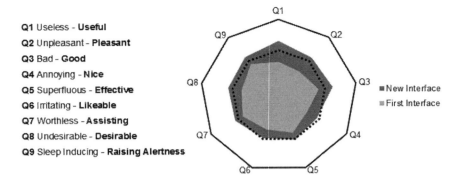

Q1 Useless - **Useful**
Q2 Unpleasant - **Pleasant**
Q3 Bad - **Good**
Q4 Annoying - **Nice**
Q5 Superfluous - **Effective**
Q6 Irritating - **Likeable**
Q7 Worthless - **Assisting**
Q8 Undesirable - **Desirable**
Q9 Sleep Inducing - **Raising Alertness**

■ New Interface
▨ First Interface

**Figure 15.1   Driver acceptance ratings of the initial and revised proximity
warning systems (adapted from Cooke and Horberry 2011c)**

'satisfying' component of the Van der Laan et al. (1997) acceptance construct. The positive ratings on seven out of the nine questions therefore indicates that drivers have mildly positive overall 'acceptance' of the revised system, are mildly positive about its 'usefulness' (positive ratings for Questions 1, 3, 5, 7 and 9) and are neither positive nor negative about its 'satisfaction' (due to positive ratings for Questions 2 and 8, but negative ratings for Questions 4 and 6).

Due to the nature of this trial, it was not possible to counterbalance the *before* and *after* conditions. As such, it is acknowledged that the procedure might have been a source of bias – for example, due to the drivers 'expecting' improvements.

As in other occupational domains, acceptance of new technology in mining is extremely important in subsequent technology utilisation. In isolated mine site environments, operators sometimes have the opportunity to choose to avoid using new technologies even when they are mandated. Therefore, if drivers do not accept proximity detection technology then its potential to prevent accidents may never be realised (Cooke and Horberry 2011c). In field usage conditions, the original interface showed several negative behavioural responses, with some drivers admitting to turning down the sound and brightness of the computer screen, in order to avoid the system as much as possible (Cooke and Horberry 2011b). The new interface made improvements to these observed deficiencies, for example, by making the auditory warning tone more salient. With improved acceptance of the new interface, it is anticipated that the drivers will be much less likely to try and avoid using the system. The underlying assumption behind this is that, because the operators are the experts, their acceptance is generally related to whether the technology aids them in driving the heavy mining vehicles (Cooke and Horberry 2011c). As mine site accidents are extremely rare, testing the effectiveness of a proximity detection system in terms of incident rates is both difficult and, potentially, unethical. Also, each mine site is very different, so undertaking the types of field operational tests that are being increasingly used in road transport would often be impractical in this domain.

Conclusions

New in-vehicle technologies, including proximity warning and collision detection systems, can help produce significant safety improvements in mining situations, especially where off-road haulage is responsible for a large number of fatalities (Horberry et al. 2010, Groves, Kecojevic and Komljenovic 2007). Mining has the opportunity to learn from other domains, such as road transport and aviation, to develop and implement technology from both a human-centred and an operational needs perspective. Therefore, rather than being introduced purely because prototype technology is available, careful consideration must be given to how it will support users' tasks, be acceptable to operators and integrate with existing work systems. It is recommended that the general Human Factors approach of systematically analysing the tasks needing to be performed, involving operators in device designs, evaluations and modifications, and using Human Factors information to develop appropriate interfaces, are of key importance to the development and deployment of successful technologies in this domain. In this regard, mining is perhaps no different from road transport; operator acceptance of technology is intricately linked to effective, user-centred design and deployment processes.

Acknowledgements

The authors would like to acknowledge the support of colleagues at the Minerals Industry Safety and Health Centre, University of Queensland, Australia. This paper was partly written with the support of an EC Marie Curie Fellowship 'Safety in Design Ergonomics' (project number 268162) held by the first author at the Engineering Design Centre, University of Cambridge, UK.

References

Australian Bureau of Statistics. 2010. *Australia's Production and Trade of Minerals.* Available at: http://www.abs.gov.au/ausstats/abs@.nsf/Latestproducts/8418.0 Main%20Features62009%20to%202010?opendocument&tabname=Summary &prodno=8418.0&issue=2009%20to%202010&num=&view [accessed 3 April 2012].

Bell, S. *Collision Detection Technology Overview.* 2009. Available at: http://www.dme.qld.gov.au/zone_files/mines_safety-health/deedi_2009_proximity_workshop_ppt_p1-18.pdf [accessed 17 December 2009].

Bersano, D., Cigna, C., Patrucco, M., Pession, J.M., Ariano, F.P., Prato, S., Romano, R. and Scioldo, G. 2010. Extractive activities start up and management: A computer assisted specially developed 'Prevention through design' approach. *International Journal of Mining, Reclamation and Environment*, 24(2): 124–37.

Burgess-Limerick, R. 2011. Avoiding collisions in underground mines. *Ergonomics Australia* – HFESA 2011 Conference Edition, 11: 7.

Cooke, T. and Horberry, T. 2011a. The operability and maintainability analysis technique: Integrating task and risk analysis in the safe design of industrial equipment. In *Contemporary Ergonomics and Human Factors 2011*. Edited by M. Anderson. Boca Raton, FL: CRC Press.

————. 2011b. Human factors in the design and deployment of proximity detection systems for mobile mining equipment. In *Contemporary Ergonomics and Human Factors 2011*. Edited by M. Anderson. Boca Raton, FL: CRC Press: UK.

————. 2011c. Driver satisfaction with a modified proximity detection system in mine haul trucks following an accident investigation. *Ergonomics Australia* – HFESA 2011 Conference Edition, 11: 30.

Dudley, J., McAree, R. and Lever, P. 2010. Bridging the automation gap. In *Automation for Success*. Mining Industry Skills Centre Report. Downloaded 30 September 2011 from http://www.miskillscentre.com.au/publications/automation-for-success.aspx.

Groves, W.A., Kecojevic, V.J. and Komljenovic, D. 2007. Analysis of fatalities and injuries involving mining equipment. *Journal of Safety Research*, 38(4): 461–70.

Horberry, T., Larsson, T., Johnston, I. and Lambert, J. 2004. Forklift safety, traffic engineering and intelligent transport systems: A case study, *Applied Ergonomics*, 35(6): 575–81.

Horberry, T., Burgess-Limerick, R. and Steiner, L. 2010. *Human Factors for the Design, Operation and Maintenance of Mining Equipment*. Boca Raton, FL: CRC Press.

Horberry, T., Lynas, D., Franks, D.M., Barnes, R. and Brereton, D. 2011. *Brave New Mine: Examining the Human Factors Implications of Automation and Remote Operation in Mining*. Second International Future Mining Conference. Sydney, Australia, 22–23 November.

Horberry, T., Burgess-Limerick, R. and Fuller, R. 2013. The contributions of human factors and ergonomics to a sustainable minerals industry. *Ergonomics*, 56(3): 556–64.

Horberry, T. and Lynas, D. 2012. Human interaction with automated mining equipment: The development of an emerging technologies database. *Ergonomics Australia*, 8: 1.

Kovalchik P.G., Matetic, R.J., Smith, A.K. and Bealko, S.B. 2008. Application of prevention through design for hearing loss in the mining industry. *Journal of Safety Research*, 39(2): 251–4.

Li, X., Powell, M.S. and McKeague, W. 2012. Unlocking processing potential by empowering our operators. In Proceedings of the 11th AusIMM Mill Operators Conference, 333–40. Hobart, Tasmania, 29–31 October.

Lynas, D. and Horberry, T. 2011. Human factors issues with automated mining equipment. *Ergonomics Open Journal*, 4 (Suppl 2-M3): 74–80.

Sanders, M.S. and Peay, J.M. 1998. *Human Factors in Mining*. Information circular IC9182. Washington, DC: US Bureau of Mines, Department of the Interior.

Simpson, G., Horberry, T. and Joy, J. 2009. *Understanding Human Error in Mine Safety*. Farnham: Ashgate.

Van der Laan, J.D., Heino, A. and De Waard, D. 1997. A simple procedure for the assessment of acceptance of advanced transport telematics. *Transportation Research Part C: Emerging Technologies*, 5(1): 1–10.

Chapter 16

Carrots, Sticks and Sermons: State Policy Tools for Influencing Adoption and Acceptance of New Vehicle Safety Systems

Matts-Åke Belin
*Swedish Transport Administration, Vision Zero Academy, Borlänge, Sweden,
and School of Health, Care and Social Welfare, Mälardalen University,
Västerås, Sweden*

Evert Vedung
*Institute for Housing and Urban Research, Uppsala University,
Uppsala, Sweden*

Khayesi Meleckidzedeck
*World Health Organization (WHO), Department of Violence and Injury
Prevention and Disability, Geneva, Switzerland*

Claes Tingvall
*Swedish Transport Administration, Borlänge, Sweden
Department of Applied Mechanics, Chalmers University, Gothenburg, Sweden*

Introduction

Road traffic fatalities and injuries are a rapidly increasing health problem in the world. The World Health Organization (WHO) has estimated that the number of traffic fatalities each year is approximately 1.2 million, while as many as 50 million people are injured. Without concerted action, the number of fatalities and injuries is estimated to increase by 65 per cent between 2000 and 2020; and in low- and middle-income countries the number of persons killed in traffic crashes is estimated to increase by as much as 80 per cent (Peden et al. 2004). According to the WHO, road traffic crashes were the ninth most common cause of death in 2004. If the trend continues, they will be the fifth most common cause of death by 2030 (WHO 2009). Despite this grim projection for road traffic crashes at the global level, there are major regional differences. For example, in high-income countries in Europe, the total number of fatalities caused by traffic crashes is the lowest in the world.

In the road transport system, consisting of users, vehicles and environment (Haddon 1980), the design of, and interaction between, these three factors determine the safety level. Vehicle safety is an area that is garnering more attention in countries with a low share of traffic fatalities and there are high ambitions for it in the field of road traffic safety. A good example is Sweden with its 'Vision Zero' policy (Belin, Tillgren and Vedung 2011). According to the European Commission, safe vehicles are a key component in creating a safe road transport network. Vehicle design is important for all road users and encompasses both crash-preventing properties and passive-damage-reducing properties in case of a crash (European Commission 2009). Electronic Stability Control, Warning and Emergency Braking Systems, Lane Support Systems, Speed Alert system are all, among others, examples of promising new vehicle safety technology (iCarSupport database 2012).

Safe designs of future vehicles are ultimately determined by the companies that produce them. For companies to be able to sell their products there needs to be a demand among consumers. Therefore, consumer demand and companies' estimation of this demand greatly influence the vehicles that are available for purchase.

Another important stakeholder is the state. The role of the state, particularly whether, when and how to intervene in the market and affairs of society, is a long-standing issue in policy research and politics (Parsons 1995). With respect to road safety, the state has various tools that it has used in other sectors that it can employ to influence the market for safe vehicles. The above relationship can be described according to Figure 16.1. In this chapter the focus will be on the first box in Figure 16.1. The purpose of this chapter is to examine the types of government actions that are being taken to influence the market for safe vehicles and how the public strategies have changed over time. This chapter is mainly descriptive and therefore will not deeply discuss which kind of policy instruments and strategies are most effective in influencing social acceptance of safe vehicle technology.

Tools of Governance: Carrots, Sticks and Sermons

Road safety research tends to focus on the effectiveness of specific measures such as helmets, seat belts and speed cameras (see, e.g., Elvik and Vaa 2004 for a summary of studies). However, behind these measures is the state, whose

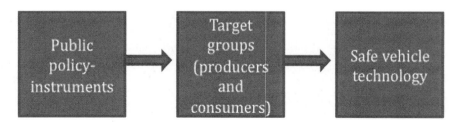

Figure 16.1   The process of influencing the market for safe vehicles

contribution ranges from formulating policy through to enforcing traffic laws to providing resources needed for road safety measures. The state carries out its role in road safety through a number of policy instruments. Literature on state instruments shows that the state has a number of policy instruments at its disposal, which can be used to ensure support and effect or prevent social change (Vedung 1998). These instruments can be classified in a number of different ways but, according to Vedung (1997 and 1998), they can be put into three main categories:

- Sticks (regulations);
- Carrots (economic means); and
- Sermons (information).

The state can either force us to do what it wants, reward us or charge us materially for doing it, or preach to us what we should do. Simply put, governments may variously use singly or in combination, the stick, the carrot or the sermon. The basis of this trifold division is the level of authority in relation to the target group. Regulations (sticks) are coercive for the target group subjected to them. Economic instruments (carrots) involve either distributing or taking away material resources. While economic instruments may be extremely controlling for the target group, there is still, in theory, a possibility for the target group not to follow the state's intentions, at the risk of being materially affected. In the case of information (sermon), the state disseminates knowledge, arguments, advice, encouraging talk and other immaterial (symbolic) assets to the target group. The target group is neither forced nor will it suffer any financial loss or enjoy any rewards if it pays heed to and follows the thrust of state information (Vedung 1997 and 1998).

In this chapter, the focus will be primarily on the role the state can play in promoting adoption of road vehicle safety technology. The basic scenario is that technology which might enhance road users' safety exists, but is not available on the market, will not reach the market quickly enough or will only be available to a small number of road users who can afford to pay for it. We shall regard this as 'failed market introduction'. Depending on the analysis of the failed market introduction, the state can choose between three strategies:

1. The first strategy is based on the assumption that the failed market introduction is due to the fact that the automotive industry, striving to maximise financial gain, estimated that there would be no demand for cars with this safety technology installed even though it would benefit the greater community of road users. The state can then force the automotive industry to equip the vehicles with the desired technology by introducing a regulation.
2. The second strategy is based on the automotive industry estimation that there is a certain demand but that the new safety technology will be too expensive to introduce, or that the technology will result in too small a profit. The state can then in various ways manipulate the market to reduce the costs or increase the profit for the automotive industry by introducing economic means.

3. In the third scenario, there is a strong potential demand and financially viable technology, but knowledge among consumers is too low. The state can then, in various ways, inform and convince both the automotive industry and consumers about positive aspects of the technology. In Figure 16.2 these three main strategies are illustrated.

Figure 16.2   Three basic tools of governance for promoting vehicle safety technology

The following sections will provide examples of how states utilise the three governance tools to stimulate the supply of certain safety technologies.

The First Strategy: Regulations and Vehicle Safety

On 9 September 1966, then-president of the United States Lyndon B. Johnson signed both the National Traffic and Motor Vehicle Safety Act and the Highway Safety Act. A federal road traffic safety institution was also established. This unique event meant that the new federal authority was given, from a contemporary perspective, a very progressive mandate to govern the automotive industry (Graham 1989). It was nothing less than a comprehensive shift in public policy, which meant that the American state abandoned a non-interventionist strategy with regards to the automotive industry in favour of an interventionist strategy (Vedung 1998:2f). A policy change of this magnitude is mostly the result of a very complex process involving many stakeholders. According to Graham (1989), the policy change can be explained in part by the automotive industry's inability to play the political game to combat federal legislation, and in part by a very strong consumer rights movement led by Ralph Nader. This policy change proved of great significance not only to road traffic safety work in the United States;

it affected other highly motorised countries such as Sweden, which also passed national vehicle legislation and, in 1968, established a national road traffic safety agency, *Statens Trafiksäkerhetsverk*. According to Stone (1982), this was also the start of an era of general consumer protection where the American Congress passed a number of laws and many federal authorities were established, such as the Environmental Protection Agency (EPA), Occupational Safety and Health Administration (OSHA) and so on. Public and government trust in the ability of companies and the market to satisfy people's safety demands was extremely low. According to Stone (1982), the federal regulation of the automotive industry was justified primarily on grounds of equity. Consumers had information deficit in relation to the automotive industry and the regulation was intended to force the automotive industry to develop and deliver vehicles that met at least one set of minimum safety requirements.

By the end of the 1970s, there were strong views against too much state intervention. The downside of an overly interventionist strategy in the form of increased bureaucracy and higher costs were emphasised and deregulation became an important watchword for Ronald Reagan when he took over as president in 1981 (Stone 1982). It wasn't just increasing costs or bureaucracy that motivated an increasing anti-regulation view. The state's ambitions to regulate the automotive industry seemed to also bring unwanted and perverse side effects (Vedung 1997), which meant that regulation essentially reduced consumers' access to safety technology rather than increasing it. According to Graham (1989), the coercive strategy that was employed when it came to passive safety systems was most likely counterproductive and delayed the introduction of airbags. Regulation in itself created strong resistance from the automotive industry, which in turn made development and the introduction of new safety technology harder.

The Second Strategy: Economic Instruments and Vehicle Safety

Economic policy instruments exist in many different forms. There is an important difference between monetary and non-monetary economic governance tools. Just like regulations, economic tools can be framed positively as incentives or negatively as disincentives (Vedung 1997 and 1998).

The state can use a number of different economic instruments to affect the introduction of safety technology in the market. In this section we will look at two examples, applied in different areas: namely, publically funded large-scale demonstration projects and public procurement.

Publically Funded Large-Scale Demonstration Projects

Large-scale demonstration projects are commonly used by the state to reduce costs and increase the utility for companies that wish to introduce new technology to the market. For example, during the period 1999–2002, the Swedish government conducted a large-scale attempt to introduce Intelligent Speed Adaptation systems (ISA) in four Swedish municipalities, at a cost of approximately US $8.4 million. A total of 10,000 drivers tested various pieces of equipment and despite it being primarily a research project it was also a step in the broader implementation of ISA systems in Sweden (Svedlund, Belin and Lie 2009). This demonstration project is an example of where the government creates the financial and other conditions for companies to showcase their new technologies. Another example of a large-scale demonstration project aimed at among other things, easing the introduction of new safety technology to the market, is the European Commission's financing of and participation in FOT-NET (FOT-NET 2012). FOT-NET are large-scale testing programs aiming at a comprehensive assessment of the efficiency, quality, robustness and acceptance of ICT solutions used for smarter, safer, cleaner and more comfortable transport solutions.

Public Procurement

Public stakeholders in many direct and indirect ways, affect the safety level of the road transport system. Public procurement has for many years been used by governments as a way of reducing environmental impacts on society. It has been used to reach environmental goals and is also used to research vehicle safety and meet road traffic safety goals. The Monash University Accident Research Centre in Melbourne, Australia, has systematically identified a number of activities that public actors perform to improve a vehicle's safety (Haworth, Tingvall and Kowadlo 2000). For example, what was then the Swedish Road Administration, introduced a travel policy that meant that internal safety requirements, in addition to contemporary legislation, were placed on vehicles that were purchased and used for work-related travel. Such a demand could be that all vehicles – company cars, private vehicles and hired vehicles – should be equipped with a specific piece of safety equipment. The Swedish Road Administration's travel policy spread to other public, nonprofit and private organisations and so, had an indirect effect on the overall market. Thus, public procurement aims to create a demand for safety technology by creating economic incentives for companies to provide vehicles with a high safety standard.

The Third Strategy: Information and Vehicle Safety

The third policy instrument strategy aims to exert influence through persuasion and information. Economic and informative means of control are similar in that, unlike regulation, they lack the element of coercion. The recipients are therefore free to follow or ignore the recommendations as they see fit. The target group neither benefits materially nor risks suffering any material losses due to their actions. All that is offered is arguments.

Vehicle Rating Program

As an alternative to government regulation, and as a response to the criticism levelled against the regulation strategy, the National Highway Traffic Safety Administration (NHTSA) launched their 1979 crash test assessment program for new vehicles. The New Car Assessment Program (NCAP) aimed to improve passenger safety through vehicle testing and evaluation in accordance with predetermined safety criteria. The results of these tests are disseminated to consumers who can then compare the safety levels of different vehicles. The aim for this sort of rating information is to encourage the automotive industry to voluntarily improve the safety of their cars (US Department of Transportation 2007). Today, there are similar programs in Europe, Japan and Australia among other places (McIntosh 2008). Although vehicle rating programs mainly can be seen as consumer information aimed at potential buyers and as a way of making their choice of safe cars easier, the programs also, to a large degree, aim to directly influence the automotive industry and increase their incentives to continuously improve the safety standards of their vehicles. The hope is that vehicles that do well in the tests and score the highest marks will also sell better in the market.

Dissemination of Scientific Results

Electronic stability control systems (ESC) have proven to be very efficient pieces of road traffic safety technology with great potential to reduce road trauma. The ESC market penetration in Sweden increased from 15 per cent to 90 per cent in 48 months (Krafft et al. 2009). A very quick diffusion process of an innovation (Rogers 1983) can be described in the steps shown in Figure 16.3: the ECSs effects are studied using scientific methods and the results are actively disseminated by governmental officials through the media; important purchasers partake of the information and some key stakeholders express their intentions of only purchasing vehicles that are equipped with ESC; and importers and producers then offer ESC as standard equipment. This is a good example of how factual information affects consumer demand in the market, which in turn affects the supply of safety technology.

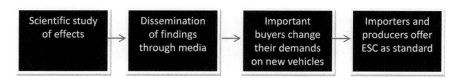

Figure 16.3   A process of influencing car importers and producers to install ESC as standard equipment

Vertical, Horizontal and Chronological Packaging Policy Instruments

Different means of control, strategies and their deployment are rarely implemented in a pure form. Most of the time, the different means of control and strategies are combined into vertical, horizontal or chronological packages. *Vertical packaging* is when a means of control is aimed at one set of actors in order to affect them to perform, in their turn, an action that affects the final target group. *Horizontal packaging* occurs when two or more means of control are aimed simultaneously at the same target group to reach the same end target. *Chronological packaging* occurs when various policy instruments are applied in certain time sequences (Bemelmans-Videc and Vedung 1998).

Packaging of regulatory, economic and informational policy instruments to affect supply and demand of road vehicle safety technology is very relevant. An example of *vertical packaging* could be the Swedish government's regulation that places requirements on government agencies in their procurement of vehicles and road transportation services. From 1 February 2009, government agencies were required (regulation) to purchase only vehicles and road transportation services that meet high environmental and road traffic safety requirements. Among other things, cars that are procured or hired by a government agency should be equipped with electronic stability control systems. This chain of influence thus leads from regulation to an economic policy instrument – namely procurement – which in turn affects the automotive industry into producing vehicles with electronic stability control systems.

Examples of *horizontal packaging* could be when the government simultaneously procures and provides information about the positive effects of electronic stability control systems. Finally, an example of *chronological packaging* could be when the government increases the market share of vehicles equipped with electronic stability control systems by providing information about their performance and, then at a later point in time, after the market share has increased voluntarily, regulate the technology. (The EU has proposed to make ESC compulsory for new cars from 2014 onward.)

Conclusion

In this chapter, different policy instruments and strategies to influence the market with safety technology in road vehicles have been analysed. In many high-income countries the number of fatalities has reached a relatively low level. Due to very demanding road safety policies such as Vision Zero, great emphasis is often placed on the use of new safety technology. Although the use of regulations are still available as a policy option for governments to promote the implementation of new vehicle safety systems, this mechanism has, in a global market, become increasingly difficult to apply. Therefore, the modern state needs to develop innovative strategies and mechanisms to amplify as well as promote societal acceptance of new traffic safety technology.

When, where and within what cultural context is a specific policy instrument or a package of several policy instruments most effective and appropriate? To answer that question, further research is needed. For example, as Belin et al. (2010) have shown in a comparative study of two different speed camera systems, even if both systems technically have the same aim – to reduce speeding – the ideas on how that should be achieved differ substantially. Interventions are based either implicitly or explicitly on theories about the way in which the interventions are supposed to work (Hoogerwerf 1990, Schneider and Ingram 1990, Vedung 1997, Rossi, Lipsey and Freeman 2004). Design and choice of policy instruments can be expected to vary with the background, roles and cognitive orientations of policymakers, as well as with contextual factors that have historically influenced their views of the instruments (Linder and Peters 1989). Therefore, the choice of different policy instruments and package strategies to influence road vehicle safety is complex and not simply a matter of choosing the most effective instrument.

References

Belin, M-Å., Tillgren P., Vedung, E., Cameron, M. and Tingvall, C. 2010. Speed cameras in Sweden and Victoria, Australia: A case study. *Accident Analysis and Prevention*, 42(6): 2165–70.

Belin, M.-Å., Tillgren, P. and Vedung, E. 2011. Vision Zero: A road safety policy innovation. *International Journal of Injury Control and Safety Promotion*, 1–9.

Bemelmans-Videc, M-L. and Vedung, E. 1998. Conclusions: Policy instruments types, packages, choices, and evaluation'. In *Carrots, Sticks and Sermons: Policy Instruments and Their Evaluation*, 249–73. Edited by M-L. Bemelmans-Videc, R.C. Rist, and E. Vedung. New Brunswick, NJ: Transaction Publishers:

Elvik, R. and Vaa, T. 2004. *The Handbook of Road Safety Measures*. Amsterdam and New York: Elsevier.

European Commission, Directorate-General Transport and Energy. 2009. eSafety. Available at: http://ec.europa.eu/transport/road_safety/specialist/knowledge/pdf/esafety.pdf.

FOT-NET. 2012. *Networking for Field Operational Tests Project*. Available at: http://www.fot-net.eu/en/welcome_to_fot-net.htm [accessed 6 December 2012]. Brussels: European Commission. Graham, J.D. 1989. *Auto safety: Assessing America's Performance*. Boston, Auburn House Publishing Company.

Haddon, W. 1980. Advance in the epidemiology of injuries as a basis for public-policy. *Public Health Reports*, 95(5): 411–21.

Haworth, N.,Tingvall, C. and Kowadlo, N. 2000. *Review of Best Practice Road Safety Initiatives in the Corporate and/or Business Environment*. Melbourne: Monash University Accident Research Centre.

Hoogerwerf, A. 1990. Reconstructing policy theory. *Evaluation and Program Planning*, 13: 285–91.

iCarSupport Database. 2012. Available at: http://www.esafety-effects-database. org/index.html [accessed 6 December 2012].

Krafft, M., Kullgren, A., Lie, A. and Tingvall, C. 2009. *From 15% to 90% ESC Penetration in New Cars in 48 Months: The swedish experience.* Twenty-first International Technical Conference on the Enhanced Safety of Vehicles. Stuttgart, Germany: US National Highway Traffic Safety Administration.

Linder, S.H. and Peters, B.G. 1989. Instruments of government: Perceptions and contexts. *Journal of Public Policy*, 9(1): 35–58.

McIntosh, L. 2008. Where do car safety assessment programs fit in with a 'Vision Zero' road safety system? 27 August. Available at: http://www.ors.wa.gov.au/ Documents/Conferences/conference-mcintosh-2008.aspx.

Parsons, D.W. 1995. *Public Policy: An Introduction to the Theory and Practice of Policy Analysis.* Brookfield, VT: Edward Elgar.

Peden, M., Scurfield, R., Sleet, D., Mohan, D., Hyder, A. and Jarawan, E. 2004. *World Report on Road Traffic Injury Prevention.* Geneva: World Health Organization.

Rogers, E.M. 1983. *Diffusion of Innovations.* New York: Free Press.

Rossi, P.H., Lipsey, M.W. and Freeman, H.E. 2004. *Evaluation: A Systematic Approach.* Thousand Oaks, CA: Sage.

Schneider, A. and Ingram, H. 1990. Behavioral assumptions of policy tools. *Journal of Politics* 52(2): 510–29.

Stone, A. 1982. *Regulation and Its Alternatives.* Washington, DC: Congressional Quarterly Press.

Svedlund, J., Belin, M-Å. and Lie, A. 2009. *ISA Implementation in Sweden: From Research to Reality.* Intelligent Speed Adaptation Conference 2009. Sydney: New South Wales Centre for Road Safety.

US Department of Transportation. 2007. *The New Car Assessment Program Suggested Approaches for Future Program Enhancements.* Washington, DC: US National Highway Traffic Safety Administration.

Vedung, E. 1997. *Public Policy and Program Evaluation.* New Brunswick, NJ: Transaction Publishers.

———. 1998. Policy instruments: Typologies and theories. In *Carrots, Sticks and Sermons: Policy Instruments and Their Evaluation*, 2158. Edited by M-L. Bemelmans-Videc, R.C. Rist and E. Vedung. New Brunswick, NJ: Transaction Publishers.

World Health Organization (WHO). 2009. *Global Status Report on Road Safety: Time for Action.* Geneva: WHO.

PART V
Optimising Driver Acceptance

Chapter 17

Designing In-Vehicle Technology for Usability

Alan Stevens
Transport Research Laboratory, UK[1]

Gary Burnett
Human Factors Research Group, Faculty of Engineering,
University of Nottingham, Nottingham, UK

Abstract

Usability of in-vehicle technology is a key contributor to drivers' acceptance of it. This chapter focuses on usability, including how usability is defined, how it can be measured and how it can be enhanced through design. The chapter describes a range of international regulations and design guidelines for information systems, warning systems and assistance systems that attempt to promote usability by incorporating best practice, both in design and in the design process. Although the technique chosen, the equipment used and the testing environment need to be carefully chosen depending on the in-vehicle system and the evaluation question being addressed, it is concluded that usability is a key contributor to drivers' acceptance of in-vehicle technology and that it can be measured.

Introduction and Scope

The Technology Acceptance Model (Davis, Bagozzi and Warshaw 1989), describes how perceived usefulness and ease of use are the main determinants of attitude towards a technology, which in turn predicts behavioural intention to use and, ultimately, actual system use. In this chapter we shall define usability and relate it to the concepts of acceptance within the Technology Acceptance Model before describing how usability can be measured and enhanced through design.

With the profusion of information and entertainment options available to drivers, the modern car has been described as 'a SmartPhone on wheels' (e.g., Toyota 2011). Information may be presented both in-vehicle and externally and needs to be relevant, timely, consistent and useful. The challenge is to provide

1   © Transport Research Laboratory, 2013

the information and services demanded by drivers that are usable without causing unsafe distraction and overload.

As well as information and entertainment, in-vehicle sensor, communications and processing technology can assist the driver by providing advice and warnings concerning the vehicle's immediate environment. Such warnings have to be perceived, understood as relevant and acted upon appropriately, if they are to be effective (Wogalter 2006). Issues such as perceived false alarm rate also have to be carefully considered through user-centred design to ensure usability and promote driver acceptance (see Chapter 9 by Jan-Erik Källhammer and colleagues earlier in this book for further discussion about warnings and false alarms).

With driver error consistently identified as a contributory factor in more than 90 per cent of vehicle crashes (Treat et al. 1977), vehicle designers are now offering systems that provide automation of specific elements of the driving task. Systems can even be designed to intervene in vehicle control to avoid or mitigate an impending collision. Nevertheless, usability issues around how the vehicle 'feels' and responds and how control is partitioned between the vehicle and the driver are crucial to achieving driver trust and acceptance.

Usability

Usability as a concept arose in the late 1970s and early 1980s as desktop computers emerged with graphical user-interfaces designed for the mass market. A range of definitions has been proposed in subsequent years, which depends predominantly on whether usability is viewed as a *property of a product/system* or an *outcome of use* (Bevan 2001). This point is specifically made in the International Organization for Standardization (ISO)/International Electrotechnical Commission (IEC) 25010 (2011: 12), which considers software quality:

> Usability can either be specified or measured as a product quality characteristic in terms of its subcharacteristics, or specified or measured directly by measures that are a subset of quality in use.

Arguably, there has been greater impact from definitions of usability that consider the outcomes that emerge from user-system interaction. In this respect, the most well-known and utilised definition is given in ISO 9241 (1998: 2), where usability is defined as

> The extent to which a product can be used by specified users to achieve specified goals with effectiveness, efficiency and satisfaction in a specified context of use

Effectiveness is essentially about whether tasks are achieved or not with a product and this, in turn, largely depends on the extent to which a product does what it was designed to do. For simple systems/functions this can be a black/white (yes/no)

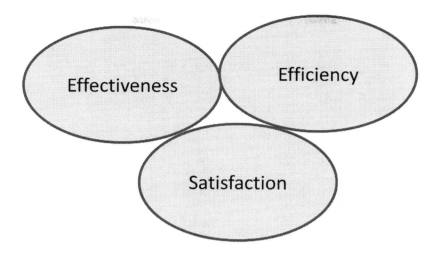

Figure 17.1   Usability Components (concept from ISO 9241 1998)

issue. For example, if a goal is to turn on an in-car entertainment system, then we may consider whether this goal was achieved or not. For more complex tasks, it may be better to think of the degree of success, as there may be partial successes. For example, with the task 'plan a route' with a navigation system, a user may be able to find and open the navigation function, enter a destination, select a route and so on, but not be able to view the complete route on a map prior to starting a journey. Although this may not be critical to achieving a planned route, the inability to preview it may impact acceptance of the navigation system for some users.

It is important to note that two systems could have the same effectiveness (i.e., can achieve goals in both), but the 'cost' to the user may be very different between the systems. For example, it could take much longer with one than the other or be more demanding physically and mentally. The usability factor, *efficiency*, considers these resources required to achieve a task. Reduced efficiency can be brought about by 'deviations from critical path', reflecting the fact that most tasks have a critical path for performance – that is, a method that requires the least steps/effort. Any deviations from this path will make the user's performance with a system less efficient. Errors are clearly of significance to efficiency, particularly when considering the safety-critical driving situation where the consequences of error may be significant.

Whereas the concepts of effectiveness and efficiency are essentially objective criteria, *satisfaction,* the third factor in the ISO 9241 definition, is largely a subjective viewpoint on usability, and is where questionnaires and interview-based techniques can be particularly useful. For instance, an evaluation team may ask questions such as 'how easy/difficult was it to use?', 'Was it enjoyable to use?', 'Were any aspects annoying?', 'What features were liked/disliked?' and so on.

Several authors have noted that satisfaction is too restrictive as a criterion for success in modern computing product design (e.g., Rogers, Sharp and Preece 2011, Jordan 2000). One may think of products, such as computer games, apps on smartphones, as well as in-vehicle entertainment systems which could be described as 'engaging', 'entertaining', 'fun', 'sociable', 'exciting' and so on. Satisfaction was conceived as a criterion for usability when software and hardware was generally considered in a work (office) context. In contrast, modern computing products are pervasive within everyday life situations, including the situation of driving a vehicle equipped with considerable computing and communications power. Consequently, a broader range of concepts become important relating to the affective needs of users, which we can think of as an extension of the satisfaction factor. Two key labels are commonly used to describe this wider view of usability: emotional design (Norman 2004) and pleasure-based design (Jordan 2000).

None of the three elements of usability within the ISO 9241 definition have a dimension of time or exposure to the system; however, this shortcoming is addressed in the work of Jordan (1998) which introduced five higher order components for usability:

- Guessability (the ability to predict without full information) is particularly important for products which have a high proportion of one-off users, for example, a hired car, or products with a number of rarely used functions. Poor guessability can put people off, and may have safety implications even if they're actually easy to use with practice.
- Learnability concerns the costs (time, effort, etc.) to a user in reaching a competent level of performance. This will be important if training time is short, or if a user is to be self-taught, as is often the case with vehicle systems.
- Experienced User Potential (EUP) is the performance of someone who has considerable experience with a product – the expert user. In other words, the likely efficiency of the interface for a proficient user (with efficiency one element of usability). This is important if a high level of knowledge (breadth and depth) and/or skills is needed, and training/time to reach it is not a significant issue, for example, basic driving skills.
- System potential is the maximum performance theoretically possible with the system – and is essentially an upper limit on EUP; for example, the minimum number of key presses required to achieve a task. It is important if it is limiting EUP (and EUP is important). For example, it may be that a user has to go through set key-presses. Shortcut options (e.g., through a command-based speech system) can raise the system potential and hence EUP if users are made aware of them and can easily access/remember them.
- Re-usability (or memorability) is the decrement in performance following a period of time away from a product – a user may forget what a function does or how to access it and so on. It is an important aspect of usability when a product or product functions are likely to be used in intermittent bursts; for example, a navigation system being used on holiday.

Although desktop computers were the subject of initial usability studies, this chapter concerns usability in the context of the driving environment where the system being examined may not be the primary focus of attention. This specific context of application was the subject of the ISO standard (ISO 17287: 2003: 5) which defined a new concept 'suitability' as 'the degree to which a [system] is appropriate in the context of the driving environment based on compatibility with the primary driving task'. Suitability focuses on two elements of product use already discussed above as important in usability: efficiency, and ease of use while learning about a new system. Moreover, the concept of suitability introduces two new elements specifically related to the driving context:

- Controllability (essentially, the effectiveness in the driving context); and
- Interference (with the driving task).

The international standard on Suitability (ISO 17287 2003) also describes a process for assessing whether a specific in-vehicle technology system or a combination of systems with other in-vehicle systems is suitable for use by drivers while driving.

Usability and Acceptability

A further perspective on usability is given by Nielsen (1993) in which it is defined in terms of five key attributes for products: learnability, efficiency, memorability, errors and satisfaction. This breakdown takes quite a narrow stance on usability, but is of particular interest here as usability is explicated in terms of broader criteria, including acceptability. Specifically, Nielsen believes the overall consideration is system acceptability (the extent to which requirements are met, both social and practical). For practical acceptability, several criteria are of relevance, including cost, reliability, compatibility and usefulness. This latter factor of usefulness is analogous to effectiveness in the ISO 9241 definition and comprises the utility of the product/system *and* the usability. The distinction between these constructs is important to any view on usability. As an example, a car without lights would be considered to be unusable according to the ISO definitions, but lacking in utility according to Nielsen.

Some key implications emerge from these alternative definitions for vehicle-based technologies and issues of acceptance. To understand usability we must specify our users and consequently understand the relevant characteristics (driving experience, technology experience, expectations and so on). We must also understand what users wish to achieve with an in-vehicle product and consider in detail the physical, social and potential organisational environment in which tasks are carried out. Without this knowledge, we cannot make any statements about whether a product is usable or more or less usable than another product.

To explore further the relationship between usability and acceptability, it is useful to consider the Technology Acceptance Model (TAM – Davis et al. 1989). This describes how perceived usefulness and ease of use are the main determinants of attitude towards a technology, which in turn predicts intention to use and ultimately, actual system use. Now, from the ISO 9241 definition of usability, there are three elements: effectiveness, efficiency and satisfaction. It is immediately clear that TAM's Usefulness aligns with the Effectiveness element of usability and TAM's Ease of Use aligns with the Efficiency element of usability. Satisfaction, the third element in the ISO 9241 definition, is likely to contribute to the perception aspects of both Usefulness and Ease of Use or (alternatively) could be regarded as a supplementary factor in an enhanced model directly influencing Intention to use and Actual System Use. So from this, it is clear that the three usability factors in the ISO 9241 definition will each have a direct impact on the overall acceptance of an in-vehicle system.

Design Guidelines

Introduction

The previous section has demonstrated how usability (and related concepts) directly influence acceptance of new technology and discussed the important role of the driver interface of those systems in determining usability.

This section reviews a range of standards and guidelines that are available to product designers that aim to promote safety and usability in the driving context. Although good design advice cannot guarantee usability or acceptance, its use is likely to lead to more usable interfaces and hence help produce technologies that are more acceptable.

International Regulations and Standards

A considerable volume of international regulation exists in relation to design requirements for motor vehicles that aims to ensure that technology within vehicles can be used safely. The Vienna Convention on Road Traffic (Vienna Convention 1968), for example, is an international treaty designed to facilitate international road traffic and to increase road safety by standardising uniform traffic rules. One of the most quoted extracts is the requirement that 'Every driver shall at all times be able to control his vehicle'. The United Nations Economic Commission for Europe (UNECE) Transport Division provides secretariat services to the World Forum for Harmonization of Vehicle Regulations. The World Forum, through its permanent Working Party 29 (WP29) provides the regulatory framework for technological innovations in vehicles to make them safer and to improve their environmental performance (UNECE 2012).

Although not legally binding, international standards provide process, design and performance advice and the following ISO groups are working in areas relevant to vehicle design and usability:

- ISO TC 22 SC13 WG8 covering basic standards for Human Factors design of in-vehicle systems;
- ISO TC 204 WG14 concerning vehicle and cooperative services (and some interface issues) including, for example, Lane Departure Warning and automatic Emergency Braking Systems; and
- ISO TC 204 WG17 concerning nomadic and portable devices for ITS services.

European Regulations

During 2010 the European Commission (EC) published a study on the regulatory situation in the Member States regarding brought-in (i.e., nomadic) devices and their use in vehicles, which highlighted the diversity of approaches across member states (European Commission 2010).

At the end of 2008, the European Commission published an action plan followed by a directive in 2010 (European Commission 2010), which has provisions for the development of specifications and standards for ITS road safety including HMI and the use of nomadic devices.

US Regulations

In the US, laws about in-vehicle distraction generally fall under the jurisdiction of individual states but with some at the national (federal) level. As an example of state provision, the State of Nevada passed a law in June 2011 concerning the operation of driverless (fully automated) cars whereby the Nevada Department of Motor Vehicles is responsible for setting safety and performance standards and for designating areas where driverless cars may be tested.

As an example of national provision, in October 2009 President Obama issued an executive order prohibiting federal employees from texting while driving. This order is specific to employees' use of government-owned vehicles or privately owned vehicles while on official government business and includes texting-while-driving using wireless electronic devices supplied by the government.

Design Guidelines for Information and Communication Systems

Europe: European Statement of Principles
The European Commission (EC) has supported the development of a document called the 'European Statement of Principles on HMI' (referred to as ESoP) which provides high-level HMI design advice (EC 2008). As an EC recommendation, it has the status of a recommended practice or code of practice (CoP) for use in Europe. The EC recommendation also contains 16 recommendations for safe use (RSU), which build on health and safety legislation by emphasising the responsibility of organisations that employ drivers to attend to HMI aspects of

their workplace. Adherence to the RSU is likely to promote greater acceptance of technology by drivers.

The design guidelines-part of the ESoP comprises 34 principles to ensure safe operation while driving. These are grouped into the following areas: Overall Design Principles, Installation Principles, Information Principles, Interactions with Controls and Displays Principles, System Behaviour Principles and Information about the System Principles.

United States: Alliance and NHTSA

US motor vehicle manufacturers have developed 'Alliance Guidelines' that cover similar, high-level, design principles to the ESoP. The guidelines (Auto Alliance 2006) consist of 24 principles organised into five groups: Installation Principles, Information Presentation Principles, Principles on Interactions with Displays/ Controls, System Behaviour Principles and Principles on Information about the System.

The US National Highway Transportation Safety Administration (NHTSA) has worked with the auto industry and the cell phone industry to develop a set of guidelines for visual-manual interfaces for in-vehicle technologies. These are based on the ESoP/Alliance guidelines and introduce some specific assessment procedures (NHTSA 2013).

The NHTSA plans to publish guidelines for portable devices in 2013 and guidelines for voice interfaces by 2014. Another suggestion has been implementation of a 'car mode' on portable devices, similar to 'airplane mode'. The idea would be to disable certain functions when the vehicle is moving.

Japan: JAMA

The Japanese Auto Manufacturers Association (JAMA) Guidelines consist of four basic principles and 25 specific requirements that apply to the driver interface of each device to ensure safe operation while driving. Specific requirements are grouped into the following areas: Installation of Display Systems, Functions of Display Systems, Display System Operation While Vehicle in Motion and Presentation of Information to Users. Additionally, there are three annexes: Display Monitor Location, Content and Display of Visual Information While Vehicle in Motion and Operation of Display Monitors While Vehicle in Motion. There is, as well, one appendix: Operation of Display Monitors While Vehicle in Motion.

Warning Guidelines

Guidelines on establishing requirements for high-priority warning signals have been under development for more than five years by the UNECE-WP29's ITS Informal Group (Warning guidelines 2011).

There has also been work in standardisation groups to identify how to prioritise warnings when multiple messages need to be presented and one 'Technical specification' (TS) has been produced:

- ISO/TS 16951: Road Vehicles – Ergonomic aspects of transport information and control systems – Procedures for determining priority of on-board messages presented to drivers.

In addition, two Technical Reports are relevant that contain a mixture of general guidance information, where supported by technical consensus, and discussion of areas for further research:

- ISO/PDTR 16352: Road Vehicles – Ergonomic aspects of transport information and control systems – MMI of warning systems in vehicles; and
- ISO/PDTR 12204: Road Vehicles – Ergonomic aspects of transport information and control systems – Introduction to integrating safety critical and time critical warning signals.

Driver Assistance System Guidelines

To help promote driver acceptance of Advanced Driver Assistance Systems (ADAS), a key issue is ensuring controllability. Controllability is determined by the possibility and driver's capability to perceive the criticality of a situation; the driver's capability to decide on appropriate countermeasures (e.g., overriding or switching off the system) and the driver's ability to perform any chosen countermeasures (taking account of the driver's reaction time, sensory-motor speed and accuracy). Drivers will expect controllability to exist in all their interactions with assistance systems

- during normal use within system limits;
- at and beyond system limits; and
- during and after system failures.

The European project RESPONSE has developed a code of practice for defining, designing and validating ADAS (Cotter, Hopkins and Wood 2007, ACEA 2009). The code describes current procedures used by the vehicle industry to develop safe ADAS with particular emphasis on the Human Factors requirements for 'controllability'.

Another European project, ADVISORS (Cotter et al. 2008), has attempted to integrate the RESPONSE code within a wider framework of user-centred design taking account of the usability of information, warning and assistance systems. There is also activity by the International Harmonized Research Activities – Intelligent Transport Systems (IHRA-ITS) Working Group to develop a set of high-level principles for the design of driver assistance systems (IHRA-ITS 2012).

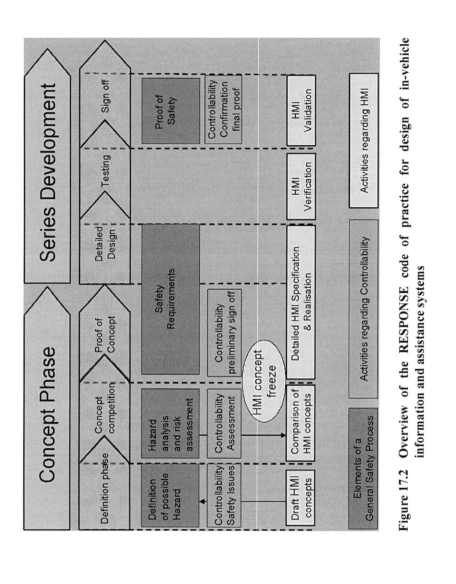

Figure 17.2 Overview of the **RESPONSE** code of practice for design of in-vehicle information and assistance systems

Methods Related to Usability Measurement and In-Vehicle Systems

A method for testing the usability of in-car systems can be seen to be a combination of three factors (Burnett 2009):

1. Which environment is the method used in (road, test track, simulator, laboratory, etc.)? Choosing an environment will be largely influenced by practical considerations, the knowledge/skills of the design/evaluation team and resource limitations. Fundamentally, there is often a trade off in choosing a method environment between the need for internal validity (control) and the ecological validity of results (Parkes 1991). For instance, road trials may have high ecological validity (we are confident that the phenomenon being observed does arise in the real world), but may have poor internal validity (we may not be able to understand clearly why such behaviour arises).
2. Which task manipulations occur (multiple task, single task loading, no tasks given, etc.)? In certain methods, there is an attempt to replicate or simulate the multiple task nature of driving. For other methods, performance and/ or behaviour on a single task may be assessed and the potential impact on other tasks is inferred from this. Other methods consider underlying opinions and attitudes (e.g., questionnaire surveys, interviews) or may not involve users, aiming instead to predict impacts or issues; for instance, through the use of expert ratings or modelling techniques.
3. Which dependent variables (operationalised as metrics) are of interest? Some will relate to drivers' performance with primary driving tasks (e.g., lane position, hazard detection) or their use of primary vehicle controls (e.g., use of brake, steering wheel). Other metrics focus on driver performance, the demand of secondary tasks (e.g., task times, errors, display glances) or various physiological parameters (ECG, EEG, EMG, etc.).

As noted by Rogers et al. (2011), in deciding on any method, the design team must consider the overall goals of the work, specific questions to be addressed, the practical and ethical issues and how data will need to be analysed and reported. By necessity, many bespoke methods (or at least specific versions of generic methods) are required that account for the particularly complex characteristics of the driving context. A recent US study (Ranney et al. 2011) assessed different methods (tests) of distraction potential in preparation for the NHTSA guidelines. Some well-known and commonly used methods which shed light on usability and hence driver acceptance, can be summarised:

Road trials – Drivers take part in a short-term (normally less than one day) focused study or a naturalistic long-term study (across many days/months). Participants may use a system in an instrumented car or their own vehicle on public roads (occasionally on test tracks). For such trials, a wide range of variables may be measured and analysed (e.g., visual behaviour, workload, vehicle control, subjective preference) depending on the aims of the study.

Simulator trials – Drivers take part in a short-term (normally less than one day) focused study using a system fitted or mocked up within a driving simulator. The faithfulness with which a simulator represents the driving task (known as its fidelity) can vary considerably.

Occlusion – This is a standardised laboratory-based method (ISO 2007) which focuses on the visual demand of in-vehicle systems. Participants carry out tasks with an in-vehicle system whilst wearing computer-controlled goggles that can open and shut in a precise manner. Consequently, by stipulating a cycle of vision for a short period of time (e.g., 1.5 seconds), followed by an occlusion interval (e.g., 1.5 seconds), glancing behaviour is mimicked in a controlled fashion (Stevens, Burnett and Horberry 2010).

Peripheral detection – This method requires drivers to carry out tasks with an in-car system (either on road or in a simulator) and to respond to changes within their periphery (e.g., the presence of lights or the modification of a shape for an object). The speed and accuracy of responses are considered to relate to the mental workload and distraction associated with secondary tasks. In a development of the method, some research has considered the potential for the use of a tactile detection task, where drivers respond to vibro-tactile stimulation (e.g., through the wrist or on the neck) whilst interacting with an in-vehicle system (Engström, Aberg and Johansson 2005, Diels 2011).

Lane change task – This standardised method (ISO 2010) occurs in a basic PC simulated environment in which drivers are requested to make various lane change manoeuvres whilst engaging with an in-vehicle system. The extent to which the profile of manoeuvre made by a driver varies from the optimum manoeuvre (the normative model) is considered to be a measure of the quality of their driving.

Keystroke Level Model (KLM) – The KLM method is a form of task analysis in which system tasks with a given user-interface are broken down into their underlying physical and mental operations; for example, pressing buttons, moving hand between controls and scanning for information. Time values are associated with each operator and summed to give a prediction of task times. In an extension of the KLM method, Pettitt, Burnett and Stevens (2007) have developed new rules that enable designers to develop predictions for a range of visual demand measures.

Conclusions

This chapter has discussed how usability can usefully be considered in terms of effectiveness, efficiency and satisfaction, all three of which contribute to a drivers' judgement of the acceptability of in-vehicle technology.

The chapter has reviewed a range of regulations, standards and design guidelines that aim to encourage better-designed in-vehicle technology that should also help to promote driver acceptance. Although basic Human Factors principles are established, the rapid development of in-vehicle technology presents a challenge for updating regulations and detailed design guidance.

Finally, the chapter has explored a range of methods through which usability can be evaluated. The technique chosen, the equipment used and the testing environment need to be carefully chosen depending on the in-vehicle system and the evaluation question being addressed. Nevertheless, it can be concluded that usability can be measured and that usability is a key contributor to drivers' acceptance of in-vehicle technology.

Acknowledgements

This chapter draws on previously published work for the European Commission. However, the content of this chapter is the responsibility of the authors and should not be construed to reflect the opinions or policies of any organisation.

References

ACEA (The European Automobile Manufacturers' Association). 2009. Available at: http://www.acea.be/images/uploads/files/20090831_Code_of_Practice_ADAS.pdf.

Auto Alliance. 2006. Available at: http://www.autoalliance.org/files/DriverFocus.pdf.

Bevan, N. 2001. International standards for HCI and usability. *International Journal of Human-Computer Systems*, 55(4): 533–52.

Burnett, G.E. 2009. On-the-move and in your car: An overview of HCI issues for in-car computing. *International Journal of Mobile Human-Computer Interaction*, 1(1): 60–78.

Cotter, S., Hopkin, J. and Wood, K. 2007. *A Code of Practice for Developing Advance Driver Assistance Systems: Final Report on Work in the RESPONSE 3 Project*. Wokingham, Berkshire: Transport Research Laboratory Ltd.

Cotter, S., Stevens, A., Mogilka, A. and Gelau, C. 2008. *Development of Innovative Methodologies to Evaluate ITS Safety and Usability: HUMANIST TF E.* Proceedings of European Conference on Human Centred Design for Intelligent Transport Systems, 55–65. Lyon, France, 3–4 April. Available at: http://www.conference.noehumanist.org/proceedings.html [accessed 13 April 2012].

Davis, F.D., Bagozzi, R.P. and Warshaw, P.R. 1989. User acceptance of computer technology: A comparison of two theoretical models. *Management Science*, 982–1003.

Diels, C. 2011. *Tactile Detection as a Real-Time Cognitive Workload Measure.* Proceedings of Institute of Ergonomics and Human Factors Annual Meeting.

Engström, J., Aberg, N. and Johansson, E. 2005. Comparison between Visual and Tactile Signal Detection Tasks Applied to the Safety Assessment of In-Vehicle Information Systems. Proceedings of the Third International Driving Symposium on Human Factors in Driver Assessment, Training and Vehicle Design. Iowa City: University of Iowa Public Policy Center.

European Commission. 2008. Commission Recommendation of 26 May 2008 on Safe and Efficient In-Vehicle Information and Communication Systems: Update of the European Statement of Principles on Human–Machine Interface. Available at: http://eur-lex.europa.eu/LexUriServ/LexUriServ.do?uri=OJ:L:20 08:216:0001:0042:EN:PDF. [accessed 30 April 2013].

European Commission. 2010. Study on the regulatory situation in the member states regarding brought-in (i.e., nomadic) devices and their use in vehicles. Available at: http://ec.europa.eu/information_society/activities/esafety/doc/studies/nomadic/final_report.pdf [accessed 30 April 2013].

European Commission Directive. 2010. http://eur-lex.europa.eu/LexUriServ/LexUriServ.do?uri=CELEX:32010L0040:EN:NOT [accessed 22 May 2013].

International Harmonized Research Activities - Intelligent Transport Systems (IHRA-ITS). 2012. Available at: http://www.unece.org/fileadmin/DAM/trans/doc/2012/wp29/ITS-20-05e.pdf [accessed 27 April 2013].

ISO. 2003. Road vehicles – Ergonomic aspects of transport information and control systems – Procedure for assessing suitability for use while driving.

ISO. 2007. Road vehicles – Ergonomic aspects of transport information and control systems – Occlusion method to assess visual distraction due to the use of in-vehicle information and communication systems. International Standard 17287.

ISO International Standard 16673. ISO. 2010. Road vehicles – Ergonomic aspects of transport information and control systems – Simulated lane change test to assess in-vehicle secondary task demand. International Standard 26022.

International Standardization Organization / International Electrotechnical Commission (ISO/IEC).

IEC. 1998. Ergonomic requirements for office work with visual display terminals (VDT)s – Part 11 Guidance on usability, 9241–11.

ISO/IEC. 2011. Systems and software engineering – Systems and software Quality Requirements and Evaluation (SQuaRE) – System and software quality models. ISO/IEC 25010:2011(E), 2011.

Jordan, P.W. 1998. *Introduction to Usability.* Taylor and Francis: London.

———. 2000. *Designing Pleasurable Products: The New Human Factors.* London: Taylor and Francis.

National Highway Traffic Safety Administration (NHTSA). 2013 Visual-manual NHTSA driver distraction guidelines for in-vehicle electronic devices. Available at: http://www.nhtsa.gov/About+NHTSA/Press+Releases/U.S.+DO T+Releases+Guidelines+to+Minimize+In-Vehicle+Distractions [accessed 30 April 2013].

Nielsen, J. 1993. *Usability Engineering*. San Diego, CA: Morgan Kaufmann.

Norman, D.A. 2004. Emotional Design: Why We Love (or Hate) Everyday Things. New York: Basic Books.

Parkes, A.M. 1991. Data capture techniques for RTI usability evaluation. In *Advanced Telematics in Road Transport: The DRIVE Conference, Vol. 2*, 1440–56. Amsterdam: Elsevier Science Publishers.

Pettitt, M.A., Burnett, G.E. and Stevens, A. 2007. *An Extended Keystroke Level Model (KLM) for Predicting the Visual Demand of In-Vehicle Information Systems*. Proceedings of the ACM Computer-Human Interaction (CHI) Conference. San Jose, May.

Ranney, T.A., Scott Baldwin, G.H., Parmer, E., Domeyer, J., Martin, J. and Mazzae, E.N. 2011. Developing a test to measure distraction potential of in-vehicle information system tasks in production vehicles. Report No. DOT HS 811 463 November.

Rogers, Y., Sharp, H. and Preece, J. 2011. *Interaction Design: Beyond Human-Computer Interaction*. Chichester, Sussex, UK: Wiley and Sons.

Stevens, A., Burnett, B. and Horberry, T.J. 2010. A reference level for assessing the acceptable visual demand of in-vehicle information systems. *Behaviour and Information Technology*, 29(5): 527–40.

Treat, J.R., Tumbas, N.S., McDonald, S.T., Shinar, D., Hume, R.D., Mayer, R.E., Stanisfer, R.L. and Castellan, N.J. 1977. *Tri-level Study of the Causes of Traffic Accidents*. Report No. DOT-HS-034-3-535-77 (TAC).

Toyota. 2011. Smartphone on wheels. Available at: http://media.toyota.co.uk/2011/11/toyota-fun-vii-a-smartphone-on-wheels/.

United Nations Economic Commission for Europe (UNECE). 2012. Available at: http://www.unece.org/trans/main/welcwp29.html.

Vienna Convention. 1968. Available at: http://www.unece.org/fileadmin/DAM/trans/conventn/crt1968e.pdf.

Warning guidelines from UNECE ITS informal Group. 2011. Available at: http://www.unece.org/fileadmin/DAM/trans/doc/2011/wp29/ITS-19-05e.pdf.

Wogalter, M.S. 2006. *Handbook of Warnings*. Routledge.

Chapter 18

The Emotional and Aesthetic Dimensions of Design: An Exploration of User Acceptance of Consumer Products and New Vehicle Technologies

William S. Green
University of Canberra, Australia

Patrick W. Jordan
University of Surrey, UK

Introduction

The technological 'level playing field' has been evident in consumer product design for more than two decades (Jordan 1997a), and even the continuous round of facelifts and upgrades to the electronics and mechatronics of our latest toys do little to differentiate the operability of one product from another. There are, arguably, no 'bad' cars. There are indeed some that will perform better than others, will last longer, will be smoother or quieter and so on, but the discrimination is often at the outer edges of the functionality envelope and the same is true of most of our consumer products. What then, are the design factors that determine the purchase impulse in the first place and the continued satisfaction with the product?

The first-century Roman architect, engineer and writer Vitruvius, in his magnum opus, *De Architectura*, identified three qualities for the critical appraisal of design. These were '*Firmitas*', '*Utilitas*' and '*Venustas*'. We may translate these as Structural Integrity, Usability and Beauty.

The first two are not very problematic. Does it work? Can we use it? There are reasonably objective metrics available for us to apply these criteria to almost any product. The third, however, has always been a slippery concept, with the common appraisal being captured in the old saying that 'beauty is in the eye of the beholder'. Unfortunately, this is not very useful for the designer or engineer charged with the task of specifying the next generation product, and indeed it is not completely accurate.

The problems facing a researcher or designer who relies on rationality are exemplified by the following excerpt from a conversation between one of the authors and a friend known for his extensive car ownership:

> When I buy a car I am absolutely meticulous. I read the test reports, the specifications, look at the safety record, the cost, the power, fuel consumption, reliability, size of the boot, even consult The Boss (referring to his partner); then I throw it away and go buy something that I like. (Anonymous 2012, in conversation with one of the authors)

What this 'like' means should be of fundamental concern to manufacturers, engineers, designers, marketers and indeed anyone involved in the production and sales of any product, because it encapsulates the entire aesthetic/emotional response to the product. The aesthetic dimension has been the subject of much discussion and debate from the early Greek philosophers (e.g., Plato, Aristotle) onwards, without ever reaching definitive conclusions, but nonetheless with the tacit understanding that there are generalisable qualities that are sufficiently robust to be useful. As an example, terms such as symmetry, balance and tension have a physical analogue that allows their application to the visual with an adequate general acceptance.

In recent years the scrutiny of the aesthetic qualities of products has been extended beyond the cerebral appreciation of form, colour, texture and so on. Current research is concerned with the emotional product experience, and this immediately amalgamates traditional aesthetic judgements with usability, reliability, longevity, value and cost. The basis of the techniques that are emerging for the assessment and formalisation of this scientifically difficult study, and some of the techniques themselves, are the subject of the following sections.

The Basics

The psychologist Abraham Maslow described what he called a 'hierarchy of human needs' (Maslow 1954). This model views the human as a 'wanting animal' that rarely reaches a state of complete satisfaction.

Indeed, if Nirvana is reached it will usually be temporary because once one desire has been fulfilled another will surface to take its place. Maslow's hierarchy is illustrated in Figure 18.1. The idea is that as soon as people have fulfilled needs lower down the hierarchy, they will turn their attention to things at the next level and look to meet those. For someone to reach a stable state of satisfaction, they would have to be able to fulfill the needs at the top of the hierarchy and all the needs underneath them on an ongoing basis.

A similar, albeit simpler, hierarchy has been proposed in the context of user needs as shown in Figure 18.2, and this relates directly to the fundamentals articulated by Vitruvius. At the bottom of the hierarchy we have the functions that the product offers the user. These need to work well and be sufficient to enable the user to achieve what they want with a product in order for it to have any value to them.

Top Level	Self-Actualisation Needs
	Esteem Needs
	Belongingness and Love Needs
	Safety Needs
Bottom Level	Physiological Needs

Figure 18.1   Maslow's hierarchy of needs (Maslow 1954)

Top Level	Pleasure
	Usability
Bottom Level	Functionality

Figure 18.2   Hierarchy of user needs (Jordan 1999)

The next level is usability. Once users know that the product offers them the functions they need, the next issue is how easy it is to use. They are likely to be discontented if functions take a lot of time and effort to use or if they struggle and make errors along the way.

Once this is sorted out, the next level is pleasure. This is a broad concept that we will discuss in more detail below, but essentially it includes things like positive emotions, good experiences and a positive self-image.

If the product also provides these, then we can expect the user to experience long-term satisfaction with it.

As an example, consider an entertainment product such as a television. In this case functionality would include things such as the size of the screen, the quality of the picture, the things that can be adjusted (e.g., volume, contrast, brightness), the facilities that are provided (HDMI, USB etc.) and the manufactured quality of the television.

Usability would include things such as how easy it is to adjust the functions, how easy it is to set the TV up and install the channels, how easy it is to navigate between functions and how easy it is to backtrack errors (the 'Oops' factor).

While pleasure in owning and using the television is likely to be heavily affected by the functionality and usability, there are also likely to be other factors that affect pleasurability and these vary with circumstance.

The perceived aesthetic qualities of the television are affected by its shape, colour, materials and design details: is it flat-screen or not, does it have a brushed metal finish or is it black plastic, are the speaker grills nicely detailed, does the remote control satisfy the user's visual and tactile expectations and so on?

The look and feel of the interface may also play a role – what are the graphics like on the menus, do the buttons on the remote control give a pleasing click when pressed, what do the buttons feel like to the touch? Haptic responses in particular become important in a control environment where visual referencing may be limited; for example, in a vehicle where directing attention to a console location may even be dangerous. In such a case the response impacts on safety and functionality as well as pleasure and satisfaction. An obvious reference is the move away from small and poorly differentiated buttons on car stereos to the current steering wheel mounted and line of sight controls.

Brand image and perceptions of status can be important. Marketing research has highly refined categorisations on the response of purchasers and users to the perceived prestige or otherwise, conferred by certain brands and the importance that varying demographics assign to them.

It must be acknowledged that the arcane world of fashion also exerts a considerable influence, but is highly complex and worthy of more consideration than is possible here. Those wishing to pursue this in the context of products could start with an essay by Jean Beaudrillard (Beaudrillard 1988) on the system of objects. Questions of status and fashionable acceptance are present at many levels of human experience but it is at the upper (pleasure) levels that they become significant. Concern with, for example, a colour being fashionable implies a level of security well beyond the need for food or shelter!

Models of the Pleasure Experience

There are a number of pleasure experience models. A good example is Norman (2004), who identifies three levels of cognitive processing at which we can experience pleasure.

Norman (2004)

Visceral
This is the most immediate level of processing. There are certain sensory aspects of a product that we perceive even before any significant level of interaction

has occurred. It gives us our instinctive first impressions of a product. It is predominantly visual but may also be olfactory and tactile.

As an example, consider the Juicy Salif citrus press designed by famous product designer Phillipe Stark for the Alessi company. The angular legs and bulbous body are visual characteristics reminiscent of a giant spider. Because spiders are frequently perceived as dangerous animals the product will immediately grab our attention, even in a cluttered kitchen.

In our previous example of the television, the size, screen type, colour, material texture and design details are the first to impact.

Behavioural

This level of processing occurs when using the product. It refers to the impression of a product that we have while we are interacting with it.

For example, when we are using products such as the Apple iPhone (and other current smartphones) we may enjoy the intuitive way they work and the bright and cheerful-looking icons on the interface.

Reflective

This level of processing is about conscious consideration and reflections on past experiences. It includes how we think about a product having had some experience of it, but when not actually using it.

For example, the Canon Ixus camera (known as the Canon Elph in some markets) had a design that was, at the time, radically different and arguably more aesthetically pleasing and interesting than other cameras on the market. This made it something of a talking point for users and also attracted attention from others. When people reflected on owning it and being seen with it, it generated a sense of pride.

Norman makes the point that pleasures at each of these three levels of processing can influence pleasure at other levels. For example, if something gives us a good impression right from the start (visceral), we may then be more inclined to feel positively about it when interacting with it (behavioural) and consequently when we think about it afterwards (reflective).

This is sometimes referred to as the 'halo' effect. It can work from any starting point; for example, if we have a great experience interacting with a product (behavioural) then this may make us think about it more positively afterwards (reflective) and have a more instinctively positive reaction the next time we see it (visceral).

The halo effect can also work in reverse; if we have a negative view of the product at any one of these stages, then we may also be inclined to take a more negative view at the other stages. Norman's model gives us an overview of the different levels at which a product can be pleasurable to own or use, but what about the different types of pleasure that a product can give?

Jordan (1999)

Based on the work of anthropologist Lionel Tiger in 1992, Patrick Jordan (1999) has identified four different types of positive experiences that we can get from products.

Physio-pleasures
These are to do with the body and the senses. They come from, for example, the visual, tactile, auditory and olfactory properties of a product. For example, the feeling in the hand of a mobile phone would come into this category or the smell of fresh coffee that comes from a coffee grinder.

Psycho-pleasures
These are to do with the mind, both cognitions and emotions. They include, for example, the pleasure of knowing how something works or the pleasure we take in finding something interesting. They include the feeling of positive emotions. For example, using a software package to produce creative imagery would come into this category as would the feeling of reassurance we get when we turn on an Apple computer and hear the resonant 'bong' when it boots up.

Socio-pleasure
This is to do with relationships. It includes both 'concrete' and 'abstract' relationships. Concrete relationships are those associated with specifically identifiable people, such as a friend, loved one or co-worker. For example, Skype brings social pleasure by allowing us to have video conversations with our friends and family at no cost.

Abstract relationships are those with society in general; they include things like status and how we are perceived by others. For example, if we wear an Armani suit it enhances our social status and may help us to be perceived as stylish. A cryptic remark by Robert Schuller sums up our concerns: 'I am not what I think I am. I am not what you think I am. I am what I think you think I am!' (Schuller 1982).

Ideo-pleasure
This is to do with our tastes and values. Tastes are generally just a matter of preference. So, for example, if we prefer blue to yellow, we may find a blue T-shirt more pleasurable to wear than a yellow one.

Values represent our morals and aspirations. If we are concerned about the environment, we may, for example, find locally sourced foods more pleasurable than foods that have been imported, thus incurring greater air miles. Meanwhile, if we aspire to be someone who is successful in their career, we may get pleasure from products associated with being a high-flyer in the business world.

Applying the Models to Vehicles and Driver Acceptance

In this section, we will combine the Four Pleasures and Three Levels models and look at how vehicles can provide pleasure in these various ways.

Physio-pleasure

Physio-visceral

These are the immediate first impression physical pleasures that we get from a vehicle. For example, the smell inside a new car would be an example of physio-visceral pleasure. This is something that we are aware of even before we have had any significant interaction with the car and it gives us a positive sense of the car's quality.

Another example is the sound that a car door makes when closing. BMW, for example, has put a lot of research into ensuring that their doors close with a deep bass thud – a sound that is associated with solidity and good build quality.

Some years ago Harley Davidson even tried (unsuccessfully) to register the sound their motorbikes make, because of the visceral association with the iconic product.

Physio-behavioural

These are the physical pleasures that we get from a vehicle when using it. It could include the benefits of comfortable seating or the pleasant feeling of the steering wheel or gearshift. For example, the feeling of the luxurious hand-stitched leather on the steering wheel and gearshift of a Bentley would be sources of physio-behavioural pleasure.

Physio-reflective

This refers to the physiological 'legacy' of making a trip in a vehicle. For example, do we have any aches and pains after the trip, do we feel tired or do we feel comfortable and fresh? The inclusion of massage seats in some high-end cars – for example, Bentley – are an example of something that helps in achieving this.

Psycho-pleasure

Psycho-visceral

These refer to our immediate psychological reactions when seeing a vehicle. Designers often use anthropomorphism – parallels with people or animals – to elicit certain reactions in us in this respect. Headlights are often used to represent eyes, grills to represent mouths. In this way cars can be made to look cute, aggressive, tough, or whatever effect the designer is going for. The reincarnation of the VW Beetle is an example of a car designed to look cute.

Psycho-behavioural
These are the psychological pleasures we get when driving the vehicle. They include the sense of control that we have over the vehicle, the feedback we are getting from the instruments and the ease of using the controls. Feeling positive emotions such as confidence or excitement when driving would also come into this category. For example, the immediate responsiveness and rapid acceleration of a Porsche 911 gives both a feeling of excitement and control and is thus a source of psycho-behavioural pleasure.

Psycho-reflective
This refers to our thoughts and emotions when reflecting on using a vehicle. For example, we may think back to what an exciting drive we have just had or reflect with quiet satisfaction on our vehicles reliability. The seven-year warranty that Kia offers on their cars is an example of this. Because the manufacturer has such a high level of confidence in their vehicles, we are likely to be confident in them too. Confidence in the vehicle is a form of psycho-reflective pleasure.

Socio-pleasure

Socio-visceral
This type of pleasure includes the immediate reaction that our car generates in others. This may have positive or negative consequences for the way that others treat us when we are driving. For example, surveys in the UK have shown consistently negative attitudes to drivers of BMW cars who are considered aggressive and inconsiderate. When some drivers see a BMW, they may react negatively and BMW drivers report being shown far less consideration by others motorists than drivers of other vehicles.

Socio-behavioural
This refers to the social reactions that people have to us and our vehicles when they see us driving and also to the social role that the car plays in our lives.
    For many of us our car plays an important role in our family lives. This includes transporting our children to and from school, various other activities, and going out together for family outings.
    Features such as DVD players in the back to keep the children entertained and wipe-clean seats to cope will spillages can be a source of socio-behavioural pleasure, as are options in, for example, the Chrysler Voyager multi-person vehicle.

Socio-reflective
This includes how others think about our vehicle and how we talk about our vehicle in the company of others. For example, if we talk about our vehicle positively to others then this would be a socio-reflective pleasure.
    Having a vehicle that is different from everyone else's can also be a socio-reflective pleasure. For example, the huge number of options that Mini offers for the

colours of the various parts of the body and the lighting and colours, give such a huge variety of potential combinations that it is possible to specify a car unique to you.

Ideo-pleasure

Ideo-visceral
This is our own instinctive response as to whether or not a vehicle appeals to our tastes and moral sensibilities. For example, some people who are concerned about the environment immediately have a negative reaction when seeing a huge 4x4 vehicle such as a Range Rover being driven on urban roads, but have an instinctively positive reaction to a hybrid vehicle such as a Toyota Prius or an electric vehicle such as G-Wiz.

Ideo-behavioural
This concerns the degree to which we perceive the vehicle as being consistent with our tastes and values when we are driving it. Many cars now give readings on how many miles or kilometres we are travelling per gallon or litre of fuel. For those concerned with environmental issues, this can be a source of ideo-behavioural pleasure. In 2012, Kawasaki introduced an 'eco' symbol on the dashboard of their motorcycles which flashes to 'congratulate' the rider when they are riding in an environmentally friendly manner.

Ideo-reflective
This refers to the degree to which we feel that a vehicle fits with our tastes and values when we reflect on it. For example, when we look at a picture of it or read about it in a magazine, do we feel proud? We may like a car because we feel that its design reflects qualities that we would like to think of ourselves as having. For example, owning a tough, rugged, vehicle like a Hummer may make the owner feel tough. 'I love my Jeep because it's tough like me' (Govers and Mugge 2004).

The Quality of the Driver Experience

When designing a car or other vehicle, it is important to consider all of the above aspects in order to maximise the quality of the driver experience. It is clear that there is much overlap between the categories, and that a single phenomenon may elicit more than one category of pleasure. To rehearse the sound of the Apple boot-up, the *reassurance* it promotes is clearly a 'psycho-pleasure' but the actual noise may be psycho-visceral.

Deciding on the nature of the driver experience is not a straightforward task. At an obvious level, there may be a considerable divergence between the perceived positive experiences of a testosterone-rich 18 year old and an elderly retiree. There are indeed some generalisable 'pleasures' but the identification of them tends to be a function of resolution. At levels of low (population) resolution, the predictability of an experience being judged pleasurable is relatively high. As the resolution increases the predictability decreases, so that for any given individual it is likely to be a far from robust judgement (Green and Kanis 1998). It is, for example, relatively safe to predict that a physio-visceral impression of good build quality as exemplified by panel fit, paint finish or detail resolution will result in positive responses from the bulk of a population. However, there may well be several individuals who ignore such markers in favour of less mainstream associations. Think of the niche fashion for 'feral' transportation generated by the *Mad Max* movies, or the appliqué mud for urban SUVs. Customisation may account for some of the inter-individual variation, but judgements still need to be made, and supported in such a way that the nature of the designed experience is as controlled as possible and not left to chance.

Once we have decided what pleasures we want the driver to experience, how do we go about delivering these through the design?

On one level much of this comes down to the judgement of the designer and their knowledge of the people they are designing for. However, there are also a number of approaches and techniques that can help with this, selected examples of which are given in the rest of the chapter. There is no basis for selection of the techniques other than to demonstrate a range of possibilities.

It is important to note that the various techniques are sometimes complex and require effort and knowledge to apply. Most are based on some elicitation of human responses to stimuli and are thus akin to techniques of user trialling and/or market research. It is not our purpose to present immediately applicable methods, but rather to illustrate some of the work that has been done in the area and to indicate further reading. References are provided for those wishing to pursue the techniques mentioned.

Kansei (Emotional) Engineering

This is a statistical approach developed by Mitsuo Nagamichi at Hiroshima University (Nagamichi 1995, 1997) whereby the design of a product is broken down into its constituent parts and a statistical analysis is used to link people's emotional responses to particular design aspects.

Each constituent part is designed to elicit the desired emotional response so that the product as a whole generates the overall emotional effect required. For example, if we were aiming to design a car that was powerful and elegant, we could look at various aspects of the design – such as colour, form and sound – and combine them in different ways and measure user responses to see which one created the overall desired affect.

Imagine that we had, for example, five colour options, five form options and five different exhaust notes to choose from. This would give us a possible 125 combinations (5 x 5 x 5) and we could put all of these in front of users to see which scored best overall on the desired characteristics of power and elegance.

This technique has been used extensively within the car industry – perhaps most notably in the design of the Mazda MX5 or Miata. *Kansei* Engineering was used to give the cars properties reminiscent of a 1970s British sports car. The MX5 went on to be a massive success and is now the best-selling two-seat convertible in history. An accessible reference is Lee, Harada and Stappers (2002).

SEQUAM

SEQUAM (SEnsorial QUality Assessment Method) has some similarities to *Kansei* Engineering in the sense that it uses statistical analyses to understand how to link product properties to emotional responses.

Where it differs from *Kansei* is that, rather than looking at the emotional responses to combinations of design elements, SEQUAM looks at responses to design elements individually. It involves plotting the properties of each element on a continuum and seeing what the response to each is.

So, for example, if trying to create a steering wheel with a high-quality feel, properties such as the roughness and hardness of the material can be plotted against perceived quality. Once the optimum level of roughness and hardness has been identified, a material using both this level of roughness and hardness can be used for the steering wheel.

Fiat has used this technique to optimise the tactile properties of their steering wheels, gear shift knob and inside door handles. Because these are usually among the first things that people touch when trying out a new car, they are important to the visceral impression that the car makes. For further detail, see Bonapace (2002).

Desmet's 'PrEmo'

Pieter Desmet (Desmet 2003, Desmet, Hekkert and Jacobs 2000) is one of a recent generation of researchers who have made significant steps forward in the formalisation of emotional responses to products, particularly vehicles. His doctoral research used animated manikins to help viewers articulate emotional responses to products. Working in the ID StudioLab in TU Delft, Desmet and Hekkert have been seminal in the study of affective design, and together with Jan Jacobs and Kees Overbeeke were prime movers in the establishment of the Design and Emotion Society, which aims to create methods and techniques for the study of emotional responses to design.

PrEmo uses 14 different animations of gender-neutral 'puppets' to depict seven positive and seven negative responses to visual stimuli. These puppets were drawn by an artist using professional models to register the required emotions. The animation begins at a neutral expression and moves to the depicted emotion in one second. The positive emotions are inspiration, desire, satisfaction, pleasant surprise, fascination, amusement and admiration. The seven negative emotions are disgust, indignency, contempt, disappointment, dissatisfaction, boredom and unpleasant surprise. For further detail of the application to automobiles, consult Desmet, Hekkert and Jacobs (2000) and Desmet (2003).

Products as Personalities

There has been considerable interest in the last decade in the concept of products as personalities. Jordan (1997b) used the Myers-Briggs personality type indicator and later, in 2002, translated this into more accessible terms. Since then, there have been numerous papers illustrating the possibilities and the difficulties of assigning anthropomorphic qualities to products and to assess the effects of doing so.

Govers and Mugge: Product Attachment

Pascalle Govers and Ruth Mugge have been active in the exploration of product attachment and the concept of product personality. Govers, a consultant with MetrixLab in Rotterdam, published her book *Product Personality* in 2004 (Govers 2004).

For her doctoral research completed in 2007 at TU Delft, Mugge considered the idea of products as personalities and the phenomenon of 'bonding' with products and has since published several related papers. Her work concentrated on the product facilitation of 'self-expression', this being one of the four factors that were drawn from the literature as being able to influence product attachment: self-expression (can I distinguish myself from others with the product?), group affiliation (does ownership of the product connect me to a group?), memories (related to the product) and pleasure (provided by the product) (Mugge, Schifferstein and Schoormans 2006). Further References are Govers, Hekkert, and Schoormans (2002), Govers (2004), Govers and Mugge (2004), Mugge, Schifferstein and Schoormans (2004), Mugge, Schoormans, and Schifferstein (2005) and Mugge, et al. (2006).

Cultural Probes: The 'Presence project' Gaver et Al.

To illustrate the diversity of techniques associated with emotional data gathering, we include the Cultural Probes technique, initiated by Gaver (Gaver, Dunne and Pacenti 1999) at the Royal College of Art and used, for example, by Wensveen (Wensveen 2005) in his doctoral thesis:

> The probes constitute a collection of evocative tasks for exploring attitudes
> and aspirations and developing an empathetic and engaging understanding of a
> particular audience. (Gaver et al. 1999)

The probes in the form of experimental designs of, primarily, communication avenues (graphic and haptic user interfaces) were eventually tested with the elderly, children and ethnic groups. Stephan Wensveen used the technique in the design of products and thus provides a bridge between the theory and product design.

Conclusion

In-vehicle technological advances have the potential to be seen as assistive/positive or restrictive/negative by differing demographics. Think, for example, of lane guidance, distance or speed control devices. In-cabin controls may be generally less controversial but recent examples of strongly criticised menu systems demonstrate that the user experience must be considered a dominant factor. A major component of the user experience is aesthetic and emotional satisfaction.

This chapter has attempted to present some of the issues confronting designers who want to move the experience of their vehicle beyond the generally accepted standards. The research and the knowledge mentioned is well known and accredited in the academic domain that generated it, and there is a wealth of published exemplars available by following the reference trail provided here. There is also no doubt that individual manufacturers and design studios have invested much time and money in determining their brand identity and have detailed information on the varying demographics that constitute their market. However, elevation of the emotional product experience to a high profile in the design process is still a work-in-progress, and translation of the academic advances into solid design parameters has much inherent uncertainty. There is at present no general theory of design for product emotion and indeed there may never be such, but some of the researchers and the techniques presented in this chapter have the potential to move the specification of a positive emotional experience to be a little more deliberate and a little less given to chance.

References

Beaudrillard, J. 1988. The system of objects. In *Design after Modernism*. Edited by J. Thackera. New York: Thames and Hudson.

Bonapace, L. 2002. Linking product properties to pleasure: The sensorial quality assessment method. In *Pleasure with Products; Beyond Usability*, 180–218. Edited by W.S. Green and P.W. Jordan. London and New York: Taylor and Francis.

Desmet, P.M.A. 2003. Measuring emotion: Development and application of an instrument to measure emotional responses to products. In *Funology: from Usability to Enjoyment*, 111–23. Edited by M.A. Blythe, A.F. Monk, K. Overbeeke and P.C. Wright. Dordrecht: Kluwer Academic.

Desmet, P.M.A., Hekkert, P. and Jacobs, J.J. 2000. When a car makes you smile: Development and application of an instrument to measure product emotions. *Advances in Consumer Research*, 27: 111–17.

Gaver, B., Dunne, T. and Pacenti, E. 1999. Design: Cultural probes. *Interactions*, 6(1).

Govers, P.C.M. 2004. *Product Personality*. Delft: Delft University of Technology.

Govers, P.C.M., Hekkert, P. and Schoormans, J.P.L. 2002. Happy, cute and tough: Can designers create a product personality that consumers understand? In *Design and Emotion: The Experience of Everyday Things*, 345–9. Edited by D. McDonagh, P. Hekkert, J. Van Erp and D. Gyi. London: Taylor and Francis:

Govers, P.C.M. and Mugge, R. 2004. *I Love My Jeep, Because Its Tough Like Me: The Effect of Product-Personality Congruence on Product Attachment.* Proceedings of the Fourth International Conference on Design and Emotion. Edited by Aren Kurtgözü. Ankara, Turkey.

Green, W.S. and Kanis, H. 1998. Product interaction theory: A designer's primer. In *Global Ergonomics*, 801–6. Edited by P.A. Scott, R.S. Bridger and J. Charteris. Amsterdam: Elsevier.

Jordan, P.W. 1997a. *Usability Evaluation in Industry: Gaining the Competitive Advantage.* Proceedings of the 13th Triennial Congress of the International Ergonomics Association, 150–52. Tampere: Finnish Institute of Occupational Health.

———. 1997b. Products as personalities. In *Contemporary Ergonomics 1997.* Edited by S. Robertson. London: Taylor and Francis.

———. 1999. Pleasure with products: Human factors for body, mind and soul. In *Human Factors in Product Design: Current Practice and Future Trends.* Edited by W.S. Green and P.W. Jordan. London: Taylor and Francis.

———. 2002. The personality of products. In *Pleasure with Products: Beyond Usability.* Edited by W.S. Green and P.W. Jordan. London and New York: Taylor and Francis.

Lee, S., Harada, A. and Stappers, P.J. 2002. Design based on Kansei. In *Pleasure with Products; Beyond Usability*, 219–30. Edited by W.S. Green and P.W. Jordan. London and New York: Taylor and Francis.

Maslow, A. 1954. *Motivation and Personality.* New York: Harper.

Mugge, R., Schifferstein, H.N.J. and Schoormans, J.P.L. 2004. *Personalizing Product Appearance: The Effect on Product Attachment.* Proceedings of the Fourth International Conference on Design and Emotion. Edited by Aren Kurtgözü. Ankara, Turkey.

Mugge, R., Schifferstein, H.N.J. and Schoormans, J.P.L. 2006. A longitudinal study on product attachment and its determinants. In *European Advances in Consumer Research, Vol. 7*, 641–7. Edited by K.M. Ekström and H. Brembeck. Duluth, MN: Association for Consumer Research.

Mugge, R., Schoormans, J.P.L. and Schifferstein, H.N.J. 2005. Design strategies to postpone consumers' product replacement: The value of a strong person-product relationship. *Design Journal*, 8(2): 38–48.

Nagamichi, M. 1995 *The Story of Kansei Engineering.* Tokyo: Kalibundo Publishing.

———. 1997. *Requirement Identification of Consumers' Needs in Product Design.* Proceedings of IEA 1997 Finnish Institute of Occupational Health, Helsinki.

Norman, D. 2004. *Emotional Design: Why We Love (or Hate) Everyday Things.* New York: Basic Books.

Schuller, R.H. 1982. *Self-Esteem: The New Reformation.* Waco: World Books.

Tiger, L. 1992. *The Pursuit of Pleasure.* Boston: Little, Brown.

Wensveen, S. 2005. *A Tangibility Approach to Affective Interaction.* Delft, the Netherlands: Delft University Press.

Chapter 19

Optimising the Organisational Aspects of Deployment: Learning from the Introduction of New Technology in Domains Other than Road Transport

Martin C. Maguire

Loughborough Design School, Loughborough University, UK

Abstract

The use of technology in any domain, such as manufacturing, finance or health care, takes place within a wider organisational environment. The road transport environment is no different in this regard from other domains. This chapter outlines how organisational aspects interact with technology deployment, providing examples from various sectors or domains. Lessons from these domains can be applied to help improve driver acceptance of new technology in the road transport domain. These are discussed and summarised. The conclusions describe general organisational strategies that can be adopted when deploying in-car technology to promote user acceptance of it.

Introduction: People's Attitudes to New Technology

When new technology is introduced into an organisation, reactions to it may vary. There may be a natural resistance to change and limited acceptance of a system that will require learning and adaptation to new procedures. There may be a feeling that the new system will increase workload, make working life more complicated or take over functions that people enjoyed doing and were skilled at (Kirk 1983, Eason 1988).

In contrast to this, people may look forward to the new system, thinking it will help them do their work more easily. They may have been part of the process that helped specify the user requirements for the system and so know what is coming. They may feel that using it will help them grow and develop new career skills and opportunities.

This chapter reviews organisational factors that are relevant to the introduction and acceptance of information technology (IT) systems. Each factor is then related to in-car technology and the driving domain. Conclusions are drawn which describe strategies for deploying in-car technology so that organisational factors help to promote rather than constrain user acceptance.

Organisational Factors and New Technology

When new technology is introduced into a work or activity setting, it sits within physical surroundings, people, procedures and other technologies that together make up a total system. This is called the 'organisational context'. The organisational context will affect how the system is then used and can have implications for the design of the user interface to it (Maguire 2013). If it is developed from a user perspective and with a consideration of how it will match the organisational context, it is more likely to be accepted and used as intended. If not, then the system may end up being only partially used, misused or not used at all.

The importance of having knowledge of organisational context is recognised in the ISO standard 9241-210 (2010) concerned with human-centred design. It states:

> The characteristics of the users, tasks and the organisational and physical environment define the context in which the system is used. It is important to understand and identify the details of this context in order to guide early design decisions, and to provide a basis for evaluation.

Figure 19.1 shows some of the key elements of the organisational environment that may interact with a new IT system when it is introduced.

There are a number of sociotechnical principles that guide system design (Clegg 2000). One of these is that the organisational context is not static and will evolve over time. When the IT system itself is introduced it may change the roles that workers occupy, their ways of working and their attitudes to technology. There are a number of well-developed and validated models for how well users receive technology. One of these is the technology acceptance model or TAM (Davis 1989, Davis, Bagozzi and Warshaw 1989), which has been successfully applied in examining adoption behaviour of various information systems (Figure 19.2).

The core idea of the TAM is that technology acceptance is based upon a person's overall attitude towards a system. This is determined by how useful or easy to use they feel that technology to be (i.e., its perceived usefulness – PU, and its perceived ease of use – PEU). This may in turn be influenced by wider aspects in the organisational context. Sometimes other features are associated with the TAM such as trust in the system and the social influence of others; for example in recommending the system or certain functions of it. These reflect the wider organisational context that influences individual user views.

Figure 19.1   Factors making up the organisational context for an IT system

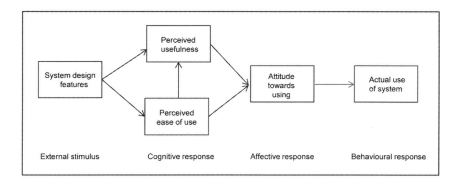

Figure 19.2   Technology acceptance model described by Davis (1989)

The TAM reflects our everyday experiences of acquiring or using technology and our perceived needs. For example, keeping a digital camera in the car to record the scene or a warning triangle as a roadside warning are both considered useful if an accident occurs.

Organisational Deployment Factors

In this section, a range of organisational factors is presented that influence the acceptance of IT and, where relevant, aspects of PU and PEU of technology. Each factor is related to the introduction of in-vehicle technology. In the context of driving, the organisational domain is the driver, passengers and other road users. Where technology may have wider implications on people or the environment, the organisation may be considered to be society itself.

It should be said that the design and installation of safe and efficient in-vehicle information and communication systems is also an important part of technology acceptance. The application of principles for achieving this, as defined by the European Commission (2008), is still needed in addition to considering how acceptance is influenced by organisational factors.

Organisational Goals

An organisation's goals are often summarised in terms of a high level vision statement of what it wishes to achieve and providing the inspiration for strategic decisions and daily operation. The goals for an organisation should lead the design of a new IT system. If they are not explicit or detailed enough, this can make it hard to specify the user requirements for new IT to support the achievement of those goals. So, for example, if the goal of a new call centre system is to provide customers with the best experience when they contact the organisation, the system needs to provide the information in a flexible way and with functions that are easily accessible to support that goal. If the call centre system supports only a rigid structure of questions in order to process customer enquiries as quickly as possible, this won't help in achieving the organisational goals. The same can be applied to in-car technology. The high level goals here may be to assist the motoring community in driving more economically, safely and comfortably, and possibly to help them maintain the car more easily. If new technology in the car helps to achieve this, it will fit in with the aspirations of the driver and passengers who will be motivated to use and accept it.

The organisational goal for the introduction of new technology may be to merge two or more systems together so that users in different roles can use the same system. This was the plan behind the proposed merger of the National Probation and HM Prison Service offender management systems in England and Wales. The details of an offender who moved back and forth between prison and probation could be kept on the same system while being accessible to both services.

System convergence goals can lead to lower system maintenance, simplification of training programmes, removal of the need to transfer data between systems and easier communication between the two services. There has been a trend to merge different systems in the car (communication, navigation, entertainment, ventilation and climate control, etc.) so that the confusing array of controls is accessed from a single control unit giving the driver a simpler user interface. In principle this is a good idea but if the integrated system contains too many functions and the controls and displays are oversimplified, this can make the user interface difficult for the driver to navigate. The BMW iDrive is one example of this (Cobb 2002, Gilbert 2004). However, the concept of a computer-oriented, integrated interface has become popular in the luxury segment of the car industry (Niedermaier et al. 2009).

Roles, Responsibilities and Skills

People within an organisation often fulfil one or more roles with accompanying responsibilities. When a new system is introduced there may be a mismatch between this and a person's current role and responsibilities. For example, a person who is only experienced in performing IT maintenance may find it difficult if they are also required, without suitable training, to provide user support. Workers usually value the skills they have built up over time and apply in their job, say in analysis, decision-making or liaison with others. However, new technology may replace these tasks, leaving the operators with more routine and less challenging roles. They may for instance be in charge of travel arrangements for staff and compiling requirements for office supplies. This role may be taken away and become less efficient if the process is devolved and travel requests and orders for office materials are sent directly by individual employees to the external supplier.

Modern cars or lorries often bring a level of automation to driving such as adaptive cruise control, vehicle platooning, driver assist lane keeping or merging into traffic, speed limiters for teenage drivers, alerts for drowsy drivers, automated parking and possibly in the future, automated driving. It can be argued that while such developments tend to deskill drivers, they may also require the driver to acquire new skills of monitoring and intervention. Methods for optimal allocation of tasks between human beings and machines have been developed, for example by Waterson (2005). These include decision criteria such as the feasibility of allocating a task to the machine and the legal requirement for a human to perform a task. In relation to control systems and automation, Sheridan developed a 10-point scale showing possible levels of computer automation, for example, the computer suggesting a path or action allowing the user to approve it, or the computer deciding on a path of action and informing the user after it has been carried out (Parasuraman, Sheridan and Wickens 2000). Deciding on what is an appropriate level of automation for particular technologies or organisational contexts depends on factors such as complexity and predictability of the control process, workload on the user, fatigue, criticality of errors and the consequences of actions and errors including ethical concerns. The potential use of automated drones in warfare has brought this topic into sharp focus (Finn 2011). Similarly, the new Google driverless car will require a rethink of what skills are required in such an environment (Nasaw 2012).

An interesting question in terms of drivers using in-car technology is how much they may try to delegate responsibility to the technology. A driver may drive close to the safety limits and trust the car to mitigate the effect of any loss of control; for example, through Active Safety systems, such as brake assist, traction control and electronic stability control. Awareness and training would be valuable to help the driver take appropriate responsibility when they come to use such systems. This would also give them confidence in engaging with such systems and being more willing to accept them.

Discretion in the use of in-car technology is likely to be preferred over a system that does not allow it. An example might be a satellite navigation system that allows the user to specify an intermediate destination part way through a journey rather than splitting it up into two separate journeys. A facility allowing a driver of a hire car to disable any special driving modes that have been set previously is likely to be helpful.

General characteristics of good job design – variety, autonomy, task identity and feedback (Hackman and Lawler 1971) – can be related to the driving environment. This could be designed to support driving as a complete task, thus matching the idea of a coherent job. In some cars feedback is offered on driving performance in terms of efficiency, fuel saving and so on. This could be translated in terms of contribution to the environment, reducing general stress on the road and saving wear and tear costs. A car promoted in these terms might be attractive to potential buyers who see driving as more than simply going from A to B.

Workflows and Procedures

Workflows and procedures describe the way things are done or ought to be done within an organisation to achieve its goals and objectives. In practice, workers may have to work around the rules to get things done. While procedures are prime candidates to be encoded in software design, there is a danger that in trying to implement them without an understanding of how they work in practice, they may produce an inefficient or non-viable system.

In the medical field, a computerised 'physician order entry system' was introduced which reduced nurse-physician interaction about critically ill infants (Harrison, Koppei and Bar-Lev 2007). In the new system, physicians initiated orders for medication and pharmacists checked them while clerks delivered them to nurses for administration. This linear workflow led to delays in delivery of orders, uncertainty over whether they had been initiated, and sometimes, divergent orders were produced. Thus nurses continued to initiate orders, to interrupt physicians to ensure that orders had been entered onto the system, and to make decisions when divergent medication orders were presented. A better understanding of the current procedures and more flexibility in the new system process would have helped to prevent some of the problems occurring and would have improved system acceptance.

In terms of procedures for setting up new technology in the car, presenting a series of prompts for each step in the process could be helpful. The car might also monitor a driver's typical behaviour in using controls and give advice on how to integrate the new technology within their driving. If carefully thought out, this would make the take up of the technology a more user-friendly experience.

Laws and Regulations

In working life, there are often laws or regulations that have to be followed; for example, about the protection of personal information, the testing of circuit boards for military use and safe practice by medical staff. These societal requirements might be in conflict with internal procedures and rules. In a hospital environment, medical staff might be issued with ID cards to access the patient records system with a regulation that they must not allow others to borrow them. However, for practical reasons, nurses may have to lend their cards so that new staff members can access the system and start work before being issued their own card. Enforcing the regulation by incorporating biometric security could prevent this practice but would then cause inconvenience for staff. Current work practices and the reasons for them need to be considered before a new system is introduced.

In terms of in-car technology, the system might disable the car phone when driving. However, this might be counterproductive if the phone is required in an emergency situation. Similarly, a police road camera might unfairly catch drivers who run through an 'always red' traffic light that has failed, or who use a usually prohibited part of the road space or pedestrian area to get around a fallen tree. Thus system design needs to take account of laws and regulations but be aware of situations where it may not be appropriate by system behaviour or restriction to enforce them strictly.

While the use of handheld phones in a car is illegal in the UK, the use of hands-free phones is allowed provided the driver is seen to be in control of the vehicle. If this is not the case, they or their employer can be prosecuted. There is evidence that hands-free phones are not necessarily safer than handheld devices since talking and listening can take up a lot of mental capacity (Nunes and Recarte 2002). Nevertheless, communication from a central office to drivers on the road remains a significant need, so there is an opportunity for creative thinking about how to facilitate it in a safe manner. For example, a light on the dashboard could warn the driver that they have a call waiting and to slow down in order to take it hands-free.

Communication and Distraction

Communication between people is a requirement for effective working in most organisations and helps to relieve boredom and work stress. IT has facilitated many new forms of communication (email, texting, instant messaging, video-telephony and conferencing), but this can create barriers to face-to-face communication or communication overload.

Some forms of technology such as machinery, manufacturing processes or even IT equipment can create noise that hampers communication between people. Partitioned work spaces can also have the effect of creating isolation which workers often view negatively (Vickers 2007). Established work procedures may involve certain kinds of communication between different groups of people; for example, nurses and doctors, managers and staff. If a new system cuts across these established channels of communication, it can create frustration and possibly result in a less efficient system (Boonstra and Broekhuis 2010).

When considering the social context of the car as including both driver and passengers, other questions arise. People within a car will normally be in communication with each other which may affect how the driver interacts with in-car technology or reduce the attention they pay to it. Passengers may also access functionality on the driver's behalf such as setting the satnav or operating the entertainment system. Within cars, soundproofing has enabled the driver and passengers to communicate freely although in-car technology can hinder such communication if they have audio output (for example, spoken directions from a satnav). Having a silent or a visual display-only mode may be helpful in making such systems more acceptable. Technology that prevents use of a cell phone while a vehicle is in motion, thus preventing use by a passenger, is another potential problem. New sensing technology which can determine whether a phone is being used by the driver or by a passenger may overcome this problem (Talbot 2012). For drivers, filtering out all but key callers for hands-free communication may be helpful.

Work Culture

Work culture is made up of the shared values, beliefs, underlying assumptions, language, attitudes and behaviours shared by a group of people (Donais 2006). Culture is the behaviour that results when a group arrives at a set of generally unspoken and unwritten rules for working together. An organisation's culture is made up of all of the life experiences each employee brings to the organisation. Culture is especially influenced by the organisation's founder, the executives and other managerial staff because of their role in decision-making and strategic direction.

If an IT system provides a way of working that does not match with the typical work culture, this can cause problems in the organisation. So if a system requires, for instance, frequent checking of emails, logging of people's movements and recording of how work hours have been spent, this might fail if it is not part of the general work culture of the group.

For in-car systems, a study of driver attitudes towards new car developments and the way they think and talk about their vehicles can be a useful input to the design process. For example, it may be found that owners of certain vehicles generally have a high level of technical knowledge and are more likely to maintain or carry out repairs for themselves. This could encourage the design of vehicles that provide diagnostic information to the owners as guidance when working on their cars.

Technical and Functional Knowledge

Technical knowledge among employees in a work situation can be an important factor in determining a user's success in the use and acceptance of new technology. If, for example, a user has some knowledge about telecommunications, they are more likely to be able to sort out the problem if an Internet connection fails. If they have knowledge about software applications, this will help them work more efficiently. Examples are the use of keyboard shortcut commands, the display of control characters when editing a word processing document and the splitting of rows and columns in a spreadsheet so that headings stay visible when scrolling through the data.

Alhussain and Drew (2009) studied employees' perceptions of using biometric equipment for identity recognition in two government departments in the Kingdom of Saudi Arabia. Fingerprint scanners were used to record and prove employees' attendance thus preventing them from signing-in for others. The survey identified a lack of trust in and acceptance of the technology by some workers, which the authors felt was due to a lack of technical knowledge and experience with technology. Employees were also unsure about their employer's motives for using the technology and many felt that this indicated a lack of trust in them. The authors advocated that managers should acquire a better understanding of biometric technology and IT so they should implement it in a more sensitive way, while employees should be made more aware of the purpose and benefits of the innovation.

In the author's own experience, a lack of knowledge by members of the public about the use of biometric technology to access ATM cash machines was a reason for reduced user acceptance of them. In a focus group study conducted for an ATM manufacturer (Maguire 2003), some participants thought that iris scanning was unsafe as it used a laser beam, and unhygienic as they had to place their eye over a tube shared with other customers. In fact, the technology is based on camera technology so the user has no physical contact with it. Other participants doubted whether face or voice recognition technology would work if someone's appearance or voice changed. However, such technologies are based on dimensions of the face and fundamental characteristics of the voice so are more reliable that people appreciated.

The same ideas can be applied to in-car technology. Knowledge of how different features of a driving system operate, could promote its use in a more appropriate way by building trust in its operation, enabling the driver to use it more efficiently and overcoming any problems when using it. An example is Adaptive Cruise Control (ACC), which uses headway sensors to continuously measure the distance to other vehicles, automatically adjusting the vehicle's speed to ensure that it does not get too close to the one in front. There are difficulties with this technology as it tends to lock onto large trucks in preference to motorcycles and 'unexpectedly' accelerates into an off-road lane when exiting into a more slowly moving traffic stream. By being aware of this, drivers can decide when it is appropriate to employ ACC and when not, and how to interact with it appropriately.

Safety Thinking

Correct operation of systems is important to ensure the safety of the workers, the public and the environment. During the development of a safety critical system, the implications of its use are assessed and, where necessary, measures are determined to meet safety needs. Careful attention to safety issues then builds up trust in the system which is shared by those who work in it and those who use it. Reason (2004) argues that in the medical domain, some organisational accident sequences could be thwarted at the last minute if those on the frontline had acquired some degree of error wisdom and appropriate mental skills. In a similar way, knowledge of dangerous situations and safe driving behaviour can do a lot to reduce the number of vehicle and pedestrian accidents. Various documents are produced by organisations, for example BVRLA (2012) and ROSPA (2012), which give guidelines on safe driving and imploring employers to disseminate them to their employees on the road.

Security and Usability

To keep a system secure, authorised users will often be required to enter one or more user codes and passwords: for example, to access the network, the operating system, and the application. Levels of user access and functions available may also be constrained depending on each person's grade or job role, so additional codes may be needed to access sensitive or confidential data.

Unfortunately, multiple passwords are hard to remember especially if they change, so tension exists between maintaining security and ease of use. System-generated passwords tend to be hard to remember while user-generated ones are open to hackers guessing them (e.g., 'secret', 'qwerty' and '1234'). Some people may write their passwords down, which compounds the problem of security. If a user forgets their password or is locked out from the system because of failed access attempts, they then have to sort out how to regain access.

In future, security access for cars may evolve and become more like IT system access, with the use of passwords or biometrics. Fingerprint car locks already exist and the use of other biometric techniques may follow. Once in the car, personal identification may be used to set the correct seating position, steering-wheel angle or to start the car. While such systems for vehicle access are attractive from a security point of view, there are dangers in locking out the driver who forgets their password or if the biometric system fails late at night in an unfamiliar location. Having a backup mechanism for such situations is likely to give future drivers and customers more confidence in car-access innovations.

Training and Support

A system developed to support a company or public organisation's work processes will normally require a training programme and user support. Careful organisation of user awareness and training sessions and matching them to individual learning styles, plays a key role in staff attitudes towards a system and enthusiasm for using it (Bostram, Olfman and Sein 1990).

Driving lessons and advanced driver training are the means by which people learn the basic controls of a car and gain driving experience. As in-car technology becomes more advanced and an increasingly significant part of driving, it may be necessary for learner or existing drivers to be trained in the safe and effective use of it. This might become part of driver training and the driving test, or be offered by the vehicle manufacturer, although similar provision may not be available within the second-hand market. Training to use vehicle technologies such as anti-collision warning, self-parking or adaptive cruise control might be conducted partly within a driving simulator so that a driver learning to use the technology can make errors and 'crash' safely. However, this should not be a substitute for the design of an intuitive user interface.

Summary of Organisational Factors That Have Implications for Vehicle Technology

The following table summarises the organisational aspects of IT implementation and the corresponding implications for the design of in-car technology.

Table 19.1   Summary of organisational context factors and how they may relate to in-car technology

Organisational factors in introduction of IT	Implications for in-car technology acceptance
Organisational goals Often defined as a high level vision statement which should lead the design of a new IT system.	Driver's goals are to get to their destination quickly, safely and comfortably. Avoiding high costs in doing this will be an underlying goal. Effectiveness of such technology will rely on design for usability.
Roles, responsibilities and skills Introduction of IT may change people's roles and responsibilities, thus affecting the tasks they perform and the skills they need.	Automation may deskill the driving task and result in mental under load. Technology to support driving as a complete task, matching the idea of a coherent job could be a way to address these problems.
Workflows and procedures Systems which fail to take account of the organisational environment and working practices may create problems for users.	Car systems should try to support existing practices and advice on use of new technology for better driving.
Laws and regulations A work system will normally be set up to abide by recognised laws, regulations and organisational policies; e.g., the protection of privacy. There may be occasions when it is necessary to work around some laws for efficient operation.	Enforcement of traffic laws may become part of in-car systems but implementation needs to be thought through in order that it does not create inappropriate barriers.
Communication and distraction Communication is a requirement for effective working in most organisations. IT facilitates new forms of communication but can create overload and hamper natural interaction.	Other passengers in the car may affect use of technology and can compound problem of information overload e.g., use of a phone on the move by passengers.
Work culture If an IT system provides a way of working that does not match with the typical work culture this can cause problems in the organisation.	Studying drivers' attitudes and thinking about new car developments can be a useful input to the design process and have implications for the technology people want and will accept.
Technical and functional knowledge Technical knowledge among employees in a work situation is an important factor in determining user attitudes to and acceptance of new technology.	Driver knowledge of how new technology in the car operates is likely to promote its use in a more appropriate way, build trust in its operation, and enable drivers to overcome problems that occur.

Safety thinking Careful attention to safety issues builds trust in the system. Error wisdom and appropriate mental skills can avert errors.	Guidelines for employers on safer driving by employees as part of safety thinking can reduce road accidents.
Security and usability Multiple user codes and passwords can make the process of system access difficult.	Technology to increase security against access to cars by intruders should avoid similar problems, for example locking drivers out in unsafe situations.
Training and support Training and user support needs to carefully match the knowledge and skills of users of the IT system.	Driving lessons and advanced driver training may need to include car information and driving assistance systems but not at the expense of an intuitive design.

Conclusions: Organisational Strategies for the Acceptance of In-Car Technologies

Knowledge of the organisational environment is important if an IT system is to integrate well with the social system and be accepted by the user community. The introduction of computer technology into vehicles can be seen in the same way. It will only be successful and accepted if matched with the social or organisational domain; that is, the driving community and broader society.

General strategies that can be employed are as follows:

1. User awareness and readiness for the new technology is paramount. If a driver is aware of what the new technology can do, they are more likely to perceive the benefits of using it. Similarly, the more they understand about how the technology operates, the easier they will find it to use. Giving drivers prior experience of in-car technology through showroom demonstrations can help to promote both of these aspects. Driving instructors could also provide training in the use of these technologies, if nothing else, to give new drivers exposure to them.

2. Drivers value the skills and experience they build up in driving. Technology within cars that makes these skills redundant is unlikely to be accepted easily. User requirements analysis can help to explore what skills drivers value, what they would be prepared to give up and what new skills they might develop when new technology is introduced.

3. As with any form of automation, problems can occur if the user of the system is not informed about automated actions that take place or given the flexibility to turn them on and off. Such systems also need to fit with the natural behaviour of the driver, which might be thought of in organisational

terms as the 'cultural context'. Knowledge of this context for particular groups of drivers will help predict how well certain in-car technologies will be accepted and used.

4. People in an organisation are normally more motivated if they can see how their particular job or role contributes to the broader success of the company or organisation. If the use of new technology is seen by the driver as contributing to desirable high-level goals, such as enhancing their driving skills, improving the experience of the passengers, and reducing costs or emissions, then it will be more attractive and acceptable.

In general terms, driving can be equated to any job or task that a person performs. Technology in the car that promotes the learning of new skills, supports the driving task, contributes to driver's aspirations and fits in with their culture and values, will greatly enhance the chances of its take-up.

References

Alhussain, T. and Drew S. 2009. Towards user acceptance of biometric technology in E-Government: A survey study in the Kingdom of Saudi Arabia. In *IFIP International Federation for Information Processing*, 26–38. Boston: Springer.

Boonstra, A. and Broekhuis, M. 2010. Barriers to the acceptance of electronic medical records by physicians from systematic review to taxonomy and interventions. *BMC Health Services Research*, 10: 1–17.

Bostrom, R.P., Olfman, L. and Sein, M.K. 1990. The importance of learning style in end-user training. *Management Information Systems Quarterly*, 14(1): 101–19.

British Vehicle Rental and Leasing Association. 2012. Driving at work guide. Available at: http://www.bvrla.co.uk/Advice_and_Guidance/Driving_at_work.aspx [accessed: 26 November 2013].

Clegg, C.W. 2000. Sociotechnical principles for system design. *Applied Ergonomics*, 31(5): 463–77.

Cobb, J.G. 2002. Menus behaving badly. *New York Times*, 12 May. [Online: 12 May]. Available at: http://www.nytimes.com/2002/05/12/automobiles/menus-behaving-badly.html [accessed: 26 November 2013].

Davis, F.D. 1989. Perceived usefulness, perceived ease of use, and user acceptance of information technology. *Management Information Systems Quarterly*, 13(3): 319–40.

Davis, F.D., Bagozzi, R.P. and Warshaw, P.R. 1989. User acceptance of computer technology: A comparison of two theoretical models. *Management Science*, 35(8): 982–1003.

Donais, B. 2006. *Workplaces That Work: A Guide to Conflict Management in Union and Non-Union Work Environments*. Aurora, Ontario: Canada Law Book.

Available at: http://www.mediate.com/articles/donaisB3.cfm [accessed: 10 September 2013].

Eason, K.D. 1988. *Information Technology and Organisational Change*. London: Taylor and Francis.

European Commission. 2008. Recommendation on safe and efficient in-vehicle information and communication systems: Update of the European Statement of Principles on human–machine interface, 26 May. Available at: http://eur-lex. europa.eu/LexUriServ/LexUriServ.do?uri=CELEX:32008H0653:EN:NOT [accessed: 26 November 2013].

Finn, P. 2011. A future for drones: Automated killing. *Washington Post*, 20 September. Available at: http://www.washingtonpost.com/national/national-security/a-future-for-drones-automated-killing/2011/09/15/gIQAVy9mgK_ story.html [accessed: 10 September 2013].

Gilbert, R.K. 2004. *BMW iDRIVE*. Prepared for Dr M. Lewis, Professor of Information Science, University of Pittsburgh. [Online: 4 March]. Available at: http://www2.sis.pitt.edu/~rgilbert/is2350/documents/iDrive.pdf [accessed: 26 November 2013].

Hackman, J.R. and Lawler, E.E. 1971. Employee reactions to job characteristics. *Journal of Applied Psychology*, 55(3): 259–86.

Harrison, M.I., Koppei, R. and Bar-Lev, S. 2007. Unintended consequences of information technologies in health care: An interactive sociotechnical analysis. *Journal of the American Medical Informatics Association*, 14: 542–9.

International Organization for Standardization (ISO). 2010. *Ergonomics of human-system interaction – Part 210: Human-centred design for interactive systems*. Geneva: ISO.

Kirk, M.S. 1983. Organisational aspects. In *Ergonomic principles in office automation: State of the art reports and guidelines on human factors in the office environment*, 115–33. Bromma, Sweden: Ericsson Information Systems AB.

Maguire, M.C. 2003. The use of focus groups for user requirements analysis. In: Focus groups: Supporting effective product development. Edited by J.D. Langford and D. McDonagh. Chapter 5, 73–96. London: Taylor and Francis.

———. 2013. Sociotechnical systems and user interface design: 21st century relevance. In 'Advances in Sociotechnical Systems Understanding and Design: A Festschrift in Honour of K.D. Eason, Applied Ergonomics'. In press. http:// dx.doi.org/10.1016/j.apergo.2013.05.011.

Nasaw, D. 2012. Driverless cars and how they would change motoring. *BBC News Magazine, Washington*, 10 May. Available at: http://www.bbc.co.uk/news/ magazine-18012812 [accessed: 26 November 2013].

Niedermaier, B., Durach, S., Eckstein, L. and Keinath, A. 2009. The new BMW iDrive: Applied processes and methods to assure high usability. In *Digital Human Modeling, Lecture Notes in Computer Science*, no. 5620: 443–52.

Nunes, L. and Recarte, M.A. 2002. Cognitive demands of hands-free-phone conversation while driving. *Transportation Research*, F5: 133–44.

Parasuraman, R., Sheridan, T.B. and Wickens, C.D. 2000. A model for types and levels of human interaction with automation. *IEEE Transactions on Systems, Man and Cybernetics* 30(3): 286–97.

Reason, J. 2004. Beyond the organisational accident: The need for 'error wisdom' on the frontline. *Quality and Safety in Health Care* (now: *BMJ Quality and Safety*), 13(ii): 28–33.

Royal Society for the Prevention of Accidents (ROSPA). 2012. Driving for work: Vehicle technology. [Online: The Royal Society for the Prevention of Accidents]. Available at: http://www.rospa.com/roadsafety/info/vehicletech.pdf [accessed: 26 November 2013].

Talbot, D. 2012. App battles driver distraction but spares passengers: A new approach inhibits dangerous phone use by detecting when a driver is on the phone. *MIT Technology Review*. Available at: http://www.technologyreview.com/news/426889/app-battles-driver-distraction-but-spares-passengers/ [accessed: 26 November 2013].

Vickers, E. 2007. Coping with speech noise in the modern workplace. Online: *The Sound Guy, Inc.* Available at: http://chatterblocker.com/whitepapers/conversational_distraction.html [accessed: 26 November 2013].

Waterson, P.E. 2005. Sociotechnical design of work systems. In *Evaluation of Human Work (3rd Edition)*, 769–92. Edited by J.R. Wilson and E.N. Corlett. London: Taylor and Francis.

Chapter 20

Adaptive Policymaking for Intelligent Transport System Acceptance

Jan-Willem van der Pas
Delft University of Technology, Faculty of Technology,
Policy and Management, the Netherlands

Warren E. Walker
Delft University of Technology, Faculty of Technology,
Policy and Management, and Faculty of Aerospace, the Netherlands

Vincent Marchau
Radboud University, Nijmegen School of Management, the Netherlands

Sven Vlassenroot
Ghent University, Belgium, and Flanders Institute for Mobility, Belgium

Introduction

Intelligent Transport System (ITS) implementation is often hindered by the uncertainties that surround implementation (see, e.g., Marchau et al. 2002, van Geenhuizen and Thissen 2002, Walta 2011, van der Pas et al. 2012). Often this uncertainty relates to general public acceptance of the ITS technology, the future acceptance of the technology or the dynamics in acceptance of the technology. (Here we refer to the uncertainty regarding future ITS acceptance among stakeholders due to, for instance, changes in the trade-offs stakeholders make among ITS outcomes and changes in the stakeholder configuration.) Transport policymakers seem paralysed in the face of this uncertainty. Often, this results in the abandonment of implementation of ITS (e.g., the implementation of road pricing in the Netherlands) or a delay in implementation due to the conclusion that more research is needed before a decision can be made (e.g., the numerous trials of Intelligent Speed Adaptation that have been held across the world – see van der Pas, Marchau and Walker 2006). But what should transport policymakers do in situations in which the future is so uncertain that analysts cannot agree upon the right model or have little understanding of what the future will look like? Hereafter, we refer to this type of uncertainty as 'deep uncertainty'.

In this chapter we introduce a policymaking approach that is especially designed to deal with deep uncertainty in developing policies. This approach is called Adaptive

Policymaking (APM). APM is a policymaking approach that was developed at the end of the 1990s at the RAND Corporation in response to the need to cope with deep uncertainty in long-term policymaking for Amsterdam Airport Schiphol (RAND Europe 1997). The approach aims at creating policies that can change over time, as the world changes, and uncertainties about the future are resolved. APM specifies a series of generic steps for decision-making under uncertainty that can be used to design an adaptive policy. The steps in APM are based on the steps of Systems Analysis (Miser and Quade 1985), and key concepts are derived from Assumption-Based Planning (ABP) (Dewar 2002). The potential of APM has been demonstrated by various researchers using transportation cases that reflect real-world policy problems (Agusdinata, Marchau and Walker 2007, Marchau et al. 2008, Agusdinata and Dittmar 2009, Taneja, Ligteringen and Schuylenburg 2010a, Taneja et al. 2010b, Kwakkel, Walker and Marchau 2010, Marchau, Walker and van Wee 2010). However, APM has seen little practical application. Only recently, almost 10 years after the first publication on APM, has attention been given to the practical use of APM (Kwakkel 2010, Walker, Marchau and Swanson 2010, van der Pas 2011).

Why is it important to include a chapter on how transport decision-makers can deal with uncertainty in a book that discusses acceptance issues for ITS technologies? ITS are highly promising when it comes to achieving transportation policy goals (e.g., less emissions, less congestion and a generally safer transport system). However, public acceptance of ITS proves crucial for its implementation and, as such, for contributing to these policy goals. Often the acceptance of the use of ITS (or policies that require the use of ITS, such as road pricing) is deeply uncertain, and in most cases policymakers do not know how, and/or traditional tools are insufficient, to cope with this uncertainty (see, e.g., van der Pas et al. 2006 and van der Pas, Kwakkel and van Wee 2011). This chapter describes a methodology that policymakers can use to overcome the uncertainties that hinder ITS implementation and that can enable them to start to implement ITS despite these uncertainties and the inherently uncertain future.

This chapter answers the question: how can transportation policymakers deal with the deep uncertainty regarding acceptance that surrounds policies aimed at implementing ITS? In particular, in this chapter, we

- explain APM;
- explain how APM can be used to deal with uncertainty regarding acceptance; and
- illustrate how to use APM to design adaptive policies using two real-world ITS examples (one based on desk research, the other based on participative research).

After reading this chapter, a reader will understand what APM is and how to use it. The chapter will also supply the reader with sources to find more information on this subject. In the next section Adaptive Policymaking is introduced and the basic principles are explained. Adaptive Policymaking is then outlined using two cases

– Personal Intelligent Travel Assistance (PITA) and Intelligent Speed Adaptation (ISA). In the final section, the main conclusions are presented.

Adaptive Policymaking

APM is a process of policy design that has five phases: Phase I sets the stage. Phases II, III and IV design the part of the adaptive policy that can be implemented at a certain moment in time (call this $t = 0$). Phase V designs the part of the adaptive policy that is to be implemented at an unspecified time after $t = 0$ (call this $t = 0+$). Figure 20.1 presents the APM process, together with the elements that comprise an adaptive policy. We briefly explain each phase, define each of their elements (policy actions), and elaborate on techniques that could be used to facilitate the

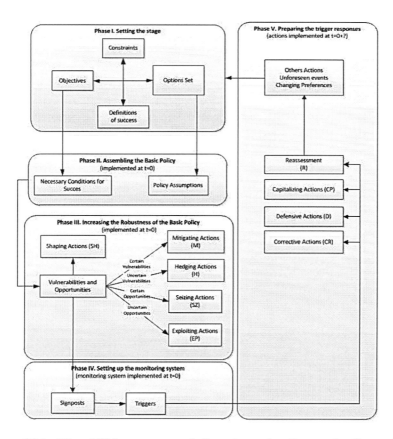

Figure 20.1   The APM process and the elements of an adaptive policy (adapted from Kwakkel 2010)

phase in a workshop setting. For more extensive descriptions and examples of the APM process, see Walker, Rahman and Cave (2001), Kwakkel et al. (2010), van der Pas (2011) and/or Marchau et al. (2008).

Setting the Stage (Phase I) and Assembling the Basic Policy (Phase II)

In this phase, the policy problem is analysed and the goals of the policy are formulated. Setting the stage is an important part of the APM process. The right policy problem has to be identified and formulated, goals and a definition of success have to be specified and a comprehensive list of policy options has to be generated.

Assembling the Basic Policy (Phase II)

Based on an ex-ante evaluation of the policy options identified in Phase I, a promising basic policy is assembled; that is, a promising starting policy. In this phase, the conditions for achieving success are also formulated. The methods in this phase are practically the same as the methods used in traditional ex-ante policy analysis to identify a promising policy (Miser and Quade 1985). In practice, there are many methods that can be used for the ex-ante evaluation of the policy options: for example, cost-benefit analysis (Sassone and Schaffer 1978), multi-criteria analysis (French, Maule and Papamichail 2009) and balanced scorecards (Kaplan and Norton 1993). These assessment techniques can be combined with the results from forecasts, scenarios, models and so on.

Increasing the Robustness of the Basic Policy (Phase III)

This phase and the following phases are designed to make the basic policy adaptive. After selecting a basic policy, the vulnerabilities and opportunities of the basic policy are identified. Vulnerabilities of the basic policy relate to ways in which the basic policy could fail (i.e., violate conditions for success). Opportunities are developments that can increase or accelerate the success of the basic policy (i.e., accelerate conditions for success). The vulnerabilities of the basic policy can be determined by examining the implicit and explicit assumptions that underlie it. Based upon the vulnerabilities and the opportunities, five types of actions can be defined that could be taken at the time the basic policy is implemented ($t = 0$), in order to increase the chances for its success:

- Mitigating actions (M) – actions aimed at reducing the relatively certain vulnerabilities of a policy;
- Hedging actions (H) – actions aimed at spreading or reducing the risk of failure from the relatively uncertain vulnerabilities of a policy;
- Seizing actions (SZ) – actions aimed at seizing relatively certain available opportunities;
- Exploiting actions (EP) – actions aimed at exploiting relatively uncertain

opportunities; and

- Shaping actions (SH) – actions aimed at reducing the chance that an external condition or event that could make the policy fail will occur, or to increase the chance that an external condition or event that could make the policy succeed will occur.

Setting-Up the Monitoring System (Phase IV)

The actions defined in Phase III are taken in advance to reduce the vulnerabilities of the basic policy and to identify opportunities to improve its chances of success. However, uncertainties about the future require the performance of the basic policy to be monitored carefully in order to know when (and if) to implement actions. This monitoring mechanism is set up in Phase IV by defining what should be monitored (signposts) and when a change in policy is needed (trigger values). Signposts are used to determine whether a defensive, corrective or capitalising action – or even a full policy reassessment – is needed (see Phase V). Implementation of a defensive, corrective or capitalising action, or a policy reassessment, occurs when a critical value of a signpost variable (trigger value) is reached.

Preparing the Trigger Responses (Phase V)

There are four different types of actions that can be triggered by a signpost:

- Defensive actions (D) – actions aimed at clarifying the basic policy, preserving its benefits or meeting outside challenges in response to specific triggers. These actions leave the basic policy unchanged;
- Corrective actions (CR) – actions aimed at adjusting the basic policy;
- Capitalising actions (CA) – actions triggered by external developments that improve the performance of the basic policy; and
- Reassessment (R) – an action that is initiated when the analysis and assumptions critical to the plan's success have clearly lost validity.

These actions are designed in Phase V. Once the basic policy and adaptive elements are agreed upon, the actions from Phases I–IV are implemented (at $t = 0$); the actions for Phase V are prepared but their implementation is suspended until a trigger event occurs.

Applying APM to the Implementation of a Personal Intelligent Travel Assistant

The Personal Intelligent Travel Assistant (PITA)

A major objective for transport policies is the efficient use by travellers of the existing transport infrastructure capacity. Although travel information through radio, television, the Internet etc., is widely available, its effectiveness is low, since those travellers that are offered alternative routes/modes generally do not accept them (Muizelaar 2011, Dicke 2012).

Therefore, a mobile phone-based travel information service has been developed that provides travellers with a full overview of travel options for travelling in the most efficient and effective way from a specific origin to a specific destination. This so-called Personal Intelligent Travel Assistant (PITA) has recently become available, but implementation is proceeding very slowly. So far, policymaking on PITA has been limited to supporting research and development designed to reduce the uncertainty in the outcomes of a PITA. In particular, the behavioural response of travellers to advanced travel information has been researched in depth (for an overview, see Chorus, Molin and van Wee 2006). Although useful, assumptions in most of these studies include the continuous availability of necessary traffic information, a perfectly functioning technology and a rational traveller. These assumptions with respect to the traveller response to PITA are unlikely to be valid. In any event, they are insufficient for PITA implementation to proceed. Instead of additional research and development on reducing the uncertainty of the outcomes, implementation of PITA could be sped up by developing an adaptive policy that takes into account the full range of uncertainty and modifies the basic policy based on what is learned over time.

Designing an Adaptive Policy Using Desk Research

The following subsections show how an adaptive policy for PITA implementation was designed using desk research and existing information. More information on methods and tools that can be used to design adaptive policies can be found in van der Pas (2011).

Phase I (Setting the Stage) and Phase II (Assembling the Basic Policy)

In Phase 1 of developing an adaptive policy for PITA implementation, important constraints would be financial and a requirement that the achievement of other transport policy objectives (e.g., safety, environmental stress) not be made more difficult due to the implementation of PITA. A definition of success might be a pre-specified improvement in (the reliability of) travel times. For instance, national policy objectives in the Netherlands include that, in 2020, 95 per cent of all movements by road should be on time during rush hours, and 90 per cent of all trains should be on time (Ministry of Transport, Public Works and Water Management 2000). Several alternative PITA options can be specified for consideration in Phase II.

In Phase II, a basic policy might be to implement PITA first for those individuals who have high demands on their time – for example, for professional drivers and business travellers (Polydoropoulou and Ben-Akiva 1998, Bovy 2001). These travellers are likely to be the most willing to adopt PITA since, by definition, they are the sub-group that is most affected by travel time losses and unreliability. Basic conditions for success include the willingness of key actors (e.g., road traffic managers, public transport operators) to provide reliable and accurate travel information, the availability of integrated models to combine multimodal travel data to meet individual preferences and the willingness of professional drivers and business travellers to buy and use PITA.

Phase III (Increasing the Robustness of the Basic Policy) and Phase IV (Setting Up the Monitoring System)

In Phase III, the several vulnerabilities of this basic policy are identified. A certain vulnerability might be a temporary lack of travel data availability for certain modes. This will likely affect the user acceptance of PITA. A mitigating action might be to include a backup travel information system that travellers can use in case of a temporary blackout. Another certain vulnerability would be that travellers resist the willingness to buy PITA because it affects their privacy; that is, it seems like 'Big Brother' watching their travel behaviour. Some travel-data encoding that avoids personal identification in relation to travel choices can be used to mitigate this vulnerability. An uncertain vulnerability involves the user acceptance of PITA – in particular, whether the PITA advice will be followed by travellers (Bonsall 2004). A signpost can be constructed that monitors the level of PITA use. As soon as the level of use drops under a predefined level (trigger), some corrective action might be initiated, such as advertising or educating travellers on the advantages of using PITA when travelling. This is related to another uncertain vulnerability – the willingness of key actors to cooperate on implementing PITA due to, for instance, too large investment risks for (public and/or private) transport operators. A hedging action might be that, at the beginning, public policymakers give some insurance for companies against potential investment losses.

Phase V (Preparing the Trigger Responses)

Once the above policy is agreed upon, the basic PITA policy plus the Phase III and Phase IV actions are implemented, and signpost information begins to be collected (see Table 20.1). In the case of a trigger event, the related prepared action is undertaken. If, for instance, the number of travellers following the PITA advice appears to be too low, some corrective action can be undertaken – for example, giving some financial incentive to those travellers who do comply with the PITA advice. For some trigger events, only a full reassessment of the basic policy might be sufficient. In case some of the key actors are not willing to participate anymore (e.g., if the returns on investment remain too low), the entire policy might come under serious pressure. However, the knowledge gathered in the initial policymaking process on outcomes, objectives, measures, preferences of stakeholders and so on would already be available and would accelerate the new policymaking process.

Table 20.1   Dealing with vulnerabilities of the basic PITA policy

Vulnerabilities	Mitigating/hedging actions	Possible signposts/triggers/ actions
Certain: (temporary) lack of travel data availability	Mitigating action: • Provide backup travel information system	
Certain: Willingness of travellers to buy PITA due to, e.g., privacy reasons, individual cost-benefit trade-offs	Mitigating actions: • Provide travel-data encoding ensuring privacy of travellers • Give (financial) incentives to travellers for buying PITA	
Uncertain: Willingness of professional drivers and business travellers to use PITA	Hedging action: • Explain advantages of PITA use to target groups	Monitor the level of PITA use. In case of too low usage (trigger), implement corrective action (e.g., provide incentives to travellers for using PITA; expand basic policy to other target groups).
Uncertain: Willingness of key actors to cooperate on implementing PITA	Hedging action: • Provide insurance for PITA companies against potential investment losses	Monitor the level to which key actors are willing to cooperate on implementing PITA. In case cooperation appears insufficient, a total reassessment of the adaptive policy is needed.

Applying APM to ISA Implementation

Intelligent Speed Adaptation in the Netherlands

Intelligent Speed Adaptation (ISA) systems are in-vehicle devices that take into account the local speed limits and warn the driver in case of speeding; some even automatically adjust the maximum driving speed to the posted maximum speed. Since speeding is the major cause of traffic accidents – roughly a third of all fatal accidents are due to inappropriate speed choice (OECD 2006) – the potential contribution of ISA to traffic safety is high. For instance, fully automatic speed control devices are estimated to produce up to a 40 per cent reduction in injury accidents (Vàrhelyi and Mäkinen 2001) and up to a 59 per cent reduction in fatal accidents (Carsten and Tate 2000). Recently, the first ISA applications have entered the market. Speed-limit information is being added to digital maps, so drivers can be warned about speeding by their navigation device using audiovisual signalling (this is called warning ISA).

So the ISA technology is available and there is experience with using it. Although expectations concerning the positive impacts of ISA are high, there still is a considerable gap between what is technologically possible and what has been implemented so far. The implementation of ISA is hindered by various deep uncertainties, including uncertainty about the way users might respond to ISA. In this case, an adaptive policy for ISA implementation was developed with ISA experts, policymakers and stakeholders during a workshop.

Phase I (Setting the Stage) and Phase II (Assembling the Basic Policy)

Important constraints for developing an adaptive policy for ISA implementation would be financial and the requirement that other transport policy objectives (e.g., safety, environmental stress) are not made more difficult to achieve due to the implementation of ISA. A definition of success in general terms would relate to the improvement of traffic safety (e.g., a reduction of 10 per cent in the number of fatalities). Based on the selected basic policy, the definition of success and the constraints have been operationalised in Table 20.2. Following interviews with policymakers from the Dutch Ministry of Infrastructure and Environment and existing policy plans, we adopted a basic policy aimed at implementing the most appropriate ISA for the most appropriate type of driver. Three types of drivers were distinguished:

- *The well-meaning driver*: This type of driver has the intrinsic motivation to stick to the speed limit;
- *The less well-meaning driver*: This type of driver lacks the intrinsic motivation to stick to the speed limit; and
- *The notorious speed offender*: Under the current regime, this type of driver would lose his or her driver's licence (and would be obliged to follow a traffic behaviour course).

In addition to different types of drivers, two different sequential phases for the implementation of ISA were identified. Phase I runs up to 2013. After 2013, a currently undefined Phase II will start. Table 20.2 presents an overview of the basic policy.

Table 20.2   Basic policy for the ISA case

Basic policy				
Type of driver	**Type of ISA**	**Measure**	**Definition of success**	**Constraints**
Phase I (2009–2012)				
Well-meaning driver	Warning ISA or speed alert	Start a campaign aimed at persuading people to turn on the speed alert functionality on their navigation device. Make agreements with companies that develop navigation devices.	Before 2013: 50% of the people that own and use a navigation device actively use the speed alert functionality.	Budget for a campaign
Less well-meaning driver (But also covers the well-meaning driver)	Free to be selected	Develop a business case with insurance companies and leasing companies.	Before 2013: 50% of the car owners and 50% of leased car drivers can choose insurance or a lease product that involves ISA.	
Notorious speed offender	Restricting	Perform a pilot test aimed at assessing the effects of implementing a restricting ISA for notorious speed offenders. Make an evidence-based decision regarding implementation of such a system for notorious speed offenders.	Before 2013: A decision has to be made on implementation of ISA for notorious speed offenders (based on, amongst others things, outcomes of the trial).	Budget/time
Phase II (2013): Phase II will be dependent on the results of Phase I. For this phase, more restricting types of ISA will be considered.				

As can be seen in Table 20.2, making a practical distinction between well-meaning and less well-meaning is not needed, because both groups are targeted with the same policies. However, it is expected that the measures would have a different effect on each of the target groups. (Notorious speeders can be defined based on past behaviour.)

Phase III (Increasing the Robustness of the Basic Policy)

The vulnerabilities and opportunities of the basic policy were specified using a Strengths, Opportunities, Weaknesses and Threats (SWOT) analysis structure

(Ansoff 1987). In our case, we considered both the opportunities and strengths to be opportunities as defined in Figure 20.1, and considered both the weaknesses and threats to be vulnerabilities as defined in Figure 20.1. This resulted in a list of more than 100 different opportunities and vulnerabilities for the basic policy. According to the participants, the most important of these relate to acceptance, technical functioning of ISA systems, the relationship between technical functioning and acceptance and the relationship between technical functioning and driver behaviour (for a full overview, see van der Pas 2011).

Next, the level of uncertainty and level of impact for each of the most important opportunities and vulnerabilities were identified and the participants were tasked to define actions for handling these. The process included ranking techniques and specially designed decision-making flowcharts (see van der Pas 2011). Table 20.3 presents a subset of Phase III actions that were generated during the workshop. (The complete set can be found in van der Pas 2011.)

Table 20.3   Increasing the robustness of the basic policy

Vulnerabilities and opportunities (certain or uncertain)	Actions: hedging (H), mitigating (M), seizing (SZ), exploiting (EP) and shaping (SH)
Implementing a restricting ISA for notorious speed offenders will damage the image of the less intervening ISA systems. ISA will be associated with punishment, not with assistance (like it is now). **(uncertain)**	**H:** Decouple the pilot from the rest of the basic policy and avoid the term 'ISA' (currently done by calling it speed-lock).
The availability of an accurate speed-limit database. Speed-limit data have to match the time (dynamic), location and vehicle. **(certain)**	This is a critical success factor, so **M:** Define who is responsible for what before starting with implementation; **M:** Tender the development of a speed-limit database (this should be arranged by public authorities); **M:** Guarantee quality through a third party that is under the supervision of the public authorities; **M:** Develop a system based on beacons that can overrule the static speed-limit information (failsafe design).
Automotive lobby to prevent large-scale implementation of ISA. **(uncertain)**	**H:** Include the automotive industry in the implementation strategy.
Speed-limit data become more and more dynamic. **(certain)**	**M:** Implement ISA systems that will function well when this happens (for instance, systems that allow for communication with the infrastructure, to transmit temporary speed limits, e.g., Bluetooth, Wi-Fi).

Cars and ISA draw lots of attention and appeal to peoples' emotions. Instead of seeing this as a threat, this can be used as an opportunity. **(uncertain)**	**SH:** Invite stakeholders that exhibit these feelings to participate in improving and implementing ISA (e.g., the presenters of Top Gear, racing drivers, etc.).
People/companies are more willing to adopt technology if they can see the technology in practice. Creating a pool of cars that are equipped can result in an uptake of the technology. **(uncertain)**	**SH:** Practice what you preach. Let the Ministry themselves equip their fleet with ISA and practice an example function. Prove that it significantly reduces the number of accidents and as such results in fewer claims.

Phase IV (Setting-Up the Monitoring System) and Phase V (Preparing the Trigger Responses)

Next, actions, signposts and triggers were designed (also using specially designed decision-making flowcharts (see van der Pas 2011). A subset of these actions, signposts and triggers is shown in Table 20.4.

Table 20.4   Contingency planning, monitoring system and trigger responses

Vulnerabilities and Opportunities	Monitoring and triggering system	Actions: reassessment (R), corrective (CO), defensive (D) and capitalising (CA)
Implementing a restricting ISA for notorious speed offenders will damage the image of the less intervening ISA systems. ISA will be associated with punishment, not with assistance (like it is now).	• Number of negative press publications • Level of acceptance of different ISA systems • Number and type of ISA-related questions asked in the politicians in the Lower House	**D:** Media campaigns to manage the perception of people regarding ISA (and the speed-lock); explain the difference between the two systems, and the need for implementing restricting ISA for this type of driver
The availability of an accurate speed-limit database. Speed-limit data have to be correct for the right time (dynamic), the right location and the right vehicle.	• Level of accuracy/reliability of speed-limit database	Changes in accuracy should be monitored over time. In addition, **D:** Start making it more accurate; **CO:** Stop implementation of certain types or combine with on/off switch and overruling possibilities; **CO:** Design the system in such a way that it only warns/intervenes in areas with certain accuracy levels.

Technology can fail: location determination can be inaccurate (e.g., in tunnels, in cities with high buildings); systems can stop functioning (sensors fail, etc.).	• Cause of accidents (relationship ISA – cause of accident) • Press releases on ISA and accidents	**D**: Make sure the market improves the systems (adjust implemented rules and regulations regarding system functioning). **R**: When large-scale failure occurs or the effects are drastic (ISA implementation leads to fatalities).
Speed limit data become more and more dynamic.	• Availability of dynamic speed limits	**D**: Make sure road authorities equip new dynamic speed limit infrastructure with infra-to-vehicle communication (so in-vehicle systems can be easily adjusted). **D**: Standardisation of communication protocol and communication standard
ISA implementation can result in larger cost savings than expected: lower and more homogeneous speeds => lower consumption costs (fuel savings + lower maintenance), resulting in higher levels of acceptance.	Monitor additional effects of implementation on: • emissions; • fuel use; and • throughput/congestion.	**CA**: Upscale the number of participating insurance companies. **CA**: Use this information in the business case for new insurance and lease companies.

The centre column of Table 20.4 can be transformed into a list of indicators that should be monitored: 'the monitoring system'. This monitoring system consists of signposts that measure the progress towards the goal (i.e., success), and signposts that are directly related to the vulnerabilities and opportunities.

Phase V (Preparing the Trigger Responses)

The workshop resulted in the development of an extensive adaptive policy, including a total of 26 mitigating actions, 16 defensive actions, three reassessment actions, two capitalising actions and two seizing actions. In practice, once the basic policy and all its adaptive elements have been agreed upon, the basic ISA implementation policy (Table 20.2) plus the Phase III and Phase IV actions would be implemented and signpost information would begin to be collected (see Tables 20.3 and 20.4). In case of a trigger event, the related (already prepared) action would be undertaken.

The Result

The designed policy was tested using wildcard scenarios, in order to determine how robust it might be. One example of a wildcard scenario is after ISA is implemented, industry starts to develop equipment that misleads the ISA systems, allowing people to speed without the system noticing. The participants were asked to think about 'what if' such a wildcard scenario were to occur. In particular, for each scenario, they were asked to answer the following questions:

- What would happen to the (road) transport system?
- What would happen to your policy, and how would the outcomes of the policy be influenced if this scenario were to occur?
- Is your adaptive policy capable of dealing with this scenario?

These wildcard scenarios led to interesting (and lengthy) discussions, which allowed the participants to reflect on the developed adaptive policy, assess its robustness and improve it.

The participants' evaluation of the workshop indicated that the resulting adaptive policy was ready to be implemented and could, if implemented, really contribute to a successful ISA policy (see van der Pas et al. 2011).

Conclusion

In this chapter, we introduced a relatively new approach that allows transportation policymakers to deal with the uncertainties that surround the implementation of ITS technologies. Based on two examples, the chapter shows that APM is an approach that allows policymakers to deal with (amongst others) issues of acceptance, and should allow them to speed-up the implementation of ITS technologies.

Two adaptive policies were designed, one for PITA based on desk research and one for ISA based on a participative workshop with ISA stakeholders. The basic PITA policy is designed for those drivers that could benefit most from PITA. The basic ISA policy is designed to implement different types of ISA for different types of drivers. Both policies would begin with the use of the ITS systems by small subsets of transport users. Both would offer the possibility of modifying the policy gradually as more information regarding acceptance becomes available (based on monitoring acceptance). This approach would allow for implementation to begin right away, for policymakers to learn over time and for the policy to be adjusted in response to new developments.

A lot of research has been performed on ISA and PITA acceptance. The time has come to begin implementation. APM is an approach that allows policymakers to deal with (amongst others) issues of acceptance, and should allow them to speed-up the implementation of ITS technologies.

References

Agusdinata, D.B. and Dittmar, L. 2009. Adaptive policy design to reduce carbon emissions, a system-of-systems perspective. *IEEE Systems Journal*, 13(4): 509–19.

Agusdinata, D.B., Marchau, V.A.W.J. and Walker, W.E. 2007. Adaptive policy approach to implementing intelligent speed adaptation. *IET Intelligent Transport Systems*, 1(3): 186–98.

Ansoff, H.I. 1987. *Corporate Strategy*. Harmondsworth, Middlesex, UK: Penguin Books.

Bonsal, P. 2004. Traveller behavior: Decision-making in an unpredictable world. *Journal of Intelligent Transportation Systems: Technology, Planning, and Operations*, 8(1): 45–60.

Bovy, P. 2001. Traffic flooding the low countries: How the Dutch cope with motorway congestion. *Transport Reviews*, 21(1): 89–116.

Carsten, O.M.J. and Tate, F.N. 2000. *Final report: Integration, Deliverable 17 of External Vehicle Speed Control Project*. Leeds: University of Leeds, Institute for Transport Studies.

Chorus, C.G., Molin, E.J.E. and van Wee, G.P. 2006. Travel information as an instrument to change car-drivers' travel choices: A literature review. *European Journal of Transport and Infrastructure Research*, 6(4): 335–64.

Dewar, J.A. 2002. *Assumption-Based Planning: A Tool for Reducing Avoidable Surprises*. Cambridge: Cambridge University Press.

Dicke, M. 2012. *Psychological aspects of travel information presentation*. Delft, the Netherlands: TRAIL Thesis Series, T2012/5.

French, S., Maule, J. and Papamichail, N. 2009. *Decision Behaviour Analysis and Support*. Cambridge: Cambridge University Press.

Kaplan, R.S. and Norton, D.P. 1993. Putting the balanced scorecard to work. *Harvard Business Review*, 71(5): 134–47.

Kwakkel, J.H. 2010. *The treatment of uncertainty in airport strategic planning*. Delft, the Netherlands: TRAIL Thesis Series, T2010/13.

Kwakkel, J.H., Walker, W.E. and Marchau, V.A.W.J. 2010. Adaptive airport strategic planning. *European Journal of Transportation and Infrastructure Research*, 10(3): 227–50.

Marchau, V.A.W.J., Walker, W.E. and van Duin, R. 2008. An adaptive approach to implementing innovative urban transport solutions. *Transport Policy*, 15(6): 405–12.

Marchau, V.A.W.J., Walker, W.E. and van Wee, G.P. 2010. Dynamic adaptive transport policies for handling deep uncertainty. *Technological Forecasting and Social Change*, 77(6): 940–50.

Marchau, V.A.W.J., Wiethoff, M., Hermans, L., Meulen, R. and Brookhuis, K.A. 2002. *Actor Analysis Intelligent Speed Adaptation*. Final Report No. AV-5157. Delft/Wijk en Aalburg.

Ministry of Transport, Public Works and Water Management. 2000. *The National Traffic and Transport Plan*. The Hague.

Miser, H.J. and Quade, E.S. 1985. *Handbook of Systems Analysis: Overview of Uses, Procedures, Applications, and Practice*. New York: Elsevier Science.

Muizelaar, T.J. 2011. *Non-recurrent traffic situations and traffic information*. Delft, the Netherlands: TRAIL Thesis Series, T2011/16.

Organization for Economic Cooperation and Development (OECD). 2006. *Speed Management*. Paris: OECD.

Polydoropoulou, A. and Ben-Akiva, M. 1998. The effect of advanced traveler information systems (ATIS) on travelers' behavior. In: Behavioral and Network Impacts of Driver Information Systems. Emmerink, R.H.M. and Nijkamp (Eds.), 317–52, Aldershot, Ashgate.

RAND Europe. 1997. *Scenarios for Examining Civil Aviation Infrastructure Options in the Netherlands*. Report DRU-1512-VW/VROM/EZ. Santa Monica, CA: RAND.

Sassone, P.G. and Schaffer, W.A. 1978. *Cost-Benefit Analysis: A Handbook*. San Diego, CA: Academic Press.

Taneja, P., Ligteringen, H. and van Schuylenburg, M. 2010a. Dealing with uncertainty in design of port infrastructure systems. *Journal of Design Research*, 8(2): 101–18.

Taneja, P., Walker, W.E., Ligteringen, H., van Schuylenburg, M. and van der Plas, R. 2010b. Implications of an uncertain future for port planning. *Maritime Policy and Management*, 37(3): 221–45.

van der Pas, J.W.G.M. 2011. *Clearing the road for ISA implementation? Applying adaptive policymaking for the implementation of intelligent speed adaptation*. Delft, the Netherlands: TRAIL Thesis Series, T2011/13.

van der Pas, J.W.G.M., Kwakkel, J.H. and van Wee, B. 2011. Evaluating adaptive policymaking using expert opinions. *Technological Forecasting and Social Change*, 79(2): 311–25.

van der Pas, J.W.G.M., Marchau, V.A.W.J. and Walker, W.E. 2006. *An Analysis of International Public Policies on Advanced Driver Assistance Systems*. Presented at the 13th World Congress and Exhibition on Intelligent Transport Systems and Services, London.

van der Pas, J.W.G.M., Marchau, V.A.W.J., Walker, W.E., van Wee, G.P. and Vlassenroot, S.H. 2012. ISA implementation and uncertainty: A literature review and expert elicitation study. *Accident Analysis and Prevention*, 48: 83–96.

van Geenhuizen, M. and Thissen, W. 2002. Uncertainty and intelligent transport systems: Implications for policy. *International Journal for Technology, Policy and Management* 2(1): 5–19.

Vàrhelyi, A. and Mäkinen, T. 2001. The effects of in-car speed limiters: Field studies. *Transportation Research Part C: Emerging Technologies*, 9(3): 191–211.

Walker, W.E., Marchau, V.A.W.J. and Swanson, D. 2010. Addressing deep uncertainty using adaptive policies: Introduction to section 2. *Technological Forecasting and Social Change*, 77(6): 917–23.

Walker, W.E., Rahman, S.A. and Cave, J. 2001. Adaptive policies, policy analysis, and policymaking. *European Journal of Operational Research*, 128: 282–9.

Walta, L. 2011. *Getting ADAS on the road – Actors' interactions in advanced driver assistance systems deployment*. Delft, the Netherlands: TRAIL Thesis Series, T2011/4.

Chapter 21

Designing Automotive Technology for Cross-Cultural Acceptance

Kristie L. Young and Christina M. Rudin-Brown

Monash University Accident Research Centre, Australia

Abstract

The rising global distribution of automobiles necessitates that the vehicle human–machine interface (HMI) is appropriate for the regions to which they are exported. Differences in cultural values across regions have been shown to be relevant to the usability and acceptance of HMI in a range of fields. This chapter examines research relevant to cross-regional automotive HMI design. Limited research specific to cross-regional automotive HMI issues currently exists. However, a large body of cross-cultural HMI work exists in other domains, such as the human–computer interaction (HCI) domain. Through a review of work from across the HCI and automotive domains, this chapter identifies a number of important cultural factors that may impact upon automotive HMI design. It is concluded that addressing cultural, as well as wider regional factors, is a vital step in the global automotive HMI design process. A failure to consider these factors could have significant implications for the acceptability and usability of in-vehicle information systems and Advanced Driver Assistance Systems in different regions.

Introduction

Like most product markets, the automotive market is becoming increasingly globalised. This trend in globalisation leads to increased interaction between diverse geographical regions and cultures. Hence, a major challenge for globalisation is how best to incorporate and accommodate cultural differences in the design of products that are destined to be used across the globe. Differences in language, social structure, education, environment and cultural values lead to vast differences in how people perceive and value objects, how they interact with them, how they want them to operate and even if they find them useful (Marcus 2009, Nisbett 2003). The extent to which a product meets the preferences and expectations of people from a particular region or culture is a critical component to its success in that region. Inadequate consideration and adaptation of a product or system to the needs and preferences of a target culture can result in issues ranging from minor frustration on the part of users through to a total failure of the product (Nielsen 1996). Indeed, there are myriad cases in the cross-cultural literature where inadequate consideration and understanding of the cultural characteristics of a region has led to devastating outcomes. In one example, members of a particular culture did not understand the meaning of the skull-and-crossbones poison symbol placed on containers of grain intended only to sow crops. The members of the village mistook the poison symbol simply as an American logo and, rather than sow the grain, consumed it with fatal consequences (Casey 1998). This example highlights how the unique life experiences and views of a particular group or culture can influence the meaning and understanding of symbols from that originally intended. It also highlights the importance of considering cultural differences in the design of any global product.

With the automotive market expanding rapidly into markets such as China and India, there is a need to ensure the design of vehicle interfaces is appropriate for the intended export region. That is, the design must meet the regions' cultural expectations and also the needs dictated by their particular traffic environment, such as driving regulations and traffic density. This is particularly true of In-Vehicle Information Systems (IVIS) and Advanced Driver Assistance Systems (ADAS), which are becoming more prevalent and also more complex in their design and functionality. Traditionally, the design and development of IVIS and ADAS has largely focused on the needs and preferences of drivers from Western markets; however, these systems are now being introduced into emerging markets unchanged or modified only slightly to suit the basic specifications of the region. With cross-cultural automotive research in its infancy, concern remains as to how drivers from emerging markets such as China, whose culture, language and traffic environment differs substantially from Western societies, will react to Western standards for IVIS and ADAS design.

As asserted in other chapters in this book, the acceptance of in-vehicle technology by drivers is a critical factor influencing the successful uptake of the technology and its effectiveness in improving road safety. A failure by drivers to accept a technology can lead to them not using it in the manner intended, or not using it at all. Sprung (1990) noted the importance of accurate adaptation of technology to a culture on acceptance. Assuming it is usable, a system that has been tailored to meet the aesthetic and linguistic preferences of a culture can have a significant competitive edge.

This chapter examines the influence of cultural characteristics on in-vehicle technology design, with a particular emphasis on its influence on user acceptance of technology. Various aspects of culture (cultural dimensions) are identified and how they impact upon technology design and, in turn, acceptability is discussed. Before reviewing research examining the influence of culture on HMI design and acceptability, it is important to briefly discuss prominent cultural theories and models and define what is meant by the term 'culture'.

Cultural Theory and Models: How Cultures Differ from Each Other

Culture has a range of definitions, most of which are complementary rather than contradictory. In theories relevant to HMI (Hofstede 1980, Hoft 1996, Trompenaars and Hampden-Turner 1998), culture is defined as a set of shared beliefs, feelings, values, customs, actions and artefacts that members of a society or group use to manage and interact with the world and one another and that are transmitted across generations through implicit learning. Cultural characteristics manifest themselves visually, for example in art and language, and non-visually, such as in preferences and interaction styles (Christaans and Diehl 2007).

There are a number of cultural theories that attempt to define those characteristics that differentiate members of one region or culture from another. Cultural models are representations of the elements or dimensions that make up a culture. They define the aspects of a culture that are observable and measureable and allow a cultural profile to be built that can then be used to compare one culture to another. With respect to HMI design, cultural models can be used to identify the characteristics of a system that are likely to be acceptable and understood by a particular group and those features that need to be adapted to better meet the needs and preferences of the group or region.

Popular cultural models include the Pyramid Model (Hofstede 1980), the Iceberg Model (Hoft 1996) and the Onion Model (Trompenaars and Hampden-Turner 1998). Each of these models considers culture to be comprised of at least an outer surface layer (directly observable aspects of culture) and a deeper, hidden layer (intrinsic aspects of culture that are beyond immediate awareness). These models also describe culture in terms of cultural variables or dimensions. Dimensions are constructs upon which cultures may differ in terms of their values, attitudes and behaviour, and include, among others, factors such as the degree of connectivity between members of a culture and the degree to which members of a culture feel in control of their life and environment.

The Onion Model

The Onion Model developed by Trompenaars and Hampden-Turner (1998) proposes that culture is comprised of three layers, each of which can be 'peeled' back to reveal the more central inner layers . The outer surface layers represent those aspects of a culture that are visible, while the inner layers are less apparent.

The outermost, surface layer represents the observable and immediately recognisable aspects of culture and includes physical characteristics and objects such as clothing, art, architecture, traditional foods and language. These cultural features are said to represent expressions of the underlying principles of the culture that exist in the innermost layers. The middle layer of the model represents the norms and values of the culture. Norms are the collective sense of right and wrong and reflect how a culture should behave, often expressed in terms of laws. Values on the other hand, comprise a sense of good and bad and represent how a culture aspires to behave. The innermost, core layer represents the most inaccessible aspects of a culture that influence all of the other layers. It comprises the implicit, fundamental principles and assumptions of the culture, such as the principle of human equality.

The Pyramid Model

The Pyramid Model, developed by Hofstede (1980), is another three-layered cultural model, which emphasises the bidirectional relationship between the model's elements. Instead of depicting culture as made up of layers, this model describes the origins of culture. The Pyramid Model summarises aspects of human behaviour, or 'mental programs', that define behaviour at different scales: the individual, all of humankind, and collective groups or cultures.

Hofstede proposes that culture is developed both from aspects of the individual and from aspects of humankind. At the lower human nature level, are those characteristics that are inherited by, and common to, all humans. The upper, personality level, refers to characteristics that are specific to the individual and are both inherited and learned. The middle level represents the culture or the 'collective', with which both of the outer layers interact in a way specific to a particular group of people. A key feature of this model is that culture is not inherited, but learned; and, therefore, is dependent on others to be passed on.

The Iceberg Model

The Iceberg Model, as described by Hoft (1996), uses an iceberg metaphor to describe culture where only the tip of the iceberg is visible above the surface. This exposed section represents consciousness and, as applied to culture, contains the observable aspects of a culture such as clothing, art and language. The layer just below the surface represents the subconscious layer and contains the unspoken rules of a culture, such as social etiquette. The largest, deepest and most hidden layer of the model represents the unconscious rules of the culture, which are not readily accessible or observable, but are intrinsically important. These unconscious rules include a culture's sense of time and preferred personal space.

Cultural Dimensions

Cultural models describe culture in terms of dimensions. Cultural dimensions are constructs upon which cultures differ in terms of their values, attitudes and behaviour; that is, they account for the manner in which culture is expressed. Dimensions are usually dichotomous and cultural groups vary in terms of their orientation towards one pole or another. For instance, the perception of time is one dimension upon which cultures can differ, with some adopting a rigid stance on time and following it precisely, while others take a far more relaxed view.

A large number of cultural dimensions exist, with most originating from cultural models and theories. Table 21.1 provides a brief description of the most commonly used cultural dimensions derived from key cultural models (e.g., Hofstede 1980, Trompenaars and Hampden-Turner 1998) that have been used in cultural studies of technology use and preferences. There have been a number of theoretical descriptions of how orientations on various cultural dimensions may contribute to or account for differences in user needs, preferences and expectations of technology across cultural and regional groups. This issue is explored in the following sections.

Table 21.1   Key cultural dimension and their definitions

Dimension	Definition
Power Distance	The extent to which members of a culture feel comfortable with hierarchy and power imbalance in relationships, both personal and business. *Low power distance* cultures are uncomfortable with inequity; *high power distance* cultures are comfortable with imbalance.
Individualism–Collectivism	The degree of connectivity between members of a culture. *Individualist* cultures show looser bonds between people and the same rules are applied to all people in all contexts. *Collectivist* cultures are characterised by strong cohesive family and social units, and tend to evaluate each situation upon its merits.
Masculinity–Femininity	Concerns the distribution and attitude towards social roles and the degree of similarity in values across genders. *Masculine* cultures demonstrate greater discrepancy between the values of males and females, are more competitive and have more traditional gender roles. *Feminine* cultures show similar values among males and females, value equality and tend to be more modest.
Uncertainty Avoidance	The extent to which members of a culture feel comfortable with ambiguity and uncertainty. Low uncertainty avoiding cultures are more tolerant of the unknown, while high uncertainty avoiding cultures are uncomfortable with ambiguity and create rules and laws to reduce uncertainty.
Long- and Short-Term Orientation	This dimension was developed to account for values identified in Asian countries. *Long-term* oriented cultures make short-term sacrifices for a better future, seen in thriftiness, perseverance and protection of reputation or 'face'. At the opposite pole, *Short-term*-oriented cultures live more for and in the present and foster virtues related to the past and the present, in particular, respect for tradition, preservation of face and fulfilling social obligations.

Attitude towards the Environment	Concerns the way in which members of a culture view themselves in relation to their environment. An internal locus of control is associated with a sense of control over life and environment. Cultures with an external locus of control feel that these factors control them.
Attitudes towards Time	Refers to the manner in which members of a culture prefer to perform tasks. *Monochronic* cultures prefer to do things sequentially, strictly adhere to plans and deadlines and value promptness and punctuality. *Polychronic* cultures prefer to do many things at once, see deadlines as objectives and are more interruptible and distractible.

The Influence of Culture on System Design and Acceptability

It is expected that culture has a range of influences on the way people interact with technology and develop expectations and preferences for how a system should look and operate. No product is culture-free: the design of all technology is influenced by the culture it was designed in and for, and this is reflected in all aspects of the system, including its appearance, functionality and purpose (Honold 2000, McLoughlin 1999).

There have been numerous attempts to explain how a group's orientations on various cultural dimensions may account for differences in user needs, preferences and expectations of technology (Choi et al. 2005, Lodge 2007, Marcus and Gould 2000, Rau, Gao and Liang 2008). Some cultural dimensions have been more commonly examined in relation to HMI design than others. For example, Hofstede's (1980) power distance, uncertainty avoidance and individualism-collectivism dimensions have been more frequently assessed than any others. However, where existing literature has used cultural dimensions to account for differences in user needs and preferences in HMI, this has been done almost exclusively in a theoretical manner, with dimensions rarely tested or used to interpret findings. As Smith and Yetim (2004) note, the existing evidence on culture and technology use is almost entirely qualitative.

This section discusses key cultural dimensions and how these have been applied in the literature to explain how and why a person's culture influences their use and acceptance of technology. Much of the existing literature comes from the wider human–computer interaction (HCI) domain and has focused on website interfaces, exploring various aspects of usability or visual preferences. Thus, the focus here is on the influence of culture on technology use and acceptability in general. Automotive technology, and how culture influences its design and acceptability, is discussed in the following section.

Power Distance

The dimension of power distance has been linked to the structure and content of information on interfaces. Cultures that score high on power distance prefer highly structured information, arranged in tall hierarchies. They also prefer information that conveys a sense of authority, expertise and security, and that reflects traditional social roles. Low power distance cultures, by contrast, tend to prefer transparency, flatter, less structured information hierarchies and dislike symbols of authority and power imbalance (Lodge 2007, Marcus and Gould 2000).

Individualism-Collectivism

Individualism is associated with information-seeking behaviour and a greater belief in technical competence (Smith et al. 2004). Highly individualistic cultures have been found to favour website designs that convey personal success, are more youth and action oriented, emphasise change and contain more controversial content than collectivist interfaces (Lodge 2007, Marcus and Gould 2000). Collectivist-oriented cultures prefer interfaces that do not distinguish the individual from the group, contain more traditional or official information, and emphasise age, wisdom and experience. Also, individualistic cultures are expected to adopt technologies which allow for personalisation, while collectivist individuals should favour those technologies which create a sense of connectivity to others, such as those with communication features or access to social networking sites (Choi et al. 2005).

Masculinity-Femininity

Interfaces designed for masculine cultures should emphasise traditional age, gender and family roles, clear definitions of tasks, and facilitate a greater sense of individual control and exploration. These interfaces should also feature the use of games and competitions and the use of graphics and sounds for practical purposes. Conversely, interfaces designed for feminine-oriented cultures should use visual and auditory features as aesthetic attractors, and should emphasise cooperation, shared roles and support (Lodge 2007, Marcus and Gould 2000).

Uncertainty Avoidance

Cultures high in uncertainty avoidance desire predictability, clear and simple directions, transparent, rigid navigation systems and redundant labels. Low uncertainty-avoiding cultures, in contrast, are comfortable with ambiguity and dislike use of redundant information. Low uncertainty-avoiding cultures also enjoy complexity and opportunity to explore and resent not feeling in control of the system (Lodge 2007, Marcus and Gould 2000). High uncertainty-avoiding cultures prefer frequent and detailed instructions, while low uncertainty-avoiding cultures view this as annoying and excessive (Heimgårtner 2005).

Attitude Towards the Environment

As applied to technology, this dimension refers to the tendency for individuals to feel they are in control of the system. Cultures with an internal locus of control believe the system should adapt to them, and any errors are the fault of the system or its designers. Externally oriented cultures, by contrast, believe systems are designed in the optimum way and errors result from their improper use or by their misunderstanding the system. Also, members of cultures with an external sense of control are less inclined to want to alter or personalise systems, as they believe they have been configured in the optimum way already (Ito and Nakakoji 1996).

Attitudes Towards Time

Time orientation affects how people perform and prioritise tasks and, therefore, should be taken into account when designing information structure. How time is perceived is also expected to impact on how people learn to operate a system. Monochronic cultures (i.e., those that like to do just one thing at a time) that value efficiency and procedure will take greater time and care to read instructions and evaluate decisions in order to avoid errors; which initially takes more time, but in the long term achieves greater efficiency. Alternatively, cultures that value freedom and autonomy prefer to learn through trial and error (Ito and Nakakoji 1996). In their review, Rau et al. (2008) found that users of a hypertext environment (electronic text with hyperlinks to other text/information) with a polychronic time orientation browsed information faster and took fewer steps than those users with monochronic time orientation. Polychronic cultures (i.e., those who like to do multiple things at one time) are also more likely than monochronic ones to engage with mobile devices in unexpected ways and are less likely to be bothered by system delays, such as long download or start-up times (Choi et al. 2005).

Internationalisation-localisation of the Interface

Central to the subject of cultural differences in HMI design is the concept of internationalisation-localisation. Internationalisation-localisation of products is a common practice in many global markets, whereby a product or interface is adapted to suit the specific needs and preferences of a target culture. The internationalisation-localisation process generally encompasses several requirements. First, it requires that the culturally specific (local) aspects of the interface are identified and removed to suit a global audience (internationalisation). Localisation then involves modifying aspects of the product or interface in a way that specifically suits the needs and cultural orientations of the target culture (Bourges-Waldegg and Scrivener 1998, Russo and Boor 1993). Internationalisation aids localisation by providing a neutral structure to which local interface features are then added (Chen and Tsai 2007).

Traditionally, internationalisation-localisation has focused on modifying the most obvious cultural artefacts relevant to the interface, such as language and time and date formats. However, successful internationalisation-localisation should also adapt a system to suit less apparent aspects of culture; that is, the process should include the functionality and interaction aspects of an interface, not just its surface features (Fernandes 1995).

Cross-Cultural Design Issues in the Automotive Context

To date, only a handful of studies have examined the influence of cross-cultural differences on IVIS or ADAS design. This may be problematic considering the increasingly globalised nature of the automotive industry and the fact that the vehicle cockpit is becoming more technically sophisticated. It is important to consider the acceptability of in-vehicle technology, as it influences the uptake of the technology and, consequently, its effectiveness in improving road safety, mobility and environmental outcomes. A failure by drivers to accept a technology can lead to them not using it in the manner intended, or not using it at all. IVIS and ADAS technology are introducing a new level of interface complexity to the vehicle cockpit and, like any system, this technology is expected to be associated with usability and acceptability issues (Rudin-Brown 2010, Young et al. 2012). As mentioned, the interaction between culture and interface design has been most extensively explored in the HCI domain. However, this knowledge might not transfer across to the automotive domain given that the addition of the driving task may result in users having a different set of needs and expectations for in-vehicle technology. Likewise, culture may impact upon these needs and expectations in different ways than they do for non-vehicle based technology. Cross-cultural design issues for automotive technology are likely to be particularly pertinent in developing regions, such as China and India, where there exist enormous markets characterised by notably different user needs and preferences to the major vehicle manufacturing countries in the West (Yang et al. 2007).

Cross-Cultural Considerations and IVIS

The potential benefits and applicability of culturally adaptive IVIS have been explored in a series of studies by Heimgärtner and colleagues (Heimgärtner 2007, Heimgärtner 2005, Heimgärtner and Holzinger 2005, Heimgärtner et al. 2007). The culturally adaptive interface proposed by Heimgärtner is designed to detect and adapt to the cultural preferences of the user, thereby optimising system usability, reducing complexity and mental workload and, ultimately, lead to greater user acceptance (Heimgärtner 2005). As part of this work, Heimgärtner has highlighted a number of cultural differences in user preferences, system navigation styles, driving styles and task management styles that are relevant to cross-cultural IVIS system design, although at present there is limited understanding of what factors are driving these differences (e.g., their causes and impact relative to other factors). Heimgärtner's (2007) results suggest there are significant differences in preferred interaction styles among users from different cultures. In particular, in his study, Chinese users showed a preference for greater information density and faster speed of information on an interface than did German or English users. The Chinese users also entered fewer characters into the system than the other two groups. Research by Knapp (2007) supports Heimgärtner's findings, having found that systems specifically designed to suit the mental models of certain cultures will significantly impact on the ability of users from different cultures to successfully perform tasks on that system.

Most recently, Young et al. (2012) examined whether there are regional differences in the needs and preferences of Australian and Chinese drivers for IVIS and to determine the impact of any differences for IVIS HMI design. A number of differences were found between the two regional groups in terms of their IVIS design preferences that could have significant implications for the appeal, acceptability and usability of IVIS in China. In particular, the Chinese drivers had more difficulty than the Australians in comprehending abbreviations used on interface labelling and showed a greater preference for the use of symbols, except for complex IVIS functions, where they preferred Chinese characters. These different preferences for use of labelling is likely to reflect differences in the languages between the two cultures, where the written Chinese language is pictorial, hence their preference for symbol- and character-based labelling. The Chinese drivers also tended to place greater value on the aesthetic appeal of an interface rather than its usability and safety aspects, preferring an interface that looks modern, sophisticated and denotes a sense of high status. These preferences may derive from the Chinese culture's typically high score on the power distance dimension, which values status in society, and their low level of uncertainty avoidance, which places value on aesthetic appeal.

Cross-Cultural Considerations and ADAS

Although similar cross-cultural issues would be expected to influence the use of both ADAS and IVIS devices, there may be differences in terms of cultural predictors of ADAS versus IVIS use. Therefore, it is important to conduct and consider research that is designed to investigate both categories of a device. Lindgren et al. (2008) examined driving culture and common traffic problems experienced in a sample of Chinese and Swedish drivers, and the consequences of any differences across the two regions for the design of ADAS. The study found that, while the traffic rules and regulations are very similar in China and Sweden, major cultural differences in terms of infrastructure and driver behaviour exist and these differences are likely to influence the acceptance and use of ADAS across the two regions. In particular, they found that the more aggressive and less law-abiding driving culture in China, which includes tailgating and constant lane switching, made some ADAS, such as adaptive cruise control and lane departure warning systems, less acceptable to drivers. They also found that Chinese drivers' preference to follow social norms rather than the road rules may also reduce the effectiveness of some ADAS in this region (e.g., the effectiveness of a follow-distance warning system may be reduced given that tailgating is firmly entrenched in Chinese driving culture). Lindgren et al. concluded that if the design of ADAS does not take such cultural differences into account, these systems may not be acceptable to Chinese drivers, but rather regarded as too intrusive and, thus, ignored or misused.

In addition to differences between countries, cross-cultural differences in attitudes towards, and preferences regarding, ADAS can exist within countries. For example, in 2008 a telephone survey was conducted of owners of vehicles equipped with electronic stability control (ESC), a system that reduces the likelihood of collisions involving loss of control (Rudin-Brown et al. 2009). Over 1,000 ESC owners were identified through the vehicle registration databases of two Canadian provinces: Québec and British Columbia (BC). (Canada's eight other provincial transport agencies chose not to participate in the project.) This allowed for the comparison of driver attitudes towards ESC, and vehicle safety features more generally, between the two regions, which differ not only in terms of weather and precipitation (Québec has a much higher annual snowfall and colder winter temperatures than BC) but also in terms of cultural background and language (Québec's population is mostly French-speaking, while BC's is mostly English). While BC owners of ESC-equipped vehicles were more likely than those from Québec to report that a vehicle's safety features, including ESC, were an important factor to consider when buying a car, they were significantly *less likely* to report: (1) having experienced ESC while driving, (2) being confident that ESC would work in an emergency, and (3) believing that ESC had made it safer to drive. The fact that BC drivers were less likely to have experienced ESC was most probably due to differences in weather patterns between the two provinces. This limited experience may have consequently contributed to the BC owners' relative lack of confidence that ESC would work in an emergency. At the same time, however, BC owners were more likely than Québec owners to report believing that vehicle safety improvements, including ESC, make it possible to drive at faster speeds, which suggests that this group of drivers may have been driving their vehicles in a more aggressive manner than those from Québec. These findings suggest the importance of considering cross-cultural differences, even those within a country's borders, when designing ADAS and other collision avoidance systems, such as ESC.

Behavioural Adaptation, Acceptance and Cross-Cultural Issues

Sometimes, the introduction of an IVIS, ADAS or other collision countermeasure within the vehicle can result in unintended driver behaviours that lead to negative safety outcomes; for example, a distracted driver over-relying on an adaptive cruise control system to maintain a safe distance to a lead vehicle. This phenomenon is known as behavioural adaptation (OECD 1990). There is a relationship between drivers' acceptance of a device or in-vehicle system and the likelihood that behavioural adaptation will develop as a consequence (Jamson, in press). If a device is not acceptable to its target user group, it will be used less often (unless its use is required by law), which would make behavioural adaptation less likely. On the other hand, a device that is widely accepted may be more prone to behavioural adaptation.

A role for device design procedures that include cross-cultural considerations to limit the likelihood of behavioural adaptation has been proposed (Rudin-Brown 2010). Individual and cultural, or group, characteristics have been shown to contribute to an individual's propensity to behaviourally adapt to ADAS. To make behavioural adaptation less likely, therefore, while at the same time producing a device that is accepted among users, designers are encouraged to use the constructs of intercultural adaptability and adaptive design to make a system that is 'user adaptive' (Jameson 2009). A system that is so-designed would use information collected regarding its user to adapt its own behaviour in some crucial way to limit the likelihood of behavioural adaptation. These concepts have been successfully applied in other domains of computer-human interaction to create interfaces that are more usable and acceptable to a broader array of users. The consequence of applying these concepts during IVIS/ADAS concept and design phases would be in-vehicle systems that are accepted across a broad variety of cultures and among a wide range of user groups, and which are associated with few negative safety outcomes.

Conclusions

As the automotive market moves further into developing regions around the world, culture is clearly becoming an important consideration for in-vehicle technology design. Currently, limited information is available to inform IVIS and ADAS design for different cultures to ensure that they are not only usable but also acceptable. Indeed, beyond concluding that culture appears to be relevant to automotive HMI design, few other reliable conclusions can be drawn at this stage. With research on cross-cultural automotive design in its infancy, there are many opportunities to explore cultural requirements for automotive interface design in greater depth across a broader range of cultures, in-vehicle systems and HMI features, particularly the interaction-level aspects of the interface.

Beyond the consideration of cultural issues, there is also a need to consider wider regional factors that are also likely to impact the design of automotive systems across countries. These include factors such as traffic regulations, traffic flow and congestion issues, and the demographic composition of the vehicle and driver fleet, all of which are likely to impact on drivers' need for, and acceptance of, certain in-vehicle technology. Indeed, Lindgren et al. (2008) found that, in addition to driving culture, wider regional factors such as the age of the vehicle fleet, traffic congestion and infrastructure, are all likely to impact on the utility and acceptance of ADAS in China.

Considering and addressing cultural and wider regional factors must be recognised as a critical part of the global IVIS and ADAS design process. Failure to do so could have significant implications for the appeal, acceptability, usability and, ultimately, the safety of IVIS and ADAS in different regions.

Acknowledgements

We thank our colleagues from the Monash University Accident Research Centre (MUARC) who were involved in our work on cross-regional HMI design: Megan Bayly, Amy Williamson and Michael Lenné.

References

Bourges-Waldegg, P. and Scrivener, S.A.R. 1998. Meaning, the central issue in cross-cultural HCI design. *Interacting with Computers*, 9: 287–309.

Casey, S. 1998. *Set Phasers on Stun: And Other True Tales of Design, Technology and Human Error*. Santa Barbara, CA: Aegean.

Chen, C.-H. and Tsai, C.-Y. 2007. Designing user interfaces for mobile entertaining devices with cross-cultural considerations. In *Usability and Internationalization, HCI and Culture, Vol. 4559*, 37–46. Edited by N. Aykin. Berlin Heidelberg: Springer.

Choi, B., Lee, I., Jeon, Y. and Kim, J. 2005. A qualitative cross-national study of cultural influences on mobile data service design. Paper presented at CHI, 2–7 April, Portland, Oregon.

Christaans, H.H.C.M. and Diehl, J.C. 2007. The necessity of design research into cultural aspects. Paper presented at the Proceedings of International Association of Societies of Design Research, Hong Kong.

Fernandes, T. 1995. *Global Interface Design: A Guide to Designing International User Interfaces*. London: Academic Press.

Heimgärtner, R. 2007. Towards cultural adaptability in driver information and assistance systems. In *Usability and Internationalization, Global and Local User Interfaces, Vol. 4560*, 372–81. Edited by N. Aykin. Berlin Heidelberg: Springer.

————. 2005. Research in progress: Towards cross-cultural adaptive human-machine-interaction in automotive navigation systems. Paper presented at the International Workshop for the Internationalisation of Products and Systems, Amsterdam.

Heimgärtner, R. and Holzinger, A. 2005. Towards cross-cultural adaptive driver navigation systems. Paper presented at the Empowering Software Quality: How Can Usability Engineering Reach These Goals? First Usability Symposium, Vienna.

Heimgärtner, R., Tiede, L.-W., Leimbach, L., Zehner, S., Nguyen-Thien, N. and Windl, H. 2007. Towards cultural adaptability to broaden universal access in future interfaces of driver information systems. In *Universal access in human–computer interaction, Part II, Vol. 6*, 383–92. Edited by C. Stephanidis. Berlin Heidelberg: Springer-Verlag.

Hofstede, G. 1991. *Cultures and Organizations: Software of the Mind.* New York: McGraw-Hill.

———. 1980. *Culture's Consequences: International Differences in Work-Related Values.* London: Sage.

Hoft, N. 1996. Developing a cultural model. In *International User Interfaces*, 41–73. Edited by E.M. del Galdo and J. Nielsen. Brisbane: John Wiley and Sons.

Honold, P. 2000. Culture and context: An empirical study for the development of a framework for the elicitation of cultural influence in product usage. *International Journal of Human-Computer Interaction*, 12(3 and 4): 327–45.

Ito, M. and Nakakoji, K. 1996. Impact of culture on user interface design. In *International User Interfaces*, 105–26. Edited by E.M. del Galdo and J. Nielsen. Brisbane: John Wiley and Sons.

Jameson, A. 2009. Adaptive interfaces and agents. In *Human-Computer Interaction: Design Issues, Solutions, and Applications*, 105–30. Edited by J. Jacko and A. Sears. Boca Raton, FL: CRC Press.

Jamson, S.L. (in press). The role of acceptability in behavioural adaptation. In *Behavioural Adaptation and Road Safety: Theory, Evidence, and Action.* Edited by C.M. Rudin-Brown and S.L. Jamson. Boca Raton, FL: CRC Press.

Knapp, B. 2007. Mental models of Chinese and German users and their implications for MMI: Experiences from the case study navigation system. In *Human-Computer Interaction, Part I, Vol. 1*, 882–90. Edited by J. Jacko. Berlin Heidelberg: Springer-Verlag.

Lindgren, A., Chen, F., Jordan, P.W. and Zhang, H. 2008. Requirements for the design of Advanced Driver Assistance Systems: The diffrences between Swedish and Chinese drivers. *International Journal of Vehicle Design*, 2(2): 41–54.

Lodge, C. 2007. The impact of culture on usability: Designing usable products for the international user. In *Usability and Internationalisation, HCI and Culture.*, Vol. 4559, 365–8. Edited by N. Aykin. Berlin Heidelberg: Springer.

Marcus, A. 2009. Global/intercultural user interface design. In *Human-Computer Interaction: Design Issues, Solutions and Applications*. Edited by A. Sears and J. Jacko. Boca Raton, FL: CRC Press.

Marcus, A. and Gould, E.W. 2000. Crosscurrents: Cultural dimensions and global Web user-interface design. *Interactions*, July–August, 32–46.

McLoughlin, C. 1999. Culturally responsive technology use: Developing an on-line community of learners. *British Journal of Educational Technology*, 30(3): 231–43.

Nielsen, J. 1996. International usability engineering. In *International User Interfaces*, 1–19. Edited by E.M. del Galdo and J. Nielsen. Brisbane: John Wiley and Sons.

Nisbett, R.E. 2003. *The Geography of Thought: How Asians and Westerners Think Differently – and Why*. New York: Free Press.

Organization for Economic Cooperation and Development (OECD). 1990. *Behavioural Adaptations to Changes in the Road Transport System*. Paris, France: OECD, Road Transport Research.

Rau, P.L.P., Gao, Q. and Liang, S.F.M. 2008. Good computing systems for everyone: How on earth? *Cultural Aspects, Behaviour and Information Technology*, 27(4): 287–92.

Rudin-Brown, C.M. 2010. 'Intelligent' in-vehicle ITS: Limiting behavioural adaptation through adaptive design. *IET Intelligent Transport Systems*, 4(4): 252–61.

Rudin-Brown, C.M., Jenkins, R.W., Whitehead, T. and Burns, P.C. 2009. Could ESC (Electronic Stability Control) change the way we drive? *Traffic Injury Prevention*, 10(4): 340–47.

Russo, P. and Boor, S. 1993. How fluent is your interface? Designing for international users. Paper presented at the INTERACT '93 and CHI '93 Conference on Human Factors in Computing Systems, Amsterdam.

Smith, A. and Yetim, F. 2004. Global human-computer systems: Cultural determinants of usability. *Interacting with Computers*, 16(1): 1–5.

Smith, A., Dunckley, L., French, T., Minocha, S. and Chang, Y. 2004. A process model for developing usable cross-cultural websites. *Interacting with Computers*, 16(1): 63–91.

Sprung, R.C. 1990. Two faces of America: Polyglot and tongue-tied. In *Designing User Interfaces for International Use, Vol. 13*, 71–102. Edited by J. Nielsen. Amsterdam: Elsevier.

Trompenaars, F. and Hampden-Turner, C. 1998. *Riding the Waves of Culture: Understanding Cultural Diversity in Global Business*. New York: McGraw-Hill.

Yang, C., Chen, N., Zhang, P.-F. and Jiao, Z. 2007. Flexible multi-modal interaction technologies and user interface specially designed for Chinese car infotainment system. In *Human-Computer Interaction, Part III, Vol. 3*, 243–52. Edited by J. Jacko. Berlin Heidelberg: Springer-Verlag.

Young, K.L., Rudin-Brown, C.M., Lenné, M.G. and Williamson, A.R. 2012. The implications of cross-regional differences for the design of in-vehicle information systems: A comparison of Australian and Chinese drivers. *Applied Ergonomics*, 43: 564–73.

PART VI
Conclusions

Chapter 22

Driver Acceptance of New Technology: Synthesis and Perspectives

Alan Stevens
Transport Research Laboratory, UK[1]

Tim Horberry
University of Queensland, Australia, and University of Cambridge, UK

Michael A. Regan
University of New South Wales, Australia

Abstract

This chapter pulls together the research findings, experiences and discussion topics documented by our book contributors to provide a synthesis of what is known and areas in which there is consensus in our understanding of driver acceptance of in-vehicle technology. We identify the perspectives of researchers, product designers and government policymakers and discuss how knowledge of acceptance of technology from related fields can be relevant to the in-vehicle environment. Finally, we try to identify the main knowledge gaps in the field and make recommendations concerning topics and methods for future research.

This Book

This book has brought together into a single volume, a body of contemporary, accumulated scientific and practical knowledge concerning driver acceptance of technology. In the four main parts, we have covered the theory behind acceptance, including definitions and models, how acceptance can be measured and how it can be improved, and we have included a number of case studies illustrating current practice. We have been fortunate to secure contributions from a range of international experts and have sought some contributions beyond the automotive domain to slightly widen the focus; for example, one chapter is on the acceptance of new technology by motorcycle riders and one is about acceptance issues with mobile machinery operators in the minerals industry. These contributions, along

---

with a perspective provided by policymakers, gives, we believe, a rich picture of acceptance within the automotive domain and provides links to a broader landscape of research and endeavour.

Identification of Key Findings from Other Chapters

In this section we identify some of the main themes and conclusions emerging from the four sections of the book. We have also tried to identify where there is consensus between views and where different views exist.

Theories and Models of Acceptance

As several chapter authors have noted, at its most basic level, acceptance of new technology can be aligned simply with long-term use of that technology. However, 'acceptance equals use' is simplistic at best, and does not help designers develop and market successful products; and nor does it help decision-makers encourage use of technology for desirable social outcomes such as increased road safety. Furthermore, it is clear throughout the book that different authors (implicitly or explicitly) think about acceptance in different ways.

One of the points made very clearly (e.g., Adell, Várhelyi and Nilsson, Chapter 2, Vlassenroot and Brookhuis, Chapter 4) is that few studies have explicitly defined acceptance and that consensus on an overall definition is lacking. This is problematic, as there is obviously a close dependence between the definition of acceptance, acceptance models and measurement of acceptance. To put it simply: if we don't have a definition of acceptance, how can we formulate models of it or measure it in a meaningful way?

Contributions to a more fundamental deconstruction of acceptance have been made by a number of authors within this book. Adell, Várhelyi and Nilsson (Chapter 2), for example, identify five categories of acceptance definitions that are based on

- using the word 'accept';
- satisfying user needs and requirements;
- the summation of attitudes;
- willingness to use; and
- actual use.

These last two categories raise the important issue of the evolution of acceptance over time. In advance of actually experiencing a new technology product, individuals will probably have a view on it, although most authors do not yet ascribe the word 'acceptance' to this initial judgement. At this point we might talk about 'acceptability' as a 'prospective judgement of measures to be introduced in the future' (Schade and Schlag 2003: 47). When a product becomes tangible, for example as part of a field trial or as a built-in feature of a particular vehicle model, drivers then have an opportunity to experience it 'for real'. At this point, and as their experience with the technology develops, drivers will, individually, form judgements including those relating to 'acceptance'.

Arguably, a consensus definition of what is meant by the term 'acceptance' cannot emerge until there is a much clearer view (supported by evidence) of the relationship between acceptance and related constructs such as usefulness, usability, trust in the technology, pleasure in use, satisfaction, desirability, impact on others, social status conferred by use, and so on. It might also be important to better recognise different types of acceptance (Adell, Várhelyi and Nilsson, Chapter 2): for example, attitudinal acceptance, behavioural acceptance, conditional/contextual acceptance and social acceptance.

The lack of a single unifying theory and definition of acceptance has been mirrored by a large number of different attempts to develop models of acceptance, and a number are reviewed and used as starting points in several chapters of this book (see, e.g., Vlassenroot and Brookhuis, Chapter 4), principal amongst these being

- the Theory of Planned Behaviour (TPB) based on the Theory of Reasoned Action;
- the Value–Belief–Norm (VBN) theory;
- the Technology Acceptance Model (TAM);
- the two dimensions direct attitudes model of Van der Laan, Heino and De Waard (1997) measuring usefulness of the system and satisfaction; and
- the Unified Theory of Acceptance and Use of Technology (UTAUT).

Our chapter contributors have found all these models of acceptance lacking or insufficient in some regard. Adell, Várhelyi and Nilsson (Chapter 3) suggest that the UTAUT is a good starting point but needs to be developed to include emotional reactions of drivers, weighting of the constructs and model reliability issues. Likewise, Vlassenroot and Brookhuis (Chapter 4) identified 14 factors possibly influencing acceptability (device specific factors, context factors and general factors) and proposed a new theoretical model of acceptance. Similarly, Ghazizadeh and Lee (Chapter 5) modified and developed the Technology Acceptance Model with a new 'trust in device' element and then combined it with a model appropriate for when an in-vehicle device provides feedback to the driver. They discuss, for example, that drivers accept mentoring or coaching more readily than monitoring of their performance.

So, although we do not have a single agreed definition, or a single model, of acceptance, there seems to be general consensus on some important issues:

- acceptance is a complex construct which has many facets and dependencies;
- acceptance is based on individual judgements, so a driver-centric view is required to measure or predict acceptance at an individual level (assessing societal acceptance requires an additional broader perspective);

- two key determinants of acceptance of new technology are usefulness and ease of use;
- acceptance depends on the individual, so issues such as gender, age, culture and personality are likely to be important;
- the context of use is also important, including the supporting 'infrastructure' (in its widest sense), whether use of the technology is voluntary and also broader social/cultural influences;
- drivers do not have to actually like a technology/system to be accepting of it (but liking it may increase use of the technology);
- acceptance should be regarded as a continuous variable, not a binary concept; and
- acceptance is not invariant; it may change (even for one individual) depending on the specific time/context in which the new technology is used and as experience with the technology develops.

To conclude this section on theories and models, we can say that a simple definition of driver acceptance is elusive and that consensus for a more complete definition is currently missing. Not unconnected with this, the situation concerning modelling of acceptance is similar: there are a number of proposed models, but consensus is lacking. It is also worth noting that many of these theories and models were developed originally for non-driving contexts. Nevertheless, seeking a better understanding of the determinants of driver acceptance appears worthwhile, as it is likely to support designers of products incorporating new technology as well as decision makers in developing implementation strategies to support desirable social goals such as increased safety.

Measurement of Acceptance

One might ask: 'why do we try to measure acceptance?' Often it is undertaken at an early stage of product concept or development with the aim of accurately predicting and optimising likely user acceptance as early as possible in the design process. Another reason might be to better understand deployment, performance issues or system failures when technology is available and in use.

As might be expected, given the diversity of definitions and models of acceptability, a range of tools and techniques are used in different environments yielding a range of metrics that measure (or purport to measure) some aspects of driver acceptance of various information, warning and assistance systems. The chapters in this book illustrate these metrics and discuss their uses and limitations.

Adell, Nilsson and Várhelyi (Chapter 6) note that the lack of a consistent way of measuring acceptance can lead to misinterpretation, and even misuse, of results. They explore a range of self-reported measures (questionnaires, focus groups, interviews, willingness to pay), driver performance measures and physiological measures. Clearly, the different tools used to derive these measures have strengths and weaknesses. Mitsopoulos-Rubens and Regan (Chapter 7) concentrate on questionnaires and focus groups and conclude that the choice of tool depends on the focus of the (acceptability) research issue. Källhammer, Smith and Hollnagel (Chapter 9) in their work with warning systems, derive benefit both from video simulations (in advance of product availability) and video analysis (during product use) to investigate the acceptability of misses and false alarms. Stevens and Burnett (Chapter 17) focus on usability aspects of acceptance and conclude that usability, at least, can be measured and that such measurement can be standardised.

As noted by Edmunds, Dorn and Skrypchuk (Chapter 8), driver acceptance of in-vehicle technology is a multidimensional concept dependent on emotional, cognitive and experiential effects. It seems likely, then, that comprehensive measurement of acceptance may need to involve multiple tools and techniques in different environments and with a representative range of users over time. Practical measurement tools will therefore need to involve compromises in cost, time and comprehensiveness but can be tailored to the research issue under investigation.

One last point is very important and is raised by a number of contributors to this book – acceptance needs to be 'operationalised' before measurement by identifying the components of it that are the focus of research. Also, since a range of tools already exist, it will often not be necessary to fundamentally develop new tools; but equally, it should be possible to build on previous work. The key points are to identify and define what is being measured and to be as consistent as possible in measurement when making comparisons of acceptability over time or between technology products.

Earlier in this chapter we revisited the distinction between acceptability and acceptance, and suggested that knowledge about acceptability (potential acceptance in advance of actual product use) may give designers and others important early feedback that can be used to optimise acceptance of technology when it is actually used. In discussing the measurement of acceptance, and in case studies reported in this book, many authors make the point that the factors determining acceptance in product use are complex. Therefore, the relationship between a product being judged as being potentially acceptable or unacceptable before introduction and its acceptance in actual use remains an area for further research.

Cases Studies and Data on Acceptance of New Technology

Various types of information can be drawn from the case studies described in the book. We see the kinds of technology that has been evaluated; the methods that have been used to evaluate acceptance of the technology and the assumptions associated with those methods; and practical data on the absolute and relative acceptance of different products. Here, we concentrate on extracting some of the insights into acceptance revealed through these studies, both from the section on case studies and from elsewhere in the book.

Vlassenroot and Brookhuis (Chapter 4), in their study of Intelligent Speed Adaptation (ISA), conclude that acceptability most strongly depends on the system doing what it is designed to do and that 'equity' between users is also a key factor (i.e., a driver is more likely to accept their speed being controlled if all drivers have their speed controlled). Källhammer, Smith and Hollnagel (Chapter 9) studied warning systems where acceptance depends on the rate and nature of misses and false alarms. They note, importantly, that 'false alarm' is a post-hoc classification; many alerts that are false alarms are actually useful and they are likely to be accepted if they provide useful, trustworthy information.

Labeye, Brusque and Regan (Chapter 11) used questionnaires, focus groups and driver logs in their studies of acceptance of electric vehicles. In their work, acceptance was judged in terms of performance, ease of use and facilitating conditions (organisational and technical infrastructure). Vilimek and Keinath (Chapter 12) note that a 'disruptive technology', like the electric vehicle, will only succeed if it meets both the requirements of early adopters and later adopters. They also advocate the necessity of a user-centred development and evaluation approach to such innovation.

Stevens and Reed (Chapter 14) used a driving simulator as a tool to investigate drivers' reaction to novel road infrastructure technology where they inferred acceptance from behaviour. They note that safe interaction between drivers requires acceptance of 'social norms of behaviour' and imply that this can be affected by acceptance of new technology. Important social factors affecting behaviour/acceptance were found to be fairness and the implementation approach when technology is deployed. Other variables were the driver's psychological characteristics (especially relevant in the case of enforcement technology) and the degree of monitoring and feedback to the driver from roadside systems.

In her work with motorcycle riders, Huth (Chapter 13) identifies that problem awareness and perceived usefulness are critically important for acceptance of new safety technology and that intervening devices are generally less acceptable than those providing warnings. She also identifies that the psychological characteristics of riders affect acceptance: with 'fun' as a motivation for riding being associated with lower acceptance of technology that removes part of this fun. She concludes that both subjective and objective evaluation of acceptance is necessary.

Horberry and Cooke (Chapter 15) looked at the minerals industry where introduction of new technology for use by operators is usually mandatory and where non-acceptance can be revealed by equipment being neglected or sabotaged. They found that improved interface design, based on best ergonomics practice, can improve operator acceptance.

Belin, Vedung, Meleckidzedeck and Tingvall (Chapter 16) examined policy instruments which aim to maximise road safety through use of new technology and where acceptance is obviously an important determinant of outcomes. They conclude that the issues are highly complex and that the choice and packaging of policies (and, by implication, acceptance of them) will depend on many contextual factors.

At first glance, these case studies may seem rather disparate as they address different users of different technology within different contexts. What unites them is that acceptance is considered important for ongoing deployment and the study findings add to the mounting body of knowledge in the area. Together, these case studies reinforce the ideas that acceptance is multifaceted and that the technology, user and contextual variables are all important in developing a fuller understanding of acceptance.

Optimisation of Acceptance

A key question for designers, and those responsible for introducing new technology, should be how to optimise its acceptance; the aim usually being to ensure that the technology/product is used 'correctly' and that its benefits are maximised. Advice on how to do so is offered in this book.

A recurring theme is the vital importance of user-centred design when trying to achieve high acceptance. Although seemingly obvious, at least to Human Factors professionals, this fundamental approach is still lacking, at least in some industrial contexts. This was the experience of Horberry and Cooke (Chapter 15), who found that investigation of actual requirements, consultation with users and user-centred design is particularly important when technology is not chosen by the driver but introduced as part of their working requirement.

In general, technology is used within a wider organisational context (including physical surroundings, people, procedures and other technology) that also evolves over time. Stevens and Burnett (Chapter 17) conclude that, for high usability, the designer needs to understand the characteristics of users and the physical, social and potential organisational environment in which tasks are carried out. Similarly, from his review of organisational factors in the introduction of technology in other domains, Maguire (Chapter 19) concludes that a range of factors may need to be considered: organisational goals; roles, responsibilities and skills; workflows; procedures; communication and distraction; work culture; technical and functional knowledge; safety thinking; security and usability trade-off; and training and support.

Horberry and Cooke (Chapter 15) identify also the need to ensure that drivers/operators have sufficient technical and non-technical skills to understand the context and benefits of the technology (to be more likely to accept it). Maguire (Chapter 19) notes that acceptability is likely to be higher when new technology promotes the learning of new skills, supports the driving task, contributes to objectives that drivers aspire to and fits in with their culture and values. Young and Rudin-Brown (Chapter 21) remind us that cultural issues are likely to have implications for usability (and hence acceptance) and that many factors that vary between countries/regions (e.g., regulations, congestion, vehicle fleet age, infrastructure) are also likely to impact on design and implementation of technology.

In terms of actual technology and product design, Stevens and Burnett (Chapter 17) review some of the many guidelines and standards available to support interface design, which have potential to make interfaces more acceptable to drivers. However, good ergonomic design is only one aspect of an acceptable product. Green and Jordan (Chapter 18) highlight that, as well as factors such as usability, reliability, longevity, value and so on, a major component of the user experience is aesthetic and emotional satisfaction. They point to the emotional product experience as an important consideration in optimising acceptance. This extends beyond the 'look-and-feel' of a technology interface to include such things as brand image and perception of status.

Finally, in terms of optimising acceptance, two chapters look at the role of government policy in driving change through new technology to achieve social outcomes such as improved road safety. Belin et al. (Chapter 16) identify three classes of policy instrument (information, economic instruments and regulations) and describe how instruments can be packaged vertically (targeted to specific stakeholders), horizontally (more than one instrument used) or chronologically (various instruments in a time sequence) in an attempt to achieve specific outcomes. van der Pas, Walker, Marchau and Vlassenroot (Chapter 20) advocate Adaptive Policymaking to help with implementation of technology in the face of uncertainty. The approaches described in these two chapters are likely to have considerable impact on acceptance of new technology, particularly where there is initial resistance or uncertainty about adoption by either drivers or wider society. The exact nature of policy implementation, however, seems to be highly dependent on the situation; sadly, no specific advice can be given as to which instruments will work best to promote acceptance in particular circumstances.

Limitations in Our Current Knowledge of Acceptance

In an ideal world, there would be a consensus definition of driver acceptance, a unifying model of the construct, and its underlying dimensions, and a range of valid tools for reliably measuring acceptance of new technology, and predicting acceptance into the future. As a corollary to this, designers and those responsible for deploying new technology would know how to optimise acceptance (or be in a position to develop alternate strategies).

In this section, we examine the extent to which this vision has been achieved and we identify gaps and deficiencies, both pertaining to this book and to the field of acceptance more generally.

As noted earlier, one fundamental point is that we are lacking a broad scientific consensus on a definition of acceptance. Similarly, despite a range of models of acceptance being available, they are often found lacking in some regard. This appears to be not because of a lack of effort on synthesis but stems from a lack of detailed knowledge concerning the factors that influence acceptance and the interaction between them. One can always make a plea for more research, but in doing so there needs to be a recognition that the research community has to get better at adopting best practice and becoming clearer in reporting its definitions, methodology and detailed findings. The reviews in several chapters in this book suggest that such detail is often lacking in publications. In most industries and contexts, there is still a continuing need for a human-centred design, operational needs analysis and a user-centred deployment process.

At the level of an individual user of new technology, there appear to be gaps in our knowledge concerning how individual characteristics, such as personal needs and motivations, contribute to acceptance. More broadly, the social context, the specific context of use and the effect of social interaction between drivers is thought to be important but not well researched. Policymakers would certainly appreciate more advice; addressing such knowledge gaps would help identify when, where and within which cultural context a specific policy instrument or package would be most effective and appropriate.

The issue of acceptance has been studied in relation to many different products and services. In this book, and for one class of user (drivers), we have reported examples of technology providing a range of services covering information, warning, assistance and automation. We also have acceptance studies in relation to other users of new technology such as motorcycle riders and operators of mobile mining machinery. This range provides a tapestry of findings concerning acceptance for specific users in specific contexts of use. Nevertheless, neither this book, nor even the wider academic literature, provides a complete picture. We are left, therefore, with a number of questions and gaps. The most important ones raised by contributors to this book in relation to the technology itself are

- differences in how acceptance is judged in prestige vehicles and more mundane vehicles;
- understanding the factors that influence driver acceptance of alerts that are false alarms;
- changes in acceptance with technology/product experience;
- acceptance of combinations of systems;
- whether measurement tools can discriminate the degree of acceptance between similar technology exemplars;
- the extent to which customisable systems might improve acceptance; and
- implications for acceptance of reliance/long-term dependency on technology support; for example, does dependency remove the choice of acceptance and is unquestioning acceptance problematic?

A note of caution should be given here to the academic community that uses existing technology for its research: often technology is a few years old, due to the traditionally steady pace of scientific research. Given the rapid nature of technology development, we need to appreciate that really new technology may have significantly different characteristics and that driver acceptance develops based on previous experience of technology.

The extent to which we can address these gaps by further harvesting knowledge from other domains is largely unknown. In this book we have concentrated on the driver of road vehicles but we have contributions also covering mobile machinery operators and motorcyclists. From Horberry and Cooke (Chapter 15), it seems that the mining industry can learn from the driving context, and perhaps the driving context can learn from other domains as suggested through the contribution of Maguire (Chapter 19). One such domain could be the aerospace industry (accepting however, that pilots are more highly skilled, trained and regulated). We have not included broader domains in this book but acceptance issues in security, medical and aerospace may be fruitful areas to explore in future.

Research Recommendations

In this section we highlight a number of areas of acceptance-related research that appear to be most valuable in terms of both theoretical and practical development. Where possible we include a suggested research approach.

The Theoretical and Practical Links Between Acceptance and Related Concepts

Exploring the theoretical and practical determinants of acceptance will be rich areas for future research, especially as new technology is increasingly deployed in road transport. Important research questions are, for example, how reliable does a technology need to be for it to be widely accepted? As Ghazizadeh and Lee (Chapter 5) explore, how do reliance and trust fit in here? Similarly, what is the effect of individual driver/rider/operator characteristics (e.g., personal needs and motivations, older and younger users) on acceptance? How does the specific context of use affect acceptance? As discussed by Burnett and Diels (Chapter 10), what is the effect of social interaction between drivers? More broadly, how do societal attitudes and prejudices affect acceptance and how do these affect policymakers, commercial organisations and individuals making judgements and plans concerning introduction of new technology?

Thus, there is a need to invest in fundamental research on the factors determining acceptability, on their definition and on modelling their interactions. As part of this effort, there is a need for the research community to adopt best practice and be more specific in reporting of findings about acceptance such that maximum use can be made of results in subsequent modelling and definitional work.

As repeatedly noted, a consensus definition of acceptance is missing. A consensus definition needs to emerge based on a body of work and from an open discussion process that is likely to hold sway in the research community. We believe that this book provides a very useful summary of the relevant work and we hope that it will provide the basis from which such ongoing discussions can be held. The forum for those discussions is not yet defined but could possibly involve national and international standardisation organisations, an expert group, a suitable professional body, or an Internet platform, perhaps associated with an international conference.

Further Development and Validation of Instruments to Measure Acceptance

As the chapters by Edmunds et al. (Chapter 8) and Green and Jordan (Chapter 18) noted, when designing new technology, a positive response from the eventual end-user is essential. So to develop and deploy successful new technology, Original Equipment Manufacturers and aftermarket technology suppliers must understand the impact of their innovations on peoples' affective and cognitive response to technology. Sadly, too few validated tools (instruments) exist in the open literature for this purpose.

As acceptance is a multifaceted concept, it is likely that a toolbox rather than a single tool needs to be the goal of this research. As noted by Huth (Chapter 13), both subjective and objective measurements need to be available. Developments in acceptance measurement need to build on the definitions and models such that there is clarity about exactly what a tool is measuring (the metric) as well as the environment in which the measurement is made and the exact measurement technique used. As noted by Adell et al., all studies of acceptance can contribute to furthering knowledge of acceptance by clearly defining what they mean by acceptance and by consequently using that definition when measuring the concept. So tools need to be developed, fully documented, validated and made widely available. Research questions to be addressed are the performance, reliability and sensitivity of different tools (e.g., can they discriminate between similar products?) and the cost of application (time, resources etc.). Further questions concern the physical environments in which different tools can operate and when during the product life-cycle their use is appropriate and valid. One critical issue for designers is the extent to which an acceptance measuring tool is predictive; that is, whether a tool applied at one point in time can assess acceptance during a later phase of a product life-cycle. As pointed out by Mitsopoulos-Rubens and Regan (Chapter 7), there is a degree of professional judgement required in knowing what combination of tools to use in measuring acceptance.

Ensuring Acceptance Research Keeps Pace with New Technology Development

An explosion in new technology design and deployment is taking place within motor vehicles: both for factory-fitted and aftermarket products. Similarly, an increase in technology deployment can be seen in highway infrastructure, industrial mobile equipment and motorcycles, as documented in the chapters by Stevens and Reed (Chapter 14), Horberry and Cooke (Chapter 15) and Huth (Chapter 13).

Many researchers note, quite rightly, that their methods and conclusions apply only to the specific technology function that they have investigated. So, there is clearly both a research and commercial need to apply existing and developing theories and measurements of acceptance to new situations.

As technology develops, the way in which drivers interact with it will also develop. This may change their perception of acceptance and the frames of reference they use to judge acceptance. The advent of 'driverless' cars is a case in point. Here, perceptions of driver satisfaction and usefulness of vehicle technology may be bound up less in the ability of the car to transport them reliably, comfortably and safely to a destination, but in it freeing up effort and time that can be used to do other things that might be more useful and satisfying: like reading, telephoning or sleeping.

Some specific technology-related research issues deserve attention:

- Some systems are multifunctional, and there might also be multiple single function devices in a vehicle, especially where aftermarket devices are purchased by drivers. How these multiple systems impact on acceptance – both now and in the future – is an area for future research.
- Exploring differences in acceptance issues in different driving contexts (e.g., depending on the sophistication of the vehicle).
- Understanding the factors that influence driver acceptance of warning systems and particularly of alerts that are false alarms.
- Further investigation of motorcycle rider information and warning acceptance, particularly from field trials of new technology products.
- How acceptance changes with technology/product experience.
- The extent to which customisable systems might improve acceptance.
- Whether there can be 'over-acceptance' of highly reliability technology leading to complacency and whether technology performance and acceptance need to be optimised to avoid dependency.

Cross-Cultural Design and Learning from Other Domains

As Young and Rudin-Brown (Chapter 21) note, research on cross-cultural automotive design is in its infancy. Numerous opportunities exist to explore requirements for automotive interface design and interaction design across different cultures. In-vehicle technology has now developed and its uptake has spread beyond a point where one interface design fits with all purposes. Given the global market for technology, especially when factory-fitted to a car, truck or motorcycle, better understanding of important cultural differences is essential for effective design.

A similarly broad issue is the extent to which the study of acceptance in the driving context can learn from other domains. The perspectives provided by Horberry and Cooke (Chapter 15) about the minerals industry and by Huth (Chapter 13) concerning motorcycle riders suggest that at least some of the issues around acceptance are universal and that there may be further insights that can be harvested from information technology, aerospace, medical and security fields. The chapter by Maguire (Chapter 19) provides a starting point and some thoughtful remarks on the similarity of challenges in these different domains.

Practical Implications of This Body of Knowledge

As has been seen throughout this book, acceptance of new technology and systems by drivers, equipment operators and wider society is becoming an increasingly important topic worldwide, especially for technology which has the potential to significantly enhance safety, efficiency or comfort.

This book provides a resource for Human Factors researchers, for industry and for policymakers to better appreciate the complexities and multifaceted nature of acceptance. The early theoretical chapters explore the factors likely to be of most importance in determining acceptance and provide some basis for designing products and services that are more likely to be accepted by the intended users. They also provide an insight into the way contemporary thinking about acceptance is developing. The case studies, measurement and optimisation chapters provide both practical experience of acceptance and advice concerning design and deployment. The case studies provide, in addition to insights about the concept of acceptance, valuable information about actual measured levels of acceptance for different products that will be of benefit to those who may be contemplating deployment of such technology.

Perhaps we have a biased view, but it seems certain that acceptance of technology by drivers will become increasingly important as the level of automation increases. We also see a need for the updating of regulations and design guidelines in this area as a result of technology developments.

Is There Anything 'Beyond' Acceptance?

The importance of driver acceptance of new technology has been emphasised throughout this book and with a general agreement that acceptance is a variable rather than a binary concept. Philosophically, one might ask whether acceptance can increase indefinitely or whether there is a 'sufficiency plateau' beyond which further improvement is of little or no value in increasing acceptance.

There has been a hint from Burnett and Diels (Chapter 10) that some scepticism about the performance of technology might be healthy at times, raising the question of whether acceptance could be too high in certain circumstances and lead to complacency. Implicitly, this suggests that there is an optimum level of acceptance which should be sought.

Other authors have considered acceptance as necessary but not sufficient in product development, implicitly supporting the idea that there is a plateau of acceptance. Green and Jordan (Chapter 18) state (also supported by Edmunds et al., Chapter 8) that emotional and aesthetic dimensions are key elements in the technology purchasing decision in the first place and in the long-term satisfaction of the driver/operator experience. For them, issues such as pleasure or joy in product use, and feelings of pride/status/comfort from owning and using the technology, are important issues that are not adequately encompassed in the term 'acceptance'.

So, as with much of the terminology around acceptance, this issue has not been resolved but serves to illustrate how concepts from different origins can contribute to enrich our understanding of the role of individual and group perceptions in the design, deployment and use of technology. The field has not yet reached a level of maturity where a simple road map or toolkit can easily be developed to help optimise driver acceptance for all technology, in all situations and for all drivers/operators. However, the research reported here, and the issues discussed in this book concerning design deployment and use of technology, are, we believe, a useful contribution to the field.

References

Schade, J. and Schlag, B. 2003. Acceptability of urban transport pricing strategies. *Transportation Research Part F: Traffic Psychology and Behaviour*, 6(1): 45–61.

Van der Laan, J. D., Heino, A. and De Waard, D. 1997. A simple procedure for the assessment of acceptance of advanced transport telematics. *Transportation Research Part C: Emerging Technologies*, 5(1): 1–10.

Index

Contributors

Nicole Abdo, BS
Morsani College of Medicine
University of South Florida
Tampa, Florida

Shiekh Aejaz Aziz, MBBS, MD, DM
Department of Medical Oncology
Sher-I-Kashmir Institute of Medical Sciences
Srinagar, India

Gul Mohd. Bhat, MBBS, MD, DM
Department of Medical Oncology
Sher-I-Kashmir Institute of Medical Sciences
Srinagar, India

Ryan J. Carlson, BS
School of Medicine
University of Washington
Seattle, Washington

Elsa Delgado-Sánchez, MD
Gynecologic Oncology Unit
La Paz University Hospital
Madrid, Spain

Gregory J. Diorio, DO
Department of Genitourinary Oncology
and
Department of Tumor Biology
Moffitt Cancer Center
Tampa, Florida

Enrique García-López, MD
Gynecologic Oncology Unit
La Paz University Hospital
Madrid, Spain

Ansar Hussain, MBBS, MD, DM
Department of Medical Oncology
Sher-I-Kashmir Institute of Medical Sciences
Srinagar, India

Andrew R. Leone, MD
Department of Genitourinary Oncology
and
Department of Tumor Biology
Moffitt Cancer Center
Tampa, Florida

Dongyou Liu, PhD
RCPA Quality Assurance Programs
Sydney, Australia

A. R. Lone, MBBS, MD
Department of Medical Oncology
Sher-I-Kashmir Institute of Medical Sciences
Srinagar, India

Paul Nghiem, MD, PhD
School of Medicine
University of Washington
Seattle, Washington

Philippe E. Spiess, MD, MS, FRCS(C)
Department of Genitourinary
    Oncology
and
Department of Tumor Biology
Moffitt Cancer Center
Tampa, Florida

Anthony Kodzo-Grey Venyo, MB, ChB, FRCS(Ed), FRCSI, FGCS Urol., LLM
Department of Urology
North Manchester General Hospital
Crumpsall, Manchester
United Kingdom

Ignacio Zapardiel, MD, PhD, MBA
Gynecologic Oncology Unit
La Paz University Hospital
Madrid, Spain

1
Introductory Remarks

1.1 Preamble

Tumor or cancer (these terms, along with *neoplasm* and *lesion*, are used interchangeably in colloquial language and publications; see the glossary) is an insidious disease that results from an uncontrolled growth of abnormal cells in parts of the body. Tumor or cancer has acquired a notorious reputation not only due to its ability to exploit host cellular machineries for its own advantages but also due to its potential to cause human misery.

With a rapidly aging world population, widespread oncogenic viruses, and constant environmental pollution and destruction, tumor or cancer is poised to exert an increasingly severe toll on human health and well-being. There is a burgeoning interest from health professionals and the general public in learning about tumor and cancer mechanisms, clinical features, diagnosis, treatment, and prognosis. The following pages in this volume, as well as those in sister volumes, represent a concerted effort to satisfy this critical need.

1.2 Tumor mechanisms

The human body is composed of various types of cells that grow, divide, and die in an orderly fashion (so-called apoptosis). However, when some cells in the body change their growth patterns and fail to undergo apoptosis, they often produce a solid tumor and sometimes a nonsolid tumor (as in the blood). A tumor is considered benign if it grows but does not spread beyond the immediate area in which it arises. While most benign tumors are not life-threatening, those found in the vital organs (e.g., brain) can be deadly. In addition, some benign tumors are precancerous, with the propensity to become cancer if left untreated. On the other hand, a tumor is considered malignant and cancerous if it grows continuously and spreads to surrounding areas and other parts of the body through the blood or lymph system.

A tumor located in its original (primary) site is known as a "primary tumor." A tumor that spreads from its original (primary) site via the neighboring tissue, bloodstream, or lymphatic system to another site of the body is called "metastatic tumor or cancer" (or secondary tumor or cancer). A metastatic

cancer has the same name and same type of cancer cells as the primary cancer. For instance, a metastatic cancer in the brain that originates from breast cancer is known as metastatic breast cancer, not brain cancer.

Typically, a tumor or cancer forms in tissues after the cells undergo genetic mutations that lead to abnormal changes known as hyperplasia, metaplasia, dysplasia, neoplasia, and anaplasia (see the glossary). Factors contributing to genetic mutations in the cells may be chemical (e.g., cigarette smoking, asbestos, paint, dye, bitumen, mineral oil, nickel, arsenic, aflatoxin, wood dust), physical (e.g., sun, heat, radiation, chronic trauma), viral (e.g., EBV, HBV, HPV, HTLV-1), immunological (e.g., AIDS, transplantation), endocrine (e.g., excessive endogenous or exogenous hormones), or hereditary (e.g., familial inherited disorders).

In essence, tumorigenesis is a cumulative process that demonstrates several notable hallmarks, including (1) sustaining proliferative signaling, (2) activating local invasion and metastasis, (3) resisting apoptosis and enabling replicative immortality, (4) inducing angiogenesis and inflammation, (5) evading immune destruction, (6) deregulating cellular energetics, and (7) genome instability and mutation.

1.3 Tumor classification, grading, and staging

Tumor or cancer is usually named for the organs or tissues from where it starts (e.g., brain cancer, breast cancer, lung cancer, lymphoma, and skin cancer). Depending on the types of tissue involved, a tumor or cancer is grouped into a number of broad categories: (1) carcinoma (involving the epithelium), (2) sarcoma (involving soft tissue), (3) leukemia (involving blood-forming tissue), (4) lymphoma (involving lymphocytes), (5) myeloma (involving plasma cells), (6) melanoma (involving melanocytes), (7) central nervous system cancer (involving the brain or spinal cord), (8) germ cell tumor (involving cells that give rise to sperm or eggs), (9) neuroendocrine tumor (involving hormone-releasing cells), and (10) carcinoid tumor (a variant of a neuroendocrine tumor found mainly in the intestinal tract).

Primary tumors of the skin encompass those affecting the keratinocytes, melanocytes, epithelial adnexal tissue (e.g., sweat glands), hematolymphoid tissue, soft tissue, and neural tissue (peripheral or autonomic nerves), in addition to inherited tumor syndromes [1].

Primary tumors of soft tissues comprise adipocytic tumors, fibroblastic or myofibroblastic tumors, so-called fibrohistiocytic tumors, smooth muscle tumors, pericytic or perivascular tumors, skeletal muscle tumors, vascular

tumors, chondro-osseous tumors, gastrointestinal stromal tumors, nerve sheath tumors, tumors of uncertain differentiation, undifferentiated or unclassified sarcomas, and congenital and inherited tumor syndromes [2].

Primary tumors of the bone range from cartilage tumors, osteogenic tumors, fibrogenic tumors, fibrohistiocytic tumors, Ewing sarcoma or primitive neuroectodermal tumors, hematopoietic tumors, giant cell tumors, notochordal tumors, vascular tumors, smooth muscle tumors, lipogenic tumors, neural tumors, miscellaneous tumors, miscellaneous lesions, and joint lesions, to congenital and inherited tumor syndromes [2].

Primary tumors affecting the urinary tract are separated into urothelial tumors, squamous neoplasms, glandular neoplasms, neuroendocrine tumors, melanocytic tumors, mesenchymal tumors, hematopoietic and lymphoid tumors, and miscellaneous tumors [3]. Primary tumors of the male genital organs include those affecting the prostate, testis, paratesticular tissue, and penis [3]. Primary tumors of the female genital organs comprise those affecting the ovary, peritoneum, fallopian tube, uterine ligaments, uterine corpus, uterine cervix, vagina, and vulva, as well as inherited tumor syndromes [4].

Primary tumors of the breast consist of epithelial tumors, myoepithelial lesions, mesenchymal tumors, fibroepithelial tumors, tumors of the nipple, malignant lymphoma, metastatic tumors, and tumors of the male breast [5].

Under the auspices of the World Health Organization (WHO), the International Classification of Diseases for Oncology, Third Edition (ICD-O-3) [6] has designed a five-digit system to classify tumors, with the first four digits being morphology code and the fifth digit behavior code [6]. The fifth-digit behavior codes for neoplasms are 0, benign; 1, benign or malignant; 2, carcinoma *in situ*; 3, malignant, primary site; 6, malignant, metastatic site; and 9, malignant, primary or metastatic site. For example, chondroma has an IDC-O code of 9220/0 and is considered a benign bone tumor; multiple chondromatosis (a subtype of chondroma) has an IDC-O code of 9220/1 and is an intermediate-grade bone tumor with the potential for malignant transformation; and central chondrosarcoma has an IDC-O code of 9220/3 and is considered a malignant bone tumor [2,6].

To further delineate tumors and cancers and assist their treatment and prognosis, the pathological stages of solid tumors are often determined by using the TNM system of the American Joint Commission on Cancer (AJCC), which incorporates the size and extent of the primary tumor (TX, T0, T1, T2, T3, and T4), the number of nearby lymph nodes involved (NX, N0, N1, N2, and N3), and the presence of distant metastasis (MX, M0, and M1) [7]. Therefore, the pathological stage of a given tumor or cancer is referred to as T1N0MX or

T3N1M0 (with numbers after each letter providing further details about the tumor or cancer). However, a much simplified clinical stage (0, I, II, III, and IV), which is based on results of clinical exam and various tests in the absence of findings during surgery, is used routinely to guide the treatment of solid tumors (see **stage** and **TNM** in the glossary) [7].

Another staging system that is more often used by cancer registries than by doctors divides tumors and cancers into five categories: (1) *in situ* (abnormal cells are present but have not spread to nearby tissue), (2) localized (cancer is limited to the place where it started, with no sign that it has spread), (3) regional (cancer has spread to nearby lymph nodes, tissues, or organs), (4) distant (cancer has spread to distant parts of the body), and (5) unknown (there is not enough information to figure out the stage).

1.4 Tumor diagnosis

As most tumors and cancers tend to induce nonspecific, noncharacteristic clinical signs, a variety of procedures and tests are employed during diagnostic workup. These involve a medical history review of the patient and relatives (for clues to potential risk factors that enhance cancer development), a complete physiological examination (for lumps and other abnormalities), imaging techniques (e.g., ultrasound, computerized tomography [CT], MRI, and positron emission tomography [PET]; see the glossary), biochemical and immunological tests (for altered substance or cell levels in blood, bone marrow, cerebrospinal fluid, urine and tissue), histological evaluation of the biopsy and tissue (using hematoxylin and eosin [H&E], immunohistochemical [IHC] stains, etc.; see the glossary), and laboratory analysis (e.g., fluorescence *in situ* hybridization [FISH] and polymerase chain reaction [PCR]; see the glossary).

1.5 Tumor treatment and prognosis

Standard cancer treatments consist of surgery (for removal of the tumor and relieving symptoms associated with the tumor), radiotherapy (also called radiation therapy or X-ray therapy; delivered externally through the skin or internally [brachytherapy] for destruction of cancer cells or impeding their growth), chemotherapy (for inhibiting the growth of cancer cells, suppressing the body's hormone production or blocking the effect of the hormone on cancer cells, etc.; usually via the bloodstream or oral ingestion), and complementary therapies (for enhancing patients' quality of life and improving well-being). Depending on the circumstances, surgery may be used in combination with

radiotherapy and/or chemotherapy to ensure that any cancer cells remaining in the body are eliminated.

The outcomes of tumor or cancer treatments include (1) cure (no traces of cancer remain after treatment and cancer will never come back), (2) remission (signs and symptoms of cancer are reduced; in a complete remission, all signs and symptoms of cancer disappearing for 5 years or more suggests a cure), and (3) recurrence (a benign or cancerous tumor comes back after surgical removal and adjunctive therapy).

Prognosis (or chance of recovery) for a given tumor is usually dependent on the location, type, and grade of the tumor; patient's age and health status, etc. Regardless of tumor or cancer types, patients with lower grade lesions generally have a better prognosis than those with higher grade lesions.

1.6 Future perspective

Tumor or cancer is a biologically complex disease that is expected to surpass heart disease to become the leading cause of human death throughout the world in the coming decades. Despite extensive past research and development efforts, tumor or cancer remains poorly understood and effective cures remain largely elusive.

The completion of the Human Genome Project in 2003 and the establishment of The Cancer Genome Atlas (TCGA) in 2005 have offered promises for better understanding of the genetic basis of human tumors and cancers, and opened new avenues for developing novel diagnostic techniques and effective therapeutic measures.

Nonetheless, a multitude of factors pose continuing challenges for the ultimate conquering of tumors and cancers. These include the inherent biological complexity and heterogeneity of tumor or cancer, the contribution of various genetic and environmental risk factors, the absence of suitable models for human tumors and cancers, and difficulty in identifying therapeutic compounds that kill or inhibit cancer cells only and not normal cells. Further effort will be necessary to help overcome these obstacles, and enhance the well-being of cancer sufferers.

Acknowledgments

Credits are due to a group of international oncologists/clinicians, whose expert contributions have greatly enriched this volume.

References

1. LeBoit PE, Burg G, Weedon D, Sarasin A. *Pathology and Genetics of Skin Tumours*. International Agency for Research on Cancer (IARC). Lyon, France: IARC Press, 2006.
2. Fletcher CDM, Bridge JA, Hogendoorn P, et al. (eds.). *WHO Classification of Tumours of Soft Tissue and Bone*. 4th ed. Lyon, France: IARC Press, 2013.
3. Eble JN (ed.). *Pathology and Genetics of Tumours of the Urinary System and Male Genital Organs*. World Health Organization, International Agency for Research on Cancer. Lyon, France: IARC Press, 2004.
4. Kurman RJ (ed.). *WHO Classification of Tumours of Female Reproductive Organs*. International Agency for Research on Cancer, World Health Organization. Lyon, France: IARC, 2014.
5. Lakhani S, Ellis I, Schnitt S, et al. *WHO Classification of Tumours of the Breast*. 4th ed. Lyon, France: IARC Press, 2012.
6. Fritz A, Percy C, Jack A, Shanmugaratnam K, Sobin L, Parkin DM, Whelan S. *International Classification of Diseases for Oncology*. 3rd ed. Geneva: World Health Organization, 2000.
7. Edge SB, Byrd DR, Compton CC, Fritz AG, Greene FL, Trotti A (eds.). *AJCC Cancer Staging Manual*. 7th ed. New York: Springer; 2010.

SECTION I
Skin, Soft Tissue, and Bone

2

Basal Cell and Squamous Cell Carcinomas

2.1 Definition

Tumors affecting the skin include the following categories: keratinocytic tumors, melanocytic tumors, appendageal tumors, hematolymphoid tumors, soft tissue tumors, neural tumors, and inherited tumor syndromes [1].

Within the keratinocytic tumor category, six types of neoplasms or lesions are recognized: (1) basal cell carcinoma (BCC) (superficial BCC, nodular [solid] BCC, micronodular BCC, infiltrating BCC, fibroepithelial BCC, BCC with adnexal differentiation, basosquamous carcinoma, and keratotic BCC), (2) squamous cell carcinoma (SCC) (acantholytic SCC, spindle cell SCC, verrucous SCC, pseudovascular SCC, and adenosquamous carcinoma), (3) Bowen disease (Bowenoid papulosis), (4) actinic keratosis (arsenical keratosis and PUVA keratosis), (5) verrucas (verruca vulgaris, verruca plantaris, and verruca plana), and (6) acanthomas (epidermolytic acanthoma, warty dyskeratoma, acantholytic acanthoma, lentigo simplex, seborrhoeic keratosis, melanoacanthoma, clear cell acanthoma, large cell acanthoma, keratoacanthoma, and lichen planus–like keratosis) [1].

Among these, BCC and SCC, which evolve from basal cells and squamous cells, represent about 75% and 15% of all skin neoplasms, respectively. Therefore, BCC and SCC are often called "common" skin cancers. Together with a number of less common skin cancers of nonmelanocyte origin, BCC and SCC are collectively referred to as nonmelanoma skin cancer (NMSC). BCC and SCC are malignant and may cause local disfiguring if not treated early. However, BCC and, to a lesser extent, SCC do not usually spread to other parts of the body [1].

Another important type of skin cancer is melanoma (of the melanocytic tumor category), which accounts for 5% of all skin neoplasms. Although melanoma is less common than BCC and SCC, it is highly aggressive with the tendency to spread to other parts of the body and cause fatality if not treated early (see Chapter 3) [2].

Other unusual types of skin cancer include Merkel cell tumor (of the neural tumor category; see Chapter 4), dermatofibrosarcoma protruberans (of the soft tissue tumor category), and microcystic adnexal carcinoma and sebaceous carcinoma (both of the appendageal tumor category) [3].

2.2 Biology

The skin is the largest organ, and it protects the body against heat, sunlight, injury, and infection; regulates body temperature; and stores water, fat, and vitamin D.

Structurally, the skin is composed of two main layers, the epidermis (outer layer) and the dermis (inner layer). The epidermis consists of squamous cells, basal cells, and melanocytes. Squamous cells (or squamous keratinocytes) are the thin, flat cells that make up most of the epidermis. Besides the skin, squamous cells are also found in the mucosa of the mouth, the esophagus, and the corneal, conjunctival, and genital epithelia. Basal cells (or basal keratinocytes) are the round cells under the squamous cells and make keratin. Basal keratinocytes gradually migrate upward and differentiate into squamous cells (squamous keratinocytes) before reaching the outermost layer of the skin (i.e., the stratum corneum) and becoming keratinazed or cornified, creating the tough outer layer of skin. For this reason, NMSCs such as BCC and SCC are sometimes called keratinocyte cancers. Melanocytes are present in the lower part of the epidermis, and make melanin (pigment) after exposure to the sun, thus causing the skin to tan, or darken. The dermis is composed of blood and lymph vessels, hair follicles, and glands.

BCC (also called basalioma, basal cell epithelioma, rodent ulcer and Jacobs' ulcer) is thought to develop in the basal cells of the epidermis. However, there is recent evidence highlighting the adnexal origin of BCC and supporting its placement under the adnexal neoplasms as trichoblastic carcinoma. SCC primarily originates in the squamous cells of the epidermis. BCC and SCC may arise sporadically and de novo, or associate with precancerous lesions or underlying conditions. Multiple BCCs often develop early in life in patients with basal cell nevus syndrome (Gorlin syndrome) or Bazex syndrome, while SCC evolves from actinic keratosis (an area of red or brown, scaly, rough skin) and abnormal moles (nevi or dysplastic nevi) in the skin.

2.3 Epidemiology

Skin cancer is the most common form of malignancies, representing at least 40% of cancer cases globally. The annual incidence of BCC in the United

States is about 40 per 100,000, and that of SCC is about 10 per 100,000. While BCC commonly occurs in Caucasians, Hispanics, Chinese, and Japanese, SCC is frequently found in blacks and Asian Indians. The peak ages for BCC and SCC are >40 years and >60 years, respectively, with a notable male predominance (1.6–2:1).

2.4 Pathogenesis

Risk factors for NMSC include ultraviolet radiation (sun exposure and tanning beds), ionizing radiation (X-rays), arsenic exposure, tars, human papillomavirus (HPV) or HIV infection, actinic keratosis, basal cell nevus syndrome (Gorlin syndrome), Bazex syndrome, xeroderma pigmentosum, epidermodysplasia verruciformis, chronic nonhealing wounds (ulcer, or osteomyelitis with draining sinuses), burn scars, PUVA treatment for psoriasis, organ transplantation, immunosuppressive medications (e.g., cyclosporin A and azathioprine), albinism (lack of pigmentation in skin), and light skin color. Molecularly, syndromic and sporadic BCC is often linked to mutations in the PTCH (patched) gene on 9q22.3 (as in basal cell nevus syndrome [Gorlin syndrome] and xeroderma pigmentosum) and p53 gene [4].

The PTCH1 gene encodes a receptor for a protein called Sonic Hedgehog (SHH), which is involved in embryonic development. When SHH binds to PTCH, it releases smoothened (SMOH), a transmembrane signalling protein, from inhibition by PTCH. SMOH in turn signals to GSK3b, which phosphorylates GLI3 (a human ortholog of the Drosophila gene cubitus interruptus), leading to subsequent activity of target genes such as WNT and NFκB genes. The development of BCC in the context of PTCH1 as well as SMOH mutations may represent uncontrolled hair follicle morphogenesis. Although p53 mutations are often observed in BCC, they appear to be secondary events that may have insignificant contribution to the tumorigenesis of BCC [5].

On the other hand, cutaneous SCC is clearly associated with mutations in p53 (40-50% of cases), RAS (10-30%), CCND1, CDKN2A, SOX2, NOTCH 1, and FBXW7 genes [6].

2.5 Clinical features

BCC appears as a pearly white or waxy bump with blood vessels (telangiectasia) on the face, ears, or neck; a flat, scaly, brown, or flesh-colored patch on the back or chest; or a white, waxy scar (morpheaform BCC). Metastases (to the lymph nodes, lung, and bone) are exceedingly rare.

SCC is a well-defined, thin, white plaque (leukoplakia) or erythematous scaly papule or nodule with ulceration and hemorrhage (bleeding). It usually occurs in areas not exposed to the sun in dark-skinned individuals, but on sun-exposed areas (e.g., the head, face, ears, and neck) in fair-skinned individuals. SCC is more likely to metastasize than BCC.

2.6 Diagnosis

BCC is a slow-growing, locally invasive, malignant skin cancer affecting the epidermis. Microscopically, BCC may appear nodular (60%, a large, rounded mass of neoplastic cells with well-defined peripheral contours and peripheral palisading in the dermis), micronodular (10%, small nodules with peripheral palisading), superficial (25%, well-delineated nests growing multifocally and radially from the epidermis into the papillary dermis), infiltrative or morpheaform (2%, tumor islands of varying size growing infiltratively with an irregular outline and spiky configuration, but poorly developed peripheral palisading), and basosquamous (metatypical) (admixture of BCC and SCC with potential to metastasize; clearly different from focal squamous differentiation in BCC). Immunohistochemically, BCC is positive for BerEP4, 34βE12, MNF116, p53, BCL2, and p63, but negative for EMA, CEA, involucrin, and CK20. Molecularly, BCC may show gains in chromosomes 5, 7, 9, 29, and 20 and LOH at 9q22.3 and trisomy 6 [7].

Differential diagnoses for BCC include actinic keratosis (which affects sun-damaged skin, is usually smaller, and retains a squamous differentiation toward the surface in comparison with Bowen disease), Bowen disease (which is a SCC in situ with full-epidermal thickness dysplasia and potential for significant lateral spread before invasion; and includes cases occurring in both sun-exposed and sun-protected skin), fibrous papule of the face, juvenile nasopharyngeal angiofibroma, malignant melanoma, melanocytic nevi, molluscum contagiosum, psoriasis, sebaceous hyperplasia, SCC, and trichoepithelioma.

SCC is slow-growing, malignant skin cancer of the epidermis that often infiltrates the dermis and shows metastatic potential (5% with 2 cm tumor; 5%–10% in transplant patients). Microscopically, SCC may be separated into well, moderately, and poorly differentiated according to degree of differentiation and keratinization. In well-differentiated SCC, tumor cells are pleomorphic or atypical, with abundant pink cytoplasm, mild to moderate atypia, and well-developed keratinization. In moderately differentiated SCC, tumor cells show well- and poorly differentiated features, with focal keratinization. In poorly differentiated SCC, tumors are pleomorphic with a high nuclear-to-cytoplasmic ratio, atypical nuclei, and no keratinization.

SCC may demonstrate a spectrum of histologic patterns, including conventional, acantholytic, clear cell, desmoplastic, lymphoepitheliomatous, spindle cell, verrucous, and warty. Immunohistochemically, SCC is positive for various keratins (34βE12, AE1/AE3, MNF116, CK5/6), EMA, and p63, but negative for CAM5.2, BerEP4, S100P, and SMA [8].

Differential diagnoses for SCC include actinic keratosis, allergic contact dermatitis, atopic dermatitis, atypical fibroxanthoma, BCC, benign skin lesions, Bowenoid papulosis, chemical burns, limbal dermoid, and pyoderma gangrenosum.

2.7 Treatment

The treatment for BCC and SCC usually involves surgical removal of the lesion (e.g., saucerization, standard full-thickness excision, Mohs micrographic surgery, curettage with or without electrodesiccation, and cryosurgery). Supplemental radiotherapy and chemotherapy may be utilized if necessary [4].

For a small BCC in a young person, the treatment with the best cure rate is Mohs surgery (a technique to remove the cancer with the least amount of surrounding tissue) or complete circumferential peripheral and deep margin assessment (CCPDMA). For an elderly frail individual with a difficult-to-excise BCC and multiple complicating medical problems, radiotherapy (slightly lower cure rate) or no treatment is prescribed. For large, superficial BCC, topical chemotherapy (5-fluorouracil or imiquimod) might be indicated for good cosmetic outcome.

For SCC, surgical excision with adequate margins, curettage, electrodessication, cryotherapy, radiotherapy (external beam radiotherapy or brachytherapy), and chemotherapy may be used.

2.8 Prognosis

BCC generally has an excellent prognosis. Unfavorable prognostic factors include certain histologic subtypes (infiltrative, morpheaform, micronodular, and basosquamous), dense fibrous stroma, loss of peripheral palisading, reduced expression of syndecan-1 and BCL2, greater expression of p53 and aneuploidy, perineurial invasion, and positive margins.

SCC has a good prognosis, with a death rate of <1%. Favorable prognostic factors consist of low stage, no or superficial dermal invasion, vertical tumor

thickness of <4 mm, short duration, and well differentiated. Unfavorable features include high-risk anatomic sites (scalp, ear, lip, nose, and eyelid), immunosuppression, tumor width of >2 cm, tumor depth of >4 mm, lymphovascular or perineural invasion, and poorly differentiated.

References

1. LeBoit PE, Burg G, Weedon D, Sarasin A. *Pathology and Genetics of Skin Tumours*. Lyon, France: IARC Press, 2006.
2. PDQ Adult Treatment Editorial Board. *Skin Cancer Treatment (PDQ®): Health Professional Version*. PDQ cancer information summaries. Bethesda, MD: National Cancer Institute, 2002.
3. PDQ Adult Treatment Editorial Board. *Melanoma Treatment (PDQ®): Health Professional Version*. PDQ cancer information summaries. Bethesda, MD: National Cancer Institute, 2002.
4. PDQ Adult Treatment Editorial Board. *Merkel Cell Carcinoma Treatment (PDQ®): Health Professional Version*. PDQ cancer information summaries. Bethesda, MD: National Cancer Institute, 2002.
5. Lupu M, Caruntu C, Ghita MA, et al. Gene expression and proteome analysis as sources of biomarkers in basal cell carcinoma. *Dis Markers*. 2016;2016:9831237.
6. Dotto GP, Rustgi AK. Squamous cell cancers: A unified perspective on biology and genetics. *Cancer Cell*. 2016;29(5):622–37.
7. PathologyOutlines.com. Basal cell carcinoma (BCC). http://www.pathologyoutlines.com/topic/skintumornonmelanocyticbcc.html. Accessed February 1, 2017.
8. PathologyOutlines.com. Squamous cell carcinoma (SCC). http://www.pathologyoutlines.com/topic/skintumornonmelanocyticscc.html. Accessed February 1, 2017.

3
Melanoma

3.1 Definition

Several categories of tumors are known to affect the skin. These include keratinocytic tumors, melanocytic tumors, appendageal tumors, hemato-lymphoid tumors, soft tissue tumors, neural tumors, and inherited tumor syndromes [1].

The melanocytic tumor category is further divided into (1) *malignant melanoma* (superficial spreading melanoma, nodular melanoma, lentigo maligna, acral lentiginous melanoma, desmoplastic melanoma, melanoma arising from blue nevus, melanoma arising in a giant congenital nevus, melanoma of childhood, nevoid melanoma, and persistent melanoma) and (2) *benign melanocytic tumors* (congenital melanocytic nevi [superficial type and proliferative nodules in congenital melanocytic nevi], dermal melanocytic lesions [Mongolian spot and nevi of Ito and Ota], blue nevus [cellular blue nevus], combined nevus, melanotic macules, simple lentigo and lentiginous nevus, dysplastic nevus, site-specific nevi (acral, genital, and Meyerson nevus), persistent (recurrent) melanocytic nevus, spitz nevus, pigmented spindle cell nevus (Reed), and halo nevus] [1].

As the third most common skin tumor after basal cell carcinoma (BCC) and squamous cell carcinoma (SCC), malignant melanoma is highly aggressive, and causes 80% of skin cancer mortality despite constituting only 5% of all skin neoplasms.

3.2 Biology

Melanocytes are melanin-producing cells located beneath basal cells and squamous cells in the epidermis. The melanin is a pigment that gives the skin its natural color. Upon exposure to the sun, melanocytes produce additional pigment to protect the nucleus of keratinocytes from ultraviolet (UV) damage, rendering the skin to tan, or darken.

Resulting from the transformation and uncontrolled proliferation of melanocytes, melanoma may occur throughout the body, including the back

(of both men and women), legs (of women), head, neck, shoulders, hips, soles of the feet, fingernails, and other areas that are not exposed to the sun.

Apart from sporadic or de novo development, melanoma is often associated with precancerous lesions, such as abnormal moles (or dysplastic nevi). Moles are flat or raised, smooth or rough ("pebbly") on the surface, round or oval in shape, and pink, tan, brown, or skin-colored growths of <0.6 cm in size. People may have as many as 100 or more moles on the body.

Despite being less common than BCC and SCC, melanoma has a propensity to spread to other parts of the body, such as nearby lymph nodes, and spread via lymphatic networks.

3.3 Epidemiology

The age-standardized incidence rate for melanoma is 49 cases per 100,000 (60 for males and 39 for females). In 2012, >232,000 cases and 55,000 deaths related to melanoma were reported globally.

Before age 50, more women are affected (1 in 155 women vs. 1 in 220 men), while after age 50, more men are affected. The average age at diagnosis is 63 years.

Melanoma often occurs on nonexposed skin with less pigment (e.g., palms, soles, mucous membranes, and nail regions) in blacks, Asians, Filipinos, Indonesians, and native Hawaiians. In addition, melanoma is responsible for about 3% of all pediatric cancers.

3.4 Pathogenesis

Risk factors for melanoma include a fair complexion (e.g., fair skin that freckles and burns easily, does not tan, or tans poorly; blue, green, or hazel eyes; and red or blond hair), exposure to natural sunlight (blistering sunburns) or artificial sunlight (tanning beds), small moles, freckles, and family history of unusual moles (atypical nevus syndrome) or melanoma. Specifically, UV induces the formation of pyrimidine dimers between adjacent thymine and cytosine bases in DNA, leading to C $\rightarrow$ T transition, and subsequent tumorigenesis.

Molecularly, melanoma often demonstrates multiple chromosomal gains and losses, which are absent in nevi. About 40%–50% of melanomas harbor the B-Raf proto-oncogene serine-threonine kinase (*BRAF*) V600 mutation. Other common mutations implicated in the tumorigenesis of melanoma

involve the *CDKN2A* tumor suppressor gene, melanocortin receptor-1 (*MC1R*) gene, seven R385C (*FBXW7*), kinase domain insert receptor Q472H variant (*KDR*), V-Ki-ras2 Kirsten rat sarcoma viral oncogene homologue G12D mutation (*KRAS*), tumor protein P53 P72R variant (*P53*), and poly-morphism of ataxia telangiectasia mutated (*ATM*) −c.8850 + 60A>G [2].

It is generally believed that melanoma undergoes step-wise progression from benign nevi (controlled proliferation of normal melanocytes; acquisi-tion of BRAF mutation), atypical/dysplastic nevi (abnormal growth of mela-nocytes in a pre-existing nevus or new location resulting in a pre-malignant lesion/flat macule of >5 mm in size with random cytologic atypia, irregu-lar borders and variable pigmentation; loss of CDKN2A and PTEN), radial growth (horizontal proliferation of melanocytes in the epidermis, leading to melanoma in situ), vertical growth (loss of E-cadherin and expression of N-cadherin allowing malignant cells to invade basement membrane and proliferate vertically in the dermis), to metastasis (spread of malignant melanocytes to nearby lymph nodes, skin, subcutaneous soft tissue, lungs and brain) [3].

3.5 Clinical features

Clinical warning signs for melanoma include change in size, shape, color, or elevation in a preexisting pigmented lesion (mole); itching or pain in a preexisting mole; and development of new pigmented lesions. An often used mnemonic for recognizing potential melanoma is ABCDE (A for asym-metrical, B for borders [irregular], C for color [variegated], D for diameter [>6 mm], and E for evolving).

Melanoma typically appears as a brown to black, or pink, red, or fleshy, flat patch with an uneven smudgy outline, ulceration, and bleeding. Nodular melanoma is raised from the start and, even in color (red-pink or brown-black), and grows quickly. It can be life threatening in 6–8 weeks if not detected and treated promptly.

3.6 Diagnosis

Malignant melanoma includes four main subtypes: superficial spreading (70%), nodular (15%), lentigo maligna (10%), and acral lentiginous (5%).

Superficial spreading melanoma often appears as a new or existing mole of variable color (red, blue, black, white) with irregular, asymmetric border and ulceration or bleeding on any site, especially the lower extremities in women and the back/trunk in men; and demonstrates prolonged radial

growth (months to years) confined to the epidermis, and rapid vertical growth after invasion of the dermis.

Nodular melanoma forms a blue or black new mole of >6 mm in diameter with irregular border, ulceration or bleeding on the trunk, head, and neck; and shows early onset, rapid growth (weeks to months, without prolonged radial growth), invasion into the dermis, vertical growth and metastases, leading to rapidly fatal outcome.

Lentigo maligna melanoma is a brown-black macule with irregular border and raised growth on the face, neck, and arms; displays slow radial growth confined to the epidermis (5–20 years); and eventually invades the dermis with rapid vertical growth.

Acral lentiginous melanoma is a dark brown, black, or blue macule/patch with irregular border, gradual thickening and ulceration on the palmar, plantar, subungual, and mucosal sites; shows prolonged radial growth confined to the epidermis (months to years); and eventually invades the dermis with rapid vertical growth. It commonly occurs in people with darker skin.

Microscopically, melanoma shows atypical melanocytes (large, epithelioid, spindle, or bizarre cells with eosinophilic and finely granular cytoplasm; nuclear pseudoinclusions, folds, or grooves; and marked atypia with pleomorphic nuclei and large eosinophilic nucleoli), junctional activity (in the dermoepidermal junction), pagetoid spread, prominent melanin pigmentation, and invasion of the surrounding tissue. Melanoma often displays various growth patterns (pseudoglandular, pseudopapillary, peritheliomatous [around blood vessels], hemangiopericytoma-like, Spitz nevus-like, trabecular, verrucous, and nevoid) [4].

Immunohistochemically, melanoma is positive for S100 (nuclear and cytoplasmic staining, 90%+ sensitive but not specific), HMB45 (cytoplasmic and weak nuclear staining, negative in desmoplastic melanoma), MelanA/Mart1 (also stains steroid-producing cells in the ovary, testis, and adrenal cortex), tyrosinase (also stains peripheral nerve sheath and neuroendocrine tumors), microphthalmia transcription factor (MITF) (also stains dermatofibroma and smooth muscle tumors; negative in spindle cell or desmoplastic melanoma), NKI-C3 and NSE (nonspecific), PHH3 and Ki-67 (may distinguish melanoma from nevi), Fontana–Masson (detects melanin granules), vimentin, Cam 5.2, CEA, EMA, α-1-antichymotrypsin, and CD68. Molecularly, melanoma often harbors genetic mutations (ARID2, BAP1, BRAF, GNAQ, HRAS,

KIT, NF1, NRAS, and PTEN) and altered pathways (RAS-RAF-MEK-ERK, p16[INK4A]-CDK4-RB, and ARF-p53) [4].

Differential diagnoses for melanoma include squamous cell carcinoma, pigmented basal cell carcinoma, metastatic tumors to the skin, pigmented spindle cell tumor, atypical fibroxanthoma (usually negative for HMB45, MelanA and S100), granular cell tumor (negative for HMB45 and MelanA), pigmented actinic keratosis, sebaceous carcinoma, histiocytoid hemangioma, mycosis fungoides, benign melanocytic lesions and nevi (blue nevus, Spitz nevus-desmoplastic type, halo nevus, activated dysplastic nevus, vulval nevus).

Key features for differentiation of melanoma from benign lesions and nevi include asymmetry, size of >6 mm, atypia, band-like chronic inflammatory infiltrate in dermis, the absence of maturation of dermal tumor cells, lateral extension of individual melanocytes, melanocytes with clear cytoplasm and finely dispersed chromatin, individual melanocyte necrosis (compared with eosinophilic hyaline bodies in Spitz nevi), mitotic figures in melanocytes (atypical melanoma), pleomorphism of tumor cells, poor circumscription of the intraepidermal component, melanin in deep cells, transepidermal migration of melanocytes, dermal lymphocytes, and the presence of chromosomal gains or losses [4].

3.7 Treatment

Surgery is curative for early-stage melanoma, but is less effective for melanoma that metastasizes to distant lymph nodes or organs [5].

Melanoma is generally poorly responsive to radiotherapy or chemotherapy. For example, the dacarbazine-based regimens induce objective response rates (ORRs) in only 15%–20% of cases, and prolong remission for 7 years. A combination of dacarbazine with cisplatin produces somewhat better results than dacarbazine alone in terms of ORR and progression-free survival, but not overall survival [2].

In view of this dilemma, the new combination of BRAF (e.g., vemurafenib and dabrafenib) and MEK (trametinib) inhibitors has been adopted for *BRAF* V600-mutated metastatic melanoma, with response rates of >70% as the first-line treatment. In *BRAF* nonmutated metastatic melanoma, immune checkpoint inhibitors (e.g., ipilimumab, pembrolizumab, and nivolumab) form the first-line therapies. The combined use of nivolumab and ipilimumab give an ORR of >75%, although at the expense of higher and more pronounced toxicities compared with immunotherapy agents used individually [6–8].

3.8 Prognosis

The estimated 5-year survival rate is 98% for patients with early melanoma, 62% when melanoma reaches the lymph nodes, and 15% when melanoma metastasizes to distant organs.

The mortality rate of melanoma is 15%–20%, and patients with advanced melanoma generally survive for 6–9 months even with chemotherapy.

Unfavorable prognostic factors include high Breslow (vertical) thickness in primary tumor, high Clark's level, vascular invasion, high TNM stage, male gender, high mitotic rate, ulceration, microscopic satellites (tumor nests of >50 μm separated from main tumor mass), higher percent tumor area per volume in sentinel node, positive nonsentinel lymph nodes, regression, tumor-infiltrating lymphocytes, and elevated lactate dehydrogenase (LDH) levels [5].

References

1. LeBoit PE, Burg G, Weedon D, Sarasin A. *Pathology and Genetics of Skin Tumours*. Lyon, France: IARC Press, 2006.
2. Simon A, Kourie HR, Kerger J. Is there still a role for cytotoxic chemotherapy after targeted therapy and immunotherapy in metastatic melanoma? A case report and literature review. *Chin J Cancer* 2017;36(1):10.
3. Yeh I. Recent advances in molecular genetics of melanoma progression: implications for diagnosis and treatment. *F1000Res.* 2016;5. pii: F1000 Faculty Rev-1529.
4. PathologyOutlines.com. Invasive melanoma—General. http://www.pathologyoutlines.com/topic/skintumormelanocyticmelanoma.html. Accessed February 3, 2017.
5. PDQ Adult Treatment Editorial Board. *Melanoma Treatment (PDQ®): Health Professional Version*. PDQ Cancer information summaries. Bethesda, MD: National Cancer Institute, 2002.
6. Muñoz-Couselo E, García JS, Pérez-García JM, Cebrián VO, Castán JC. Recent advances in the treatment of melanoma with BRAF and MEK inhibitors. *Ann Transl Med* 2015;3(15):207.
7. Diamantopoulos P, Gogas H. Melanoma immunotherapy dominates the field. *Ann Transl Med* 2016;4(14):269.
8. Hashim PW, Friedlander P, Goldenberg G. Systemic therapies for late-stage melanoma. *J Clin Aesthet Dermatol* 2016;9(10):36–40.

4
Merkel Cell Carcinoma

Ryan J. Carlson and Paul Nghiem

4.1 Definition

Merkel cell carcinoma (MCC) is an aggressive neuroendocrine cancer that is associated with the Merkel cell polyomavirus. MCC gets its name because it has ultrastructural features similar to normal Merkel cells present in the skin that are involved in mediating the sensation of fine touch. MCC expresses neuroendocrine markers characteristic of the normal Merkel cell, a mechanoreceptor present in the skin. Although MCC is much rarer than melanoma, it nonetheless displays higher rates of recurrence (both local and regional), sentinel lymph node involvement, and overall mortality. Due to its association with Merkel cell polyomavirus (MCPyV) and/or ultraviolet (UV)-induced mutations, MCC appears to be an immunogenic malignancy. Patients with advanced MCC show striking responses to newly available immunotherapeutic treatments, which will likely transform management of the advanced stages of this cancer [1].

4.2 Biology

First described by Friedrich Sigmund Merkel, the Merkel cell is a mechanoreceptor located in the skin that aids in sensing light touch. In combination with other cells in the skin, Merkel cells allow for two-point discrimination and perception of texture, shape, and curvature. Normal Merkel cells are at their highest density in touch-sensitive areas, such as the hairless portions of the hands and feet, as well as "touch domes" in hairy skin. While they have sensory functions, recent research has shown that Merkel cells are not developmentally derived from neural crest tissue, but are instead descended from the epidermal lineage. It is currently unclear if MCC arises from an earlier epidermal progenitor or from a more differentiated, Merkel-like cell [2].

4.3 Epidemiology

MCC is a rare cancer, with approximately 1600 new patients diagnosed each year in the United States. However, its incidence has been rising steadily, and has roughly tripled over the last three decades. This increase is most likely

due to improved diagnostic techniques and growing numbers of higher risk individuals (e.g., the elderly and immunosuppressed persons). MCC incidence is at least eightfold higher in Caucasian individuals than in African-Americans and occurs somewhat more often in men (60%) than in women.

4.4 Pathogenesis

Risk factors for MCC include immunosuppression (e.g., solid organ transplantation, HIV infection, chronic leukemia or lymphoma, and autoimmune disease requiring immunosuppression), a history of sun exposure or other UV-associated conditions (leading to mutations in RB1, TP53, and other genes), and MCPyV infection [3,4].

MCPyV, discovered in 2008, is a small, nonenveloped, double-stranded DNA virus, but is unique in that it is clonally integrated into cancer cells in ~80% of MCC cases. The MCPyV genome encodes two viral proteins that are important to the pathogenesis of MCC, the large and small T (tumor) antigens. The primary oncogenic action of the large T antigen is its ability to inhibit the retinoblastoma tumor suppressor protein, promoting E2F activity and cell cycle progression. In addition, its C-terminus contains several regions that are essential for viral replication, such as its ATPase, helicase and origin-binding domains. In order for clonal integration of MCPyV to progress to MCC, a large T-antigen truncation mutation is required to abolish viral replication. The role of the MCPyV small T antigen in the pathogenesis of MCC is controversial, but it has been shown to possess several oncogenic functions. Surprisingly, MCPyV is an extremely common infection, and has been found to be present on 60%–80% of forehead swabs from healthy control individuals (and 90% of MCC patients). The rarity of MCC can be attributed to the low probability of both the clonal integration of MCPyV and the specific truncation mutation necessary to cause MCC, as well as the presence of effective immune surveillance in the majority of people [5,6].

Approximately 80% of MCC is virus positive but in some parts of the world (Australia) as few as 25% of MCC cases are caused by MCPyV. As with all viral cancers, the immune response is heavily implicated in the development, progression, and treatment of MCC. For example, circulating antibodies against MCPyV have been found in the majority of MCC patients, to both viral capsid proteins and the T antigens. However, these antibodies are not thought to protect against MCC progression. Instead, cellular immunity is thought to be vital to tumor clearance, and T cell responses against MCC have been observed in patients. Virus-specific CD8+ T cells are present among tumor infiltrating lymphocytes in MCC tumors, but they often fail to

respond to MCPyV peptide due to exhaustion, as suggested by expression of exhaustion markers PD-1 and TIM-3 [3].

In virus-negative MCC, a higher mutational burden is observed, including nearly ubiquitous mutation of the tumor suppressor genes *RB1* and *TP53*. Many of the mutations seen in these virus-negative cancers are characteristic of chronic UV exposure [4].

4.5 Clinical features

MCC typically manifests as a painless mass on the skin with a cystic or nodular appearance. Sometimes, it can be surrounded by small satellite lesions, and may have a plaque-like quality. MCC often expands rapidly in the months after first being noticed by a patient, and can commonly invade the subcutis, while rarely extending to the overlying epidermis. The most common clinical features of MCC are summarized by the mnemonic AEIOU: *asymptomatic* (or a lack of tenderness), *expanding* rapidly, *immunosuppressed, older* than 50 years, and arising on an *UV*-exposed site. In a study of MCC patients' clinical presentations, 89% of patients exhibited three or more of these criteria [7].

4.6 Diagnosis

Diagnosis of MCC typically requires biopsy because MCC tumors often have a non-specific, minimally concerning appearance. Pathologically, MCC tumors appear as an intradermal mass with tumor cells forming sheets, nests, or anastomosing aggregates (less common). MCC cells have round to oval nuclei, with a "salt and pepper" appearance of the chromatin (Figure 4.1). Cells exhibit significant mitotic activity, and have very scant cytoplasm. Typically, necrosis or apoptotic bodies can also be seen, along with lymphocytic infiltrates. However, these morphological features are not sufficient to differentiate MCC from other malignancies, or to distinguish between virus-positive and virus-negative cases.

MCC is confirmed through a series of immunohistochemical assays, of which cytokeratin 20 (CK20) positivity is the most discriminating feature (in 89%–100% of MCC), and only rarely positive in small cell lung carcinoma [SCLC]). Tumors may also be screened for markers that are common to SCLC, lymphoma, and melanoma in order to rule out these more common cancers. MCC cells can express both neuroendocrine and epithelial proteins, and proliferation markers are frequently seen.

A serology test for T-antigen-specific antibodies has been developed for MCC and has significant clinical value for both diagnosis and surveillance.

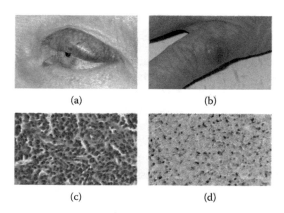

Figure 4.1   Characteristic clinical presentations and histological features of MCC. (a) Primary MCC lesion located on the upper eyelid of an 85-year-old man with a history of chronic lymphocytic leukemia. The red-purple color is common for MCC, as is its location on sun-exposed skin. (b) Primary MCC on the left small finger of a 70-year-old man. As is common for MCC, neither of these lesions were tender or pruritic. (c) Hematoxylin and eosin staining of an MCC tumor showing the large nuclei and scant cytoplasm associated with MCC. (d) This MCC tumor stained positively with anti-CK20, and the stain appears in the typical perinuclear dot-like pattern.

A positive result confirms virus-positive status, and informs the physician that this test can be used to measure the patient's disease burden over the course of his or her treatment and to detect early recurrences. A negative result indicates a higher risk of recurrence and that the patient should undergo increased surveillance, typically with imaging studies. More than 40% of patients diagnosed with virus-positive MCC will develop recurrent disease.

MCC staging is performed using the eighth edition staging system set by the American Joint Committee on Cancer (AJCC). Patients are grouped into both clinical and pathological stages based on the size of their primary tumor, their nodal involvement, and sites of distant metastases (see Table 4.1). This staging system aids patients and physicians in understanding the severity of MCC, and can guide treatment options and prognostic predictions [8].

4.7 Treatment

The standard of care for MCC involves surgical excision, often with wide margins (2 cm), strong recommendation for sentinel lymph node biopsy (SLNB), and adjuvant radiation in higher risk cases (Figure 4.2). SLNB is an important prognostic indicator, as patients with microscopic nodal involvement have at least 20% lower survival than similar patients without microscopic nodal disease [9,10].

Table 4.1 Current American Joint Committee on Cancer (AJCC) Staging System for Merkel Cell Carcinoma

Stage	Primary Tumor	Lymph Node	Metastasis
0	*In situ* (within epidermis only)	No regional lymph node metastasis	No distant metastasis
I Clinical[a]	≤ 2 cm maximum tumor dimension	Nodes negative by clinical exam (no pathological exam performed)	No distant metastasis
I Pathological[b]	≤ 2 cm maximum tumor dimension	Nodes negative by pathologic exam	No distant metastasis
IIA Clinical	> 2 cm tumor dimension	Nodes negative by clinical exam (no pathological exam performed)	No distant metastasis
IIA Pathological	> 2 cm tumor dimension	Nodes negative by pathological exam	No distant metastasis
IIB Clinical	Primary tumor invades bone, muscle, fascia, or cartilage	Nodes negative by clinical exam (no pathological exam performed)	No distant metastasis
IIB Pathological	Primary tumor invades bone, muscle, fascia, or cartilage	Nodes negative by pathologic exam	No distant metastasis
III Clinical	Any size / depth tumor	Nodes positive by clinical exam (no pathological exam performed)	No distant metastasis
IIIA Pathological	Any size / depth tumor	Nodes positive by pathological exam	No distant metastasis
IIIA Pathological	Not detected ("unknown primary")	Nodes positive by clinical exam, and confirmed via pathological exam	No distant metastasis
IIIB Pathological	Any size / depth tumor	Nodes positive by clinical exam, and confirmed via pathological exam OR in-transit metastasis[c]	No distant metastasis
IV Clinical	Any	+/− regional nodal involvement	Distant metastasis detected via clinical exam

(Continued)

Table 4.1 (Continued) Current American Joint Committee on Cancer (AJCC) Staging System for Merkel Cell Carcinoma

Stage	Primary Tumor	Lymph Node	Metastasis
IV Pathological	Any	+/− regional nodal involvement	Distant metastasis confirmed via pathological exam

Source: Adapted from Harms, et al. (*Annals of surgical oncology,* 2016) by Aubriana Ard & Paul Nghiem, see http://www.merkelcell.org/staging

[a] Clinical detection of nodal or metastatic disease may be via inspection, palpation, and/or imaging.

[b] Pathological detection/confirmation of nodal disease may be via sentinel lymph node biopsy, lymphadenectomy, or fine needle biopsy; and pathological confirmation of metastatic disease may be via biopsy of the suspected metastasis.

[c] In transit metastasis: a tumor distinct from the primary lesion and located either (1) between the primary lesion and the draining regional lymph nodes or (2) distal to the primary lesion.

Surgical excision with adjuvant radiotherapy has been associated with better survival and fewer recurrences than excision alone, and is often recommended except for very low-risk cases. In contrast, no survival benefit has been demonstrated from adjuvant chemotherapy. MCC rapidly develops resistance to chemotherapy, which is also immune suppressive, and can be quite toxic in this older patient population.

For patients with unresectable disease, primary radiation therapy (RT) can often provide excellent control. The recommended dose for traditional RT is 55–60 Gy for unresected disease and 46–50 Gy for adjuvant therapy, typically given over 5–6 weeks in ~25 doses. Single-dose radiotherapy (8 Gy) has also proven to be effective in the palliative treatment of metastatic MCC, and should be considered for patients with few distant metastases due to its financial and logistical benefits. While studies have reported 50%–65% initial response rates with systemic chemotherapy, these treatments rarely cure MCC, and typically patients develop progressive disease only 3 months after beginning chemotherapy.

Immunotherapeutic approaches are also showing exciting early success in MCC. Like other virus-associated cancers, MCPyV-positive MCC is highly immunologically responsive due to viral antigens expressed by tumor cells. In addition, virus-negative MCC has a large number of UV-associated mutations that are often immunogenic. Several ongoing clinical trials are using methods such as checkpoint inhibition, cytokine supplementation, and adoptive T-cell therapy to successfully treat patients with MCC. For example, pembrolizumab, an anti-PD-1 therapy, has recently been shown to induce striking responses in advanced MCC.

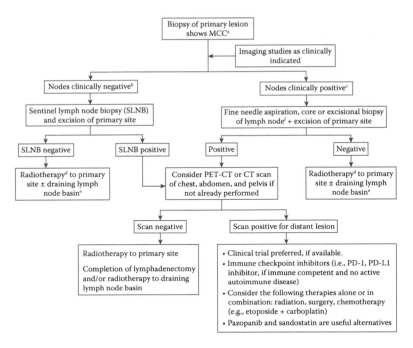

Figure 4.2   Primary cutaneous MCC: simplified evaluation and treatment. [a]Consider baseline MCPyV serology for prognostic significance and to track disease. [b]No pathologically enlarged nodes in physical exam and imaging study. [c]Pathologically enlarged nodes in physical exam or imaging study. [d]Radiotherapy is indicated in most patients; the exception is for low-risk disease (e.g., primary is ≤1 cm and of the extremities or trunk, no lymphovascular invasion, and negative margins; no immunosuppression). [e]Consider RT to the nodal basin in high-risk patients. [f]Consider excisional biopsy primarily or after negative needle or core biopsy to rule out false-negative biopsy result. See NCCN guidelines (https://www.nccn.org/) and http://www.merkelcell.org/ for further information, including surveillance guidance.

Treatment resulted in an objective response rate of 56%, with 86% of responding patients exhibiting continued responses well beyond those observed for chemotherapy. These responses occurred in both virus-positive and virus-negative cancers [10].

4.8 Prognosis

The 5-year overall survival rates for clinical stages I, IIA, and IIB are 45%, 30.9%, and 27.3%, respectively (note: OS includes deaths both from MCC and non-MCC causes, which can be very significant in this older population). The 5-year overall survival rates for pathological stages are 62.8%

(stage I), 54.6% (IIA), 34.8% (IIB), 40.3% (IIIA), 26.8% (IIIB), and 13.5% (IV) [8]. Immune function is critical for preventing the occurrence and halting the advancement of MCC, with survival in immune-suppressed patients being markedly lower [3].

Acknowledgments

R.J.C. was funded by the University of Washington MSTP training grant (5T32GM007266) and by the UW Merkel Cell Carcinoma Gift Fund. We would like to thank Aubriana McEvoy and Dr. Song Park for their assistance with the Figure 4.2 in this work.

References

1. Ramahi E, Choi J, Fuller CD, Eng TY. Merkel cell carcinoma. *Am J Clin Oncol* 2013;36(3):299–309.
2. Morrison KM, Miesegaes GR, Lumpkin EA, Maricich SM. Mammalian Merkel cells are descended from the epidermal lineage. *Dev Biol* 2009;336(1):76–83.
3. Triozzi PL, Fernandez AP. The role of the immune response in Merkel cell carcinoma. *Cancers (Basel)* 2013;5(1):234–54.
4. Wong SQ, Waldeck K, Vergara IA, et al. UV-associated mutations underlie the etiology of MCV-negative Merkel cell carcinomas. *Cancer Res* 2015;75(24):5228–34.
5. Chang Y, Moore PS. Merkel cell carcinoma: A virus-induced human cancer. *Annu Rev Pathol* 2012;7:123–44.
6. Feng H, Shuda M, Chang Y, Moore PS. Clonal integration of a polyomavirus in human Merkel cell carcinoma. *Science* 2008;319(5866):1096–100.
7. Heath M, Jaimes N, Lemos B, et al. Clinical characteristics of Merkel cell carcinoma at diagnosis in 195 patients: The AEIOU features. *J Am Acad Dermatol* 2008;58(3):375–81.
8. Harms KL, Healy MA, Nghiem P, et al. Analysis of prognostic factors from 9387 Merkel cell carcinoma cases forms the basis for the new 8th edition AJCC staging system. *Ann Surg Oncol* 2016;23(11):3564–71.
9. Miller NJ, Bhatia S, Parvathaneni U, Iyer JG, Nghiem P. Emerging and mechanism-based therapies for recurrent or metastatic Merkel cell carcinoma. *Curr Treat Options Oncol* 2013;14(2):249–63.
10. Nghiem PT, Bhatia S, Lipson EJ, et al. PD-1 blockade with pembrolizumab in advanced Merkel-cell carcinoma. *N Engl J Med* 2016;374(26):2542–52.

5
Angiosarcoma

5.1 Definition

Tumors of soft tissues consist of the following categories: adipocytic tumors, fibroblastic or myofibroblastic tumors, so-called fibrohistiocytic tumors, smooth muscle tumors, pericytic or perivascular tumors, skeletal muscle tumors, vascular tumors, chondro-osseous tumors, gastrointestinal stromal tumors, nerve sheath tumors, tumors of uncertain differentiation, and undifferentiated or unclassified sarcomas [1].

In turn, vascular tumors are divided into four groups: (1) *benign* (hemangioma [capillary, cavernous, arteriovenous, venous, intramuscular, and synovial], epithelioid hemangioma, angiomatosis, and lymphangioma, which usually do not recur), (2) *intermediate (locally aggressive)* (kaposiform hemangioendothelioma, which often recurs but does not metastasize), (3) *intermediate (rarely metastasizing)* (retiform hemangioendothelioma, papillary intralymphatic angioendothelioma, composite hemangioendothelioma, and Kaposi sarcoma, which often recur and metastasize in <2% cases), and (iv) *malignant* (epithelioid hemangioendothelioma, and angiosarcoma of soft tissue, which commonly recur and have a high risk of metastasis) [1,2].

Benign soft tissue tumors have an annual incidence of 3000 cases per million, of which lipomas account for 33%, fibrohistiocytic and fibrous tumors for 33%, vascular tumors for 10%, and nerve sheath tumors for 5%. By contrast, soft tissue sarcomas (STSs) have an annual incidence of 30 cases per million, of which sarcomas showing endothelial differentiation (i.e., angiosarcoma and epithelioid hemangioendothelioma) account for <1% of all sarcoma cases.

As a rare malignant mesenchymal tumor, angiosarcoma (synonyms: lymphangiosarcoma, hemangiosarcoma, hemangioblastoma, malignant hemangioendothelioma, and malignant angioendothelioma) typically shows anastomosing vascular channels lined by atypical and proliferative active endothelial cells, or solid sheets of high-grade epithelioid or spindled cells without clear vasoformation [3].

5.2 Biology

The musculoskeletal system is composed of hard and soft tissues. While the hard tissue consists of bones and cartilages (articular cartilages), the soft tissues include fat, muscle (smooth, skeletal, and cardiac), fibrous tissue (tendons and ligaments), synovial tissue (joint capsules and ligaments), blood vessels, lymph vessels, and peripheral nerves. The main functions of soft tissues are to connect, support, or protect other structures and organs of the body.

Not surprisingly, STSs are biologically diverse, with >50 subtypes involving various soft tissues and cells, including adipocytes (liposarcoma), fibroblastic or myofibroblastic tissues (desmoid-type fibromatosis), so-called fibrohistiocytic tissue (giant cell tumor of soft tissue), pericytic or perivascular tissues (malignant glomus tumor), peripheral nerve tissues (malignant peripheral nerve sheath tumor), smooth muscle (leiomyosarcoma), skeletal (striated) muscle (rhabdomyosarcoma), vascular tissue (angiosarcoma), chondroosseous tissue (extraskeletal osteosarcoma), gastrointestinal stromal tissue (gastrointestinal stromal tumor), and cells of uncertain differentiation or origins (e.g., synovial sarcoma and pleomorphic sarcoma).

Soft tissue angiosarcoma mostly arises de novo from endothelial cells of blood vessels (e.g., aorta, inferior vena cava, and pulmonary artery), although some may occur in the setting of preexisting benign vascular tumors (e.g., lymphangioma), "port wine" stains, and benign and malignant peripheral nerve sheath tumors, as well as familial cancer syndromes. Angiosarcoma has also been described as a complication of renal transplantation, arteriovenous fistulae, varicose ulceration, xeroderma pigmentosum, and gouty tophus.

Typically occurring in the soft tissue and skin (with or without lymphedema) angiosarcoma may be occasionally present in the liver, (associated with arsenic, thorotrast, PVC and radiation to other sites), kidney (primary renal angiosarcoma), breast (3-12 years after radiation therapy for carcinoma), spleen, lung, and rarely bone (<1% of all angiosarcomas).

5.3 Epidemiology

STSs are a broad group of mesenchymal neoplasms, which account for <1% of all diagnosed malignancies, about 2% of adult and 15% of pediatric malignancies. The reported incidences of STS range from 1.8 to 5.0 per 100,000 annually.

Representing 2% of sarcomas, angiosarcoma is a rare, aggressive, vascular tumor commonly found in the skin and soft tissue. The estimated incidence of angiosarcoma in the United States is 2 cases per 1 million, with a mean age of 65 years at diagnosis, and a slight male predilection. In addition, a small number of angiosarcoma cases involve neonates and toddlers.

5.4 Pathogenesis

Risk factors for angiosarcoma include exposure to vinyl chloride (for the production of synthetic rubber), radiation, sun, thorium dioxide (thorotrast for cerebral angiography), and AsO_3-containing insecticides (vineyard workers); chronic lymphedema (Stewart–Treves syndrome); and other syndromes (e.g., Klippel–Trenaunay and Maffucci syndromes). Pediatric angiofibroma is often associated with tuberous sclerosis.

Molecularly, angiosarcoma is noted for upregulation of vascular-specific receptor tyrosine kinases (e.g., *TIE1*, *KDR*, *TEK*, and *FLT1*), *KDR* mutations (encoding for VEGFR2), high-level *MYC* amplification on 8q24 (a hallmark of radiation-induced and lymphedema-related angiosarcoma), and *FLT4* (encoding for VEGFR3) coamplification on 5q35 (in 25% of secondary angiosarcoma). Both *MYC* and *FLT4* gene abnormalities appear to be unrelated to radiation-induced atypical vascular lesions. Other alterations in angiosarcoma relate to gains of 5p11, 8p12, and 20q12 and losses of 4p and 7p15, and elevated expression of TP53 and MDM2 proteins (in 60% of cases) [3].

MYC, encoded by the Myc (v-Myc avian myelocytomatosis viral oncogene homolog) gene, is a multifunctional, nuclear phosphoprotein that binds on enhancer box sequences, recruits histone acetyltransferases, and thus regulates expression of 15% of all genes, including those involved in cell cycle progression, apoptosis and cellular transformation. Mutation, overexpression, rearrangement and translocation of the Myc gene lead to the unregulated expression of many genes and contribute to the formation of a variety of hematopoietic tumors, leukemias, lymphomas, and carcinomas of the cervix, colon, breast, lung and stomach. Through upregulation of the miR-17–-92 cluster, which subsequently downregulates thrombospondin-1 (THBS1), a potent endogenous inhibitor of angiogenesis, MYC amplification plays a crucial role in the angiogenic phenotype of angiosarcoma.

5.5 Clinical features

Soft tissue angiosarcomas often present as enlarging masses, which may lead to disfigurement, chronic pain, recurrent infections, and organ dysfunction. In one-third of patients, other symptoms (e.g., coagulopathy, anemia, persistent hematoma, or bruisability) may be also noted.

As the most prevalent form, cutaneous angiosarcoma is mainly found on the scalp and upper forehead, as multifocal bruise-like areas with an indurated border, which may become elevated, nodular, and occasionally ulcerated at the advanced stages. In the pediatric population, angiofibroma is usually a benign, red papule on the face.

5.6 Diagnosis

Soft tissue angiosarcomas at the early stages are small, sharply demarcated, asymptomatic, multiple red nodules. At the late stages, they are fleshy, grayish white masses of several centimeters in diameter with hemorrhage, necrosis, and deep invasion.

Microscopically, angiosarcoma in soft tissue shows epithelioid and spindled areas. Epithelioid areas contain large rounded cells arranged in sheets, small nests, cords, or rudimentary vascular channels. These vascular channels intercommunicate sinusoidally and infiltrate surrounding tissues destructively, offering a useful diagnostic mark for angiosarcoma. Other notable features include brisk mitotic activity, coagulative necrosis, significant nuclear atypia, necrosis, extensive hemorrhage, and involvement of subcutaneous tissue. Angiosarcoma predominated by large rounded "epithelioid" endothelial cells with abundant amphophilic or eosinophilic cytoplasm, large vesicular nuclei and prominent nucleoli is defined as "epithelioid angiosarcoma." Immunohistochemically, angiosarcoma is positive for von Willebrand factor (Factor VIII), other vascular markers (CD31, CD34 [nonspecific], Fli1 [nuclear stain], ERG, and Ki-67), thrombomodulin, c-kit (50%), VEGFR3 (variable), keratin (epithelioid tumors), podoplanin (D2-40 and lymphatic marker), EMA, Cam5.2, and AE1/3. Immunostaining for Kaposi sarcoma herpes virus is negative [4].

Molecularly, angiosarcoma shows high-level amplifications of MYC (particularly in radiation-induced angiosarcoma, and secondary tumors), and FLT4-VEGFR3 amplifications (<50% of cases).

Differential diagnoses for angiosarcoma include atypical vascular lesions (circumscribed, no atypia, and no mitotic figures), hemangioma (usually

<2 cm, well circumscribed, fibrous septa, thick-walled vessels, and noninvasive), Kaposi sarcoma (less prominent cytological atypia and no endothelial multilayering), and Kaposi-like hemangioendothelioma (no cytological atypia and absence of infiltrative anastomosing channels).

In addition, epithelioid angiosarcoma should be distinguished from epithelioid hemangioendothelioma. The former is noted for its greater cellularity, larger cells, more prominent mitotic activity, greater nuclear and nucleolar pleomorphism, and more frequent tightly cohesive cell clusters without a myxohyaline matrix (immunoreactivity for vimentin and cytokeratin). The latter tends to affect younger patients (20–40 years), has an indolent course, and contains specific gene fusion at WWTR1 (WW domain–containing transcription regulator protein 1) and CAMTA1 (calmodulin-binding transcription activator 1) due to the nonrandom reciprocal t(1;3)(p36;q25) translocation, turning the affected endothelial cells to tumors [5]

5.7 Treatment

Treatment options for soft tissue angiosarcoma include surgery, chemotherapy, and/or radiotherapy [6,7].

Localized angiosarcoma is cured by aggressive surgery. Multimodal treatment with surgery, systemic chemotherapy (e.g., doxorubicin, ifosfamide and dacarbazine), and radiation therapy is useful for metastatic angiosarcoma, despite being rarely curative. A combination of bevacizumab, a monoclonal antibody against vascular endothelial growth factor, with systemic chemotherapy has been reported for treating childhood angiosarcoma secondary to malignant transformation from infantile hemangioma [8].

Doxorubicin represents a standard first-line cytotoxic chemotherapy for advanced soft tissue sarcomas, resulting in a median survival of approximating 1 year in most studies.

Ifosfamide is an alkylating agent that has response rates comparable to those seen with doxorubicin for soft tissue sarcomas.

Dacarbazine has a slightly lower activity than doxorubicin or ifosfamide, but is useful as second-line therapy for soft tissue sarcomas. A combination of dacarbazine and gemcitabine appears to improve overall survival and progression-free survival in comparison with dacarbazine alone [8].

5.8 Prognosis

Soft tissue angiosarcomas are aggressive malignancies with a mortality of >50% within the first year and local recurrences in about 20% of patients. The 5-year overall disease-specific survival (DSS) rate for patients with localized, resected tumors is 44%, and that for patients with metastases at diagnosis is only 16%. Poor prognostic factors include older age, retroperitoneal location, large size (>5 cm), deep invasion (>3 mm), positive surgical margins, frequent mitosis (>3 high-power field [HPF]), high Ki-67 values, and tumor recurrence.

References

1. Fletcher CDM, Bridge JA, Hogendoorn P, et al. (eds.). *WHO Classification of Tumours of Soft Tissue and Bone*. 4th ed. Lyon, France: IARC Press, 2013.
2. International Society for the Study of Vascular Anomalies. *ISSVA Classification for Vascular Anomalies*. Melbourne, Australia: International Society for the Study of Vascular Anomalies, 2014.
3. Antonescu C. Malignant vascular tumors—An update. *Mod Pathol* 2014;27(Suppl 1):S30–8.
4. PathologyOutlines.com. Angiosarcoma. www.pathologyoutlines.com/topic/softtissueangiosarcoma.html. Accessed February 6, 2017.
5. Flucke U, Vogels RJ, de Saint Aubain Somerhausen N, et al. Epithelioid hemangioendothelioma: Clinicopathologic, immunohistochemical, and molecular genetic analysis of 39 cases. *Diagn Pathol* 2014;9:131.
6. PDQ Pediatric Treatment Editorial Board. *Childhood Soft Tissue Sarcoma Treatment (PDQ®): Health Professional Version*. PDQ cancer information summaries. Bethesda, MD: National Cancer Institute, 2002.
7. PDQ Pediatric Treatment Editorial Board. *Childhood Vascular Tumors Treatment (PDQ®): Health Professional Version*. PDQ cancer information summaries. Bethesda, MD: National Cancer Institute, 2002.
8. Ryan CW, Desai J. The past, present, and future of cytotoxic chemotherapy and pathway-directed targeted agents for soft tissue sarcoma. *Am Soc Clin Oncol Educ Book* 2013:386.

6
Leiomyosarcoma

6.1 Definition

Grouped along other tumors of soft tissues (including adipocytic tumors, fibroblastic or myofibroblastic tumors, so-called fibrohistiocytic tumors, pericytic or perivascular tumors, skeletal muscle tumors, vascular tumors, chondro-osseous tumors, gastrointestinal stromal tumors (GISTs), nerve sheath tumors, tumors of uncertain differentiation, and undifferentiated or unclassified sarcomas), smooth muscle tumors of soft tissues and nonuterine visceral organs encompass leiomyoma (angioleiomyoma, gastrointestinal [GI] leiomyoma, Mullerian leiomyoma, and pilar or cutaneous leiomyoma) and leiomyosarcoma (peripheral or extremity leiomyosarcoma, retroperitoneal leiomyosarcoma, and cutaneous leiomyosarcoma) [1]. In contrast to leiomyoma, which is noted for its absent or minimal atypia, absent or rare mitotic activity, small size, lack of hemorrhage, and necrosis, leiomyosarcoma is characterized by its nuclear atypia, mitotic activity, large size, frequent hemorrhage, and necrosis [2,3].

6.2 Biology

Consisting of both hard and soft tissues, the musculoskeletal system plays a fundamental role in connecting, supporting, and protecting other vital organs and structures of the body. While the hard tissue is composed of bones and cartilages (articular cartilages), the soft tissue is made up of fat, muscle (smooth, skeletal, and cardiac), fibrous tissue (tendons and ligaments), synovial tissue (joint capsules and ligaments), blood vessels, lymph vessels, and peripheral nerves.

Smooth (visceral) muscle (so called due to its lack of striations) is one of the three types of muscle tissue within the body (i.e., visceral, cardiac, and skeletal). Smooth muscle appears very smooth (in contrast to the banded or striated appearance of cardiac and skeletal muscles) and uniform under microscope and is present in the walls of hollow organs, like the urinary bladder, uterus, stomach, intestines, bronchi, and blood vessels. Smooth muscle is also found in the eye (for changing the size of the iris and altering

the shape of the lens) and skin (for making hair stand erect in response to cold temperature or fear).

Smooth muscle cells (also known as smooth muscle fibers or myocytes) are fusiform or spindle shaped and measure 20–200 μm in length and 3–10 μm in thickness. These cells have a single, centrally located nucleus and an eosinophilic cytoplasm or sarcoplasm (which is filled with actin and myosin, along with dense bodies in the cell membrane or sarcolemma), and often group in branching bundles. The actin filaments stretch between dense bodies in the cytoplasm and attachment plaques at the cell membrane, whereas the myosin filaments lie between the actin filaments. The presence of intermediate filaments, such as desmin and vimentin, aids in pulling the sarcolemma toward the fiber's middle, and provides further support to the cell function.

There is a special type of smooth muscle cell called myofibroblast located in alveolar septa of the lung and scar tissue, which has additional qualities of fibrocytes. Involved in the production of connective tissue proteins, such as collagen and elastin, myofibroblast is thus also called fixed or stationary connective tissue cell.

Smooth muscle is controlled by the unconscious part of the brain (and thus also called involuntary muscle) and is responsible for making organs contract to move substances through the organ, and for sealing orifices (e.g., pylorus and uterine orifices).

Leiomyosarcoma develops from immature smooth muscle cells, in five distinct anatomic sites: (1) retroperitoneum; (2) deep extremity (thighs); (3) uterus; (4) blood vessels (e.g., inferior vena cava, saphenous vein, femoral vein, pulmonary artery, femoral artery); and (5) superficial dermis. Soft tissue leiomyosarcomas are mostly of high or intermediate grade, with strong potential (40%–50%) for metastasis, and only rarely of low grade (e.g., cutaneous leiomyosarcoma). Purely cutaneous (dermal) leiomyosarcoma (sometimes called atypical smooth muscle tumor) shows a 30% local recurrence rate but no metastases [2,3].

6.3 Epidemiology

Soft tissue sarcomas are a heterogeneous group of mesenchymal neoplasms that represent 2% of adult and 15% of pediatric cancers.

Soft tissue leiomyosarcoma is the most common sarcoma, and accounts for >20% of all soft tissue sarcomas. This tumor affects people of 21–98 years in age (median 59 years).

6.4 Pathogenesis

Predisposing factors for leiomyosarcoma include Epstein-Barr virus (EBV) infection, in the setting of severe immunosuppression (eg, AIDS, or post kidney, cardiac, and liver transplantation), retinoblastoma syndrome (loss of retinoblastoma protein), Li–Fraumeni syndrome (TP53/p53 germline mutations), and Budd–Chiari syndrome.

Molecularly, leiomyosarcoma demonstrates multiple and variable gene alterations and very complex karyotypes. These include gains from 1q12-q31, 1q21, 1p3, 5p15, 6q, 8q24 (47%), 15q12-15, 15q25-q26, 16p, 17p11.2 (encompassing MAP2K4, FLCN and MYOCD genes), 17q25.1 (41%), 19q13.12 (53%), 20q, 22q, and Xp; losses from 1q42-qter, 1p36, 2p, 2q, 2p15-pter, 3p21-p23, 4q, 8p21-pter, 9p21.3 (41%), 10p, 10q23-qter (encompassing PTEN gene), 11p, 11q23-qter, 13q14 (encompassing RB gene), 13q14-21, 13q32-qter, 16q, and 18p11; and regions of amplification at 1q21, 5p14-pter, 8q, 12q13-15, 13q31, 17p11 (encompassing p53 gene), 19p13, and 20q13.

Interestingly, small tumors (<5 cm in diameter) tend to have fewer (and sometimes unique) cytogenetic gains or losses than larger tumors (>5 cm but <20 cm).

The MYOCD gene (on 17p12) encodes myocardin (MYOCD), a transcriptional cofactor of serum response factor, that regulates smooth muscle differentiation and cell migration. It is the most overexpressed in leiomyosarcomas of the retroperitoneum. Further, gains at 1q21.3, 11q12.2-q12.3, 16p11.2, and 19q13.12 correlate to increased risk of death. Some leiomyosarcomas demonstrate high expression levels of muscle-associated genes, including CALD1, SLMAP, ACTG2, CFL2, MYLK, ACTA2, MBNL1, TPM1, PPP1R12A, DTNA, FZD6, PPP1R12A, CLIC4, CDC42EP3, BARD1, TPM1, RAB27A, MAP1B, and EDIL [4,5].

6.5 Clinical features

Leiomyosarcoma often manifests as pain or unusual swelling (or lump) in the body, limited mobility, trouble breathing, bloating, vomiting or constipation, blood clots, and a change in menstruation (or vaginal bleeding after menopause).

6.6 Diagnosis

Leiomyosarcoma is a smooth muscle tumor that may affect the skin (cutaneous variant), retroperitoneum (superficial or deep soft tissues;

usually in women) and immunocompromised individuals (multifocal lesions, in association with EBV in HIV patients). Leiomyosarcoma may appear as either a hard mass with a white whorled cut surface (low grade, resembling leiomyoma) or a large, soft mass of >10 cm with necrosis, hemorrhage, and cystic degeneration (high grade). Imaging approaches (CT/MRI) usually reveal a nonspecific soft tissue mass, but are helpful in delineating the relationship to adjacent structures within the retroperitoneum. Histologically, leiomyosarcoma demonstrates a fascicular growth pattern (bundles intersecting at right angles); palisading of spindle cells with eosinophilic fibrillary cytoplasm, focal granularity, and cigar-shaped and blunt-ended nuclei containing cytoplasmic vacuoles at both ends; variable mitosis; hemangiopericytoma-like vasculature; myxoid change; osteoclast-like giant cells; focal pleomorphism (mimicking pleomorphic sarcoma or malignant fibrous histiocytoma [MFH]); and infiltration into adjacent tissue [6].

Use of mitotic figures (MF) and other parameters (tumor size, location, and necrosis) enables estimation of the relative malignancy of leiomyosarcomas, including soft tissue leiomyosarcoma (1–2 MF/10 high-power field [HPF] and deep), skin or subcutaneous leiomyosarcoma (1–2 MF/10 HPF), retroperitoneal leiomyosarcoma (5 MF/10 HPF, or 1–4 MF/10 HPF plus necrosis plus size of >7.5 cm), and vascular leiomyosarcoma (1–4 MF/10 HPF plus large size plus necrosis).

Immunohistochemically, leiomyosarcoma is positive for HHF35 (90%), α-smooth muscle actin (90%), vimentin and desmin (75%), heavy caldesmon, smooth muscle myosin, phosphotungstic acid–hematoxylin (PTAH) (myofibrils), keratin (30%), ER (uterine and female retroperitoneal), S100 (occasionally weak), EMA (focal), and CD34 (occasionally), but it is negative for CD117.

Differential diagnoses for leiomyosarcoma consist of dedifferentiated liposarcoma (usually trunk, well-differentiated component, and MDM2 and CDK4 amplification), leiomyoma (no or rare mitotic activity, no or minimal atypia, small size, no hemorrhage, and no necrosis), and myofibroblastic sarcoma (negative for heavy caldesmon and smooth muscle myosin). Retroperitoneal leiomyosarcoma is frequently intermixed with GIST, which was historically classified as GI leiomyosarcoma and which reacts with c-kit and Ano-1/DOG1, but not desmin [6].

The grade of leiomyosarcoma (other than uterine type) is often determined according to the Federation Nationale des Centres de Lutte Contre le Cancer (FNCLCC) grading system, which distinguishes three malignancy grades based on differentiation, necrosis and mitotic rate. For uterine

leiomyosarcoma, the revised 2008 International Federation of Gynaecology and Obstetrics (FIGO) staging system is used to predict patient outcome.

6.7 Treatment

Treatment options for leiomyosarcoma include surgery (excision with clear margins), radiotherapy, and chemotherapy (e.g., gemcitabine-docetaxel, doxorubicin, ifosfamide, trabectedin, and pazopanib) [7,8].

Gemcitabine is a nucleoside analog with modest activity against soft tissue sarcomas. Gemcitabine and docetaxel appear to demonstrate synergy against leiomyosarcoma (especially uterine leiomyosarcoma) despite docetaxel's apparent lack of activity as a single-agent in soft tissue sarcomas. Although their side-affects include neutropenia and anemia, gemcitabine and docetaxel ofter an important therapeutic option for advanced leiomyosarcomas. Trabectedin is a marine alkaloid that is believed to involve multiple DNA cellular pathways (including secondarily inhibiting the TLS-CHOP oncogene), leading to changes in the tumor microenvironment. Its chief toxicities include cytopenias, hepatitis, and fluid retention. Pazopanib is a vascular endothelial growth factor (VEGF) inhibitor, which is especially valuable for treating uterine leiomyosarcoma. Its toxicities are related to diarrhea (secondary to colitis), nausea, vomiting, hypertension, myocardial dysfunction, thromboembolism, and hepatitis, [9,10].

6.8 Prognosis

The 1-, 5-, and 10-year metastasis-free survival (MFS) rates for patients with leiomyosarcomas are 83%, 51%, and 45%, respectively. The 1-, 5-, and 10-year overall survival (OS) rates are 95%, 63%, and 49%, respectively.

Poor prognostic factors for leiomyosarcomas include retroperitoneal, mesenteric, or other deep location, size of >5 cm, age of >62 years, high grade, tumor disruption by prior incisional biopsy or incomplete excision, and possibly nuclear c-myc expression.

References

1. Fletcher CDM, Bridge JA, Hogendoorn P, et al. (eds.). *WHO Classification of Tumours of Soft Tissue and Bone.* 4th ed. Lyon, France: IARC Press, 2013.
2. Miettinen M. Smooth muscle tumors of soft tissue and non-uterine viscera: Biology and prognosis. *Mod Pathol* 2014;27(Suppl 1):S17–29.

3. Baheti AD, Jagannathan JP, O'Neill A, Tirumani H, Tirumani SH. Current concepts in non-gastrointestinal stromal tumor soft tissue sarcomas: A primer for radiologists. *Korean J Radiol* 2017;18(1):94–106.

4. Italiano A, Lagarde P, Brulard C, et al. Genetic profiling identifies two classes of soft-tissue leiomyosarcomas with distinct clinical characteristics. *Clin Cancer Res* 2013;19(5):1190–6.

5. Lee YF, Roe T, Mangham DC, Fisher C, Grimer RJ, Judson I. Gene expression profiling identifies distinct molecular subgroups of leiomyosarcoma with clinical relevance. *Br J Cancer* 2016;115(8):1000–7.

6. PathologyOutlines.com. Leiomyosarcoma—General. http://www.pathologyoutlines.com/topic/softtissueleiomyosarcoma.html.

7. PDQ Adult Treatment Editorial Board. *Adult Soft Tissue Sarcoma Treatment (PDQ®): Health Professional Version.* PDQ cancer information summaries. Bethesda, MD: National Cancer Institute, 2002.

8. PDQ Pediatric Treatment Editorial Board. *Childhood Soft Tissue Sarcoma Treatment (PDQ®): Health Professional Version.* PDQ cancer information summaries. Bethesda, MD: National Cancer Institute, 2002.

9. Duffaud F, Ray-Coquard I, Salas S, Pautier P. Recent advances in understanding and managing leiomyosarcomas. *F1000Prime Rep.* 2015;7:55

10. Gordon EM, Sankhala KK, Chawla N, Chawla SP. Trabectedin for soft tissue sarcoma: Current status and future perspectives. *Adv Ther.* 2016;33(7):1055–71.

7
Liposarcoma

7.1 Definition

Classified along with fibroblastic or myofibroblastic tumors, so-called fibrohistiocytic tumors, smooth muscle tumors, pericytic or perivascular tumors, skeletal muscle tumors, vascular tumors, chondro-osseous tumors, gastrointestinal stromal tumors, nerve sheath tumors, tumors of uncertain differentiation, and undifferentiated or unclassified sarcomas under tumors of soft tissues [1], adipocytic tumors are divided into (1) *benign* (lipoma, lipomatosis, lipomatosis of nerve, lipoblastoma or lipoblastomatosis, angiolipoma, myolipoma, chondroid lipoma, extrarenal angiomyolipoma, extra-adrenal myelolipoma, spindle cell or pleomorphic lipoma, and hibernoma), (2) *intermediate (locally aggressive)* (atypical lipomatous tumor [ATL] or well-differentiated liposarcoma [LPS]), and (3) *malignant* (dedifferentiated LPS, myxoid LPS, pleomorphic LPS, and LPS not otherwise specified [NOS]) [1].

The major LPS subtypes are well-differentiated LPS, dedifferentiated LPS, myxoid LPS, and pleomorphic LPS.

Well-differentiated LPS (or ATL) is an intermediate, locally aggressive mesenchymal neoplasm with a predilection for the extremity, retroperitoneum, inguinal, and paratesticular regions. Specifically, tumors found in the retroperitoneum or mediastinum are referred to as well-differentiated LPS, and tumors arising in the extremities are known as ATLs. Accounting for 40% of all LPS cases, well-differentiated LPS has a tendency to progress to dedifferentiated LPS (25%–40%) and is chemoresistant [2].

Dedifferentiated LPS is a malignant neoplasm that usually forms a deep-seated mass in the retroperitoneum. Representing 25% of all LPS cases, dedifferentiated LPS demonstrates a propensity for distant lung metastases (10%–15%) and recurrence, and is chemoinsensitive.

Myxoid LPS is a malignant neoplasm that tends to occur in the extremities. Accounting for 30% of all LPS cases, this tumor may develop distant metastases (10%–20%) and recur, and is chemosensitive. High-grade myxoid LPS is now considered synonymous to round cell LPS, which has an increasing round cell component (>5%) and worsening prognosis.

Pleomorphic LPS is the least common (representing 5% of all LPS cases) and most aggressive subtype of LPS. It commonly occurs in the extremities, shows local recurrence (especially in the lung, bone, and liver) in 30%–35% of patients, and is chemoresponsive [3].

7.2 Biology

The musculoskeletal system is composed of hard and soft tissues. While the hard tissue includes bones and cartilages (articular cartilages), the soft tissue encompasses fat, muscle (smooth, skeletal, and cardiac), fibrous tissue (tendons and ligaments), synovial tissue (joint capsules and ligaments), blood vessels, lymph vessels, and peripheral nerves. The main functions of soft tissues are to connect, support, or protect other structures and organs of the body.

Adipose tissue (or fat) is a loose connective tissue located beneath the skin (at the subcutaneous layer), between muscles, around the kidneys and heart, and behind the eyeballs and abdominal membranes. Two types of adipose tissue exist in the body, i.e. white adipose tissue (WAT) and brown adipose tissue (BAT). WAT is mostly made up of fat cells called adipocytes, which synthesize and store fat as lipid droplets in the middle, and whose nuclei are displaced to the side. BAT contains smaller brown fat cells, which derive their color from the high concentration of mitochondria for energy production and vascularization of the tissue. The main functions of adipose tissue are to store excess dietary fat (called triglycerides) inside the adipocytes, to cushion or absorb physical trauma, and to insulate or maintain the core body temperature.

LPS (also known as adipocyte sarcoma) usually arises de novo from adipocytes at deep-seated, well-vascularized stroma rather than submucosal or subcutaneous fat. While myxoid LPS and pleomorphic LPS commonly occur in the extremities (limbs) and rarely in the retroperitoneum (abdomen), well-differentiated LPS (or ALT) and dedifferentiated LPS develop frequently in the retroperitoneum. However, in rare cases, LPS may produce dome-shaped or polypoid lesions in the cutis and subcutis.

7.3 Epidemiology

Soft tissue sarcomas (STSs) represent about 2% of adult and 15% of pediatric cancers. Being one of the most common STSs (second only to leiomyosarcoma followed by malignant fibrous histiocytoma, and synovial sarcoma), LPS constitutes 15% of all STSs (but about 4% of pediatric STSs), 24% of all extremity STSs, and 45% of all retroperitoneal STSs.

LPS has an annual incidence of 2.5 cases per million. The mean age at onset is 50 years, and a slight male predominance is noted.

7.4 Pathogenesis

Well-differentiated LPS (or ALT) and dedifferentiated LPS frequently exhibit 12q13-15 amplification involving cyclin-dependent kinase 4 (CDK4) and murine double minute 2 (MDM2). MDM2 is a ubiquitin ligase that binds to the transactivation domain of p53 and promotes its degradation. Myxoid LPS harbors a recurring unique chromosome rearrangement, t(12;16)(q13;p11), leading to FUS (fused in sarcoma)-DDITR (TLS-CHOP) fusion. Another translocation fusion, EWS-CHOP oncogene t(12;21)(q13;q12), is occasionally observed in myxoid LPS. Pleomorphic LPS demonstrates a diverse mix of chromosomal rearrangements and genomic profiles without unified alterations, with the most common mutations found in p53. The role of p14^{ARF} methylation in the origination and growth of pleomorphic LPS is also hypothesized [3,4].

7.5 Clinical features

LPS is usually a slow-growing and painless mass in the early stage. However, late-stage LPS may lead to pain (chest pain or abdominal pain); swelling; numbness; constipation; diarrhea; bloody bowel movements; enlargement of varicose veins; fatigue; nausea; vomiting; trouble urinating, swallowing, speaking, and breathing; decreased range of motion in the limbs; and weight loss.

7.6 Diagnosis

Well-differentiated LPS (or ALT) is an intermediate, locally aggressive neoplasm, with a predilection for the extremities, retroperitoneum, inguinal, and paratesticular regions; and a notable absence in the dermis or subcutis of the posterior neck, upper back or shoulders . Macroscopically, the tumor is a large mass (>10 cm) with thick or nodular septations and a nodular nonlipomatous component. Microscopically, the tumor shows scattered lipoblasts with a large, irregular nucleus and dense, smudgy chromatin surrounded by large intracytoplasmic vacuoles in a background of adipocytes. Focal nuclear atypia and hyperchromasia are present. Upon undifferentiation into the dedifferentiated form, this locally aggressive tumor may metastasize. Molecularly, it is mostly positive for MDM2 and CDK4.

Dedifferentiated LPS mostly develops de novo, and sometimes progresses from ALT (25%–40%), and usually presents as a large, predominantly

fatty mass with a >1 cm nonlipomatous component. Commonly occurring in the retroperitoneum, the tumor is characterized by the transition from an adipocyte-rich, well-differentiated region to a nonlipogenic spindle cell–rich region. Despite its morphologic and molecular similarities to ALT, dedifferentiated LPS behaves aggressively and has a sixfold higher risk of death. Recurrence can be local, in the form of a nonlipomatous lesion with or without a fatty component, or as peritoneal sarcomatosis or distant metastases to the liver or lungs [2]. Molecularly, dedifferentiated LPS is positive for MDM2 (95%) and CDK4 (92%).

Myxoid LPS is a malignant neoplasm with a predilection for the extremities. Despite the median age of 50–65 years at diagnosis, myxoid LPS demonstrates a higher incidence in children and adolescents than other LPS subtypes. The tumor with a >5% round cell component is considered high-grade LPS or myxoid round cell LPS, which is more aggressive than myxoid LPS. Microscopically, the tumor is characterized by spindle or oval primitive nonmesenchymal cells organized in a myxoid stroma, with signet ring or multivacuolated lipoblasts and a distinctive plexiform vasculature. Molecularly, myxoid LPS harbors t(12;16) or t(12;22); and may be positive for MDM2 and CDK4 (4-20%).

Pleomorphic LPS is a malignant neoplasm with common occurrence in the extremities, high local recurrence, and distant metastasis. Macroscopically, the tumor is an indeterminate heterogeneous soft tissue mass with minimal fat. Its lack of septations and presence of T2 heterogeneity are useful for differentiation from high-grade myxoid LPS. Microscopically, the tumor is characterized by a variable number of pleomorphic lipoblasts with sharply defined cytoplasmic vacuole and indent and distort nucleus in a setting of a nonlipogenic high-grade sarcoma [3]. Molecularly, pleomorphic LPS is negative for MDM2 and CDK4.

Differential diagnoses for dedifferentiated LPS include lipomatous hemangiopericytoma (cytologically bland, no areas of atypical lipomatous tumor, circumscribed), non-lipogenic sarcoma and GI stromal tumor (infrequent pleomorphism, CD117, DOG1 85-95%); those for myxoid LPS include lipomatous tumors lipoblastoma/ipoblastomatosis - abnormalities involving 8q(11-13), spindle cell lipoma (prominent spindle cell component, lacking "chicken wire" vascular pattern, dense collagen bundles, CD34 positivity, abnormalities of chromosomes 13 or 16), hibernoma (lacking "chicken wire" vascular pattern, 11q13 rearrangement), chondroid lipoma (chondroid matrix, lacking "chicken wire" vascular pattern), myxolipoma (lacking "chicken wire" vascular pattern, no lipoblasts, CD34 positivity in myxoid areas)], other myxoid neoplasms [aggressive angiomyxoma (large,

disorganized vessels, no lipoblasts), intramuscular myxoma, juxta-articular myxoma (no lipoblasts, hypovascular pattern), low grade myxofibrosarcoma (no lipoblasts, no consistent abnormalities), extraskeletal myxoid chondro-sarcoma (no lipoblasts, lacking arborizing pattern, t(9;22) or t(9;17))] and small round blue cell tumor (no lipoblasts) [5,6].

7.7 Treatment

Surgical resection (limb-sparing surgery) with negative margins (opti-mally 5 cm) is the mainstay of treatment for localized or operable LPSs. Neoadjuvant doxorubicin-based chemotherapy and/or external beam radia-tion therapy may improve the efficacy of limb-sparing surgery in patients with large lesions or those adjacent to vital structures (e.g., retroperito-neal LPS). Myxoid LPS appears highly radiosensitive, and shows dramatic responses to preoperative radiation. On the other hand, pleomorphic LPS often responds to radiotherapy with an increase in size secondary to intra-tumoral hemorrhage and necrosis [7].

Systemic treatment, particularly chemotherapy, is limited in unresectable or metastatic LPSs. LPS is generally resistant to chemotherapy, although myxoid LPS has a higher sensitivity (48%) to cytotoxic drugs (e.g., doxoru-bicin, ifosfamide, and anthracyclines) than pleomorphic LPS (33%), dedif-ferentiated LPS (25%), high-grade (round cell) myxoid LPS (17%), and well-differentiated LPS (0%) [3].

Trabectedin, a newly available TLS-CHOP inhibitor, is effective for myxoid and dedifferentiated LPS. By binding to DNA within the minor groove of the double helix, trabectedin causes a conformational change that alters its interactions with DNA-binding proteins. In addition, through interaction with a transcription-coupled nucleotide excision repair complex, trabect-edin causes double-stranded DNA breaks [8].

7.8 Prognosis

Five-year disease-specific survival rates for LPSs are 100% in well-differentiated LPS, 88% in myxoid LPS, and 56% in pleomorphic LPS.

Local recurrence may occur in two-thirds of all LPS patients, and metastases in 50% of patients. Patients with metastases of extremity sarcomas to the lungs have a 5-year survival rate of 60%.

Factors impacting on the prognosis of LPS include tumor site, size, grade, and histologic subtype. In general, limb tumor has a more favorable

prognosis than retroperitoneal tumor; tumor of <10–15 cm in size fairs better than larger one; and patients with a low-grade, well differentiated subtype have a more encouraging outcome than those with a high-grade, poorly differentiated subtype.

References

1. Fletcher CDM, Bridge JA, Hogendoorn P, et al. (eds.). *WHO Classification of Tumours of Soft Tissue and Bone*. 4th ed. Lyon, France: IARC Press, 2013.
2. Baheti AD, Jagannathan JP, O'Neill A, Tirumani H, Tirumani SH. Current concepts in non-gastrointestinal stromal tumor soft tissue sarcomas: A primer for radiologists. *Korean J Radiol* 2017;18(1):94.
3. De Vita A, Mercatali L, Recine F, et al. Current classification, treatment options, and new perspectives in the management of adipocytic sarcomas. *Onco Targets Ther* 2016;9:6233.
4. Abbas Manji G, Singer S, Koff A, Schwartz GK. Application of molecular biology to individualize therapy for patients with liposarcoma. *Am Soc Clin Oncol Educ Book* 2015:213.
5. Surgpathcriteria. Dedifferentiated liposarcoma. http://surgpathcriteria. stanford.edu/softfat/dedifferentiated_liposarcoma/differentialdiagnosis. html, Accessed March 10, 2017.
6. Surgpathcriteria. Myxoid liposarcoma. http://surgpathcriteria. stanford.edu/softfat/myxoid_liposarcoma/differentialdiagnosis.html, Accessed March 10, 2017.
7. PDQ Pediatric Treatment Editorial Board. *Childhood Soft Tissue Sarcoma Treatment (PDQ®): Health Professional Version*. PDQ cancer information summaries. Bethesda, MD: National Cancer Institute, 2002.
8. Nassif NA, Tseng W, Borges C, Chen P, Eisenberg B. Recent advances in the management of liposarcoma. *F1000Res* 2016;5:2907.

8
Pleomorphic Sarcoma

8.1 Definition

Soft tissues are affected by a diversity of tumors, including adipocytic tumors, fibroblastic or myofibroblastic tumors, so-called fibrohistiocytic tumors, smooth muscle tumors, pericytic or perivascular tumors, skeletal muscle tumors, vascular tumors, chondro-osseous tumors, gastrointestinal stromal tumors, nerve sheath tumors, tumors of uncertain differentiation, and undifferentiated or unclassified sarcomas [1].

Pleomorphic sarcoma (formerly malignant fibrous histiocytoma [MFH]) was previously considered part of the so-called fibrohistiocytic tumors, but has been recently classified inclusively as undifferentiated pleomorphic sarcoma (UPS) or pleomorphic undifferentiated sarcoma (PUS), among undifferentiated and unclassified sarcomas [1].

Of the former MFH subtypes (i.e., storiform-pleomorphic [50–60%], myxoid [25%], giant cell [5–10%], and inflammatory [5–10%]), storiform-pleomorphic MFH is now recognized as "undifferentiated high-grade pleomorphic sarcoma," giant cell MFH as "undifferentiated pleomorphic sarcoma with giant cells," inflammatory MFH as "undifferentiated pleomorphic sarcoma with prominent inflammation," and myxoid MFH as myxofibrosarcoma (of the myofibroblastic tumor category) [1,2].

Pleomorphic sarcoma (UPS) is an aggressive tumor with frequent local recurrence and occasional metastasis to distant sites. Due to its absence of the lineage with specific differentiation and its morphological similarity to other tumors (e.g., leiomyosarcoma, liposarcoma, and rhabdomyosarcoma), diagnosis of pleomorphic sarcoma is largely by exclusion, that is, searching for other components to rule out a dedifferentiated tumor or evidence of specific differentiation other than fibroblasts or myofibroblasts [2,3].

8.2 Biology

The main components of the musculoskeletal system are hard and soft tissues. While the former consist of bones and cartilages (articular cartilages),

the latter encompasses fat, muscle (smooth, skeletal, and cardiac), fibrous tissue (tendons and ligaments), synovial tissue (joint capsules and ligaments), blood vessels, lymph vessels, and peripheral nerves. The main functions of soft tissues are to connect, support, or protect other structures and organs of the body.

Pleomorphic sarcoma (UPS) is thought to arise from a primitive mesenchymal cell capable of differentiating into histiocytes, fibroblasts, myofibroblasts, and osteoclasts. The uncertain evolutionary origin of UPS is reflected by the story of MFH, which was coined in 1963 to refer to a group of soft tissue tumors characterized by a storiform or cartwheel-like growth pattern. Demonstration of ameboid movement and phagocytosis of explanted tumor cells in early tissue culture studies suggested the histiocytic origin of MFH. However, subsequent immunohistochemical investigations highlighted the close relationship of MFH phenotype with a fibroblast rather than a histiocyte. The ongoing difficulty in precisely subclassifying MFH as lineage-specific sarcomas has further exacerbated the problem. Therefore, the current consensus is to apply the term *MFH* synonymously with *UPS* when immunohistochemistry reveals no definable line of differentiation and electron microscopy demonstrates fibroblastic or myofibroblastic features [2,3].

Soft tissue UPS commonly occurs in the lower extremity (mainly thighs), followed by the upper extremity (arms), retroperitoneum, viscera, head, and neck (in childhood). Osseous UPS often affects the distal femur, proximal tibia, proximal femur, and humerus. Most UPSs recur locally, and metastases to distant sites such as the lungs (90%), bones (8%), and liver are common.

8.3 Epidemiology

Soft tissue sarcomas (STSs) are a heterogeneous group of mesenchymal neoplasms that account for 2% of adult and 15% of pediatric cancers.

Pleomorphic sarcoma (UPS) is the fourth most common STS and makes up about 5% of adult STS. The annual incidence of UPS is 1–2 cases per 100,000, mostly involving adults in the sixth and seventh decades of life. There is a male preponderance (2:1) among UPS patients, and whites appear more susceptible than blacks or Asians. As this tumor rarely occurs in children, adolescents, and young adults, diagnosis of UPS should be made with caution in patients who are <20 years of age.

8.4 Pathogenesis

Risk factors for pleomorphic sarcoma (UPS) include prior radiotherapy, secondary lesion from bone infarct, Paget disease, non-Hodgkin lymphoma, Hodgkin lymphoma, multiple myeloma, and malignant histiocytosis.

Molecularly, UPS is linked to gains at 20q13.33 (75%); 1q21.3-q23.1 (60%); 7q22.1 (60%); 9q34.11 and 20p11.21 (45%); and 1q21.1-q21.2, 8p11.21, 11q13.1, and 16p13.3 (40%); as well as losses at 3p26.3 (60%) [4–6].

8.5 Clinical features

Pleomorphic sarcoma (UPS) usually appears as a painless, slowly enlarging mass (typically 5–10 cm in diameter) in the lower extremity (especially thighs) and upper extremity. Clinical symptoms include swelling, pain, anorexia, fatigue, decreased range of motion (limp), pathological fracture, fevers, chills, night sweats, and weight loss.

8.6 Diagnosis

Pleomorphic sarcoma (UPS) is usually a large and deep-seated tumor of 20 cm (retroperitoneal location) with a fibrous or fleshy cut surface, necrosis, or myxoid change. Histologically, the classic form of UPS shows frequent transitions from storiform to pleomorphic areas. Storiform areas consist of plump spindle cells arranged in short fascicles in a cartwheel, or storiform, pattern (i.e., cells emanating from a central focus) around slit-like vessels. Pleomorphic areas contain plumper fibroblastic cells and more rounded histiocyte-like cells arranged haphazardly with no particular orientation to vessels. High cellularity, marked nuclear pleomorphism, abundant mitotic activity (including atypical mitoses), chronic inflammatory cells, and necrosis (in high-grade lesions) are observed. Immunohistochemically, pleomorphic sarcoma is positive for vimentin, α-1-antitrypsin, α-1-antichymotrypsin, factor XIIIa, CD68, CD10, CD34, and CD99 (35%), but negative for keratin, melanocytic markers, CD45, S100, and muscle markers [7].

Undifferentiated high-grade pleomorphic sarcoma (formerly storiform-pleomorphic MFH) is usually located in the deep fascia or substance of the skeletal muscle of the extremities and sometimes in the retroperitoneum. Appearing as large, multinodular, gray-white masses, the tumor contains pleomorphic tumor cells; bizarre multinucleated cells; a storiform pattern;

inflammatory cells composed of lymphocytes, plasma cells, eosinophils, and xanthoma cells; and foci of metaplastic bone and cartilage [8].

UPS with giant cells (formerly giant cell MFH; also known as malignant giant cell tumor of soft parts) is a superficial, multinodular tumor (subcutis or fascia), usually located in the skeletal muscle of the extremities. Histologically, the tumor displays focal osteoid or bone formation present at the periphery, stromal hemorrhage, cellular pleomorphism, and mitotic figures [8].

UPS with prominent inflammation (formerly inflammatory MFH) is usually located in the retroperitoneum. Histologically, the tumor shows a dense, diffuse neutrophilic infiltrate unassociated with tissue necrosis, admixed with xanthoma cells and tumor cells; phagocytosis of neutrophils by the tumor cells; occasional bizarre atypical nuclei; focal storiform fibrous areas; and multinucleate or Reed–Sternberg-like cells. Immunohistochemically, the tumor is positive for vimentin, α-1-antitrypsin and α-1antichymotrypsin, and CD74, but negative for S-100, HMB-45, CD34, desmin, actin, cytokeratin, and melanocytic markers [8].

Differential diagnoses for pleomorphic sarcoma include dedifferentiated liposarcoma (lipoblasts, often S100+ or smooth muscle actin+), pleomorphic leiomyosarcoma (smooth muscle differentiation), rhabdomyosarcoma (skeletal muscle differentiation), anaplastic large cell lymphoma (CD30+), atypical fibroxanthoma (cutaneous, small and superficial), synovial sarcoma (specific t(X;18) (p11.2;q11.2)), aggressive fibromatosis, benign fibrous tumor, metastatic renal cell carcinoma (keratin+), dermatofibrosarcoma protuberans (characteristic t(17;22)(q22;q13)), and histiocytoid leprosy (prominent histiocytes but no prominent atypia or atypical mitotic figures, presence of mycobacteria).

Like other soft tissue sarcomas, pleomorphic sarcoma can be staged using the AJCC (American Joint Commission on Cancer) system, with stage I tumor being any size, any depth, no metastasis; stage II tumor being <5 cm/ any depth or >5cm/superficial, no metastasis; stage III tumor being >5cm, deep location, no metastasis; and stage IV tumor being any size, any depth, metastasis.

8.7 Treatment

For operable pleomorphic sarcoma (UPS), treatment involves neoadjunctive chemotherapy, wide excision (or amputation), postoperative chemotherapy, and radiotherapy (for high-grade, large [>5 cm], deep-seated tumors; in limb-sparing surgeries; incomplete or questionable margins; typical doses of 45–65 Gy) [9,10].

For nonoperable pleomorphic sarcoma (UPS), chemotherapy (e.g., doxo-rubicin, ifosfamide, trabectedin, dacarbazine, and pazopanib) is preferred. Radiotherapy provides another option [9,10].

8.8 Prognosis

Pleomorphic sarcomas (UPS) are high-grade lesions, and have a local recurrence rate of 19%–31%, a metastatic rate of 31%–35%, and a 5-year survival of 65%–70%. Favorable prognostic factors include age of <60 years, small tumor (<5 cm), superficial or extremity location, low grade, absence of metastatic disease, and a myxoid subtype. Unfavorable prognostic factors are older age, large tumor (>5 cm), deeply seated location, and high grade.

References

1. Fletcher CDM, Bridge JA, Hogendoorn P, et al. (eds.). *WHO Classification of Tumours of Soft Tissue and Bone*. 4th ed. Lyon, France: IARC Press, 2013.
2. Matushansky I, Charytonowicz E, Mills J, Siddiqi S, Hricik T, Cordon-Cardo C. MFH classification: Differentiating undifferentiated pleomorphic sarcoma in the 21st century. *Expert Rev Anticancer Ther* 2009;9(8):1135–44.
3. Goldblum JR. An approach to pleomorphic sarcomas: Can we sub-classify, and does it matter? *Mod Pathol* 2014;27(Suppl 1):S39–46.
4. Jain S, Xu R, Prieto VG, Lee P. Molecular classification of soft tis-sue sarcomas and its clinical applications. *Int J Clin Exp Pathol* 2010;3(4):416–28.
5. Kelleher FC, Viterbo A. Histologic and genetic advances in refining the diagnosis of "undifferentiated pleomorphic sarcoma." *Cancers (Basel)* 2013;5(1):218–33.
6. Silveira SM, Villacis RA, Marchi FA, et al. Genomic signatures predict poor outcome in undifferentiated pleomorphic sarcomas and leio-myosarcomas. *PLoS One* 2013;8(6):e67643.
7. PathologyOutlines.com. Undifferentiated pleomorphic sarcoma. http://www.pathologyoutlines.com/topic/softtissuemfhpleo.html. Accessed March 1, 2017.
8. Pathology of pleomorphic undifferentiated sarcoma (PUS) or undif-ferentiated pleomorphic sarcoma. http://www.histopathology-india. net/malignantfibroushistiocytoma.htm. Accessed March 1, 2017.

9. PDQ Adult Treatment Editorial Board. *Adult Soft Tissue Sarcoma Treatment (PDQ®): Health Professional Version.* PDQ cancer information summaries. Bethesda, MD: National Cancer Institute, 2002.

10. PDQ Pediatric Treatment Editorial Board. *Osteosarcoma and Malignant Fibrous Histiocytoma of Bone Treatment (PDQ®): Health Professional Version.* PDQ cancer information summaries. Bethesda, MD: National Cancer Institute, 2002.

9
Rhabdomyosarcoma

9.1 Definition

Tumors of soft tissues include a number of categories, such as adipocytic tumors, fibroblastic or myofibroblastic tumors, so-called fibrohistiocytic tumors, smooth muscle tumors, pericytic or perivascular tumors, skeletal muscle tumors, vascular tumors, chondro-osseous tumors, gastrointestinal stromal tumors, nerve sheath tumors, tumors of uncertain differentiation, and undifferentiated or unclassified sarcomas [1].

Skeletal muscle tumors are subdivided into *benign* (rhabdomyoma [adult type, fetal type, and genital type]) and *malignant* (embryonal rhabdomyosarcoma [RMS], alveolar RMS, spindle cell or sclerosing RMS, and anaplastic or pleomorphic RMS) [1].

Embryonal RMS represents 84% of RMS cases and occurs mainly in younger patients. Alveolar RMS accounts for 12% of cases, and is characterized by a monomorphic round cell cytology and a variable alveolar pattern, and mainly affects older children. Spindle cell or sclerosing RMS constitutes 2.5% of cases and shows predilection for paratesticular and head and neck sites in children. Anaplastic (pleomorphic) RMS is rare and shows anaplastic cells in aggregates or diffuse sheets throughout the tumor. [2,3].

9.2 Biology

Hard and soft tissues are the key components of the musculoskeletal system. While the hard tissues consist of bones and cartilages (articular cartilages), the soft tissues comprise fat, muscle (smooth, skeletal, and cardiac), fibrous tissue (tendons and ligaments), synovial tissue (joint capsules and ligaments), blood vessels, lymph vessels, and peripheral nerves. The main functions of soft tissues are to connect, support, or protect other structures and organs of the body.

Of the three types of muscles (i.e., visceral, cardiac, and skeletal) within the body, visceral muscle is an involuntarily controlled, smooth muscle located inside various organs (e.g., the stomach, intestines, and blood vessels); cardiac muscle is an autorhythmic or intrinsically controlled striated muscle

found in the heart; and skeletal muscle is a voluntarily controlled, striated muscle mostly attached to two bones through tendons.

Skeletal muscle cells (fibers) possess cell membrane (the sarcolemma) and cytoplasm (the sarcoplasm). Within the sarcoplasm are transverse tubules (T tubules, which connect and carry electrochemical signals from the cell membrane into the middle of the muscle fiber) and myofibrils (which are the contractile structures made up of proteins fibers [thick myosin filaments and thin actin filaments] arranged into repeating subunits called sarcomeres). Skeletal muscles move by shortening their length, pulling on tendons, and moving bones closer to each other.

RMS is soft tissue sarcoma (STS) that arises from primitive mesenchymal cells/myogenic precursor cells and demonstrates evidence of skeletal muscle differentiation. Given the wide distribution of skeletal muscle cells, RMS can develop in almost any part of the body, including the genitourinary system, head and neck, extremities, retroperitoneum, and other sites (e.g., thoracic cavity, axillary region, sacral region, and biliary tract). Specifically, embryonal RMS typically involves the head and neck (46.7%) and genitourinary system (16.1%); alveolar RMS and anaplastic (pleomorphic) RMS are commonly found in the extremities (>65%) [3].

9.3 Epidemiology

STSs are responsible for about 2% of adult and 15% of pediatric cancers. As the most common STS in children and adolescents and the fourth most common pediatric solid tumor, RMS accounts for 5%–8% of all pediatric malignancies (83.2% of which involve children of <10 years) and 50% of pediatric STS. Interestingly, children <5 years of age tend to have RMS in the genitourinary system, bladder, and prostate; those aged 5–9 years have the tumor in the head and neck; and older children and adolescents have the tumor in the extremities, trunk, and paratesticular region. Further, >65% of embryonal RMS cases are detected in the first decade of life and >90% in the first two decades. Alveolar RMS often affects people in the first three decades of life. Anaplastic (pleomorphic) RMS usually occurs in middle-aged men involving the extremities (especially the thigh) and occasionally in children <6 years, involving the lower extremity, retroperitoneum, head and neck. On the whole, a male predominance (1.5–2:1) is noted among RMS patients.

9.4 Pathogenesis

Besides certain chemotherapeutic agents, ionizing radiation, and parental recreational drug use, RMS is associated with several familial syndromes

(e.g., Li–Fraumeni syndrome [involving germline mutations of the p53 tumor suppressor gene], Beckwith–Wiedemann syndrome [involving abnormalities on 11p15], Costello syndrome, Noonan syndrome, and neurofibromatosis type 1) and aberrant DNA methylation [4–6].

Embryonal RMS commonly contains loss of heterozygosity at 11p15.5 and gains of chromosomes 2, 8, and 12 in varying combinations.

Alveolar RMS carries specific chromosomal translocations t(2;13)(q35;q14) or t(1;13)(p36;q14) that lead to PAX3-FOXO1 or PAX7-FOXO1 fusion genes in 55% or 1–20% of cases, respectively. In addition, some cases of alveolar RMS harbor t(2;2)(q35;p23), which has similar transactivation properties as PAX3-FOTO1. The remaining alveolar RMS (20%) is PAX gene fusion-negative (PFN), and demonstrates a similar clinical course to embryonal RMS [6].

Spindle cell or sclerosing RMS within the pediatric age group may be distinguished into three molecular subsets. The first subset occurs at birth or within 1 year of age with predilection in the trunk and is associated with NCOA2 gene rearrangements or gene fusions (e.g., *VGLL2*, *TEAD1*, and *SRF*) and a favorable clinical outcome, lacking metastatic potential. The second subset contains MYOD1 mutation, with or without accompanying *PIK3CA* mutations, and follows a highly aggressive course with high mortality. The third subset lacks gene fusions or MYOD1 mutations and often occurs intra-abdominally or in the genitourinary area, with a favorable clinical outcome [4,5].

Anaplastic (pleomorphic) RMS shows numerical and unbalanced structural abnormalities, including gains (chromosomes 1, 5, 8, 14, 18, 20, and 22) and losses (chromosomes 2, 5, 6, 10, 11, 13–19, and Y) [5]. Further, anaplastic (pleomorphic) shows amplification of JUN (1p31), MYC (8q24), CCND1 (11q13), INT2 (11q13.3), MDM2 (12q14.3–q15), and MALT (18q21) [6].

9.5 Clinical features

Depending on the location, clinical symptoms of RMS include persistent lump or swelling; fever; headache; nausea; vomiting; abdominal pain; constipation; trouble urinating; blood in the urine; bleeding from the nose, throat, vagina, or rectum; bulging or swollen eye (proptosis) (orbital RMS, about 10% of cases); painless scrotal mass (paratesticular tumor); protruding grapelike mass in the vagina (botryoidal RMS); weakness; and weight loss.

9.6 Diagnosis

RMS is a relatively large tumor (>5 cm in 47% cases), with lymph node involvement (in 19% cases) and metastasis (in 34% cases).

Diagnosis and identification of RMS subtypes (e.g., alveolar, anaplastic, embryonal, sclerosing, and mixed) rely on hematoxylin and eosin (H&E), immunohistochemistry, and/or electron microscopy.

Embryonal RMS comprises sheets of rhabdomyoblasts (malignant cell of RMS) with irregularly distributed dense chromatin. The botryoid variant embryonal RMS is less common (around 6% of all cases), has a classic gross morphology (resembling a cluster of grapes), and shows a cambium (or layer) of rhabdomyoblasts beneath an intact epithelial layer.

Alveolar RMS is responsible for the majority of disease-related deaths despite its relatively low incidence. Histologically, alveolar RMS shows segmentation of rhabdomyoblasts (with uniform, popcorn configuration nuclei; coarse chromatin; and less myogenic differentiation than embryonal RMS) in clusters and cleft-like spaces lined with tumor cells (similar to lung alveoli with loss of the centrally located cells).

Spindle cell RMS contains spindle-shaped rhabdomyoblasts with varying degrees of myogenic differentiation. Sclerosing RMS is characterized by cords or microalveolar patterns, embedded in a prominent sclerotic or hyalinized stroma (stromal sclerosis).

Anaplastic (pleomorphic) RMS is more common in the adult population than in children, and contains rhabdomyoblasts with large, lobulated, and hyperchromatic nuclei; atypical mitoses; and focal or diffuse anaplasia (in 13% of embryonal RMS and alveolar RMS).

Mixed RMS may have separate, discrete alveolar RMS and embryonal RMS foci. While tumors with >50% alveolar components are considered alveolar RMS, fusion-positive tumors are exceptions.

Immunohistochemically, RMS is positive for myogenin (particularly alveolar RMS), sarcomeric actin (90%), desmin (95%; reliable for the solid variant of alveolar RMS, tumors with round or strap cell rhabdomyoblasts, and smooth muscle tumors), myoglobin, MyoD1 (sclerosing RMS shows strong and diffuse expression of MyoD1, but weak, patchy myogenin expression), and vimentin (not specific), but negative for FLI1 [6].

9.7 Treatment

Treatment options for RMS consist of surgery, radiotherapy, and chemotherapy [2,8–10].

Surgical removal of the entire tumor (with a surrounding margin of normal tissue and sampling of possibly involved lymph nodes in the draining

nodal basin, and without functional and cosmetic impairment) represents the optimal local control management for RMS.

A second operative procedure (primary re-excision) may be considered for patients with microscopic residual tumor after their initial excisional procedure, and adjunct chemotherapy (e.g., cyclophosphamide/vincristine/dactinomycin) may be given to achieve a complete remission.

Radiotherapy offers another effective local control method for patients with microscopic or gross residual disease after biopsy, initial surgical resection, or chemotherapy, with proton-beam treatment plans helping spare more normal tissue adjacent to the targeted volume than IMRT plans.

Treatment for recurrent rhabdomyosarcoma may also involve surgery, radiotherapy (for patients who have not already received radiotherapy in the area of recurrence or for whom surgical excision is not possible) and chemotherapy (e.g., carboplatin/etoposide, ifosfamide/carboplatin/etoposide, cyclophosphamide/topotecan, irinotecan with or without vincristine) [9].

9.8 Prognosis

Patients with localized RMS have a 5-year survival rate of >70%, following a multimodal approach that includes chemotherapy, radiotherapy, and surgery; those with metastasis have a 5-year overall survival rate of <20%–30% [3].

Spindle cell RMS has a superior prognosis, embryonal RMS has an intermediate prognosis (metastatic rate of 30.9% and 5-year survival rate of 61%–73%), and alveolar RMS has a poor prognosis (metastatic rate of 54.5% and 5-year survival rate of 40%–48%).

RMSs in the orbit, nonparameningeal head and neck sites (e.g., cheek or ear lobe), and male (paratesticular) or female (vagina, vulva, cervix, or uterus) genital tracts are considered favorable in comparison with RMSs in other sites.

References

1. Fletcher CDM, Bridge JA, Hogendoorn P, et al. (eds.). *WHO Classification of Tumours of Soft Tissue and Bone*. 4th ed. Lyon, France: IARC Press, 2013.
2. Hiniker SM, Donaldson SS. Recent advances in understanding and managing rhabdomyosarcoma. *F1000Prime Rep* 2015;7:59.

3. Ma X, Huang D, Zhao W, et al. Clinical characteristics and prognosis of childhood rhabdomyosarcoma: A ten-year retrospective multicenter study. *Int J Clin Exp Med* 2015;8(10):17196.

4. Rudzinski ER, Anderson JR, Hawkins DS, Skapek SX, Parham DM, Teot LA. The World Health Organization classification of skeletal muscle tumors in pediatric rhabdomyosarcoma: A report from the children's oncology group. *Arch Pathol Lab Med* 2015;139(10):1281.

5. Seki M, Nishimura R, Yoshida K, et al. Integrated genetic and epigenetic analysis defines novel molecular subgroups in rhabdomyosarcoma. *Nat Commun* 2015;6:7557.

6. Sun X, Guo W, Shen JK, Mankin HJ, Hornicek FJ, Duan Z. Rhabdomyosarcoma: Advances in molecular and cellular biology. *Sarcoma.* 2015;2015:232010.

7. PathologyOutlines.com. Rhabdomyosarcoma—General. http://www.patholog youtlines.com/topic/softtissuerhabdomyosarcoma.html. Accessed February 20, 2017.

8. PDQ Adult Treatment Editorial Board. *Adult Soft Tissue Sarcoma Treatment (PDQ®): Health Professional Version.* PDQ cancer information summaries. Bethesda, MD: National Cancer Institute, 2002.

9. PDQ Pediatric Treatment Editorial Board. *Childhood Rhabdomyosarcoma Treatment (PDQ®): Health Professional Version.* PDQ Cancer Information Summaries [Internet]. Bethesda (MD): National Cancer Institute (US); 2002-. 2017 Apr 6.

10. Hettmer S, Li Z, Billin AN, Barr FG, et al. Rhabdomyosarcoma: current challenges and their implications for developing therapies. *Cold Spring Harb Perspect Med.* 2014;4(11):a025650.

10
Synovial Sarcoma

10.1 Definition

Tumors of soft tissues encompass adipocytic tumors, fibroblastic or myofibroblastic tumors, so-called fibrohistiocytic tumors, smooth muscle tumors, pericytic or perivascular tumors, skeletal muscle tumors, vascular tumors, chondro-osseous tumors, gastrointestinal stromal tumors, nerve sheath tumors, tumors of uncertain differentiation, and undifferentiated or unclassified sarcomas [1].

Tumors of uncertain differentiation may be differentiated into (1) *benign* (intramuscular myxoma [including cellular variant], juxta-articular myxoma, deep ["aggressive"] angiomyxoma, pleomorphic hyalinizing angiectatic tumor, and ectopic hamartomatous thymoma), (2) *intermediate* (rarely metastasizing) (angiomatoid fibrous histiocytoma, ossifying fibromyxoid tumor [including atypical and malignant], and mixed tumor [myoepithelioma and parachordoma]), and (3) *malignant* (synovial sarcoma, epithelioid sarcoma, alveolar soft part sarcoma, clear cell sarcoma of soft tissue, extraskeletal myxoid chondrosarcoma ["chordoid" type], primitive neuroectodermal tumor [PNET]/extraskeletal Ewing tumor, desmoplastic small round cell tumor, extra-renal rhabdoid tumor, malignant mesenchymoma, neoplasms with perivascular epithelioid cell differentiation [PEComa], clear cell myomelanocytic tumor, and intimal sarcoma) [1].

Synovial sarcoma (formerly tendosynovial sarcoma, synovial cell sarcoma, malignant synovioma, and synovioblastic sarcoma) is an intermediate- to high-grade malignant soft tissue tumor of the extremities (often in close proximity to joint capsules and tendon sheaths) in young patients. Histologically, synovial sarcoma is a mesenchymal spindle cell tumor with variable epithelial differentiation (including glandular formation) and may be subdivided into four subtypes: biphasic (BPSS), monophasic fibrous or monophasic epithelial (MPSS), and poorly differentiated (PDSS, round cell).

10.2 Biology

The musculoskeletal system is made up of hard and soft tissues. The former includes bones and cartilages (articular cartilages), and the latter consists of fat, muscle (smooth, skeletal, and cardiac), fibrous tissue (tendons and ligaments), synovial tissue (joint capsules and ligaments), blood vessels, lymph vessels, and peripheral nerves.

The joint is the point where two or more bones meet or articulate. Typically, a joint consists of cartilage (a thin layer of hyaline tissue covering the surface of a bone at a joint), synovial membrane (also known as synovium or stratum synoviale; a thin, loose vascular connective tissue lining the interior surface of the joint cavity), ligaments (tough, elastic bands of connective tissue surrounding the joint), tendons (tough connective tissue on each side of a joint connecting muscles to bones), bursas (fluid-filled sacs between bones, ligaments, or other adjacent structures), and synovial fluid (a clear, sticky fluid secreted by the synovial membrane).

Synovial sarcoma (a name reflecting its microscopic resemblance to normal synovium) likely arises from undifferentiated mesenchymal stem cells instead of synovial cells, as it stains for epithelial markers (e.g., epithelial membrane antigen and cytokeratin), which synovium does not. With a tendency to occur in the soft tissue surrounding larger joints, synovial sarcoma is typically found near the joints in the lower limb (60%–70%), upper limb (15%–25%), and head and neck (5%). Metastasis to the lungs (70%), lymph nodes (15%), bone (10%), and liver (4.5%) occurs in about 50% of cases.

10.3 Epidemiology

Soft tissue sarcomas (STSs) contain a heterogeneous group of mesenchymal neoplasms, which account for 2% of adult and 15% of pediatric cancers.

Synovial sarcoma constitutes 8%–10% of all STS, and is the fourth most common STS after liposarcoma, malignant fibrous histiocytoma (MFH), and rhabdomyosarcoma, and the third most common STS in adolescents and young adults. Synovial sarcoma has an estimated incidence of 2.75 per 100,000, median age of 35 years (range 15–40 years), and a mild male predilection (male-to-female ratio of 1.2:1).

10.4 Pathogenesis

Risk factors for synovial sarcoma include Li–Fraumeni syndrome, neurofibromatosis type 1, past radiotherapy, and exposure to certain chemical carcinogens.

Molecularly, synovial sarcoma is characterized by t(X;18)(p11.2;q11.2), which is a fusion between the SS18(SYT) gene on chromosome 18 and one of the SSX genes on the X chromosome, leading to SS18-SSX1, SS18-SSX2, or SS18-SSX4 chimeric genes, and inappropriate transcription of SSX sequences. The resulting SYT-SSX can interfere with assembly of BAF (BRG- or BRM-associated factor) complexes and thus affect the integration of a tumor suppressor component and consequent SRY (sex-determining region Y)-box 2 (SOX2) activation, which stimulates cell proliferation. Interestingly, the monophasic subtype comprises vimentin-expressing spindle cells, and usually carries the SS18-SSX2 translocation, whereas the biphasic subtype comprises a mixture of vimentin-expressing spindle cells and keratin-expressing glandular epithelial cells, and harbors the SS18-SSX1 or SS18-SSX2 translocation [2–5].

10.5 Clinical features

In the early stages, synovial sarcoma may be asymptomatic. In the late stages, the growing tumor may produce a lump or swelling, along with a limited range of motion, numbness, pain, and fatigue.

10.6 Diagnosis

Synovial sarcoma is usually a deep-seated, circumscribed or infiltrative, firm, gray-pink mass of <1 cm around large joints with hemorrhage, necrosis, and dystrophic calcification (in 30% of cases).

Radiographically, synovial sarcoma appears as a soft tissue mass of heterogeneous density and enhancement on CT, iso- (slightly hyper-) intense to muscle and heterogeneous on T1, mostly hyperintense and markedly heterogeneous (areas of very high signal due to necrosis and cystic degeneration, relatively high-signal soft tissue components, and areas of low signal intensity due to dystrophic calcifications and fibrotic bands) on T2, and diffuse (40%), heterogeneous (40%), or peripheral (20%) enhancement on T1 C + (Gd).

Histological, synovial sarcoma is divided into four subtypes: biphasic (20%–30%), monophasic fibrous (50%–60%), monophasic epithelial (very rare), and poorly differentiated (15%–25%). The biphasic subtype contains both fibrous cells (also known as spindle or sarcomatous cells, which are relatively small and uniform), forming sheets or fascicles, and large pale or columnar epithelial cells, forming glands, tubules, or papillae (rare). The monophasic fibrous or poorly differentiated subtype consists only of

sheets of spindle cells with hyalinization and distinct lobulation accompanied by mast cells, occasional osseous or cartilaginous metaplasia, focal whirling, and a hemangiopericytomatous vascular pattern. The monophasic epithelial subtype contains only epithelial cells.

Immunohistochemically, synovial sarcoma is positive for cytokeratin 7, 8/18, 19 (both components), AE1/AE3 (70% of monophasic fibrous, 46% of poorly differentiated), EMA (epithelial areas, 100% of monophasic fibrous, 92% of poorly differentiated), CD99/O13 (Ewing or PNET marker, 90%–100% of monophasic fibrous or poorly differentiated), vimentin (spindle cells), CEA, BCL2 (both components, 90% of monophasic fibrous or poorly differentiated), TLE1 (97%), CD57 (neural marker in 72%), E-cadherin (50%), S100 (30%–40%), c-kit (children), nuclear β-catenin, mucin in spindle cell areas, positive periodic acid–Schiff (PAS) in the epithelium, and reticulin (highlighting biphasic pattern), but negative for CD34 (94% monophasic fibrous, 100% poorly differentiated), desmin (98% monophasic fibrous, 100% poorly differentiated), myogenin, h-caldesmon, CD141, WT1, and FLI-1. Electron microscopy shows glandular formation of epithelioid tumor cells with sparse luminal microvilli [6,7].

Molecularly, identification of SYT-SSX1 and SYT-SSX2 genes related to t(X;18)(p11.2; q11) by polymerase chain reaction (PCR) is definitional (in 95% of cases), and is required for the diagnosis of a pure poorly differentiated subtype and useful for some monophasic subtypes. A p16INK4A gene deletion is present in 74% cases, and the poorly differentiated subtype is distinguished from monophasic and biphasic subtypes by its high expression of EZH2 [6].

Differential diagnoses for biphasic synovial sarcoma include malignant peripheral nerve sheath tumor (MPNST) (usually negative for CK7 and CK19, and negative for SYT-SSX fusion products), carcinoma, (rare spindle cell component, no SYT-SSX gene fusion) and diffuse type tenosynovial giant cell tumor (keratin negative, no SYT-SSX gene fusion); those for monophasic spindled synovial sarcoma include fibrosarcoma (keratin and EMA negative, no SYT-SSX gene fusion), leiomyosarcoma (actin positive, no SYT-SSX gene fusion), epithelioid sarcoma (CD34 50% positive , CA125 90% positive, no SYT-SSX gene fusion), clear cell sarcoma (melanin and HMB45 80% positive, frequent t(11;12)), and palmar or plantar fibromatosis (no SYT-SSX fusion); and those for poorly differentiated synovial sarcoma include MPNST, fibrosarcoma, hemangiopericytoma, and primitive peripheral neuroectodermal tumor [7].

10.7 Treatment

Surgery represents the mainstay of treatment for synovial sarcoma, with chemotherapy and radiotherapy used as adjuvant treatment. Chemotherapy with doxorubicin or ifosfamide appears to marginally improve the survival of patients with advanced, poorly differentiated disease. Preoperative radiotherapy is valuable where an adequate clear margin cannot be achieved [8].

Therapeutic agents under clinical trials for synovial sarcoma include inhibitors of the Hedgehog and Notch pathways and Wnt signaling pathways, blockers of immune checkpoints, peptide targeting SS18-SSX oncoprotein, and chimeric antigen receptor T cells (for recognizing and killing tumor cells independent of MHC antigen presentation) [9].

10.8 Prognosis

In general, synovial sarcoma has a 5-year survival rate of 50%–60% and 10-year survival rate of 40%–50%. Synovial sarcoma has a local recurrence rate of 30%–50% and distant metastases of 40%–70%, most commonly to the lungs (80%), bones (15%), regional lymph nodes (10%), and chest wall or abdomen (7.5%).

Favorable prognostic factors include small size (<5 cm), negative resection margin, location in the extremity (e.g., hands, feet, and ankles), age of <50 years, female gender, solid homogenous mass, presence of calcification, biphasic histologic pattern, and SYT-SSX gene fusion. Poor prognostic factors include large size (>5 cm), location in the trunk or head and neck, older patients, cystic or hemorrhagic components, marked heterogeneity, poorly differentiated histology, extensive tumor necrosis, high nuclear grade, p53 mutations, and high mitotic rate (>10 mitoses per 10 high-power field).

References

1. Fletcher CDM, Bridge JA, Hogendoorn P, et al. (eds.). *WHO Classification of Tumours of Soft Tissue and Bone.* 4th ed. Lyon, France: IARC Press, 2013.
2. Lee YF, John M, Edwards S, et al. Molecular classification of synovial sarcomas, leiomyosarcomas and malignant fibrous histiocytomas by gene expression profiling. *Br J Cancer* 2003;88(4):510.
3. Jain S, Xu R, Prieto VG, Lee P. Molecular classification of soft tissue sarcomas and its clinical applications. *Int J Clin Exp Pathol* 2010;3(4):416.

4. Saito T. The SYT-SSX fusion protein and histological epithelial differentiation in synovial sarcoma: relationship with extracellular matrix remodeling. *Int J Clin Exp Pathol* 2013;6(11):2272–9.

5. de Necochea-Campion R, Zuckerman LM, Mirshahidi HR, Khosrowpour S, Chen CS, Mirshahidi S. Metastatic biomarkers in synovial sarcoma. *Biomark Res* 2017;5:4.

6. PathologyOutlines.com. Synovial sarcoma. http://www.pathology-outlines.com/topic/softtissuesynovialsarc.html. Accessed February 25, 2017.

7. Stanford School of Medicine. *Surgical Pathology Criteria.* Synovial Sarcoma. Stanford, CA: http://surgpathcriteria.stanford.edu/softmisc/synovial_sarcoma/, Accessed February 25, 2017.

8. PDQ Pediatric Treatment Editorial Board. *Childhood Soft Tissue Sarcoma Treatment (PDQ®): Health Professional Version.* PDQ cancer information summaries. Bethesda, MD: National Cancer Institute, 2002.

9. Nielsen TO, Poulin NM, Ladanyi M. Synovial sarcoma: recent discoveries as a roadmap to new avenues for therapy. *Cancer Discov* 2015;5(2):124–34.

11
Chondrosarcoma

11.1 Definition

Tumors of the bone include cartilage tumors, osteogenic tumors, fibrogenic tumors, fibrohistiocytic tumors, Ewing sarcoma or primitive neuroectodermal tumors, hematopoietic tumors, giant cell tumors, notochordal tumors, vascular tumors, smooth muscle tumors, lipogenic tumors, neural tumors, miscellaneous tumors, miscellaneous lesions, and joint lesions.

Out of these, cartilage tumors are further divided into five types: (1) osteochondroma, (2) chondroma (enchondroma, periosteal chondroma, and multiple chondromatosis), (3) chondroblastoma, (4) chondromyxoid fibroma, and (5) chondrosarcoma (central, primary, and secondary; peripheral; dedifferentiated; mesenchymal; and clear cell) [1].

Chondrosarcoma is a malignant tumor of cartilage characterized by the production of cartilage matrix by tumor cells accompanied by diverse histopathology and clinical behavior. As the second most frequent primary malignant bone tumor after osteosarcoma (of osteogenic tumors; see Chapter 14), chondrosarcoma accounts for nearly 20% of primary malignant bone neoplasms. While 90% of clinical cases are due to central and peripheral chondrosarcomas, the rest are caused by dedifferentiated, clear cell, and mesenchymal chondrosarcomas.

11.2 Biology

The musculoskeletal system is made up of hard and soft tissues, which are types of connective tissues in the body. The hard tissue consists of bone and cartilage, whereas the soft tissue comprises fat, muscle (smooth, skeletal, and cardiac), fibrous tissue (tendons and ligaments), synovial tissue (joint capsules and ligaments), blood vessels, lymph vessels, and peripheral nerves.

Like other connective tissues, cartilage is composed of cells and extracellular matrix. The cells of cartilage include chondroblasts and chondrocytes (chondro = cartilage). Chondroblasts secrete matrix and fibers, become

trapped inside and mature into chondrocytes. Chondrocytes divide and form "nests" of two to four cells in the enclosed compartments (called lacunae or little lakes or small pits) in the matrix. The extracellular matrix of cartilage comprises aggregating GAG chondroitin sulfate (or aggrecan) and collagen or collagen-elastin fibers. Most cartilage is covered by a dense irregular connective tissue (called perichondrium), consisting of collagen-producing fibroblasts (in the outer layer) and chondroblasts in the inner layer. In contrast to highly vascularized and rigid bone, cartilage is avascular (and thus relies on long-range diffusion from nearby capillaries in the perichondrium for nourishment), flexible, semirigid, and resistant to compressive forces. Cartilage is present in the ear, walls of airways (nose, trachea, larynx, and bronchi), and joints (articular cartilage).

Chondrosarcoma is a tumor of the cartilage (articular cartilage) located on the joints (the ends of the bone), with common occurrence in the pelvis, proximal femur (thighs, accounting for almost one-third of all chondrosarcomas), proximal humerus (upper arms), distal femur, shoulder blades (scapula), and ribs.

Chondrosarcoma often arises de novo in the medulla of bone (85%), or as a secondary tumor (15%) from a preexisting condition, including enchondroma, osteochondroma, exostosis (solitary or multiple), and chondrodysplasia. Based on the osseous location, chondrosarcoma is separated into central (intramedullary, arising in the medullary cavity) and peripheral subtypes (which together are referred to as conventional chondrosarcomas by some authors). According to the histologic characteristics, chondrosarcoma is divided into mesenchymal, clear cell, and dedifferentiated variants.

Approximately 90% of all chondrosarcoma (e.g., central and clear cell chondrosarcoma) are of low to intermediate grades, contain normal-looking cartilage cells with small to moderate amounts of abnormality, and occur in the shoulder, pelvis, or proximal femur. A small number of chondrosarcoma (e.g., dedifferentiated and mesenchymal chondrosarcomas) are of high grade, possess many abnormal cartilage cells, and have the ability to grow rapidly and spread.

11.3 Epidemiology

Chondrosarcoma represents about 20% of primary malignant bone neoplasms, and has an estimated annual incidence of 1–10 per 200,000.

Chondrosarcoma commonly affects people of age 45 years (range 40–70) and shows a male predilection (2:1). About 16% of chondrosarcoma cases

occur in patients of <20 years, and <10% involve children with a rapidly fatal outcome [2].

11.4 Pathogenesis

Risk factors for chondrosarcomas include enchondroma (benign cartilage tumor of the hands), Ollier disease (a cluster of enchondromas), Maffucci syndrome (a combination of multiple enchondromas), multiple hereditary exostosis (MHE) (osteochondromatosis or osteochondroma, and an overgrowth of cartilage and bone near the end of the growth plate), Wilms tumor, Paget disease, and angioma (benign tumor of blood vessels), as well as diseases in children that require chemotherapy or radiotherapy.

Molecularly, low-grade chondrosarcomas tend to display reduced levels of genomic imbalances compared with high-grade chondrosarcomas. Central chondrosarcomas contain more frequent abnormalities of chromosome 9p and extra copies of chromosome 22 than peripheral chondrosarcomas. More than 50% of primary central chondrosarcomas and around 87% of benign enchondromas have somatic mutations at the *IDH* genes. High-grade chondrosarcomas are linked to a p53 mutation. Further, signaling pathways (e.g., Hedgehog, Src, PI3k–Akt–mTOR) and high levels of platelet-derived growth factor isoform AA and PDGF-α receptor may also be implicated in the pathogenesis of chondrosarcoma [2,3].

11.5 Clinical features

About 95% of patients with a chondrosarcoma report insidious, slowly worsening pain (most pronounced at night). Some patients have a palpable soft tissue mass (or boney bump), while others (~20%) present as a pathologic fracture. The affected area may be swollen and tender to touch.

11.6 Diagnosis

Chondrosarcoma (central and peripheral) is usually a large, painful, pearly white or light blue tumor (4–10 cm) of the long bone or ribs that may recur at a higher histologic grade.

Radiographically, chondrosarcoma shows fluffy calcification, poorly defined margins, erosion or thickening of the cortex, and no periosteal new bone formation. On CT, the nonmineralized portion is hypodense to muscle;

the mineralized portion shows the characteristic "ring-and-arc" chondroid matrix. On MRI, chondrosarcoma gives a low to intermediate signal on T1; the nonmineralized portion of hyaline cartilage is hyperintense (bright) (owing to the high water content of hyaline cartilage), while the mineralized areas or fibrous septa are hypointense (dark) on T2. Postcontrast sequences reveal enhancement of the septa in a ring-and-arc pattern [2].

Based on nuclear size, staining pattern (hyperchromasia), mitotic activity, and degree of cellularity in chondrocytes, chondrosarcoma may be well, moderately, or poorly differentiated (corresponding to low, intermediate or high grades; or grades 1–3), as grade 4 is a spindled tumor representing either dedifferentiated chondrosarcoma or chondroblastic osteosarcoma (with direct osteoid or bone formation).

Well-differentiated chondrosarcoma has low cellularity, with hyperchromatic plump nuclei of uniform size and cytology very similar to benign cartilaginous tumors such as enchondroma and displays a lobulated architecture with an abundant cartilaginous matrix separated by narrow fibrovascular bands. The tumor may permeate existing trabecular bone and fill marrow space, or stay in the lacunar space surrounding the hyaline cartilaginous matrix. As a low grade lesion, well-differentiated chondrosarcoma rarely metastasizes.

Moderately differentiated chondrosarcoma shows increased cellularity, hypercromasia, distinct nucleoli, and foci of myxoid alteration. As an intermediate grade lesion, moderately differentiated chondrosarcoma metastasizes in 10-15% of cases.

Poorly differentiated chondrosarcoma demonstrates marked hypercellularity, nuclear atypia (nuclei 10× the usual size, hyperchromatic, and binucleated to vesicular), neoplastic chondrocytes arranged in lobular cords and clumps (causing characteristic endosteal scalloping), abundant necrosis, frequent mitotic figures, destruction of the cortex, and formation of a soft tissue mass. As a high grade lesion, poorly differentiated chondrosarcoma metastasizes in up to 70% of cases.

Immunohistochemically, chondrosarcoma is positive for S100 (nuclear and cytoplasmic) and p53 (high-grade tumor), but negative for cadherin (neural type). Molecularly, chondrosarcoma is often 20q+ or 8q+ [4].

Dedifferentiated chondrosarcoma is a grade 4 tumor characterized by a bimorphic histology with a well-differentiated cartilaginous component and a dedifferentiated, noncartilaginous component. The tumor stains positive for SOX9 and has IDH1 and IDH2 overexpression.

Mesenchymal chondrosarcoma is a highly malignant lesion characterized by varying amounts of differentiated cartilage admixed with undifferentiated small round cells. The tumor shows minimal pleomorphism, and no or rare mitotic figures; stains positive for CD57/Leu7, CD99/MIC2 (in small round blue cells), neuron-specific enolase, and vimentin, but negative for actin, cytokeratin, desmin, EMA, S100 (positive only in chondroid areas), and synaptophysin; and contains occasional chromosomal abnormality of der(13;21)(q10;q10) and p53 overexpression [4].

Clear cell chondrosarcoma is a low-grade malignant tumor characterized by tumor cells with clear, empty cytoplasms. The tumor may harbor extra copies of chromosome 20 and loss or rearrangements of 9p, as well as expression of parathyroid hormone-like hormone (PTHLH) and platelet derived growth factor (PDGF) [5].

11.7 Treatment

The preferred treatment for chondrosarcoma is surgical excision, as neither chemotherapy nor radiotherapy is effective (with the exception of mesenchymal chondrosarcoma, which is sensitive to doxorubicin-based combination chemotherapy). Physical therapy may help one to regain strength and use of the affected area after surgery [6,7].

Low-grade tumors confined to the bone may be managed by extensive intralesional curettage to minimize functional disability. High-grade tumors require wide, en bloc local excision with negative margins and reconstruction of the affected bone with graft or metal (or curettage with thermal ablation if excision is impossible) [7].

11.8 Prognosis

In general, low-grade chondrosarcomas have an excellent prognosis, with a small chance of recurrence and no chance of disease spread. High-grade chondrosarcomas such as dedifferentiated tumors have a high risk of recurrence and metastasis and a poor prognosis.

The 5-year survival rate for chondrosarcoma for grades 1, 2, and 3 is 90%, 50%, and 25%, respectively. The overall survival for patients with chondrosarcoma is 90% and 80% at 5 and 10 years, respectively. Within chondrosarcoma subtypes, the 5-year survival rate is 76% for conventional, 71% for myxoid, 87% for juxtacortical, and 52% for mesenchymal [7].

References

1. Fletcher CDM, Bridge JA, Hogendoorn P, et al. (eds.). *WHO Classification of Tumours of Soft Tissue and Bone.* 4th ed. Lyon, France: IARC Press, 2013.
2. Mosier SM, Patel T, Strenge K, Mosier AD. Chondrosarcoma in childhood: The radiologic and clinical conundrum. *J Radiol Case Rep* 2012;6(12):32–42.
3. Evola FR, Costarella L, Pavone V, et al. Biomarkers of Osteosarcoma, Chondrosarcoma, and Ewing Sarcoma. *Front Pharmacol* 2017;8:150.
4. PathologyOutlines.com. *Chondrosarcoma—General.* http://www.pathologyoutlines.com/topic/bonechondrosarcoma.html. Accessed February 28, 2017.
5. Gelderblom H, Hogendoorn PC, Dijkstra SD, et al. The clinical approach towards chondrosarcoma. *Oncologist* 2008;13(3):320–9. Erratum in *Oncologist* 2008;13(5):618.
6. Polychronidou G, Karavasilis V, Pollack SM, Huang PH, Lee A, Jones RL. Novel therapeutic approaches in chondrosarcoma. *Future Oncol* 2017;13(7):637–48.
7. Bindiganavile S, Han I, Yun JY, Kim HS. Long-term outcome of chondrosarcoma: A single institutional experience. *Cancer Res Treat* 2015;47(4):897–903.

12
Ewing Sarcoma

12.1 Definition

The bone is affected by a diverse range of tumors, including cartilage tumors, osteogenic tumors, fibrogenic tumors, fibrohistiocytic tumors, Ewing sarcoma or primitive neuroectodermal tumor (PNET), hematopoietic tumors, giant cell tumors, notochordal tumors, vascular tumors, smooth muscle tumors, lipogenic tumors, neural tumors, miscellaneous tumors, miscellaneous lesions, and joint lesions [1].

Ewing sarcoma is an aggressive neoplasm of bone and soft tissue characterized by sheets of small round cells (containing round to oval nuclei, finely dispersed chromatin without nucleoli, and variable amounts of cytoplasm) that may form Homer–Wright rosettes. Along with soft tissue PNET (showing neuroendocrine differentiation), Askin tumor, and neuroepithelioma, Ewing sarcoma is often grouped into Ewing sarcoma family tumors (ESFTs), which typically share a nonrandom t(11;22)(q24;q12) chromosome rearrangement. In classic Ewing sarcoma, the cytoplasm is clear and contains glycogen, which may be highlighted with a periodic acid–Schiff stain, and the tumor cells are tightly packed and grow in a diffuse pattern without structural organization [1].

Commonly referred to as a small, round, blue cell tumor of childhood, Ewing sarcoma (n = 4,870) presents the third most common primary malignant bone tumor after osteosarcoma (n = 11,961) and chondrosarcoma (n = 9,606), according to the epidemiologic data from the National Cancer Data Base of the American College of Surgeons between 1985 and 2003.

12.2 Biology

The musculoskeletal system is composed of specifically designated connective tissues, namely, hard and soft tissues. The hard tissue includes bone and cartilage, while the soft tissue contains fat, muscle (smooth, skeletal, and cardiac), fibrous tissue (tendons and ligaments), synovial tissue (joint capsules and ligaments), blood vessels, lymph vessels, and peripheral nerves.

The 206 pieces of bones within the body are classified into long, short, and flat types. Mature long bone contains three zones: epiphysis (the region at the polar ends, commonly associated with joint surfaces and composed of compact bone shell with bony struts or trabecular bone for support of the cortical shell), metaphysis (located on both sides of diaphysis as a transitional zone between the epiphysis and diaphysis, and characterized by a thinner cortical wall with dense trabecular bone), and diaphysis (the middle of the long bone characterized by a thick cortical bone with minimal trabecular bone). In addition, the epiphyseal plate (physis) is present between the epiphysis and metaphysis during the developmental stage, and becomes a scar in adulthood. Examples of lone bones include the femur, tibia, fibula, humerus, radius, ulna, metacarpals, metatarsals, and phalanges. Short bone goes through a cartilaginous model similar to that of long bone, but has a unique shape and function. Flat bone stems from mesenchymal tissue sheets, which condense and ossify. It is often broad and flat, consisting of a cortical shell and a cancellous interior. Examples of flat bone include the skull and scapula (for muscular attachment).

Structurally, bone (or osseous tissue) is made up of cells and an extracellular matrix, much like other types of connective tissue. The cells in bone consist of osteoblasts, osteocytes (osteoblasts that have been incorporated into cortical bone), osteoclasts, and bone-lining cells (old osteoblasts that no longer play a role in synthesis). The extracellular matrix in bone comprises an organic matrix (30%, including proteoglycans, glycosaminoglycans [GAGs], glycoproteins, osteonectin [which anchors bone mineral to collagen], osteocalcin [calcium-binding protein], and collagen fibers [type I and some type V]) and hydroxyapatite (70%, bone mineral). Bone is evolved from osteoid (bone-like) tissue, in which the extracellular matrix becomes calcified with the deposition and crystallization of calcium and phosphate ions in the spaces between the collagen fibers. Bone often has a thick, well-organized outer shell (cortex or cortical bone, 80%), a less dense mesh of bony struts in the center (trabecular bone, 20%), and a soft tissue envelope (called periosteum, for perfusion and nutrient supply to the outer third of the bone). Bone is highly vascularized, and its calcified matrix makes it hard and rigid, ideally suited for forming the skeletal structure of the body.

Ewing sarcoma is thought to arise from cells of the neural crest, possibly mesenchymal stem cells in red bone marrow. The tumor affects both long (60%, including the femur [25%], tibia [11%], and humerus [10%]) and flat (40%, including the pelvis [14%], scapula, and ribs [6%]) bones. Within the long bones, metadiaphysis accounts for 44%, middiaphysis for 33%, metaphysis for 15%, and epiphysis for 1%–2%. Among various anatomic

locations, Ewing sarcoma typically occurs in the lower limb 45% (femur most common), pelvis 20%, upper limb 13%, spine and ribs 13% (sacrococcygeal region is the most common), and skull or face 2%. Extraosseous primary tumors may form in the trunk (32%), extremity (26%), head and neck (18%), retroperitoneum (16%), and other sites (9%).

12.3 Epidemiology

Ewing sarcoma represents the third most common bone tumor (after osteosarcoma and chondrosarcoma), the second most common bone tumor in children and adolescents, and the most lethal bone tumor.

Ewing sarcoma has an annual incidence of 1–2 cases per million, 0.3 cases per million in children under 3 years, 4.6 cases per million in adolescents aged 15–19 years, 2.6 cases per million in females, and 3.3 cases per million in males. Overall, 27% of cases occur in the first decade of life, 64% of cases in the second decade, and 9% of cases in the third decade. With a mean age of 15 years at diagnosis and a slight male predilection (male-to-female ratio of 1.6:1), Ewing sarcoma appears to affect whites more frequently than Asians (particularly Japanese) and blacks.

12.4 Pathogenesis

Ewing sarcoma is shown to contain chromosomal translocation of EWS-ETS t(11;22) (q24;q12) (in 85% of cases), EWS-ERG t(21;22)(q22;q12) (in 10% of cases), EWS-ETV t(7;22)(p22;q12) and t(17;22)(q12;q12), and EWS-FEV t(2;22) (q35;q12), as well as TLS-ERG t(16;21)(p11;q22) and TLS-FEV t(2;16)(q35;p11). In addition, patients often harbor the Ewing sarcoma susceptibility gene (EGR2), located within the chromosome 10 susceptibility locus (10q21.3), which is regulated by the EWSR1 (Ewing sarcoma breakpoint region 1 of the TET family on chromosome 22)-FLI1 (friend leukemia insertion of the ETS family on chromosome 11) fusion oncogene via a GGAA microsatellite. Additional numerical and structural aberrations in Ewing sarcoma include gains of chromosomes 2, 5, 8, 9, 12, and 15; the nonreciprocal translocation t(1;16)(q12;q11.2); deletions on the short arm of chromosome 6; and trisomy 20. Further, STAG2 mutations are observed in 15%–20% of the cases, CDKN2A deletions in 12%–22% of cases, TP53 mutations in 6%–7% of cases, and intrachromosomal X-fusion in 4% of cases, leading to altered BCOR (encoding the BCL6 corepressor) and CCNB3 (encoding the testis-specific cyclin B3). In contrast, small, round, blue cell tumors of bone and soft tissue do not have rearrangements of the EWSR1 gene [2].

12.5 Clinical features

Clinical symptoms of Ewing sarcoma may include localized pain, stiffness, swelling, or tenderness; back pain (suggestive of a paraspinal, retroperitoneal, or deep pelvic tumor); palpable mass (a lump near the surface of the skin that may feel warm and soft to the touch); fracture (without an injury); fever (that does not go away); anemia; leukocytosis; and weight loss (suggestive of metastatic disease, in 25% of cases).

12.6 Diagnosis

Ewing sarcoma (usually an undifferentiated bone tumor) is a white, fleshy, ill-defined tumor with extensive involvement of the medulla and cortex, periosteal elevation, and necrosis.

Radiographically, Ewing sarcoma is a destructive, lytic tumor with reactive periosteal bone (lamellated, resembling onion skin) and widening of the medullary canal. It gives a low to intermediate signal on T1, a heterogeneously high signal (with hair-on-end low-signal striations) on T2, and heterogeneous but prominent enhancement on T1 C+ (Gd) [2].

Histologically, Ewing sarcoma displays sheets of small, round, uniform cells (10–15 μm, larger than lymphocytes) with scant clear cytoplasm (divided into irregular lobules by fibrous strands), indistinct cell membranes, round nuclei with indentations, and small nucleoli. The tumor cells may show Homer–Wright rosettes (central fibrillary space) or pseudorosettes (cells arranged around vessels), hemorrhage, necrosis, prominent vasculature, and variable mitotic figures. Sometimes, large pleomorphic cells, an organoid pattern, or a filigree pattern (large areas of perivascular tumor necrosis with "ghost cells") may be observed [3].

Immunohistochemically, Ewing sarcoma stains positive for CD99 (O13 or MIC2), PAS+ diastase sensitive (glycogen), NSE, Leu7/CD57, FLI1 protein, and vimentin; is variable for low-molecular-weight keratin and synaptophysin; but negative for S100, CD45/LCA, muscle markers, and vascular markers. Molecularly, t(11,22)(q24;q12) is present in 85% of Ewing sarcoma (22q12 EWS and 11q24 FL1), and t(21;22)(q22;q12) in 5%–10% (ERG and EWS) [3].

Differential diagnoses for Ewing sarcoma include PNET (neural morphologically or a large soft tissue component with extension into the bone; also PAS+/CD99+ and translocation EWSR1), Askin tumor (chest wall), small cell or telangiectatic osteosarcoma (deposition of bone; alkaline phosphatase

(ALP) not elevated in Ewing sarcoma), and other small, blue, round cell tumors (e.g., neuroblastoma [CD99– and CD56+; MYCN amplification], non-Hodgkin lymphoma [PAS– and CD45+], poorly differentiated synovial sarcoma [CD99+ and BCL-2+; translocation SS18], alveolar rhabdomyosarcoma (MYF4+ and desmin+; translocation FKHR], desmoplastic small round cell tumor [keratin+ and desmin+; translocation EWSR1], and mesenchymal chondrosarcoma [deposition of cartilage]).

12.7 Treatment

The standard care for patients suffering from Ewing sarcoma relies on systemic chemotherapy, together with surgery and/or radiotherapy for local tumor control [4–6].

Chemotherapy consists of alternating courses of (1) vincristine, doxorubicin, and cyclophosphamide, and (2) ifosfamide and etoposide for 6–12 months, in addition to neutrophil support and red blood cell and platelet support . This multiagent chemotherapy leads to long-term event free survival in 70% of patients with no clinically detectable metastatic disease, and in 20-50% of patients with lung or bone metastasis. For patients with metastatic disease, the five drug chemotherapy at a higher dose may be complemented by autologous stem cell rescue [4,7].

New therapeutic agents under investigation for Ewing sarcoma include topotecan, irinotecan, temozolomide, and insulin-like growth factor receptor antibodies (IGFR-1) such as ganitumab [8].

12.8 Prognosis

The overall survival rate for patients with localized disease (which represents 80% of cases) is 60%–70%, and that for patients with metastatic disease (usually to the lung, skull, pleura, or central nervous system [CNS]) is <25%, after undergoing chemotherapy. Compared with osteosarcoma (53.9%) and chondrosarcoma (75.2%), the relative 5-year survival rate for Ewing sarcoma is 50.6%.

Unfavorable prognostic factors include high stage, direct extension into soft tissue, aneuploidy, metastases, grossly viable tumor postchemotherapy, possible filigree pattern (bicellular strands of tissue separated by filmy vascular stroma), increased serum lactate dehydrogenase (LDH) levels, overexpression of p53 protein, Ki-67 expression, loss of 16q, and high expression of microsomal glutathione S-transferase.

References

1. Fletcher CDM, Bridge JA, Hogendoorn P, et al. (eds.). *WHO Classification of Tumours of Soft Tissue and Bone*. 4th ed. Lyon, France: IARC Press, 2013.
2. Jain S, Xu R, Prieto VG, Lee P. Molecular classification of soft tissue sarcomas and its clinical applications. *Int J Clin Exp Pathol* 2010;3(4):416–28.
3. PathologyOutlines.com. *Ewing Sarcoma/Primitive or Peripheral Neuroectodermal Tumor (PNET)*. http://www.pathologyoutlines.com/topic/boneewing.html. Accessed February 25, 2017.
4. PDQ Pediatric Treatment Editorial Board. *Ewing sarcoma treatment (PDQ®): Health Professional Version. PDQ Cancer Information Summaries*. Bethesda, MD: National Cancer Institute, 2002.
5. Biswas B, Bakhshi S. Management of Ewing sarcoma family of tumors: Current scenario and unmet need. *World J Orthop* 2016;7(9):527–38.
6. Yu H, Ge Y, Guo L, Huang L. Potential approaches to the treatment of Ewing's sarcoma. *Oncotarget* 2017;8(3):5523–39.
7. Meyers PA. Systemic therapy for osteosarcoma and Ewing sarcoma. *Am Soc Clin Oncol Educ Book* 2015:e644–7.
8. Pishas KI, Lessnick SL. Recent advances in targeted therapy for Ewing sarcoma. *F1000Res* 2016;5. pii: F1000 Faculty Rev-2077.

13
Fibrosarcoma

13.1 Definition

Tumors of the bone include cartilage tumors, osteogenic tumors, fibrogenic tumors, fibrohistiocytic tumors, Ewing sarcoma or primitive neuroectodermal tumors, hematopoietic tumors, giant cell tumors, notochordal tumors, vascular tumors, smooth muscle tumors, lipogenic tumors, neural tumors, miscellaneous tumors, miscellaneous lesions, and joint lesions [1]. Of these, fibrogenic tumors consist of desmoplastic fibroma and fibrosarcoma.

Fibrosarcoma (or fibroblastic sarcoma) is a malignant tumor that originates in the connective fibrous tissue at the ends of bones, and then spreads to other surrounding soft tissues. Clinically, fibrosarcoma may display two disease forms: infantile or congenital fibrosarcoma and adult fibrosarcoma. Infantile or congenital fibrosarcoma (also known as congenital-infantile fibrosarcoma) is a low-grade tumor that appears as a rapidly growing mass at birth or shortly after (under 1 year of age). Although this form of disease is locally aggressive and may recur locally, it is rarely metastatic. Adult fibrosarcoma occurs in older children and adolescents (between 10 and 15 years), as well as adults, and often presents as a deep-seated tumor with frequent metastasis.

13.2 Biology

The musculoskeletal system contains two special types of connective tissue, namely, hard and soft tissues, whose main functions are to connect, support, or protect other structures and organs of the body. The hard tissue comprises bone and cartilage (articular cartilage), whereas the soft tissue encompasses fat, muscle (smooth, skeletal, and cardiac), fibrous tissue (tendons and ligaments), synovial tissue (joint capsules and ligaments), blood vessels, lymph vessels, and peripheral nerves.

Fibrous tissue (tendons and ligaments) is composed of collagen, glycosaminoglycan, and glycoprotein, which are synthesized by fibroblasts. In tendons, dense, tightly arranged collagen fibers (containing fibrocytes or fibroblasts)

run in parallel and connect the ends of bones to muscles, facilitating the transmission of large forces from muscles to bones.

Fibrosarcoma is a tumor of mesenchymal cell origin containing immature proliferating fibroblasts or undifferentiated anaplastic spindle cells in a storiform/cartwheel pattern. Fibrosarcoma may arise de novo or emerge as a secondary tumor from a preexisting benign lesion. Primary fibrosarcoma is mostly central (intramedullary) and occasionally peripheral (periosteal or parosteal). Although fibrosarcoma commonly occurs at the ends of bones in the legs, arms, or pelvis, it may sometimes form a soft tissue mass in the thigh and posterior knee. Infantile or congenital fibrosarcoma usually produces a subcutaneous lesion in the distal extremities, and less commonly in the trunk, head and neck, mesentery, retroperitoneum, and orbit. By contrast, adult fibrosarcoma is often located in the proximal extremities.

13.3 Epidemiology

Fibrosarcoma represents about 10% of musculoskeletal sarcomas and <5% of all primary bone tumors.

Infantile fibrosarcoma is typically found in children of <10 years. Adult fibrosarcoma of bone has an incidence of 1 case per 2 million, and mainly occurs in the fourth decade of life, with a slight male predominance. Adult fibrosarcoma of soft tissue commonly affects people aged between of 35–55 years.

13.4 Pathogenesis

Risk factors for fibrosarcoma include previous radiotherapy, bone infarct, surgically treated fracture or joint reconstruction, preexisting benign lesions (e.g., enchondroma, bizarre parosteal osteochondromatous proliferation, chronic osteomyelitis, giant cell tumor, or fibrous dysplasia), known malignant lesion (e.g., low-grade chondrosarcoma), Paget disease, and inherited genetic conditions (e.g., Gardner syndrome, Li–Fraumeni syndrome, retinoblastoma, and multiple neurofibromas).

Molecularly, 70% of infantile or congenital fibrosarcoma contain a characteristic t(12;15)(p13;q26), which leads to an ETV6-NTRK3 gene fusion product (ETS variant gene 6 and neurotrophic tyrosine kinase receptor type 3).

13.5 Clinical features

Fibrosarcoma of bone may cause pain or soreness (due to suppressed nerves and muscles), swelling, loss of motion (limp or difficulty using legs,

feet, arms, and hands), and pathologic fracture (following trivial trauma). New pain and/or swelling at a sight of known fibrous dysplasia is indicative of malignant transformation. Adult fibrosarcoma of soft tissue is generally a large, painless mass deep to fascia with an ill-defined margin.

13.6 Diagnosis

Adult fibrosarcoma is a deep-seated, well-circumscribed, nonencapsulated, fleshy, hemorrhagic, necrotic, white-tan mass of >4 cm in size. Histologically, the tumor contains a monomorphic population of spindle cells arranged in fascicles and intersecting at acute angles, giving a herringbone appearance; cells with scant cytoplasm; tapering elongated dark nuclei; increased granular chromatin; variable nucleoli; mitotic activity; and variable collagen. Immunohistochemically, the tumor is positive for reticulin (abundant fibers wrapped around each cell), phosphotungstic acid–hematoxylin (cytoplasmic fibrils), vimentin, type I collagen, p53, and high Ki-67, but negative for S100 (neural marker), keratin, smooth muscle markers (desmin, smooth muscle actin, and HHF-35), CD31 (neutrophils), CD34 (hematopoietic cells), histiocytic markers, and basal lamina. Cytogenetically, fibrosarcoma is aneuploid [2].

Infantile or congenital fibrosarcoma resembles adult fibrosarcoma morphologically, but shows locally aggressive behavior with an increased rate of local recurrence. The tumor varies from a few centimeters to 30 cm in size, with a fleshy, gray-tan cut surface and areas of necrosis and hemorrhage, in addition to myxoid and cystic areas. Microscopically, the tumor shows relatively uniform spindle-shaped cells (with large hyperchromatic nuclei) arranged in fascicles, which are more rounded and immature than those in the adult type, along with prominent hemangiopericytoma-like areas, dystrophic calcification, and extramedullary hematopoiesis. Cytogenetically, the tumor harbors a t(12;15)(13p;q26) translocation, in addition to trisomy 8, 11, 17, and 20.

Compared with fibrosarcoma of bone, *desmoplastic fibroma*—the other fibrogenic bone tumor—is the bony counterpart of soft tissue fibromatosis commonly occurring in the metaphysis of long bones (56%), mandible (26%), and pelvis (14%). Macroscopically, the tumor is a white-gray, fibrous rubbery mass with variable bony spicules and cysts. Radiographically, the tumor is a lytic and honeycombed ("soap bubble") lesion with cortical thinning, soft tissue extension, local destruction, and no metastases. Microscopically, the tumor contains mature, bland fibroblasts separated by abundant collagen with thin-walled, dilated vascular

channels, infiltration into soft tissue, and the absence of necrosis, pleo-morphism, atypia, and mitotic activity. Cytogenetically, the tumor is tri-somy 8 and 20.

Fibrosarcoma of soft tissue is distinct from other fibroblastic or myofi-broblastic tumors, such as *myxofibrosarcoma* (superficial multiple myx-oid nodules or a single deep mass with infiltrative margins; multinodular tumor composed of pleomorphic spindle cells in myxoid background; curvilinear vessels [thick walled with broad arc] with characteristic con-densation of round to spindled cells around vessels, incomplete fibrous septa, myxoid stroma [in 10% of cases], and infiltrating immature den-dritic cells; high-grade tumors showing more cellular with atypical mitotic figures, hemorrhage, necrosis, and possibly bizarre multinucleated giant cells; positivity for muscle markers [HHF35, α-smooth muscle actin, calponin, desmin, and vimentin]), *low-grade fibromyxoid sarcoma* (also called Evans tumor, hyalinizing spindle cell tumor with giant rosettes; a well-circumscribed, slow-growing, painless mass of 6 cm [range 1–18 cm] with a fibromyxoid cut surface and gross infiltration; low cellularity, bland fibroblastic cells, and curvilinear or arcuate vessels; recurrence in 64%, metastasis in 45%, death in 42%; t(7;16)(q33;p11) in 66%, supernumer-ary ring chromosome in 25%), and *sclerosing epithelioid fibrosarcoma* (a grossly well-circumscribed, lobulated, bosselated, or multinodular, painful mass of 9 cm [range 4–22 cm] in deep soft tissue of muscle asso-ciated with fascia or periosteum, firm, gray-white cut surface; nests and cords of small to medium size, round to ovoid, relatively uniform epitheli-oid cells; hemangiopericytoma-like vasculature, vascular invasion, incon-spicuous mitoses 4/10 high-power field [HPF]; histologically low grade but clinically aggressive with local recurrence in 50% and distal metas-tases in 43%; negativity for CD34, CD45 [leukocyte common antigen (LCA)], HMB45, desmin, α-smooth muscle actin, and CD68).

Other differential diagnoses for fibrosarcoma consist of nodular fasciitis (undulating pattern, fine, pale, even chromatin instead of herringbone pat-tern, coarse, granular, irregular chromatin), extra-abdominal desmoid fibro-matosis (mild to moderate cellularity, cells separated by collagen instead of high cellularity, cells touching each other), monophasic synovial sarcoma (keratin+, EMA+, frequent hemangiopericytomatous vessels, SYT-SST gene fusion instead of keratin–, EMA–, no hemangiopericytomatous vessels, no SYT-SSX gene fusion), malignant fibrous histiocytoma (histiocytic differen-tiation instead of no histiocytic differentiation), leiomyosarcoma (desmin+ lack of t(12;15) instead of desmin–, t(12;15) in infantile case), dermatofibro-sarcoma protuberans (CD34+, storiform pattern, t(17;22)(q22;q13) instead of CD34–, fascicular, herringbone pattern, no t(17;22)(q22;q13)), GI stromal

tumor (mostly CD34+, CD117+ instead of CD34–, CD117–), malignant peripheral nerve sheath tumor (50% S100+, lack of t(12;15) instead of S100–, t(12;15) in infantile case), and fibroma of tendon sheath (2 cm, lack of t(12;15) instead of >4 cm, t(12;15) in infantile case).

13.7 Treatment

Treatment options for fibrosarcoma consist of radical excision with an adequate margin, radiation if residual tumor or positive margins, and chemotherapy if high grade [3,4].

In limb-sparing surgery, either a prosthesis (a metal replacement bone) or a bone graft (bone taken from another part of the body) is inserted to replace what has been taken out.

Radiotherapy may be given either post-operatively (to destroy any remaining cancer cells) or pre-operatively (to shrink the tumor and make the surgery easier). Radiotherapy may also form the only treatment for fibrosarcoma, depending on the location and size.

Similarly, chemotherapy (with cytotoxics) may be administered prior to surgery (neoadjuvant treatment) and after surgery (adjuvant chemotherapy).

13.8 Prognosis

Primary fibrosarcoma of the bone has a 5-year survival rate of 65% and 10-year survival rate of <30%. Secondary fibrosarcoma has a 10-year survival rate of <10%.

Patients with high-grade medullary fibrosarcoma have a 5-year survival rate of about 30%; those with low-grade and surface fibrosarcomas have a 5-year survival rate of 50%–80%.

Infant or congenital fibrosarcoma in children has a better prognosis, with a 5-year survival rate of >80%.

References

1. Fletcher CDM, Bridge JA, Hogendoorn P, et al. (eds.). *WHO Classification of Tumours of Soft Tissue and Bone*. 4th ed. Lyon, France: IARC Press, 2013.

2. PathologyOutlines.com. Fibrosarcoma of soft tissue—Adult. http://www.pathologyoutlines.com/topic/softtissuefibrosarcoma.html. Accessed February 24, 2017.

3. PDQ Pediatric Treatment Editorial Board. *Childhood soft tissue sarcoma treatment (PDQ®): Health professional version. PDQ Cancer Information Summaries.* Bethesda, MD: National Cancer Institute, 2002.

4. PDQ Adult Treatment Editorial Board. *Adult Soft Tissue Sarcoma Treatment (PDQ®): Health Professional Version. PDQ Cancer Information Summaries.* Bethesda, MD: National Cancer Institute, 2002.

14
Osteosarcoma

14.1 Definition

Neoplasms affecting the bone encompass cartilage tumors, osteogenic tumors, fibrogenic tumors, fibrohistiocytic tumors, Ewing sarcoma or primitive neuroectodermal tumors, hematopoietic tumors, giant cell tumors, notochordal tumors, vascular tumors, smooth muscle tumors, lipogenic tumors, neural tumors, miscellaneous tumors, miscellaneous lesions, and joint lesions.

Among osteogenic tumors, three types are recognized: (1) osteoid osteoma, (2) osteoblastoma, and (3) osteosarcoma including primary central or medullary osteosarcoma (conventional [chondroblastic, fibroblastic, and osteoblastic], telangiectatic, small cell, and low-grade central), primary surface or peripheral osteosarcoma (parosteal, periosteal, and high-grade surface), and secondary osteosarcoma [1].

Defined as the primary malignant mesenchymal bone tumor where the malignant tumor cells directly form the osteoid or bone or both, osteosarcoma represents the most common primary osseous tumor excluding malignant neoplasms of marrow origin (myeloma, lymphoma, and leukemia), and accounts for 60% of the primary bone malignancies diagnosed within the first two decades of life, and 20% of all bone cancers.

14.2 Biology

The musculoskeletal system is made of specifically designated connective tissues known as hard and soft tissues. Whereas the hard tissue consists of bone and cartilage, the soft tissue includes fat, muscle (smooth, skeletal, and cardiac), fibrous tissue (tendons and ligaments), synovial tissue (joint capsules and ligaments), blood vessels, lymph vessels, and peripheral nerves.

Inside the human body, there are 206 pieces of bones of long, short, and flat types, which are composed of cells (osteoblasts, osteocytes, osteoclasts,

and bone-lining cells) and extracellular matrix (organic matrix [30%] and hydroxyapatite [70%]).

Osteosarcoma is a histologically diverse tumor arising from cells of osteoblastic lineage, including primitive mesenchymal stem cells and well-differentiated osteocytes. It usually develops within the medullary cavity and extends to the cortex. In rare cases, osteosarcoma may involve tissues of mesenchymal origin in the absence of bone (e.g., extraskeletal osteosarcoma). Regardless of its developmental origin, osteosarcoma invariably produces osteoid, which provides a diagnostic hallmark [2].

Osteosarcoma commonly occurs as an enlarging and palpable mass with progressive pain in the metaphysis (91%) of long bones of the extremities, especially the distal femur (43%), proximal tibia (23%), proximal humerus (10%), and proximal femur. About 9% of osteosarcomas are found in the diaphysis. Flat bones or short bones are seldom affected [2].

Common sites of osteosarcoma metastasis include the lung (80%), other bones (37%), pleura (33%), and heart (20%), and rarely lymph nodes, gastrointestinal (GI) tract, liver, and brain.

14.3 Epidemiology

Bone tumors make up <1% of adult cancers and 3%–5% of childhood cancers. As the most common bone tumor, osteosarcoma is bimodally distributed by age with peaks in adolescence (<20 years) and in the elderly (>65 years). The annual incidence of osteosarcoma is 2.1–2.6 cases per 1 million among 5–9 years, 7–8.3 cases per 1 million among 10–14 years, 8.2–8.9 cases per 1 million among 15–19 years, and 1.3–1.9 cases per 1 million among 25–59 years. Osteosarcoma appears to affect more males than females, with a male-to-female ratio of 1.28:1.

14.4 Pathogenesis

Osteosarcoma is the most common radiation-induced sarcoma, and has been linked to Thorotrast administration, chemotherapy in children and genetic syndromes and diseases (e.g., Li–Fraumeni syndrome [p53 gene on 17p13.1], Bloom syndrome [BLM or RecQL3 gene on 15q26.1], Diamond–Blackfan anemia, Paget disease [LOH18CR1 gene on18q21-qa22, 5q31, or 5q35-qter], retinoblastoma [RB1 gene on 13q14.2], Rothmund–Thomson syndrome or poikiloderma congenitale [RTS or RecQL4 gene on 8q24.3], and Werner syndrome [WRN or RecQL2 gene on 8p12-p11.2]). In addition,

mutations in P16INK4A, P19ARF, ATRX, and DLG2, and amplification of 12q13 (containing CDK4 and MDM2 genes) are also implicated in the pathogenesis of osteosarcoma [3].

14.5 Clinical features

Clinical symptoms of osteosarcoma include pain (usually around the knee or in the upper arm, which is sufficiently intense to wake the patient from sleep), swelling (due to a lump or mass), limp (if the tumor is in a leg bone), and bone fractures (breaks, particularly with telangiectatic osteosarcoma).

14.6 Diagnosis

Diagnosis of bone tumors such as osteosarcoma involves physical examination, x-ray, CT scan, MRI, and PET scan, followed by microscopic confirmation on biopsy specimen.

Osteosarcoma is a bulky, gritty, hemorrhagic tumor with well-defined proximal and distal margins, and cystic degeneration. A large quantity of cartilage may be seen in 25% of cases. The tumor often spreads within the medullary cavity, destroys cortical bone, elevates periosteum, invades soft tissue, and forms satellite nodules (skip metastases).

Radiographically, osteosarcoma appears as a large, destructive, lytic or blastic mass with permeative margins and a sunburst pattern (due to new bone formation in soft tissue) [3]. Histologically, osteosarcoma is a high-grade spindle cell tumor that produces characteristic neoplastic osteoid (eosinophilic, homogenous, glassy matrix with irregular contours and osteoblastic rimming) or neoplastic bone (basophilic thin trabeculae). The tumor often causes trabecular destruction, vascular invasion, and necrosis, as well as atypical mitoses. According to the predominant type of stroma, osteosarcoma is distinguished into various subtypes (e.g., osteoblastic, chondroblastic, fibroblastic, and giant cell rich), with osteoblastic, fibroblastic predominance showing pure spindle cell growth with minimal matrix, and chondroblastic predominance showing malignant-appearing cartilage with peripheral spindling and osteoid production. Immunohistologically, osteosarcoma is positive for alkaline phosphatase and vimentin; variable for smooth muscle actin, desmin, S100 (if chondroid differentiation), and vWF; but negative for keratin and EMA. Molecularly, osteosarcoma is aneuploid or hyperploid except periosteal and well-differentiated (usually diploid) subtypes, and 20% have p53 mutations [4].

Differential diagnoses include osteoid osteoma and osteoblastoma. Osteoid osteoma is a benign, small, circumscribed bone-forming tumor (circumscribed nidus) of <1.5 cm (any larger size is considered an osteoblastoma) associated with pain (due to production of prostaglandin E2 or nerve fibers in the reactive zone), arthritis, or joint dysfunction. This tumor also affects children, adolescents, and young adults (mainly 5–24 years), and commonly occurs in long bones (femur and tibia [50%]). Histologically, this tumor shows anastomosing, irregular trabeculae or solid, sclerotic nidus of woven bone with variable mineralization, which is rimmed by a single layer of osteoblasts plus frequent osteoclasts. Other features include loose, fibrovascular stroma, surrounding thick sclerotic bone, and lymphoplasmacytic synovitis with juxta-articular tumors. Molecularly, the tumor may have structural alterations involving 22q13.1 [5].

Osteoblastoma (also called giant osteoid osteoma) is a benign, well-demarcated, hemorrhagic, bone-forming tumor comprised of anastomosing trabeculae of osteoid and woven bone rimmed by osteoblasts, with one-third occurring in the spine and sacrum. Clinically, the tumor manifests with progressive pain (less intense than that in osteoid osteoma), neurological findings and scoliosis (in spinal tumors), fever, weight loss, and generalized periostitis (toxic osteoblastoma). Radiography reveals demarcated bone tumor with intralesional ossification, mostly nondestructive growth, and occasional central nidus or nidi. Histologically, the tumor contains anastomosing trabeculae of osteoid and woven bone rimmed by a single layer of benign activated osteoblasts (which are large with eccentric cytoplasm and perinuclear hof and large vesicular nuclei with prominent nucleoli), osteoclastic giant cells, loose fibrovascular stroma between bone trabeculae, intralesional hemorrhage, secondary aneurysmal bone cyst (ABC), low mitotic rate, rare cartilaginous matrix, and degenerative atypia (pseudomalignant osteoblastoma) [6].

14.7 Treatment

Similar for Ewing sarcoma, curative therapy for osteosarcoma involves a combination of systemic therapy and local control of all macroscopic tumors.

Systemic therapy for osteosarcoma is based on a multiagent protocol consisting of high-dose methotrexate, doxorubicin, and cisplatin, (so called MAP combination). Addition of liposomal muramyl tripeptide to chemotherapy may improve overall survival.

Surgical resection (limb-sparing surgery with adequate margins) represents a valuable local control approach, which is usually supported by

preoperative or postoperative chemotherapy (depending on the extent of tumor necrosis).

If complete surgical resection is not feasible (e.g., metastatic osterosarcoma) or if surgical margins are inadequate, radiotherapy (doses of 70 Gy or higher) may improve the local control rate, disease-specific survival, and overall survival [7–9].

14.8 Prognosis

Patients with nonmetastatic osteosarcoma of the extremities have an expected 5-year survival rate of 70%–80%. Patients with metastases at diagnosis or recurrent osteosarcoma have a 5-year survival rate of 20%–30%.

Within osteosarcoma subtypes, the 5-year survival rate is 52.6% for high grade, 85.9% for parosteal, and 17.8% for Paget subtypes. For osteosarcoma patients, the 5-year survival rate is 60% for those under 30 years, 50% for those aged 30–49 years, and 30% for those over 50 years.

Poor prognostic factors include Paget disease, telangiectatic histology, elevated serum alkaline phosphatase, minimal postchemotherapy tumor necrosis, involvement of craniofacial bones (not jaw) or vertebrae, multifocal tumor, and loss of heterozygosity of the RB gene.

Favorable prognostic factors include young age (child or young adult), female, tumor on an arm or leg (instead of the hip), complete resectable tumor, normal blood alkaline phosphatase and lactate dehydrogenase (LDH) levels, and good response to chemotherapy.

References

1. Fletcher CDM, Bridge JA, Hogendoorn P, et al. (eds.). *WHO Classification of Tumours of Soft Tissue and Bone.* 4th ed. Lyon, France: IARC Press, 2013.
2. Abarrategi A, Tornin J, Martinez-Cruzado L, et al. Osteosarcoma: Cells-of-origin, cancer stem cells, and targeted therapies. *Stem Cells Int* 2016;2016:3631764.
3. Kundu ZS. Classification, imaging, biopsy and staging of osteosarcoma. *Indian J Orthop* 2014;48(3):238–46.
4. PathologyOutlines.com. *Osteosarcoma—General.* http://www.pathologyoutlines.com/topic/boneosteosarcomageneral.html. Accessed March 1, 2017.

5. PathologyOutlines.com. *Osteoid osteoma.* http://pathologyoutlines. com/topic/boneosteoidosteoma.html. Accessed March 1, 2017.

6. PathologyOutlines.com. *Osteoblastoma.* http://www.pathologyoutlines. com/topic/boneosteoblastoma.html. Accessed March 1, 2017.

7. PDQ Pediatric Treatment Editorial Board. *Osteosarcoma and Malignant Fibrous Histiocytoma of Bone treatment (PDQ®): Health Professional Version. PDQ Cancer Information Summaries.* Bethesda, MD: National Cancer Institute, 2002.

8. Isakoff MS, Bielack SS, Meltzer P, Gorlick R. Osteosarcoma: Current treatment and a collaborative pathway to success. *J Clin Oncol* 2015;33(27):3029–35.

9. Meyers PA. Systemic therapy for osteosarcoma and Ewing sarcoma. *Am Soc Clin Oncol Educ Book* 2015:e644–7.

SECTION II
Urogenitory System

15
Bladder Cancer

15.1 Definition

Tumors affecting the urinary tract consist of eight categories: (1) *urothelial tumors* (infiltrating urothelial carcinoma [with squamous differentiation, with glandular differentiation, with trophoblastic differentiation, nested, microcystic, micropapillary, lymphoepithelioma-like, lymphoma-like, plasmacytoid, sarcomatoid, giant cell, and undifferentiated] and noninvasive urothelial neoplasias [urothelial carcinoma *in situ*, noninvasive papillary urothelial carcinoma—high grade, noninvasive papillary urothelial carcinoma—low grade, noninvasive papillary urothelial neoplasm of low malignant potential, urothelial papilloma, and inverted urothelial papilloma]), (2) *squamous neoplasms* (squamous cell carcinoma [SCC], verrucous carcinoma, and squamous cell papilloma), (3) *glandular neoplasms* (adenocarcinoma [enteric, mucinous, signet ring cell, and clear cell] and villous adenoma), (4) *neuroendocrine tumors* (small cell carcinoma, carcinoid, and paraganglioma), (5) *melanocytic tumors* (malignant melanoma and nevus), (6) *mesenchymal tumors* (rhabdomyosarcoma, leiomyosarcoma, angiosarcoma, osteosarcoma, malignant fibrous histiocytoma, leiomyoma, and hemangioma), (7) *hematopoietic and lymphoid tumors* (lymphoma and plasmacytoma), and (8) *miscellaneous tumors* (carcinoma of Skene, Cowper, and Littre glands; metastatic tumors and tumors extending from other organs) [1].

Out of these, infiltrating urothelial carcinoma (formerly invasive transitional cell carcinoma) of the urothelial tumor category is most common and accounts for >90% of all bladder cancers. This is followed by SCC of the squamous neoplasm category and adenocarcinoma of the glandular neoplasm category, which represent 5% and 2% of all bladder cancers, respectively.

Depending on the involvement of the muscularis propria (or the detrusor muscle), infiltrating urothelial carcinoma can be further separated into muscle-invasive bladder cancers (MIBCs) (high-grade stages T2–T4; 25% of cases) and non-muscle-invasive bladder cancers (NMIBCs) (low-grade

stages Tis/carcinoma in situ [CIS], Ta, and T1; 75% of cases). SCC and adenocarcinoma are high-grade urothelial carcinomas with squamous and glandular differentiation, respectively.

15.2 Biology

The urinary tract comprises the kidneys, ureters, bladder, and urethra. Located in the anterior pelvis, the adult bladder is a round, hollowed organ of 300–600 mL in capacity. The body of the bladder receives inferior support from the pelvic diaphragm in females or prostate in males, lateral support from the obturator internus and levator ani muscles, and superior support from the medial umbilical ligament or the urachal remnant. With extraperitoneal fat and connective tissue covering outside, detrusor, neck, and pelvic floor muscles in the middle, and the mucosa lining inside, the bladder stores urine produced in the renal tubules, and collected via the renal pelvis of each kidney and the ureters. The urine eventually leaves the bladder through the urethra.

The bladder mucosa is formed by the transitional epithelium (which is also found in the lower part or the renal pelvises of the kidneys, the ureters, and the proximal urethra) and is connected by the lamina propria (or the bladder submucosa) to the muscular bladder wall. Most cancers of the bladder, renal pelvises, ureters, and proximal urethra arise through an accumulation of somatic mutations from transitional cells in the transitional epithelium, and are thus called transitional cell carcinomas or urothelial carcinomas. As women typically have a thinner detrusor muscle than men, extravesical tumor growth in women is much faster.

15.3 Epidemiology

Bladder cancer is the sixth most common cancer, the second most common genitourinary cancer, the fourth most common malignancy in men, and the eighth most common malignancy in women globally.

Invasive bladder cancer has an annual incidence of 20 per 100,000, and an annual mortality of 7 per 100,000 in Europe. In developing countries, 75% of bladder cancer is SCC, due mostly to endemic schistosomiasis.

Bladder cancer primarily affects people of >65 years. Men are at a three- to fourfold higher risk of developing bladder cancer than women. However, more women present with advanced or aggressive disease stages at the primary diagnosis and have worse outcomes than men [2].

15.4 Pathogenesis

Up to 80% of bladder cancer is associated with cigarette smoking; arsenic exposure; occupational exposure to aluminum, metal, aromatic amines, polycyclic aromatic hydrocarbons, oil, leather, arylamines, aniline dye, and paint; chronic urinary tract infections (e.g., *Schistosoma haematobium* and human papillomavirus [HPV]); and family disease syndromes (e.g., Costello syndrome [*HRAS*], Facio–Cutaneous skeletal syndrome [*HRAS*], and Cowden syndrome [*PTEN/MMAC1*].

In muscle-invasive bladder cancers (MIBCs), frequently mutated genes include *TP53* (41%), *KDM6A* (28%), *ARID1A* (22%), *PIK3CA* (18%), *MLL2* (17%), *CREBBP* (15%), *RB1* (15%), *STAG2* (13%), *FGFR3* (13%), *EP300* (13%), *TSC1* (8%), and *HRAS* (8%). Most MIBCs contain mutations in chromatin remodeling genes (86%), including histone methyltransferases (58%), histone lysine demethylases (54%), SWI-SNF complexes (40%), and histone acetyltransferases (32%). *TP53* mutations occur more frequently in MIBCs (41%) than in NMIBCs (8%) [3–5].

In non-muscle-invasive bladder cancers (NMIBCs), class 1 tumors (which have a lower risk of progression and a better prognosis than classes 2 and 3 tumors) show upregulation of early-cell-cycle genes (*CCND1*, *ID1*, and *RBL2*), class 2 tumors display upregulation of late-cell-cycle genes (*CDC20*, *CDC25A*, *CDKs*, and *PLK1*) and overexpression of cancer stem cell markers (*ALDH1A1*, *ALDH1A2*, *PROM1*, *NES*, and *THY1*), and class 2 and/or class 3 tumors show increased expression of the keratin (*KRT*) gene family. Additionally, NMIBCs show a high frequency (~80%) of activating mutations in the fibroblast growth factor receptor 3 (FGFR3) signaling pathway. The *FGFR3* and *RAS* gene mutations are mutually exclusive in bladder cancer [3–5].

Bladder cancer often harbors deletions of 9ptr-p22, 9q22.3, 9q33, or 9q34; amplifications of 6p22.3 (*E2F3*), 8p12 (*FGFR1*), 8q22.2 (*CMYC*), 11q13 (*CCND1*, *EMS1*, and *INT2*), and 19q13.1 (*CCNE*); and homozygous deletions at 9p21.3, 8p23.1, and 11p13. Epigenetic regulation (e.g., DNA methylation, histone modification, microRNA regulation, and nucleosome positioning) is also distorted in bladder cancer [3–5].

15.5 Clinical features

Bladder cancer typically causes macroscopic hematuria (blood in the urine), pain or burning during urination, frequent urination, nocturia, dysuria, and general weakness.

15.6 Diagnosis

Patients suspected of urothelial carcinomas should undergo cytoscopic examination, which is more sensitive than CT scan and ultrasound for detecting bladder cancer.

For staging and surveillance, upper urinary tract is assessed by ureteroscopy, retrograde pyelograms during cystoscopy, intravenous pyelograms, or CT urograms.

Infiltrating or invasive urothelial carcinoma (formerly invasive transitional cell carcinoma) is typically a large, flat to papillary infiltrative mass of the bladder (70%) that has penetrated the basement membrane and invaded the lamina propria or deeper. Histologically, high-grade lesions often have foci of squamous differentiation with focal or extensive keratinization and intracellular bridges; scattered syncytiotrophoblasts; bizarre nuclear pleomorphism; focal clear cells or choriocarcinomatous areas; spindle cells, osteoclasts, glandular or benign stromal elements, plasmacytoid cells, and lipid cells; focal pseudosarcomatous stroma; necrosis; and papillary (70%), sessile or mixed (20%), or nodular (10%) growth patterns. Immunohistochemically, the tumor is positive for CK7, CK20 (50%), HMW keratin (80%), uroplakin (40%–60%, specific), thrombomodulin, p63 (variable), MUC1, CEA, p53, GATA3, S100, CA125 (variable), HER2 (variable), CD31, CD34, and podoplanin (D2-40), but negative for prostatic markers (PSA, P501S, PSMA, NKX3.1, and pPSA), WT1, MUC2, MUC5AC, HPV, and Leu7/CD57. Molecularly, the tumor shows p16(INK4a) expression, monosomy 9, 9p– (p16 INK4/TS1), 9q–, 13q– (retinoblastoma gene), 14q–, 17p– (p53), and polysomy 1 and 17 (more common in pT1 than in pTa) [6].

Differential diagnoses include pseudocarcinomatous epithelial proliferations, high-grade or poorly differentiated prostatic adenocarcinoma (foamy and pale cytoplasm; oval nuclei with smooth borders; fine, powdery, evenly distributed nuclear chromatin; large prominent nucleolus; no or rare mitotic figures; no or rare necrosis; no intraductal growth; positivity for PSA or PAP and CD57/Leu7; negativity for CK7, CK20, 34βE12, Uroplakin III, thrombomodulin, and p53), vasculitis, and low-grade lesions resembling Brunn nests or cystitis glandular or cystica [6].

The stage of bladder cancer is often determined in accordance with the TNM system of AJCC, in which T (X, 0, a, is, 1, 2a, 2b, 3a, 3b, 4a, 4b) refers to tumor size and location, N (X, 0, 1, 2, 3) refers to degree of lymph node involvement, and M (0, 1) refers to absence or presence of metastasis. Thus, bladder cancer stage 0a is defined as Ta, N0, M0; stage 0is as Tis, N0, M0; stage I as T1, N0,

M0; stage II as T2a or T2b, N0, M0; stage III as T3a, T3b, or T4a, N0, M0; stage IV as T4b, N0, M0; or any T, N1 to N3, M0; or any T, any N, M1.

15.7 Treatment

NMIBC is often treated by tumor removal via a transurethral approach, and introduction of chemotherapy or other therapies into the bladder with a catheter [7–9].

MIBC is treated with radical cystectomy and bilateral lymph node dissection. Neoadjuvant chemotherapy with various cisplatin-based regimens may be used after radical cystectomy, providing clear survival benefit. Adjuvant chemotherapy may be considered for patients who have not received neoadjuvant therapy [10].

Locally advanced and metastatic bladder cancer may be treated with chemotherapy (e.g., methotrexate, vinblastine, doxorubicin, and cisplatin [MVAC]; gemcitabine and taxanes; and more recently, gemcitabine + cisplatin [GC]) [10].

15.8 Prognosis

Patients with bladder cancer have 5-, 10-, and 15-year survival rates of 77%, 70%, and 65%, respectively.

Patients with NMIBC or superficial bladder cancer have a better prognosis than patients with MIBC. The latter tend to recur (30%–80% of cases) and progress to a higher grade (up to 45% of cases) within 5 years.

Carcinoma with rhabdoid features, micropapillary carcinoma, plasmacytoid carcinoma, sarcomatoid carcinoma, small cell carcinoma, and undifferentiated carcinoma have a poor prognosis. Other unfavorable prognostic factors include decreased expression of p63, loss of E-cadherin expression, and increased CK20 expression.

References

1. Eble JN (ed.). *Pathology and Genetics of Tumours of the Urinary System and Male Genital Organs*. Lyon, France: IARC Press, 2004.
2. Marks P, Soave A, Shariat SF, Fajkovic H, Fisch M, Rink M. Female with bladder cancer: What and why is there a difference? *Transl Androl Urol* 2016;5(5):668–2.

3. van Rhijn BW, Musquera M, Liu L, et al. Molecular and clinical support for a four-tiered grading system for bladder cancer based on the WHO 1973 and 2004 classifications. *Mod Pathol* 2015;28(5):695–705.
4. Li HT, Duymich CE, Weisenberger DJ, Liang G. Genetic and epigenetic alterations in bladder cancer. *Int Neurourol J* 2016;20(Suppl 2):S84–94.
5. Zhao M, He XL, Teng XD. Understanding the molecular pathogenesis and prognostics of bladder cancer: An overview. *Chin J Cancer Res* 2016;28(1):92–8.
6. PathologyOutlines.com. *Urothelial Carcinoma—Invasive.* http://www.pathologyoutlines.com/topic/bladderurothelialinvasivegen.html.
7. Klaile Y, Schlack K, Boegemann M, Steinestel J, Schrader AJ, Krabbe LM. Variant histology in bladder cancer: How it should change the management in non-muscle invasive and muscle invasive disease? *Transl Androl Urol* 2016;5(5):692–701.
8. Baumann BC, Sargos P, Eapen LJ, et al. The rationale for post-operative radiation in localized bladder cancer. *Bladder Cancer* 2017;3(1):19–30.
9. PDQ Adult Treatment Editorial Board. *Bladder Cancer Treatment (PDQ®): Health Professional Version. PDQ Cancer Information Summaries.* Bethesda, MD: National Cancer Institute, 2002.
10. Pinto IG. Systemic therapy in bladder cancer. *Indian J Urol* 2017;33(2):118–26.

16
Breast Cancer

16.1 Definition

Tumors of the breast include (1) *epithelial tumors* (invasive ductal carcinoma [IDC] not otherwise specified [NOS] [mixed-type carcinoma, pleomorphic carcinoma, carcinoma with osteoclastic giant cells, carcinoma with choriocarcinomatous features, carcinoma with melanotic features], invasive lobular carcinoma [ILC], tubular carcinoma, invasive cribriform carcinoma, medullary carcinoma, mucinous carcinoma and other tumors with abundant mucin [mucinous carcinoma, cystadenocarcinoma and columnar cell mucinous carcinoma, signet ring cell carcinoma], neuroendocrine tumors [solid neuroendocrine carcinoma, atypical carcinoid tumor, small cell or oat cell carcinoma, large cell neuroendocrine carcinoma], invasive papillary carcinoma, invasive micropapillary carcinoma, apocrine carcinoma, metaplastic carcinomas [pure epithelial metaplastic carcinomas—squamous cell carcinoma, adenocarcinoma with spindle cell metaplasia, adenosquamous carcinoma, mucoepidermoid carcinoma; mixed epithelial or mesenchymal metaplastic carcinomas], lipid-rich carcinoma, secretory carcinoma, oncocytic carcinoma, adenoid cystic carcinoma, acinic cell carcinoma, glycogen-rich clear cell carcinoma, sebaceous carcinoma, inflammatory carcinoma, lobular neoplasia [lobular carcinoma *in situ*, or LCIS], intraductal proliferative lesions [usual ductal hyperplasia, flat epithelial atypia, atypical ductal hyperplasia, ductal carcinoma *in situ*], microinvasive carcinoma, intraductal papillary neoplasms [central papilloma, peripheral papilloma, atypical papilloma, intraductal papillary carcinoma, intracystic papillary carcinoma], benign epithelial proliferations [adenosis including variants—sclerosing adenosis, apocrine adenosis, blunt duct adenosis, microglandular adenosis, adenomyoepithelial adenosis; radial scar or complex sclerosing lesion; adenomas—tubular adenoma, lactating adenoma, apocrine adenoma, pleomorphic adenoma, ductal adenoma]), (2) *myoepithelial lesions* (myoepitheliosis, adenomyoepithelial adenosis, adenomyoepithelioma, malignant myoepithelioma), (3) *mesenchymal tumors* (hemangioma, angiomatosis, hemangiopericytoma, pseudoangiomatous stromal hyperplasia, myofibroblastoma, fibromatosis—aggressive, inflammatory myofibroblastic tumor, lipoma [angiolipoma], granular cell tumor, neurofibroma, schwannoma, angiosarcoma, liposarcoma, rhabdomyosarcoma, osteosarcoma,

leiomyoma, leiomyosarcoma), (4) *fibroepithelial tumors* (fibroadenoma, phyllodes tumor [benign, borderline, malignant], periductal stromal sarcoma—low grade, mammary hamartoma), (5) *tumors of the nipple* (nipple adenoma, syringomatous adenoma, Paget disease of the nipple), (6) *malignant lymphoma* (diffuse large B-cell lymphoma, Burkitt lymphoma, extranodal marginal-zone B-cell lymphoma of mucosa-associated lymphoid tissue [MALT] type, follicular lymphoma), (7) *metastatic tumors*, and (8) *tumors of the male breast* (gynecomastia, carcinoma [invasive, in situ]) [1,2].

Among these, invasive ductal carcinoma (IDC) and invasive lobular carcinoma (ILC) of the epithelial tumor category constitute about 80% and 10% of all breast malignancies, respectively, while invasive carcinoma of other special types (e.g., apocrine, adenoid cystic, acinic cell, and oncocytic carcinoma) accounts for about 10% of breast malignancies [3].

16.2 Biology

The breast (or mammary gland) sits on the pectoralis (chest) muscle (atop the rib cage), and extends horizontally from the edge of the sternum (the firm flat bone in the middle of the chest) out to the midaxillary line (the center of the axilla, or underarm). Covered by the skin with sweat glands and hair follicles, the breast tissue is encircled underneath the skin and on top of the pectoralis muscle by a thin layer of connective tissue called fascia.

Each breast consists of 15–20 sections called lobes, and each lobe has many smaller sections called lobules. The lobules end in dozens of tiny bulbs of 1–4 mm in size (also called the terminal duct lobular units [TDLUs]), which constitute the effective secretory units of the breast and are composed of extralobular terminal duct, intralobular terminal duct, and lobule. Thin tubes that connect the lobes, lobules, and bulbs are called ducts, which lead out to the nipple (surrounded by areola).

The epithelium of the mammary gland comprises luminal and basal or myoepithelial cells. Lining the ductal lumen, luminal cells secrete milk upon terminal differentiation into lobuloalveolar cells. Located just below the luminal cells, basal or myoepithelial cells ensure ductal contractility to release milk. In addition, breast ducts are infiltrated with stem cells that are tightly regulated to produce all cellular elements for breast ducts and contribute to normal gland development and cycling.

Between the lobes, lobules, bulbs, and ducts is stroma (a connective tissue, consisting of adipocytes, fibroblasts, and endothelial cells), which provides support for the extensive system of ducts and alveoli.

Structurally, the breast in children and males is nearly identical to that of adult females, except that the former lacks the specialized lobules, without physiologic need for milk production.

Breast carcinomas (including ductal carcinoma and lobular carcinoma) arise either inside or just proximal to the lobule of the TDLU. IDC begins in the cells of the ducts that carry milk to the nipple and is characterized by malignant ductal proliferation and stromal invasion. ILC starts in the lobes or lobules that are the milk-making glands, and typically contains single cells arranged individually, in single file, or in sheets.

16.3 Epidemiology

Breast cancer is the most common malignancy in women, with an annual incidence of 18–90 per 100,000 women worldwide. Occasionally, breast cancer may affect men (e.g., gynecomastia or breast enlargement) and children or adolescents (e.g., secretory carcinoma or juvenile carcinoma, and juvenile fibroadenoma), accounting for 1% of all breast cancer cases.

16.4 Pathogenesis

Risk factors for breast cancer include (1) a personal history of benign (noncancerous) breast disease, *in situ* carcinoma, or invasive carcinoma; (2) a family history of breast cancer in a first-degree relative (mother, daughter, or sister); (3) inherited mutational changes in the *BRCA1* or *BRCA2* genes or other breast cancer–related genes; (4) exposure of breast tissue to estrogen made by the body (e.g., menstruating at an early age, older age at first birth or never having given birth, and starting menopause at a later age); (5) hormone replacement therapy (e.g., estrogen combined with progestin for symptoms of menopause); (6) exposure to diethylstilbestrol (DES) (to prevent miscarriage); (7) past radiotherapy in the breast or chest; (8) alcohol intake; (9) obesity; and (10) older age.

Molecularly, breast cancer demonstrates gains in chromosomes 1q, 8q, 17q, 20q, and 11q and losses in 8p, 13q, 16q, 18q, and 11q, with mutations most frequently centered on BRCA1, BRCA2, p53, HER2-neu, cyclin D1, and cyclin E [4].

16.5 Clinical features

Clinical signs of breast cancer include (1) a lump or thickening in or near the breast or in the underarm area; (2) a change in the size or

shape of the breast; (3) a dimple or puckering in the skin of the breast; (4) a nipple turned inward into the breast; (5) fluid, other than breast milk, from the nipple, especially if it is bloody; (6) scaly, red, or swollen skin on the breast, nipple, or areola (the dark area of skin around the nipple); and (7) dimples in the breast that look like the skin of an orange, called peau d'orange. Some patients with breast cancer may be asymptomatic.

16.6 Diagnosis

IDC is a large, firm, poorly circumscribed mass with hemorrhage, necrosis, and cystic degeneration. Histologically, IDC contains sheets, nests, cords, or individual cells; prominent tubular formations in well-differentiated tumors, but absent in poorly differentiated tumors; desmoplastic stroma, calcification (60%), and variable necrosis; prominent mitotic figures; and occasional eosinophils and intraluminal crystalloids. IDC stains positive for CK8/18, CK19, CK7, EMA, E-cadherin, ER (70%), milk fat globule, lactalbumin, CEA, B72.3, BCA225, glycogen (60%), mucin (20%), cytokeratin 5/6 (30%), S100 (10%–45%), HER2 (15%–30%), RCC Ma (renal cell carcinoma marker), and CD5 clone 4C7, but negative for myoepithelial markers p63 (positive in benign lesions), CD10, and calponin [5].

ILC often contains small, uniform, round cells in single file (linear, Indian file) or a targetoid pattern of noncohesive cells encircling ducts; evenly disbursed chromatin and no nucleoli (like LCIS cells); signet ring cells, intracellular lumina, and intracellular mucin; variable dense fibrous stroma with periductal and perivenous elastosis; dense lymphoid infiltrate; <10 mitoses/10 high-power field (HPF); and no necrosis. ILC is positive for ER, PR, HMW keratin, mucicarmine (intracellular mucin), GCDFP-15 (30%), and PLEKHA7, but negative for E-cadherin, p53, HER2, and Ki-67. Molecularly, ILC shows inactivation of E-cadherin, loss of heterozygosity, or methylation [6].

After histologic and imaging assessments, breast cancer is often graded according to the Nottingham histologic score system (which is the Elston-Ellis modification of Scarff-Bloom-Richardson grading system). This system takes three factors into consideration: (i) glandular (acinar)/tubular differentiation (how well the tumor cells try to recreate normal glands), (ii) nuclear pleomorphism (how "ugly" the tumor cells look), and (iii) mitotic count (how much the tumor cells are dividing). With each of these features scored from 1-3, the final added scores range from 3-9, with Grade 1 (low grade or well differentiated) having a score

of 3, 4 or 5; Grade 2 (intermediate grade or moderately differentiated) having a score of 6 or 7; and Grade 3 (high grade or poorly differentiated) having a score of 8 or 9.

The stage of breast cancer is determined on the basis of to the TNM system, with stage 0 (noninvasive tumor or carcinoma *in situ*), stage I (tumor of <2 cm or invasion of surrounding breast tissue), stage II (tumor of 2–5 cm or invasion of lymph nodes in the armpit, or both), stage III (tumor of 2–5 cm attaching to the skin or surrounding tissues, and invasion of lymph nodes in the armpit and near the breastbone), and stage IV (tumor of any size spreading beyond the breast and nearby lymph nodes to other organs of the body) [7].

Once a breast cancer is diagnosed, further tests are conducted with cancer tissue to determine the growth potential, likelihood to spread through the body, chance of recurrence, and best treatment options.

In particular, based on the presence of hormone receptors (ER and PR) and the quantity of HER2 detected by immunohistochemical (IHC) and other related procedures, invasive breast cancer can be classified as: (i) hormone receptor-positive, (ii) hormone receptor-negative, (iii) HER2-positive, (iv) HER2-negative, (v) triple-negative, (vi) triple-positive. The availability of this information facilitates the design of tailor-made treatments for invasive breast caner types/subtypes, and provides valuable insights on the prognosis of invasive breast cancer.

Gene expression profiling helps predict likelihood of cancer spread to other parts of the body or recurrence as well as potential response to chemotherapy.

16.7 Treatment

Five types of standard treatment are used for patients with breast cancer, that is, surgery, chemotherapy, radiation therapy, hormone therapy, and targeted therapy [7,8].

Surgical approaches consist of breast-conserving surgery, total mastectomy (or simple mastectomy), and modified radical mastectomy.

Chemotherapy includes preoperative therapy (or neoadjuvant therapy) and postoperative therapy (or adjuvant therapy).

Radiation therapy comprises external radiation therapy and internal radiation therapy.

Hormone therapy is most often used as an adjuvant therapy to help reduce the postsurgery relapse risk, as well as metastases.

Targeted therapy relies on drugs or other substances to identify and attack specific cancer cells without harming normal cells.

16.8 Prognosis

Breast cancer has a 5-year overall survival rate of 93% for stage 0 (carcinoma *in situ*), 88% for stage I, 74%–81% for stage II, 41%–67% for stage III, and 15% for stage IV. The recurrence rate of breast cancer is ~30%.

References

1. Lakhani S, Ellis I, Schnitt S, et al. *WHO Classification of Tumours of the Breast*. 4th ed. Lyon, France: IARC Press, 2012.
2. Sinn HP, Kreipe H. A brief overview of the WHO Classification of Breast Tumors, 4th Edition, focusing on issues and updates from the 3rd Edition. *Breast Care (Basel)* 2013;8(2):149–54.
3. Makki J. Diversity of breast carcinoma: Histological subtypes and clinical relevance. *Clin Med Insights Pathol* 2015;8:23–31.
4. Dai X, Xiang L, Li T, Bai Z. Cancer hallmarks, biomarkers and breast cancer molecular subtypes. *J Cancer* 2016;7(10):1281–94.
5. PathologyOutlines.com. *Ductal Carcinoma, NOS—General*. http://www.pathologyoutlines.com/topic/breastmalignantductalNOS.html. Accessed March 5, 2017.
6. PathologyOutlines.com. *Classic Infiltrating Lobular Carcinoma*. http://www.pathologyoutlines.com/topic/breastmalignantlobularclassic.html. Accessed March 5, 2017.
7. PDQ Adult Treatment Editorial Board. *Breast Cancer Treatment (PDQ®): Health Professional Version*. PDQ Cancer Information Summaries. Bethesda, MD: National Cancer Institute, 2002.
8. Tang Y, Wang Y, Kiani MF, Wang B. Classification, treatment strategy, and associated drug resistance in breast cancer. *Clin Breast Cancer.* 2016;16(5):335–43.

17
Cervical Cancer

Elsa Delgado-Sánchez, Enrique García-López,
and Ignacio Zapardiel

17.1 Definition

Cervical tumors can be divided into several histological types: (1) squamous cell carcinoma, (2) adenocarcinoma, (3) mixed carcinomas, (4) neuroendocrine carcinomas, and (5) others [1].

Squamous cell carcinoma is the most common type, although its incidence is declining in recent years, while adenocarcinoma is increasing. This can be explained by an increased prevalence of human papillomavirus (HPV) infection despite the fact that there is a higher rate of early-stage detection of squamous cell carcinoma.

Mixed adenosquamous carcinomas are rare (including glassy cell carcinoma). Neuroendocrine tumors include small and big cell carcinomas. They are extremely aggressive, and their diagnosis can be confirmed with neuroendocrine markers.

The other more rare types of cervical adenocarcinoma include sarcoma, mesonephric adenocarcinoma, and lymphoma.

17.2 Biology

The cervix is the inferior portion of the uterus. It can be divided into two parts: the ectocervix portion, covered by a multilayered epithelium, and the endocervix, lined by a single layer of cylindrical cells. The origin of cervical cancer is almost always located in the junction of two epithelia (transition zone) from a precursor injury.

HPV infection is considered necessary for development of the disease, and most risk factors are related to HPV infection (especially early onset of sexual activity, high-risk sexual partners, multiple sexual partners, and history of sexually transmitted diseases), or the body's ability to respond to such infection, such as immunosuppression and cigarette smoking (only for squamous cell cancer).

Oral contraceptive use has also been declared to be a risk factor for cervical cancer, and IUDs have been reported to accelerate clearance of the virus in infected patients (probably by enhancing local immune response).

A precursor injury is a microscopic lesion that, if left to evolve, can turn into cancer. The resulting cancer is named cervix intraepithelial neoplasia (CIN), and it can be divided histologically into

1. CIN 1: Dysplastic lesion limited to the basal third of the epithelium.
2. CIN 2: The injury spreads to the basal and medium thirds of the epithelium.
3. CIN 3: The alteration exceeds two-thirds of the basal epithelium. This should be considered equivalent to carcinoma *in situ* (CIS) [2].

17.3 Epidemiology

Cervical cancer is the second most common cancer in women worldwide, with an estimation of 528,000 new cases per year and 266,000 deaths. Its incidence is especially high in developing countries, accounting for >83% of reported cases.

Knowledge of the natural history of the disease and the introduction of screening techniques in developed countries have contributed to a large decrease in mortality rates, being more than three times lower than in developing countries.

Age at diagnosis of this disease is lower than that of other gynecologic cancers, reaching the median age at diagnosis between 40 and 59 years. For example, in Spain, cervical cancer is ranked the fifteenth most common cause of death from cancer in general terms, but the third considering the age group between 15 and 44 years, surpassed only by lung and breast cancers.

17.4 Pathogenesis

HPV infection is considered necessary for development of the disease. Although most infected women quickly eliminate the virus, those with persistent infection may end up developing a preinvasive lesion, most cases in the squamocolumnar junction. Not all HPV types have the same oncogenic power. HPV 16 and 18 are the most prevalent in cervical cancers, and HPV 31, 33, 35, 45, 52, and 58 may also play a relevant role.

Other types of tumors different from squamous cell carcinoma and adenocarcinoma, such as clear cell tumor or mesonephric adenocarcinoma, are not related to HPV.

HPV can be integrated into the human genome, causing cytological changes that are detected as low-grade squamous intraepithelial lesions (LSILs) in the Papanicolaou test. Later, there may be the transformation of normal cells into tumoral cells, especially in relation to HPV E6 and E7 proteins, which cause the degradation of tumor suppressor protein of retinoblastoma and p53, respectively.

However, this does not occur in all patients infected with HPV, suggesting that certain not well-clarified immune, environmental, or genetic mechanisms may play a part in the disease process.

Regarding cancer spread, it can occur through three different ways: local, lymphatic, or hematogenous dissemination. In local dissemination, it may involve the uterine corpus, vagina, parametria, peritoneal cavity, bladder, or rectum. In lymphatic extension, the sentinel node technique has shown that any pelvic node can be affected in the first place, although it is more common in obturator lymph nodes, moving from the pelvic nodes to the para-aortic group [2].

17.5 Clinical features

Cervical cancer at the initial stages is usually an asymptomatic illness. When present, the most common symptoms are abnormal genital bleeding, postcoital bleeding, or malodorous discharge. Other symptoms related to advanced stages could be pelvic pain, rectal tenesmus, or leg lymphedema [3].

17.6 Diagnosis

It is necessary to obtain a biopsy for the final diagnosis.

In case of macroscopic lesion, the biopsy should be taken avoiding the central part of the lesion, because in some cases this part is only necrosis [4].

If there is no macroscopic lesion, a colposcopy should be done to take the biopsy. If there is no colposcopic lesion, endocervical brushing is needed.

The International Federation of Gynecology and Obstetrics (FIGO) classification is based on tumor size, vaginal or parametrial involvement, bladder or rectum extension, and distant metastasis. It requires radiological imaging techniques such chest X-ray, computerized tomography (CT), MRI, or positron emission tomography/computerized tomography (PET/CT) [5,6]. A comparison between FIGO staging and TNM classification is shown in Table 17.1.

Table 17.1 ESMO Clinical Practice Guidelines for Diagnosis, Treatment, and Follow-Up

TNM Categories	FIGO Stages	Comments
TX		Primary tumor cannot be assessed
T0		No evidence of primary tumor
Tisb		CIS (preinvasive carcinoma)
T1	I	Cervical carcinoma confined to uterus (extension to corpus should be disregarded)
T1ac	IA	Invasive carcinoma diagnosed only by microscopy; stromal invasion with a maximum depth of 5.0 mm measured from the base of the epithelium and a horizontal spread of ≤7.0 mm (vascular space involvement, venous or lymphatic, does not affect classification)
T1a1	IA1	Measured stromal invasion of ≤3.0 mm in depth and ≤7.0 mm in horizontal spread
T1a2	IA2	Measured stromal invasion of >3.0 mm and ≤5.0 mm with a horizontal spread of ≤7.0 mm
T1b	IB	Clinically visible lesion confined to the cervix or microscopic lesion of >T1a/IA2
T1b1	IB1	Clinically visible lesion of ≤4.0 cm in greatest dimension
T1b2	IB2	Clinically visible lesion of >4.0 cm in greatest dimension
T2	II	Cervical carcinoma invades beyond uterus but not to the pelvic wall or lower third of the vagina
T2a	IIA	Cervical carcinoma invades beyond uterus but not to the pelvic wall or lower third of the vagina
T2a1	IIA1	Clinically visible lesion of ≤4.0 cm in greatest dimension
T2a2	IIA2	Clinically visible lesion of >4.0 cm in greatest dimension
T2b	IIB	Tumor with parametrial invasion
T3	III	Tumor extends to pelvic wall and/or involves lower third of the vagina and/or causes hydronephrosis or nonfunctioning kidney
T3a	IIIA	Tumor involves lower third of the vagina, no extension to pelvic wall
T3b	IIIB	Tumor extends to the pelvic wall and/or causes hydronephrosis or nonfunctioning kidney
T4	IV	The carcinoma has extended beyond the true pelvis or has involved (biopsy proven) the mucosa of the bladder or rectum; a bullous edema, as such, does not permit a case to be allotted to stage IV
T4a	IVA	Spread of the growth to adjacent organs
T4b	IVB	Spread to distant organs

17.7 Treatment

Depending on the stage, primary treatment consists of surgery, radiotherapy, or a combination of radiotherapy and chemotherapy [7–9].

See the following for the initial stages of the disease:

Stage IA1 with no lymphovascular space invasion (LVSI): Conization or extrafascial hysterectomy.

Stage IA1 with LVSI, IA2, IB1, and IIA1: Intraoperative sentinel lymph node biopsy.

- Negative: Pelvic lymphadenectomy and radical hysterectomy. If the patient has gestational desire, a radical trachelectomy can be done (not recommended for neuroendocrine tumor, tumors of >2 cm, with LVSI, distant to the internal orifice of the uterus, or remaining cervix of >1 cm).
- Positive: Consider like an advanced stage.

Recurrence risk groups

- High risk: Positive surgical margin, node affection or microscopic parametrial affection (≥1 criteria). The patient would benefit from adjuvant radiochemotherapy.
- Medium risk: Tumor of >4 cm or deeper infiltration of the stroma (≥2 criteria). The patient would benefit from adjuvant radiotherapy.
- Low risk: Any previous criteria. No adjuvant treatment.

For advanced stages of the disease, IB2, IIA2, IIB, II, IVA, and initial stages with positive sentinel node: Paraaortic lymphadenectomy and debulking of suspicious pelvic nodes of >2 cm (laparoscopic extraperitoneal dissection if possible) to determine radiation fields. After the surgery, the patient will receive radiochemotherapy and brachytherapy.

In the case of surgical contraindication, initial stages can be treated only with radiotherapy.

For stages IB2 and IIA2, neoadjuvant treatment is a good option for further surgery (radical hysterectomy or trachelectomy).

In the case of distant disease, the treatment is chemotherapy.

17.8 Prognosis

The 5-year survival rates for cervical cancer are 93% for stage IA, 80% for IB, 63% for IIA, 58% for IIB, 35% for IIIA, 32% for IIIB, 16% for IVA, and 15% for IVB, according to data from the National Cancer Data Base (Commission on Cancer of the American College of Surgeons and the American Cancer Society).

References

1. Kurman RJ (ed.). *WHO Classification of Tumours of Female Reproductive Organs.* Lyon, France: IARC Press, 2014.
2. Oncoguia SEGO. *Cáncer de cuello uterino 2013. Guías de práctica clínica en cáncer ginecológico y mamario.* Madrid, Spain: Publicaciones SEGO, Octubre 2013.
3. Colombo N, Preti E, Landoni F, Carinelli S, Colombo A, Marini C, Sessa C. Cervical cancer: ESMO Clinical Practice Guidelines for diagnosis, treatment and follow-up. ESMO Guidelines Working Group. *Ann Oncol* 2013;(Suppl 6):vi33–8.
4. Scottish Intercollegiate Guidelines Network. *Management of Cervical Cancer. Guideline 99.* Edinburgh: Scottish Intercollegiate Guidelines Network, January 2008.
5. Federation Internationale de Gynecologie et d'Obstetrique (FIGO), Committee on Gynecology Oncology. Revised FIGO staging for carcinoma of the vulva, cervix and endometrium. *Int J Gynecol Obstet* 2009;105:103–104.
6. American Joint Committee on Cancer (AJCC). http://www.cancer-staging.org. Accessed on July 3, 2016.
7. National Comprehensive Cancer Network (NCCN). *Clinical practice guidelines in oncology (NCCN guidelines®). Versión 1.2015.* Fort Washington, PA: NCCN. http://www.nccn.org. Accessed on July 3, 2016.
8. Landoni F, Maneo A, Colombo A, et al. Randomised study of radical surgery versus radiotherapy for stage IB-IIA cervical cancer. *Lancet* 1997; 350:535–40.
9. PDQ Adult Treatment Editorial Board. *Cervical Cancer Treatment (PDQ®): Health Professional Version. PDQ Cancer Information Summaries.* Bethesda, MD: National Cancer Institute, 2002.

18
Fallopian Tube, Broad Ligament, and Peritoneal Cancer

18.1 Definition

Tumors of the fallopian tube consist of (1) *epithelial tumors* (malignant [serous adenocarcinoma, mucinous adenocarcinoma, endometrioid adenocarcinoma, clear cell adenocarcinoma, transitional cell carcinoma, squamous cell carcinoma, undifferentiated carcinoma], borderline tumor of low malignant potential [serous borderline tumor, mucinous borderline tumor, endometrioid borderline tumor], carcinoma *in situ*, benign tumors [papilloma, cystadenoma, adenofibroma, cystadenofibroma, metaplastic papillary tumor, endometrioid polyp], tumorlike epithelial lesions [tubal epithelial hyperplasia, salpingitis isthmica nodosa, endosalpingiosis], (2) *mixed epithelial-mesenchymal tumors* (malignant Müllerian mixed tumor [carcinosarcoma, metaplastic carcinoma], adenosarcoma), (3) *soft tissue tumors* (leiomyosarcoma, leiomyoma), (4) *mesothelial tumors* (adenomatoid tumor), (5) *germ cell tumors* (teratoma—mature or immature), (6) *trophoblastic disease* (choriocarcinoma, placental site trophoblastic tumor, hydatidiform mole, placental site nodule), (7) *lymphoid and hematopoietic tumors* (malignant lymphoma, leukemia), and (8) *secondary tumors* [1].

Tumors of the broad ligament and other uterine ligaments include (1) *epithelial tumors of Müllerian type* (serous adenocarcinoma, endometrioid adenocarcinoma, mucinous adenocarcinoma, clear cell adenocarcinoma, borderline tumor of low malignant potential, adenoma and cystadenoma), (2) *miscellaneous tumors* (Wolffian adnexal tumor, ependymona, papillary cystadenoma [with von Hippel–Lindau syndrome], uterus-like mass, adenosarcoma), (3) *mesenchymal tumors* (malignant, benign), and (4) *secondary tumors* [1].

Tumors of the peritoneum comprise (1) *peritoneal tumors*, (2) *mesothelial tumors* (diffuse malignant mesothelioma, well-differentiated papillary mesothelioma, multicystic mesothelioma, adenomatoid tumor), (3) *smooth muscle tumor* (leiomyomatosis peritonealis disseminate), (4) *tumor of*

uncertain origin (desmoplastic small round cell tumor), and (5) *epithelial tumors* (primary peritoneal serous adenocarcinoma, primary peritoneal borderline tumor) [1].

Among primary fallopian tube cancer (PFTC), papillary serous adenocarcinoma is an epithelial tumor that is responsible for >95% of clinical cases. Another epithelial tumor (transitional cell carcinoma) and a soft tissue tumor (leiomyosarcoma) are occasionally encountered [2].

The most common solid tumor of the broad ligament is leiomyoma, which is a benign mesenchymal tumor involving smooth muscle cells. Rare primary tumors of the broad ligament include leiomyosarcoma (smooth muscle), Ewing sarcoma (bone and soft tissue), steroid cell tumor, and papillary cystadenoma (in patients with von Hippel–Lindau syndrome) [3]).

Primary peritoneal cancer (PPC) is a neoplasm in the peritoneum, with a close relationship to epithelial ovarian cancer and a tendency to occur in women [4].

18.2 Biology

Connecting the upper, outermost part of the uterus with the ovary, the fallopian tubes are a pair of thin tubes that transport a female egg released from one of the ovaries to the uterus for fertilization. Structurally, the fallopian tubes are composed of the serosa (adventitia layer), the muscularis mucosa (muscle layer), and the mucosa (containing ciliated columnar cells [~25%], secretory cells [~60%], and intercalated cells or peg cells [<10%]).

The broad ligament is a double-layered sheet of mesothelial cells that attaches the uterus, fallopian tubes, and ovaries to the pelvis. Located on each side of the fused Müllerian ducts, the broad ligament is formed by a double fold of the peritoneum.

The peritoneum is the serous membrane (containing a layer of mesothelial tissue and a thin layer of connective tissue) that lines the abdominal cavity and covers most of the intra-abdominal organs. The peritoneum serves as a conduit for the blood vessels, lymph vessels, and nerves within the abdominal organs.

PFTC usually starts as a dysplasia or carcinoma *in situ*, with subsequent transition into adenocarcinoma. PFTC may spread by local invasion, by transluminal migration, and via the lymphatics and the bloodstream to the para-aortic lymph nodes and retroperitoneum.

Broad ligament leiomyoma is a benign smooth muscle tumor, which may arise primarily from the broad ligament hormone-sensitive smooth muscle or evolve secondarily from the uterine smooth muscle and subsequently invade the broad ligament.

Peritoneal cancer is thought to arise from the mesothelium of the peritoneum (malignant mesothelioma), or from a mucus-producing tumor situated in the appendix (pseudomyxoma peritonei [PMP]). Close histological and clinical relationships between peritoneal cancer and epithelial ovarian cancer suggest their common origin. Apart from the peritoneum (30%), malignant mesothelioma also occurs on the serous surfaces of the pleura (>65%), tunica vaginalis testis, and pericardium (1%–2%) [5].

18.3 Epidemiology

PFTC makes up about 0.5% (range 0.14%–1.8%) of all malignancies of the female genital tract. The annual incidence of PFTC is 3.6 per 1 million women in the United States. PFTC commonly affects women of 40–65 years (mean age 55, peak age 60–64, range 17–88).

Broad ligament leiomyoma accounts for <1% of all leiomyomas and shows a decline after menopause.

Peritoneal cancer (e.g., malignant mesothelioma and PMP) has an annual incidence of 1–2 cases per 1 million. Malignant mesothelioma is mainly diagnosed in men of >60 years. The incidence of PMP is slightly higher in women than in men.

18.4 Pathogenesis

Chromosomal changes (BRCA1 or BRCA2 mutations) may be a contributing factor for PFTC.

Broad ligament leiomyoma is possibly associated with pregnancy, long-term use of oral contraceptives, and granulosa cell tumors of the ovary. Deletion of portions of 7q, trisomy 12, and rearrangements of 12q15, 6p21, or 10q22 may be also implicated.

Risk factors for PPC include older age, asbestos exposure, BRCA1 and BRCA2 genetic mutations, altered expressions of HER-2/neu, p53, Wilms tumor suppressor protein (WT1), and estrogen and progesterone receptors.

18.5 Clinical features

Early-stage fallopian tube cancer may be asymptomatic. As the disease advances, it may cause abdominal vaginal bleeding (especially after menopause, 47.5% of cases), abdominal pain (pain with intercourse) or pressure, abnormal vaginal discharge (white, clear, or pinkish, 20% of cases), and abdominal or pelvic mass, as well as nulliparity, subfertility, and pelvic inflammatory disease.

Broad ligament leiomyoma is asymptomatic in 50% of patients. When present, symptoms include abnormal uterine bleeding, menstrual disturbances, pelvic pain, bloating, urinary frequency, or bowel disturbances. The growing tumor may sometimes push the uterus to the contralateral side, and compress the ureter, leading to hydronephrosis.

Peritoneal cancer may manifest as abdominal pain (35% of cases), abdominal swelling (31%), anorexia, marked weight loss, ascites, shortness of breath, night sweats, hypercoagulability, fever, and intestinal obstruction.

18.6 Diagnosis

Papillary serous adenocarcinoma of the fallopian tubes is a small, solid, lobulated mass. The tumor usually appears hypointense on T1 and homogeneously hyperintense on T2, and has marked enhancement on T1 C+. Histologically, the tumor shows a filigree pattern of small, uniform, elongated, stroma-poor or stroma-free papillae (length >5× width), arising from large fibrotic, edematous, or myxoid papillary stalks or from cyst walls, or a cribriform pattern of epithelial cells lining the stalks or cyst walls, or both. The noninvasive subtype may be cribriform due to fusion, with micropapillae covered by cuboidal or columnar cells containing scant cytoplasm, variable psammoma bodies, no cilia, and no or rare mitotic figures. The invasive subtype has destructive infiltrative growth (>3 mm), with stromal invasion characterized by micropapillae and solid epithelial nests surrounded by a cleft or space and fibroblastic stroma; variable psammoma bodies; and rare or atypical mitotic figures. Useful markers for this tumor include CA 125 (serum, 80%), MOC-31 (98%), BG8 (73%), and BerEP4 (83%–100%) [6].

Broad ligament leiomyoma is a solid, well-circumscribed adnexal mass, which is separate from both the uterine body and ovary. The tumor gives iso to low signal on T1, typically a low signal on T2, and enhancement on T1 C+. Histopathologically, the tumor shows intersecting fascicles of spindle

cells with cigar-shaped blunt spindle nuclei and eosinophilic cytoplasm, and various degenerative changes (hyaline [63%], myxomatous [13%], calcified [8%], mucoid [6%], red [3%], cystic [4%], and fatty [3%]). Leiomyoma difffers from leiomyosarcoma by latter's >5 mitotic figures, hypercellularity, and nuclear atypia.

Peritoneal cancer typically presents as small nodules (2–5 mm) in the right diaphragm, small pelvis, omentum, and surface of the intestine. Malignant mesothelioma shows three basic histologic forms: epithelioid (most frequent), sarcomatoid (25%), or mixed (biphasic). Useful markers for this tumor include calretinin (85%–100%) and D2-40 (93%–96%). Pseudomyxoma peritonei (PMP) is divided into two subtypes (peritoneal adenomucinosis and peritoneal mucinous carcinoma). Typically, PMP contains the tumor cells that do not infiltrate the peritoneal and omental tissue, but rather dissect into the tissue.

18.7 Treatment

The treatment of choice for PFTC is surgery (involving abdominal total hysterectomy, bilateral salpingo-oophorectomy, omentectomy, and selective pelvic and para-aortic lymphadenectomy). Postoperative adjuvant chemotherapy consists of intravenous taxol and cisplatin (TP), cisplatin and cyclophosphamide (PC), or cisplatin, cisplatin, adriamycin, and cyclophosphamide (PAC) [7].

Broad ligament leiomyoma is treated by surgery e.g., laparotomy, laparoscopy, or hysteroscopy). Myomectomy is indicated if preservation of fertility is desired.

Peritoneal cancer may be treated with palliative or curative intent. The former consists of pain relief, removal of ascites, systemic chemotherapy (oxaliplatin, irinotecan, and 5-fluorouracil [5-FU]), surgery to resolve bowel obstruction, and nutritional support. The latter includes systemic chemotherapy, extensive surgery, and hyperthermic intraperitoneal chemotherapy (HIPEC) involving administration of heated chemotherapy into the peritoneum after surgery [4,8].

18.8 Prognosis

PFTC has a 5-year survival rate of 22%–57% (including 68%–76% for stage I disease, 27%–42% for stage II disease, and 0%–6% for stage III and IV diseases).

Broad ligament leiomyoma is a benign neoplasm with a favorable prognosis. Myomectomy for broad ligament leiomyoma is reported to result in pregnancy rates up to 70%.

Peritoneal cancer has a median survival time of 6 weeks without treatment.

References

1. Kurman RJ (ed.). *WHO Classification of Tumours of Female Reproductive Organs*. Lyon, France: IARC Press, 2014.
2. Cobb LP, Gaillard S, Wang Y, Shih IM, Secord AA. Adenocarcinoma of Mullerian origin: review of pathogenesis, molecular biology, and emerging treatment paradigms. *Gynecol Oncol Res Pract*. 2015;2:1.
3. Veloso Gomes F, Dias JL, Lucas R, Cunha TM. Primary fallopian tube carcinoma: review of MR imaging findings. *Insights Imaging*. 2015;6(4):431–9.
4. Meinhold-Heerlein I, Fotopoulou C, Harter P, et al. Statement by the Kommission ovar of the AGO: The new FIGO and WHO classifications of ovarian, fallopian tube and primary peritoneal cancer. *Geburtshilfe Frauenheilkd*. 2015;75(10):1021–7.
5. Hjerpe A, Ascoli V, Bedrossian C, et al. Guidelines for cytopathologic diagnosis of epithelioid and mixed type malignant mesothelioma. Complementary statement from the International Mesothelioma Interest Group, also endorsed by the International Academy of Cytology and the Papanicolaou Society of Cytopathology. *Cytojournal*. 2015;12:26.
6. PathologyOutlines.com. *Serous Micropapillary Carcinoma*. http://www.pathologyoutlines.com/topic/ovarytumorserousmicropapillary.html. Accessed March 5, 2017.
7. PDQ Adult Treatment Editorial Board. *Ovarian Epithelial, Fallopian Tube, and Primary Peritoneal Cancer Treatment (PDQ®): Health Professional Version. PDQ Cancer Information Summaries*. Bethesda, MD: National Cancer Institute, 2002.
8. Berek JS, Crum C, Friedlander M. Cancer of the ovary, fallopian tube, and peritoneum. *Int J Gynecol Obstetr*. 2015; 131 (suppl 2): S111–22.

19
Gestational Trophoblastic Neoplasia

Ansar Hussain, Shiekh Aejaz Aziz, Gul Mohd. Bhat, and A. R. Lone

19.1 Definition

Encompassing both benign and malignant lesions arising from the anomalous growth of trophoblastic tissue within the uterus, gestational trophoblastic diseases (GTDs) are classified into (1) trophoblastic neoplasms (choriocarcinoma, placental site trophoblastic tumor [PSTT], epithelioid trophoblastic tumor [ETT]), (2) molar pregnancies (hydatidiform mole [HM]—complete, partial, invasive, metastatic), and (3) nonneoplastic, nonmolar trophoblastic lesions (placental site nodule [PSN] and plaque, exaggerated placental site [EPS]) [1].

The most common form of GTD is HM (80% of cases), which can behave in a malignant or benign fashion. This is followed by invasive mole (chorioadenoma destruens, 10%), choriocarcinoma (5%), PSTT (1%), and ETT (<1%), which represent the malignant forms of GTD. Along with malignant HM, invasive mole, choriocarcinoma, PSTT, and ETT are capable of invading locally or metastasizing, and are collectively referred to as gestational trophoblastic neoplasia (GTN) [2].

Other uncommon benign or premalignant lesions include EPS and PSN, which are not considered part of GTN.

19.2 Biology

The placenta is a disc-shaped, fetomaternal organ of 22 cm in length, 2–2.5 cm in thickness, and 500 g in weight that attaches to the wall of the uterus, and connects with the developing fetus via the umbilical cord (55–60 cm in length). Composed of the fetal placenta (chorion frondosum, which develops from the same blastocyst that forms the fetus) and the maternal placenta (decidua basalis, which develops from the maternal uterine tissue), the placenta provides a vital source of nutrient and gas uptake for the developing fetus, and also carries away waste from the fetus.

Placenta villi are made up of three layers, with different cell types in each: (1) the outer surface is covered by villous trophoblasts (i.e., undifferentiated cytotrophoblasts and fully differentiated syncytiotrophoblasts, which are a specialized layer of epithelial cells); (2) the villous core stroma comprises mesenchymal cells, mesenchymal-derived macrophages (Hofbauer cells), and fibroblasts; and (3) the inner layer contains fetal vascular cells (e.g., vascular smooth muscle cells and pericytes) and endothelial cells.

GTD appears to evolve from defective differentiation of the trophoblast as a rare complication of pregnancy. While HM and choriocarcinoma arise from villous trophoblast, PSTT involves interstitial trophoblast. Metastases of malignant GTN may sometimes involve the lungs, lower genital tract, brain, liver, kidney, and gastrointestinal (GI) tract.

19.3 Epidemiology

The reported incidence of GTD varies from 23 cases per 100,000 pregnancies in Paraguay, to 120 cases per 100,000 pregnancies in the United States, to 1,299 cases per 100,000 pregnancies in Indonesia.

Accounting for 80% of all GTD cases, HM has an incidence of 60–110 cases per 100,000 pregnancies in North America. The incidence of choriocarcinoma, the most aggressive form of GTD, is 2–7 cases per 100,000 pregnancies in the United States.

Women >45 years of age or <16 years of age are at an increased risk of HM (or molar pregnancy).

19.4 Pathogenesis

Maternal age and history of HM (including familial recurrent hydatidiform mole [FRHM]) are implicated in the pathogenesis of GTD.

Complete HM is caused by the fertilization of an oocyte without maternal chromosomes by a haploid sperm. Subsequent duplication of paternal DNA results in an egg of exclusively parthenogenetic origin (diploid [46,XX] karyotype). In rare cases (<10%), complete HM is caused by the fertilization of an oocyte without genetic material by two separate sperm (dispermy), giving rise to an egg of exclusively androgenetic origin (46,XX or 46,XY karyotype). Partial HM is caused by the fertilization of a normal egg by two sperm, resulting in a zygote with a triploid (69,XXY or 69,XXX) diandric karyotype [2].

The incidence of GTN after complete HM ranges from 18% to 29%, in comparison with 0.5%–11% after partial HM [2].

19.5 Clinical features

Clinical symptoms of GTD consist of vaginal or uterine bleeding (84%), increased uterine volume (50%), theca lutein cysts (40%), high serum levels of β-human chorionic gonadotropin (β-hCG) (50%), early preeclampsia, hyperemesis gravidarum, and hyperthyroidism.

In metastatic cases, additional symptoms may include purple to blue-black papules or nodules in the lower genital tract, abdominal tenderness (in liver or GI metastases), abdominal guarding and rebound tenderness (hemoperitoneum), hemorrhagic shock, neurologic deficits (e.g., lethargy and coma in brain metastases), and jaundice (biliary obstruction).

19.6 Diagnosis

Diagnosis of GTN involves laboratory tests (e.g., serum quantitative hCG, complete blood count [CBC], and liver enzymes), imaging studies (pelvic ultrasonography, computerized tomography [CT], and MRI), and histological characterization.

As outlined by the International Federation of Gynecology and Obstetrics (FIGO), the criteria for GTN diagnosis include (1) a plateau in the serum β-hCG level lasting for >3 weeks (days 1, 7, 14, and 21) and an elevated serum β-hCG level for >2 weeks (days 1, 7, and 14); (2) histopathological diagnosis of choriocarcinoma; and (3) an elevated serum β-hCG level for >6 months after uterine evacuation [3].

HM is defined as products of conception that show a botryoid structure induced by abnormal trophoblast hyperplasia, stromal hypercellularity, stromal karyorrhectic debris, and collapsed villous blood vessels. It is considered malignant when the serum hCG levels plateau or rise during the follow-up period and an intervening pregnancy is excluded.

A partial or incomplete mole is an HM with a fetus or fetal tissue and a triploid karyotype; it shows patchy villous hydrops with scattered abnormally shaped irregular villi, trophoblastic pseudoinclusions, and patchy trophoblast hyperplasia. Partial moles also have marginal malignant potential (2%–3%) [3].

Invasive mole (chorioadenoma destruens) is a locally invasive, rarely metastatic lesion characterized by trophoblastic invasion of the myometrium,

hyperplasia of the cytotrophoblastic and syncytial elements, and persistence of the villous structures. It may be preceded by either complete or partial molar pregnancy, and shows more aggressive behavior than either complete or partial HMs. Resembling choriocarcinoma in histologic appearance, invasive moles are treated similarly to choriocarcinoma (with chemotherapy). However, unlike choriocarcinoma, it may regress spontaneously [3].

Choriocarcinoma is a malignant tumor of the trophoblastic epithelium that has no villi, but contains columns and sheets of trophoblasts, hemorrhage, and necrosis. It often spreads to distant sites, such as the lungs, brain, liver, pelvis, vagina, spleen, intestines, and kidney. Choriocarcinoma is aneuploid and can be heterozygous. About 50% of choriocarcinomas are preceded by an HM, 25% by an abortion, 3% by ectopic pregnancy, and 22% by a full-term pregnancy [3].

PSTT is a rare form of GTN and shows infiltration of trophoblastic cells into the myometrium without causing tissue destruction, along with vascular invasion. PSTT has lower growth rates than choriocarcinoma, and shows a delayed presentation after a full-term pregnancy. About 35% of PSTT may metastasize to the lungs, pelvis, and lymph nodes [3].

ETT (formerly atypical choriocarcinoma) is an extremely rare gestational trophoblastic tumor and appears less aggressive than choriocarcinoma and closer to PSTT clinically. Histologically, it shows a monomorphic cellular pattern of epithelioid cells. Clinically, ETT demonstrates a spectrum of behavior (from benign to malignant), and about 33% of patients develop metastases in the lungs [2,3].

The stage of GTN is determined according to the official FIGO staging of GTN as stage I (confined to the uterus), stage II (extending into the pelvis), stage III (lung and/or vagina metastases), and stage IV (other metastases, including liver, kidney, spleen, and brain) [4–6].

The prognostic score is obtained by adding the individual scores for each prognostic factor (Table 19.1), with 0–6 being low risk and ≥7 being high risk [4].

19.7 Treatment

Patients with nonmetastatic GTN or metastatic low-risk GTN are treated with single-agent chemotherapy (e.g., methotrexate or actinomycin D for patients with poor liver function) [7,8].

Table 19.1   FIGO 2000 Scoring System for GTN

Prognostic Factor	Score			
	0	1	2	4
Age (years)	<40	≥40	—	—
Antecedent pregnancy (AP)	Mole	Abortion	Term	—
Interval (end of AP to chemotherapy in months)	<4	4–6	7–12	>12
hCG (IU/L)	$<10^3$	$10^3–10^4$	$10^4–10^5$	$>10^5$
Number of metastases	0	1–4	5–8	>8
Site of metastases	Lung	Spleen, kidney	GI tract	Brain, liver
Largest tumor mass	—	3–5 cm	>5 cm	—
Prior chemotherapy	—	—	Single drug	2 drugs

Patients with high-risk metastatic GTN are subdivided into two groups. Patients with a FIGO2000 score of <7 can be treated with single-agent chemotherapy (methotrexate or actinomycin D). Patients with a FIGO2000 score of >7 have a high risk of developing resistance to single-agent systemic chemotherapy. Therefore, a polychemotherapy regimen with etoposide, methotrexate (MTX), and actinomycin D (EMA), alternating with cyclophosphamide plus vincristine (EMA/CO), as first-line therapy is recommended and well accepted [6–8].

In cases of relapse (about 20%–40% of cases), additional multiagent, platinum-based chemotherapy is administered [7,8].

19.8 Prognosis

Nonmetastatic GTN and metastatic low-risk GTN have a cure rate of close to 100% with chemotherapy treatment, while metastatic high-risk GTN has a cure rate of approximately 75% with chemotherapy treatment. After 12 months of normal hCG levels, <1% of patients with GTN have recurrences.

Fertility appears largely unaffected by methotrexate-folinic acid (*MTX-FA*) or EMA/CO treatment, as 83% of women become pregnant after such chemotherapy. Moreover, there is no obvious increase in the incidence of congenital malformations. The hCG level should be rechecked at 6 and 10 weeks after the pregnancy to ensure no recurrence or new disease.

Factors affecting the prognosis of GTN are related to histologic type (invasive mole or choriocarcinoma), extent of spread of the disease/largest tumor size, level of serum β-hCG, duration of disease from the initial pregnancy

event to start of treatment, number and specific sites of metastases, nature of antecedent pregnancy, and extent of prior treatment.

Selection of treatment depends on these factors plus the patient's desire for future pregnancies. The β-hCG is a sensitive marker to indicate the presence or absence of disease before, during, and after treatment. Given the extremely good therapeutic outcomes of most of these tumors, an important goal is to distinguish patients who need less intensive therapies from those who require more intensive regimens to achieve a cure.

References

1. Kurman RJ (ed.). *WHO Classification of Tumours of Female Reproductive Organs.* Lyon, France: IARC Press, 2014.
2. Stevens FT, Katzorke N, Tempfer C, et al. .Gestational trophoblastic disorders: An update in 2015. *Geburtshilfe Frauenheilkd* 2015;75(10):1043–50.
3. Lima LL, Parente RC, Maestá I, et al. Clinical and radiological correlations in patients with gestational trophoblastic disease. *Radiol Bras* 2016;49(4):241–50.
4. FIGO Oncology Committee. FIGO staging for gestational trophoblastic neoplasia. *Int J Gynecol Obstet* 2000;77:285–7.
5. Benedet JL, Bender H, Jones H 3rd, Ngan HY, Pecorelli S. FIGO staging classifications and clinical practice guidelines in the management of gynecologic cancers. FIGO Committee on Gynecologic Oncology. *Int J Gynaecol Obstet* 2001;70:209–62.
6. Kohorn EI. The new FIGO 2000 staging and risk factor scoring system for gestational trophoblastic disease: Description and critical assessment. *Int J Gynecol Cancer* 2001;11:73–7.
7. Berkowitz RS, Goldstein DP. Current management of gestational trophoblastic diseases. *Gynecol Oncol* 2009;112:654–62.
8. PDQ Adult Treatment Editorial Board. *Gestational Trophoblastic Disease Treatment (PDQ®): Health Professional Version. PDQ cancer Information Summaries.* Bethesda, MD: National Cancer Institute, 2002.

20
Ovarian Cancer

20.1 Definition

The ovary is affected by a diversity of primary and secondary tumors. Of these, primary tumors account for 95% of all ovarian malignancies (including surface epithelial-stromal tumors [65%], germ cell tumors [15%], sex cord-stromal tumors [10%], and miscellaneous tumors [5%]), whereas secondary tumors account for 5% of all ovarian malignancies [1].

Using histological criteria, surface epithelial-stromal tumors are divided into serous (55%), mucinous (13%), endometrioid (17%), clear cell (8%), mixed (5%), and undifferentiated and unclassified tumors (2%). Based on clinical and molecular characteristics, epithelial ovarian carcinomas (EOCs) may be separated into types 1 and 2. Type 1 tumors include low-grade serous, endometrioid, mucinous, clear cell, and transitional cell carcinomas; rarely contain TP53 mutations; often present at earlier stages; and have an indolent clinical course. Accounting for 75% of EOCs and the vast majority of ovarian cancer deaths, type 2 tumors include high-grade serous, endometrioid carcinosarcomas and undifferentiated carcinomas; harbor TP53 mutations; and behave aggressively [2–4].

Germ cell tumors comprise dysgerminoma (30%–50%), immature and mature teratoma (20%), endodermal sinus tumor (20%), yolk-sac tumor, embryonal carcinoma (rare), choriocarcinoma (very rare), and mixed germ cell tumors (10–15%; including combination of dysgerminoma and endodermal sinus tumor, or choriocarcinoma and immature teratoma, but rarely endodermal sinus tumor and emryonal carcinoma), of which dysgerminoma predominates [5].

Sex cord-stromal tumors consist of pure stromal, pure sex cord, and mixed sex cord-stromal tumors. Of these, granulosa cell tumor (GCT) within the pure sex cord tumor category is most common, representing 5% of all malignant ovarian tumors [6].

20.2 Biology

The ovaries are a pair of oval-shaped, unevenly surfaced, grayish organs of 4 cm in length, 2 cm in width, 8 mm in thickness, and about 3.5 g in

weight (during childbearing), located on either side of the uterus within the broad ligament below the uterine (fallopian) tubes. Histologically, the ovary is covered by a single-cell mesothelial layer (called the ovarian surface epithelium [OSE]; also known as the germinal epithelium of Waldeyer). Within the ovary exist the outer cortex and inner medulla. Composed of connective tissue (spindle-shaped fibroblasts, which are able to generate the female sex hormones estrogen and progesterone on demand), the outer cortex houses the follicles and oocytes. The medulla comprises primarily loose stromal tissue that includes ovarian vasculature.

The ovaries contain 1–2 million eggs (ova) at birth and undergo two distinct phases of development during a 28-day cycle. In the follicular phase (days 1–14), a follicle develops, grows, matures (containing nucleated cells called the membrana granulose), and releases an egg at the time of ovulation (around day 14); in the luteal phase (days 14–28), the remaining immature follicles degenerate up until day 28. The released egg is picked up by the fimbriae of the uterine tube, and transported toward the uterus. If fertilization does not occur, the egg degenerates and menstruation occurs. A new cycle will begin. During a woman's lifetime, only 300 eggs will ever become mature and be released for the purpose of fertilization.

Due to the presence of a wide spectrum of cell types in the ovaries, ovarian cancer is notably diverse. EOC arises from the ovarian surface epithelium (OSE), which is derived from the coelomic epithelium, not the Müllerian ducts, and which also covers the serosa of the fallopian tubes, uterus, and peritoneal cavity. Type 1 EOC evolves from atypical proliferative borderline tumors, benign cystic lesions, or endometriosis. Type 2 EOC may develop from the precursor lesion present in the fallopian tube. Interestingly, OSE expresses the mesenchymal markers vimentin and N-cadherin instead of cancer antigen 125 (CA125) or E-cadherin, which are markers of the mature, differentiated epithelium.

Sex cord-stromal tumors are derived from stromal cells (theca cells, fibroblasts, and Leydig cells) and gonadal primitive sex cords (granulosa cells and Sertoli cells), which are engaged in steroid hormone production (androgens, estrogens, and corticoids).

20.3 Epidemiology

Ovarian cancer is responsible for 3.6% of all neoplasms among women globally, and ranks fifth in incidence (after breast, colorectum, lung, and corpus uteri) and sixth in mortality among all women's neoplasms (after breast, colorectum, lung, pancreas, and stomach) in Europe.

Germ cell tumor (particularly dysgerminoma) has a mean age of 13 years at diagnosis (range 4–27 years). Adult GCT (95%) has a median age of 52 at diagnosis, while juvenile GCT (5%) usually occurs in women <30 years.

20.4 Pathogenesis

About 90%–95% of ovarian cancer is sporadic, affecting mostly older women; the remainder is linked to previous ovarian cancer, breast–ovarian cancer syndrome (mutations in the BRCA1 and BRCA2 genes), nulliparity, early menarche, and late menopause.

Type 1 EOC rarely contains TP53 mutations, but carries genetic mutations in the phosphatase and tensin homologue (PTEN), v-Ki-ras2 Kirsten rat sarcoma viral oncogene homology (KRAS), and v-raf murine sarcoma viral oncogene homologue B1 (BRAF). Type 2 EOC tends to harbor the TP53 mutation [2,3].

Dysgerminoma has gains of 1p, 6p, 12p, 12q, 15q, 20q, 21q, and 22q, as well as chromosomes 7, 8, 17, and 19, but loss of chromosome 13. About 5% of dysgerminoma occurs in women with abnormal gonads (e.g., 46XY [bilateral streak gonads], 46X/46XY [unilateral streak gonads, contralateral testis], or 46XY [testicular feminization]) [5].

Sex cord-stromal tumor may be associated with Peutz–Jeghers syndrome and mutations in the DICER1 (14q32.13) gene. Further, GCT has characteristic recurrent cytogenetic aberrations of trisomy 12 (isochromosome 12p) and 14, and monosomy 22. A FOXL2 mutation (3q23) is present in most malignant GCT [6].

20.5 Clinical features

Symptoms of ovarian cancer may range from abnormal uterine bleeding, abdominal pain (due to torsion, rupture, or hemorrhage), and fever to iso-sexual precocity in children.

Sex cord-stromal tumors of the ovary are associated with hyperestrogenic and occasionally hyperandrogenic signs.

20.6 Diagnosis

Ovarian serous tumor may be benign (60%), borderline (15%), or malignant (30%). The cystic form is called serous cystadenoma, while tumor with a prominent fibrous component is called an adenofibroma or

adenocarcinofibroma. Histologically, the tumor shows tall, columnar, ciliated epithelial cells filled with clear serous fluid, and frequent psammoma bodies (concentric calcifications). The tumor stains positive for CK7, CK8/18, CK19, EMA, B72.3, S100, amylase (25%), ER (50%), PR (50%), androgen receptor (50%), and N-cadherin, but negative for CEA, CDX2, and CK20. Cancer antigen (CA) 125 has high sensitivity and high specificity in postmenopausal women, but high sensitivity and low specificity in premenopausal women [3].

Dysgerminoma is a solid, nodular, small to huge, gray-pink mass with a smooth surface. Histologically, dysgerminoma shows nests of round and ovoid cells separated by fibrous stroma with T lymphocytes. The tumor stains positive for OCT4 (strong nuclear staining in 90% cells), c-kit (87%), Cam5.2 (20%), and AE1-AE3 (8%), but negative for CK7, CK20, EMA, HMW keratin, CD30, and vimentin. Molecularly, 12p abnormalities are detected in 81% [5].

Granulosa cell tumor (GCT) of the pure sex cord tumor category typically presents as unilateral, gray or yellow, solid to cystic masses (average 12 cm) with a smooth lobulated surface. Histologically, GCT includes small, bland, cuboidal to polygonal cells with a coffee bean–like longitudinal nuclear groove and a microfollicular structure (the Call–Exner body). GCT demonstrates microfollicular, macrofollicular, trabecular, insular, diffuse, and watered-silk (gyriform) growth patterns. Microfollicular GCT contains pathognomonic Call–Exner bodies (small rings of granulosa cells surrounding eosinophilic fluid and basement membrane material). Macrofollicular GCT contains one or more large cysts lined with granulosa cells. Trabecular and insular GCT have granulosa cells organized into nests and bands, with an intervening fibrothecomatous stroma present in trabecular GCT. Diffuse GCT contains sheets of cells arranged in no pattern. Watered-silk GCT contains cells arranged in single file in line. Adult GCT displays large, pale, ovoid or angular nuclei with nuclear grooves; mild nuclear atypia; few mitotic figures; little cytoplasm; and occasional luteinization. In contrast, juvenile GCT shows round hyperchromatic nuclei without nuclear grooves, severe nuclear atypia, more mitotic figures, and more cytoplasm (which is dense). GCT stains positive for inhibin-α, vimentin, calretinin, CD99, smooth muscle actin, desmoplakin, S100 (50%), keratin (30%–50%), anti-Müllerian hormone (focally), desmin (35%), and silver stains (reticulin surrounding cluster of cells), but negative for EMA. Molecularly, GCT shows monosomy 22 (~40%), trisomy 12 (~30%), trisomy 14 (~30%), monosomy X (~10%), and monosomy 17 (5%), although most tumors (80%) are diploid or near diploid [6].

Staging of ovarian cancer is based on the TNM (tumor, node, metastasis) or the International Federation of Gynecology and Obstetrics (FIGO) system. In the FIGO system, stage I describes tumors confined to ovaries, stage II has pelvic extension or primary peritoneal cancer, stage III indicates spread to the peritoneum outside the pelvis and/or metastasis to the retroperitoneal lymph nodes, and stage IV contains distant metastasis [4].

20.7 Treatment

Stage I OEC is treated with surgery alone (e.g., total abdominal hysterectomy, bilateral salpingo-oophorectomy); later-stage OEC is treated with surgery and chemotherapy (e.g., cisplatin or carboplatin with paclitaxel) [7,8].

Malignant ovarian germ cell tumors (e.g., dysgerminoma, endodermal sinus tumor, embryonal carcinoma) are generally treated with surgery and chemotherapy, but not radiotherapy. For conservation of fertility, only unilateral oophorectomy and surgical staging are carried out. Current use of bleomycin, etoposide, and cisplatin (BEP) leads to a 5-year survival rate of up to 100% for dysgerminomas and 85% for nondysgerminomatous tumors.

Granulosa cell tumor (GCT) is treated with surgery (transabdominal hysterectomy or bilateral salpingo-oophorectomy for women beyond childbearing age, and unilateral oophorectomy for younger women), in combination with chemotherapy when appropriate PVB (cisplatin + vinblastin + bleomycin) and BEP (bleomycin + etoposide + cisplatin) are considered effective chemotherapies for patients with residual foci, a high risk of recurrence (tumor rupture, stage Ic or greater, a poorly differentiated tumor, or a tumor diameter of >10–15 cm); or actual recurrence.

20.8 Prognosis

The prognosis of ovarian cancer is poor, with a 5-year overall survival rate of only 45%, given that a majority of ovarian cancer is diagnosed in advanced stages, which are often refractory to the current intervention measures.

The 5-year survival rates for dysgerminomas and nondysgerminomatous tumors are 100% and 85%, respectively, following BEP chemotherapy. Poor prognostic indicators for ovarian germ cell tumors include large size, unfavorable histological type, advanced stage at presentation, and elevation of both AFP and hCG levels.

Patients with stage I or II GCT have a 5-year survival rate of 95%, and those with stage III or IV GCT have a rate of 59%. Juvenile GCT may recur within 3 years and is rapidly fatal.

References

1. Kurman RJ (ed.). *WHO Classification of Tumours of Female Reproductive Organs.* Lyon, France: IARC Press, 2014.
2. Ramalingam P. Morphologic, immunophenotypic, and molecular features of epithelial ovarian cancer. *Oncology (Williston Park)* 2016;30(2):166–76.
3. Tanaka YO, Okada S, Satoh T, et al. Differentiation of epithelial ovarian cancer subtypes by use of imaging and clinical data: A detailed analysis. *Cancer Imaging* 2016;16:3.
4. Foti PV, Attinà G, Spadola S, et al. MR imaging of ovarian masses: Classification and differential diagnosis. *Insights Imaging* 2016;7(1):21–41.
5. Low JJ, Ilancheran A, Ng JS. Malignant ovarian germ-cell tumours. *Best Pract Res Clin Obstet Gynaecol* 2012;26(3):347–55.
6. Horta M, Cunha TM. Sex cord-stromal tumors of the ovary: A comprehensive review and update for radiologists. *Diagn Interv Radiol* 2015;21(4):277–86.
7. PDQ Adult Treatment Editorial Board. *Ovarian Epithelial, Fallopian Tube, and Primary Peritoneal Cancer Treatment (PDQ®): Health Professional Version. PDQ Cancer Information Summaries.* Bethesda, MD: National Cancer Institute, 2002.
8. Berek JS, Crum C, Friedlander M. Cancer of the ovary, fallopian tube, and peritoneum. *Int J Gynecol Obstetr* 2015; 131 (suppl 2): S111–22.

21
Penile Cancer

Andrew R. Leone, Nicole Abdo, Gregory J. Diorio, and Philippe E. Spiess

21.1 Definition

Tumors of the penis comprise several histologic types, ranging from squamous cell carcinoma (SCC), melanoma, sarcomas, and lymphomas to secondary metastasis.

Accounting for >95% of the clinical cases, penile SCC consists of several common subtypes, including usual, basaloid, warty, verrucous, papillary, sarcomatoid, adenosquamous, and mixed. Basaloid, sarcomatoid, and adenosquamous subtypes are more aggressive (high grade and invasive) than verrucous, papillary, and warty subtypes (low grade and superficial involvement) [1]. While basaloid and warty subtypes demonstrate the highest percentages of human papillomavirus (HPV) DNA, keratinizing SCC and verrucous subtypes contain the lowest percentage [2].

In addition, basal cell carcinoma (BCC), melanoma, sarcoma, and lymphoma, as well as secondary metastasis (usually originating from prostatic and colorectal primary tumors), are responsible for the remaining malignant lesions of the penis [1]. Although Kaposi sarcoma rarely presents on the penis, case reports of this occurrence even in HIV-negative males have been published. Thus, Kaposi sarcoma should be considered in the differential diagnosis [3].

21.2 Biology

Located in the pubic region superior to the scrotum and inferior to the umbilicus along the body's midline, the penis is the male external excretory and sex organ that has three main regions: the root (connecting the penis to the bones of the pelvis via several tough ligaments), body (large masses of erectile tissue), and glans (containing the urethral orifice for semen and urine to exit the body).

The outside of the penis is covered with skin. Underneath the skin is a layer of subcutaneous tissue containing blood vessels and protein fibers.

Further down is a tough and elastic layer of fibrous connective tissue (called the tunica albuginea). Inside the tunica albuginea are three columnar masses of erectile tissue (two corpora cavernosa and corpus spongiosum), which surround the urethra, in addition to the columns' enveloping fascial layers, nerves, lymphatics, and blood vessels.

Penile SCC typically arises from the epithelium in the inner prepuce or the glans, and behaves similarly to SCC at other sites. such as the anus, oropharynx, and female genitalia [1]. Malignancy may evolve through precursor lesions. such as penile intraepithelial neoplasia (PeIN) or carcinoma *in situ* (CIS). Although both pathways are associated with HPV, most PeIN lesions contain HPV DNA, while most CIS originates from inflammatory conditions, such as phimosis, balanitis, and lichen sclerosis [4].

21.3 Epidemiology

Penile carcinoma is a rare malignancy in North America and Europe, but its devastating clinical course and challenging management have raised concern particularly in South America, Asia, and Africa [5]. In the United States, Hispanic males demonstrate the highest incidence, followed by African-American males, white non-Hispanic males, Native American males, and Asians or Pacific Islanders, in order of decreasing incidence. The incidence of penile cancer increases with age, and while the disease does occur in young men, its incidence reaches its peak during the sixth decade of life [1]. The median age at diagnosis is 60–62 years, and the majority is diagnosed at the stage of local lymph node involvement. Indeed, Hispanic and African-American males are diagnosed with more advanced systemic lymph node involvement [4].

21.4 Pathogenesis

Major risk factors for the development of penile cancer include premalignant lesions, chronic inflammation, poor hygiene, low socioeconomic status, photochemotherapy (PUVA), tobacco use, lack of circumcision, and HPV infection [1,4].

Two main mechanisms of pathogenesis leading to precursor lesions (PeIN and CIS) and then penile cancer are inflammation and infection (HPV, human herpesvirus [HHV], and HIV) (Table 21.1). As HPV infections are becoming more prevalent, particularly in the young sexually active population, the relevance of HPV as it relates to the development of penile SCC has provided insight into its pathogenesis [1,4].

Table 21.1   Precursor Lesions to Penile Cancer

Origin	Precursor Lesion	Associated Malignancy
Inflammatory	Bowen disease	Invasive SCC
	Erythroplasia of Queyrat	Invasive SCC
	Bowenoid papulosis	SCC sporadically
	Phimosis	Invasive SCC
	Balanitis	Invasive SCC
	Lichen sclerosis	SCC sporadically
	Paget disease	Invasive SCC
Infectious	HPV	Invasive SCC
	HHV-8	Kaposi sarcoma
	HIV	Kaposi sarcoma
Neoplastic	Buschke–Lowenstein tumor (HPV origin)	Invasive SCC
	Intraepithelial neoplasia grade III	Invasive SCC

Source: Hakenberg OW, et al., *Eur. Urol.*, 2014, 1.

Chronic inflammatory conditions (e.g., phimosis, balanitis, and lichen sclerosis) demonstrate varying degrees of invasiveness [1]. Phimosis is highly correlated with the development of invasive penile cancer, due to its associated risk of chronic infection [1]. In contrast, while there is a significant connection between the incidence of lichen sclerosis in males and penile cancer, this inflammatory condition does not increase the rate of penile carcinogenesis [1].

Although HPV types 6 and 11 are considered "low risk," they can escalate the formation of giant condyloma, or "Buschke–Lowenstein" tumor. HPV types 16, 18, 31, and 33 are recognized as "high risk," with the ability to cause premalignant and malignant lesions [1,4].

Penile CIS may arise as cutaneous lesions, such as erythroplasia of Queyrat or Bowen disease, and up to one-third of these lesions transform into invasive SCC [1,4]. Erythroplasia of Queyrat presents as erythematous, nontender, solitary or multiple plaques on the glans penis or foreskin, usually in uncircumcised men in their fifties, and ulceration of these plaques correlates with progression from CIS to invasive SCC [4].

21.5 Clinical features

The median age at presentation is 55–60 years, and disease usually manifests as a tumor, ulcer, or inflammatory lesion on the glans penis, but the prepuce, coronal sulcus, and shaft may also be affected [5,6]. In general,

penile carcinoma is an aggressive malignancy due to its early lymph node involvement and potential for systemic dissemination, but inguinal lymphadenopathy on physical exam demonstrates low positive and negative predictive values [4]. Thus, although 72% of patients with penile carcinoma have lymphadenopathy on physical exam, the absence of palpable lymph nodes does not exclude a diagnosis of penile carcinoma with lymph node involvement [1,4].

21.6 Diagnosis

Diagnosis of penile cancer involves a history review and physical exam (for visible lesion and palpable inguinal lymphadenopathy), radiographic imaging (computerized tomography [CT], MRI, and fluorodeoxyglucose [FDG]–positron emission tomography [PET]), and biopsy (e.g., fine-needle aspiration [FNA] and excisional biopsy).

If FNA is negative, the National Comprehensive Cancer Network (NCCN)© recommends excisional biopsy in order to address a possible sampling error, and if the FNA is positive, full lymph node dissection is recommended. CT scan is the standard method for staging penile tumors presenting as >T1, and abdominal and chest CT scan is recommended for poorly differentiated tumors or those with >N2 stage. MRI and FDG-PET can be used to detect lymph node metastasis as well, but pelvic CT scan remains the standard modality for staging. In patients without palpable inguinal lymphadenopathy, the NCCN protocol suggests surveillance for low-risk (<T1G1) patients and sentinel lymph node biopsy in high-risk (>T1G1) patients [7]. Dynamic sentinel node biopsy (DSNB) offers staging in patients with non-palpable inguinal lymph nodes who still exhibit poor prognostic factors within the primary tumor. DSNB is a low-morbidity surgical staging technique that localizes the sentinel node via visual or gamma emission. This process consists of injecting blue dye or technetium-labeled colloid next to a lesion and allowing the lymphatic system to transport the tracer to a specific node.

Macroscopically, a penile tumor is a slow-growing, irregular mass of 2–5 cm located on the glans or inner foreskin near the coronal sulcus. The tumor may show superficial spreading, vertical growth (basaloid and sarcomatoid carcinomas), verruciform (verrucous, warty, papillary, or cuniculatum carcinomas), or mixed patterns. Microscopically, the tumor is usually keratinized with moderate differentiation (e.g., intraepithelial neoplasia and squamous hyperplasia), and often has variable stromal lymphoplasmacytic infiltrate and foreign-body-type giant cells (particularly in highly keratinized tumors). The tumor stains positive for epidermal growth factor receptor (EGFR).

The stage of penile cancer is determined as 0, I, II, III, or IV on the basis of the American Joint Committee on Cancer (AJCC) TNM system, which incorporates features related to tumor (TX, T0, Tis, Ta, T1a, T1b, T2, T3, T4), lymph node (NX, N0, N1, N2, N3), and metastasis (M0, M1). Specifically, stage 0 is defined as Tis or Ta/N0/M0; stage I as T1a/N0/M0; stage II as T1b, T2, or T3/N0/M0; stage IIIa as T1, T2, or T3/N1/M0; stage IIIb as T1, T2, or T3/N2/M0; and stage IV as T4/any N/M0, or any T/N3/M0, or any T/any N/M1 [4].

21.7 Treatment

Guided by the TNM staging, treatment for penile cancer consists of excision of a primary lesion, and if metastatic disease is present, chemotherapy, radiotherapy, and surgery, as well as palliation [5–8].

For small, low-grade primary tumors (CIS, Ta, T1) and high-grade, invasive tumors (T1b, T2–T3), penile-sparing treatment options, such as wide local excision, circumcision, glansectomy, glans resurfacing, laser ablation, Mohs microsurgery, partial amputation, and topical therapy, are preferred. Topical chemotherapy consisting of 5-fluorouracil and imiquimod, combined with circumcision, is reported to be effective first-line treatment for penile CIS [1]. Topical treatment requires daily application for 3-4 weeks in order to achieve a sequence of inflammation, sloughing, and then healing.

For metastatic disease, lymphadenectomy is the appropriate protocol for palpable inguinal lymph nodes or nonpalpable inguinal lymph nodes with a concerning primary tumor, and the extent of the procedure (inguinal lymph node dissection vs. pelvic lymph node dissection) is determined by the number of nodes involved. Patients with local metastasis who display clinical response to chemotherapy may be treated with neoadjuvant chemotherapy (e.g., cisplatin, paclitaxel, 5-fluorouracil, irinotecan, docetaxel, or ifosfamide), followed by surgery. Radiotherapy (via external beam or brachytherapy) is appropriate for treating primary tumors of stages T1 and T2 with involvement of the glans penis and <4 cm of the coronal sulcus. Neoadjuvant or adjuvant radiation therapy has a palliative role in node-positive patients, but it does not necessarily improve survival.

21.8 Prognosis

The most predictive prognostic factor for penile cancer relates to inguinal and pelvic lymph node involvement, as penile metastases yield a poor prognosis. The 3-year disease-specific survival rate of penile cancer is 89.6%

(75.9–100) for pN1, 65.9% (51.0–80.8) for pN2, 33.6% (13.4–53.8) for pN3, 86.7% (76.5–96.9) for pN+ unilateral, and 36.8% (21.5–52.1) for pN+ bilateral.

The 10-year survival rates for verrucous, adenosquamous, mixed, papillary, and warty carcinomas are much higher (100%, 100%, 97%, 92%, and 90%, respectively) than those for the usual and basaloid types (78% and 76%, respectively), and sarcomatoid tumors carry the worst prognosis, as 75% of patients die within a year of diagnosis.

References

1. Hakenberg OW, Comperat E, Minhas S, Necchi A, Protzel C, Watkin N. Guidelines on penile cancer. *Eur Urol* 2014:1–38.
2. La-Touche S, Lemetre C, Lambros M, et al. DNA copy number aberrations, and human papillomavirus status in penile carcinoma. Clinicopathological correlations and potential driver genes. *PLoS One* 2016;11(2).
3. Buechner SA. Common skin disorders of the penis. *BJU Int* 2002;90(5):498–506.
4. Lynch D, Pettaway C. Tumors of the penis. In *Campbell's Urology*, ed. PC Walsh, AB Retik, ED Vaughan Jr. 8th ed. Philadelphia, PA: Saunders, 2002.
5. PDQ Adult Treatment Editorial Board. *Penile Cancer Treatment (PDQ®): Health Professional Version. PDQ Cancer Information Summaries.* Bethesda, MD: National Cancer Institute, 2002.
6. Diorio GJ, Leone AR, Spiess PE. Management of penile cancer. *Urology* 2016;96:15–21.
7. (NCCN) National Comprehensive Cancer Network. *Clinical Practice Guidelines in Oncology: Penile Cancer.* Version 1.2014. Fort Washington, PA: NCCN.
8. Sonpavde G, Pagliaro LC, Buonerba C, Dorff TB, Lee RJ, Di Lorenzo G. Penile cancer: Current therapy and future directions. *Ann Oncol* 2013;24(5):1179–89.

22
Prostatic Cancer

22.1 Definition

The prostate is affected by primary tumors of six histological categories: (1) *epithelial tumors* (glandular neoplasms [acinar adenocarcinoma—atrophic, pseudohyperplastic, foamy, colloid, signet ring, oncocytic, lymphoepithelioma-like; carcinoma with spindle cell differentiation—carcinosarcoma, sarcomatoid carcinoma; prostatic intraepithelial neoplasia [PIN]—PIN grade III; ductal adenocarcinoma—cribriform, papillary, solid]; urothelial tumors [urothelial carcinoma]; squamous tumors [adenosquamous carcinoma, squamous cell carcinoma]; basal cell tumors [basal cell adenoma, basal cell carcinoma]), (2) *neuroendocrine tumors* (endocrine differentiation within adenocarcinoma, carcinoid tumor, small cell carcinoma, paraganglioma, neuroblastoma), (3) *prostatic stromal tumors* (stromal tumor of uncertain malignant potential, stromal sarcoma), (4) *mesenchymal tumors* (leiomyosarcoma, rhabdomyosarcoma, chondrosarcoma, angiosarcoma, malignant fibrous histiocytoma, malignant peripheral nerve sheath tumor, hemangioma, chondroma, leiomyoma, granular cell tumor, hemangiopericytoma, solitary fibrous tumor), (5) *hematolymphoid tumors* (lymphoma, leukemia), and (6) *miscellaneous tumors* (cystadenoma, nephroblastoma [Wilms tumor], rhabdoid tumor, germ cell tumors [yolk sac tumor, seminoma, embryonal carcinoma and teratoma, choriocarcinoma], clear cell carcinoma, melanoma), in addition to a number of metastatic tumors [1].

The most common primary cancer of the prostate is acinar adenocarcinoma, which accounts for about 95% of all prostatic malignancies and is thus called prostatic or prostate cancer (carcinoma). Other rare tumors of the prostate include ductal carcinoma (4%), urothelial carcinoma, squamous cell carcinoma, and small cell carcinoma [2].

22.2 Biology

Located posterior to the pubic symphysis, superior to the perineal membrane, inferior to the bladder, and anterior to the rectum, the prostate is a walnut-sized, conical organ ($4 \times 3 \times 2$ cm in dimension and 20 g in weight) surrounding

the urethra that empties urine from the bladder. The prostate is enclosed by a capsule composed of collagen, elastin, and smooth muscle, and includes a base, an apex, an anterior, a posterior, and two lateral surfaces. The anterior, lateral, and posterior aspects are covered by three distinct layers of fascia.

Historically, the prostate consists of approximately 70% glandular tissue and 30% fibromuscular stroma, and is divided into three zones: (1) transition zone (two pear-shaped lobes surrounding the proximal urethra, accounting for 5% of the glandular tissue and 10% of prostate cancer, often ductal carcinoma), (2) central zone (surrounding the transition zone to the angle of the urethra to the bladder base, accounting for 25% of the glandular tissue and 5% of prostate cancer), and (3) peripheral zone (from the apex posterior to the base, surrounds the transition and central zones, accounting for 70% of the glandular tissue and 80% of prostate cancer, including 75% adeno-carcinoma). The prostate contains several cell types, such as secretory (along the glandular lumen), basal (low cuboidal epithelium and columnar mucus-secreting cells from the basement membrane), neuroendocrine (irregularly distributed), and urothelial (lining proximal 2 mm of prostatic ducts) cells.

The main function of the prostate is to produce fluid for semen, which transports sperm during the male orgasm.

Adenocarcinoma mostly arises from the glands and ducts in the prostate, with 75% occurring in the peripheral zone, 15% in the central zone, and 10% in the transition zone.

22.3 Epidemiology

Prostate cancer is the second most frequent male malignancy (after skin cancer), and the fifth most common cancer worldwide. The incidence rates during 2008–2012 were 131.5 cases per 100,000 and 21.4 deaths per 100,000. African-Americans have a higher incidence of prostate cancer than the Caucasian population. The median age at diagnosis is 66 years.

22.4 Pathogenesis

Risk factors for prostate cancer are advancing age (>50 years), ethnicity or race, family history, and possibly diet.

Molecularly, prostate cancer is associated with somatic copy number alterations (e.g., gains of Xp11.22–q13.1, 1p12–q43, 1q32.1–q32.3, 3q26.1, 7p22.3–q36.3, 8p12–q24.3, and 9q31.3, and losses of 6q14.3–15, 8p23–p11, 10p13, 10q11.21, 10q22–q24, 11p13–p12,

12p13,   13q12.3–q14.2,   15q25.1–q26.3,   16q11.2–q24.3,   17p13.1,
17p13.3-p11.2, 17q21.31, 17q24.2–q25.3, 18q22.3, and 21q22.3), point
mutations (*TP53, PTEN, RB1*, BRCA2, *PIK3CA, KRAS,* and *BRAF*), and
structural rearrangements (e.g., *TMPRSS2:ERG*) [3–5].

Occurring in almost 50% of all primary prostate tumors, *TMPRSS2:ERG*
rearrangement places the growth-promoting activity of the *ERG* oncogene under the control of the regulatory elements of androgen-responsive
TMPRSS2, and results in new fusion protein with altered function.

In addition, prostate cancer often overexpresses prostate specific antigen
(PSA; also called kallikrein III, seminin, seminogelase, γ-seminoprotein and
antigen P-30; a 34 kD glycoprotein with serine-protease activity encoded by
the gene localized on chromosome 19q13), and androgen receptor.

22.5 Clinical features

Clinical presentations of prostate cancer include no symptoms (47%),
urinary frequency (38%), urinary urgency (10%), decreased urine stream
(23%), hematuria (1.4%), nocturia, erectile dysfunction, pain or burning
during urination, discomfort when sitting, pain in back and bone, fatigue,
and weight loss.

22.6 Diagnosis

Diagnosis of prostate cancer involves digital rectal exam (DRE), transurethral ultrasound (with a sensitivity of 70%), prostate-specific antigen (PSA)
detection (above 4 ng/dL or increasing over time), and histological confirmation as well as molecular urine assay.

Prostatic acinar adenocarcinoma is a gritty, firm, gray-yellow, poorly delim-
ited, easily felt mass. Histologically, acinar adenocarcinomas show cuboidal
and/or columnar-shaped malignant cells in the neoplastic tissue forming
acini and tubules; finely granular cytoplasm, nuclear enlargement, hyperchromasia, and prominent nucleoli (>3 μm); rare mitotic figures; perineural
invasion; glomerulation; and mucinous fibroplasia [6].

Immunohistochemically, prostatic acinar adenocarcinoma is positive for
PSA (a Kallikrein-related serine protease produced by secretory epithelium
and drained into the ductal system), prostatic acid phosphatase (PAP),
α-methylacyl coenzyme A (CoA) reductase (AMACR), androgen receptor
(AR), and cytokeratin 7 (CK7), but negative for proto-oncogene (p63), high-
molecular-weight CK (clone 34βE12), and CK5/6 [6].

Among various subtypes, ductal adenocarcinoma (formerly endometrioid carcinoma of the prostate) shows a papillary or cribriform pattern with slit-like lumina (86%) or discrete glands lined by tall, pseudostratified epithelium with abundant, amphophilic cytoplasm (14%); pale or clear cytoplasm; and stromal fibrosis (67%), and stains positive for PSA and PAP.

Atrophic adenocarcinoma contains round acini, scant cytoplasm, nuclear enlargement, prominent nucleoli, and an infiltrative growth pattern.

Foamy gland adenocarcinoma shows abundant foamy cytoplasm, small hyperchromatic nuclei, minimal or no atypia, and pink luminal secretions, and stains positive for colloidal iron, Alcian blue, and P504S, but negative for mucicarmine, PAS, and lipid.

Pseudohyperplastic adenocarcinoma shows papillary infoldings (100%), crowded glands, large atypical glands (95%), nuclear enlargement (95%), pink amorphous secretions (70%), occasional to frequent nucleoli (45%), branching (45%), crystalloids (45%), and corpora amylacea (20%). It stains positive for P504S (70% of cases).

Mucinous (colloid) adenocarcinoma contains tumor cells floating in pools of mucin; has microglandular, cribriform, comedo, solid, and hypernephroid patterns; and stains positive for PSA, PAP, and MUC2. Signet ring carcinoma displays solid, acinar, single-line patterns [6].

It should be noted that variants of prostate cancer (e.g., pseudohyperplastic, foamy gland, atrophic, microcystic, with Paneth cell-like changes, with collagenous micronodules, with glomeruloid formations, oncocytic) do not have any known prognostic significance, which relies on the Gleason system or the National Comprehensive Cancer Network (NCCN) guideline for determination (see below).

Differential diagnoses for prostate cancer include benign lesions and patterns (e.g., partial atrophy, postatrophic hyperplasia, basal cell hyperplasia, clear cell cribriform hyperplasia, adenosis or atypical adenomatous hyperplasia, sclerosing adenosis, nephrogenic adenoma, seminal vesicle and ejaculatory duct, and Cowper gland) [7].

The stage of prostate cancer is often determined according to the TNM system of AJCC. While the clinical stage is based on the pre-surgery results of DRE, biopsy, x-rays, CT and/or MRI as well as PSA level and size of cancer, the pathologic stage is ascertained with information found during surgery, plus laboratory/pathology results (see TNM in Glossary).

The grade of prostate cancer is also assessed by using the Gleason scoring system. On the basis of how much the cancer looks like healthy tissue under a microscope, the Gleason scoring system classifies prostate cancer into Gleason group I (former Gleason 6, with cancer cells being well differentiated, and looking similar to healthy cells), Gleason group II (former Gleason 3 + 4 = 7, with cancer cells being moderately differentiated, and looking somewhat similar to healthy cells), Gleason group III (former Gleason 4 + 3 = 7, with cancer cells being moderately differentiated, and looking somewhat similar to healthy cells), Gleason group IV (former Gleason 8, with cancer cells being poorly differentiated or undifferentiated, and looking very different from healthy cells), and Gleason group V (former Gleason 9 or 10, with cancer cells being poorly differentiated or undifferentiated, and looking very different from healthy cells).

To help plan the best treatment and predict how successful treatment will be, prostate cancer may be assessed as very low risk, low risk, intermediate risk, high risk or very high risk in accordance with the NCCN guideline, which takes into account of PSA level, prostate size, needle biopsy findings, and the stage of cancer.

22.7 Treatment

Treatment options for prostate cancer are watchful waiting (for low-grade, localized tumors or limited life expectancy), radical prostatectomy (not warranted if positive pelvic nodes), external beam radiation therapy/brachytherapy, focal laser ablation, cryotherapy, androgen deprivation therapy, chemotherapy, and combination therapy. External beam radiotherapy chemotherapy, and brachytherapy are potentially curative for prostate cancer [8].

In recent years, immune-based treatment strategies are increasingly applied to both advanced and recurrent prostate cancer. These range from passive (utilizing either immune check point blockade or specific monoclonal antibodies) to active immnotherapy (via tumor specific antigen vaccination) [9].

22.8 Prognosis

The 5-year relative survival rate for patients with local or regional prostate cancer is 80%–100%, that for patients with locally advanced disease is 75%, and that for patients with distant disease is 28%. Recurrence may occur 40 months after radical prostatectomy.

Unfavorable prognostic factors are a Gleason score of >6, PSA of >40 ng/mL, stage 3 or higher, Caucasian, patients of <20 years, and poor response to treatment. A tissue-based neuropeptide called Pro-NPY (pro-neuropeptide) has been recently identified as a useful prognostic indicator for prostate cancer [10].

References

1. Eble JN (ed.). *Pathology and Genetics of Tumours of the Urinary System and Male Genital Organs*. Lyon, France: IARC Press, 2004.
2. Chang AJ, Autio KA, Roach M 3rd, Scher HI. High-risk prostate cancer-classification and therapy. *Nat Rev Clin Oncol* 2014;11(6):308–23.
3. Schoenborn JR, Nelson P, Fang M. Genomic profiling defines subtypes of prostate cancer with the potential for therapeutic stratification. *Clin Cancer Res* 2013;19(15):4058–66.
4. Ahmed M, Eeles R. Germline genetic profiling in prostate cancer: Latest developments and potential clinical applications. *Future Sci OA* 2015;2(1):FSO87.
5. Sharma P, Zargar-Shoshtari K, Pow-Sang JM. Biomarkers for prostate cancer: Present challenges and future opportunities. *Future Sci OA* 2015;2(1):FSO72.
6. PathologyOutlines.com. *Adenocarcinoma of Peripheral Ducts and Acini*. http://www.pathologyoutlines.com/topic/prostateadenoNOS. html. Accessed March 8, 2017.
7. Stanford School of Medicine. *Prostatic Adenocarcinoma. Surgical Pathology Criteria*. Stanford, CA: Stanford School of Medicine. http:// surgpathcriteria.stanford.edu/prostate/adenocarcinoma/printable. html. Accessed March 8, 2017.
8. PDQ Adult Treatment Editorial Board. *Prostate Cancer Treatment (PDQ®): Health Professional Version. PDQ Cancer Information Summaries*. Bethesda, MD: National Cancer Institute, 2002.
9. Alberti C. Prostate cancer immunotherapy, particularly in combination with androgen deprivation or radiation treatment. Customized pharmacogenomic approaches to overcome immunotherapy cancer resistance. *G Chir* 2017;37(5):225–35.
10. Grozescu T, Popa F. Prostate cancer between prognosis and adequate/ proper therapy. *J Med Life* 2017;10(1):5–12.

23
Renal Cancer

23.1 Definition

Tumors of the kidney come under nine histological categories: (1) *renal cell tumors* (clear cell renal cell carcinoma [ccRCC], multilocular ccRCC, papillary renal cell carcinoma [pRCC], chromophobe renal cell carcinoma [crRCC], carcinoma of the collecting ducts of Bellini [cdRCC], renal medullary carcinoma, Xp11 translocation carcinomas, carcinoma associated with neuroblastoma, mucinous tubular and spindle cell carcinoma, renal cell carcinoma [RCC] unclassified, papillary adenoma, oncocytoma), (2) *metanephric tumors* (metanephric adenoma, metanephric adenofibroma, metanephric stromal tumor), (3) *nephroblastic tumors* (nephrogenic rests, nephroblastoma, cystic partially differentiated nephroblastoma), (4) *mesenchymal tumors* (occurring mainly in children [clear cell sarcoma, rhabdoid tumor, congenital mesoblastic nephroma, ossifying renal tumor of infants], occurring mainly in adults [leiomyosarcoma—including renal vein, angiosarcoma, rhabdomyosarcoma, malignant fibrous histiocytoma, osteosarcoma, angiomyolipoma and epithelioid angiomyolipoma, leiomyoma, hemangioma, lymphangioma, juxtaglomerular cell tumor, renomedullary interstitial cell tumor, schwannoma, solitary fibrous tumor]), (5) *mixed mesenchymal and epithelial tumors* (cystic nephroma, mixed epithelial and stromal tumor, synovial sarcoma), (6) *neuroendocrine tumors* (carcinoid, neuroendocrine carcinoma, primitive neuroectodermal tumor, neuroblastoma, pheochromocytoma), (7) *hematopoietic and lymphoid tumors* (lymphoma, leukemia, plasmacytoma), (8) *germ cell tumors* (teratoma, choriocarcinoma), and (9) *metastatic tumors* [1].

Accounting for approximately 90% of kidney malignancies, renal cell tumors (carcinomas and cancers) comprise a diverse group of solid tumors originating from renal parenchyma. Common RCC histological subtypes include ccRCC (75%), pRCC (10%), crRCC (5%), and cdRCC (1%). In children, 90% of renal cancer cases are due to Wilms tumor (WT) (or nephroblastoma) [2,3].

23.2 Biology

Located along the posterior muscular wall of the abdominal cavity, the kidneys are a pair of bean-shaped organs of 10–12 cm in length, 5–7 cm in width,

2–3 cm in thickness, and 135–150 g in weight, which are covered by a thin layer of fibrous connective tissue (the renal capsule). Underneath the renal capsule is the soft, dense, vascular renal cortex, and further down is the renal medulla, which consists of seven cone-shaped renal pyramids separated by the cortical tissue called renal columns (of Bertin). The bases of the renal pyramids are oriented outward toward the renal cortex, and their apexes point inward toward the center of the kidney. Each apex connects to a minor calyx (a small hollow tube), which merges to form three larger major calyces, and then form the hollow renal pelvis at the center of the kidney. The renal pelvis exits the kidney at the renal hilus (on the concave side), where urine drains into the ureter.

Structurally, each kidney contains around 1 million individual nephrons (the functional units), which are made of renal corpuscle and renal tubule. The renal corpuscle comprises the capillaries of the glomerulus that is surrounded by the glomerular capsule (or Bowman capsule, a cup-shaped double layer of simple squamous epithelium with a hollow space between the layers). The glomerulus contains podocytes and a basement membrane, allowing water and certain solutes to be filtered across. Podocytes form a thin filter with the endothelium of the capillaries to separate urine from blood passing through the glomerulus. The outer layer of the glomerular capsule keeps the urine separated from the blood within the capsule. At the far end of the glomerular capsule is the mouth of the renal tubule, which carries urine from the glomerular capsule to the renal pelvis.

RCC arises from the cells in the parenchyma of the kidneys. Specifically, about 85% of RCC are adenocarcinomas (including ccRCC, pRCC, crRCC) of the proximal convoluted tubules epithelium (renal cortex) presenting a predominantly expansile growth pattern, while the remainder are mainly transitional cell carcinomas of the renal pelvis (renal medulla). Most clear cell RCCs (95%) are sporadic, and the remaining 5% are linked to hereditary syndromes (e.g., von Hippel–Lindau [VHL] disease and tuberous sclerosis). In addition, 25-30% of RCC patients present with metastatic disease (to the lungs, liver and bones) at diagnosis, and about 30% of patients relapse after treatment for local RCC.

WT evolves from pluripotent renal precursors that undergo excessive proliferation, leading to undifferentiated stromal components, blastemal cells, and primitive epithelial structures; and failure to mature to normal renal parenchyma.

23.3 Epidemiology

RCC constitutes the seventh most common malignancy in men, and the ninth in women, amounting to 2% of the total human cancer burden globally.

The reported incidence rates are 0.6–14.7 per 100,000, with about 190,000 new cases diagnosed each year. Most patients are in the fifth to seventh decades of life (median age at diagnosis of 66 years and median age at death of 70 years). RCC occurs slightly more commonly in blacks than in whites, and shows a male predilection (2–3:1).

Childhood renal tumors (mainly WT or nephroblastoma) account for 7% of all pediatric cancers, with an annual incidence of 7.1 cases per 1 million children of <15 years. The median age at diagnosis of WT is 3 years, and the male-to-female ratio in unilateral cases of WT is 0.92 to 1.00, but in bilateral cases, it is 0.60 to 1.00.

23.4 Pathogenesis

Risk factors for RCC include cigarette smoking; obesity; diuretic use; exposure to petroleum products, chlorinated solvents, cadmium, lead, asbestos, and ionizing radiation; high-protein diets; hypertension; kidney transplantation; and HIV infection.

RCC is associated with VHL disease, hereditary leiomyomatosis and renal cell cancer (HLRCC), Birt–Hogg–Dubé (BHD) syndrome, and hereditary papillary renal cancer (HPRC) [4,5].

Specific genetic mutations include *PBRM1*, *BAP1*, *SETD2*, *KDM5a*, *ARID1a*, *UTX*, *PIK3CA*, *PTEN*, and *MTOR* in clear cell RCC; *MET* and the fumarate hydratase (*FH*) gene in papillary RCC; and losses of whole chromosomes 1, 2, 6, 10, 13, 17, and 21, and mutations in *PTEN* (at 10q23) and *TP53* (at 17p13) in chromophobe RCC [4,5].

WT may occur as a part of a genetic predisposition syndrome in 5%–10% of cases, including WAGR (WT, aniridia, genitourinary anomalies, and mental retardation), Denys–Drash syndrome, Beckwith–Wiedemann syndrome, asymmetric overgrowth, or family history of WT [6].

23.5 Clinical features

Patients with RCC are usually asymptomatic, although some may show flank or abdominal pain.

Children with WT may present with abdominal pain, hematuria, fever, hypertension, hemihyperplasia, cryptorchidism, and hypospadias.

23.6 Diagnosis

ccRCC is a solid, yellowish lesion with variable degrees of internal necrosis, hemorrhage, and cystic degeneration. On CT, ccRCC usually displays intense contrast uptake in the corticomedullary phase (120–140 HU) and typical washout in the nephrographic phase (90–100 HU). On MRI, ccRCC has signal intensity similar to that of the renal cortex at T1-weighted images, and hypersignal at T2-weighted images. Histologically, the tumor is defined by clear, eosinophil granular cytoplasm (due to the lipid- and glycogen-rich cytoplasmic content), with nested clusters of cells surrounded by a dense endothelial network [2]. ccRCC carries a worse prognosis than pRCC and crRCC.

pRCC is usually a peripheral lesion with necrosis, hemorrhage, and calcification. On CT, pRCC presents a mean density of between 50–60 HU and 65–75 HU in the nephrographic phase (progressive uptake). On MRI, pRCC appears as peripheral lesions with intense hyposignal (due presumably to the intratumoral hemosiderin content or to their architectural arrangement) on T2-weighted images. Histologically, pRCC shows cells organized in a spindle-shaped pattern and possible areas of internal hemorrhage and cystic alterations. Of the two subtypes recognized, subtype 1 (basophilic, 73% of cases) is usually of low grade, and consists of papillae lined by a single layer of small cells with scanty clear-to-basophilic cytoplasm and hyperchromatic nuclei; subtype 2 (eosinophilic) is of high grade, and consists of papillae lined by cells with abundant granular eosinophilic cytoplasm and prominent nucleoli [2]. Further confirmation of pRCC is achieved through its positivity for cytokeratin 7 and AMACR and polysomy of chromosomes 7 and 17 and loss of chromosome Y.

crRCC is a less aggressive variant, showing orange colored turning grey after fixation. On CT, crRCC demonstrates contrast uptake (80–100 HU in the corticomedullary phase), and is less intense than ccRCC, and more intense than pRCC. On MRI, crRCC presents a slight hyposignal or intermediate signal intensity at T2-weighted sequences. Histologically, crRCC contains large pale cells with reticulated cytoplasm, characteristic perinuclear clearing (halo), and low mitotic rates [2]. crRCC appears to be closely related to oncocytoma (a benign tumor of intercalated type B cells of the cortical collecting ducts), as both are thought to evolve from collecting ducts intercalated cells and constitute common findings of Birt-Hogg-Dubé syndrome.

cdRCC (collecting duct renal cell carcinoma) is a rare variant that is possibly originated from the medulla (implying a differential diagnosis with transitional cell carcinoma), but may invade the cortex. cdRCC appears hypovascular on multidetector CT, and heterogeneous with extremely variable

signal intensity on T2-weighted sequences, depending on the amount of hemorrhage, necrosis, cystic component and calcification. Histologically, cdRCC is characterized by irregular arranged, infiltrating cells in the collecting duct walls, with remarkable desmoplasia. cdRCC is highly aggressive and causes 60-70% of mortality within a two-year period.

WT usually forms well-circumscribed, large, solid, pale gray to slightly pink or yellow-gray masses with soft consistency and a heterogeneous cut surface (including areas of viable tumor, hemorrhage, and necrosis). Histologically, WT displays a classic triphasic pattern of epithelial, stromal, and blastemal components. The blastemal component is the least differentiated and contains small round blue cells with overlapping nuclei and brisk mitotic activity in diffuse, serpentine, nodular, or basaloid patterns. The epithelial component may show a spectrum of differentiation, as exemplified by squamous epithelial islands and mucinous epithelium. The stromal component may include densely packed undifferentiated mesenchymal cells or loose cellular myxoid areas. Biphasic and monophasic variants may also occur. In addition, chemotherapy-induced change (especially in primitive, highly proliferative blastemal component) includes areas of necrosis, hemorrhage and fibrosis and areas with foamy and/or hemosiderin-laden macrophages. Anaplastic WT (in 5%–8% of all WT cases) may show large, atypical multipolar mitotic figures and significantly enlarged and hyperchromatic nuclei, and often expresses p53 [7].

The stage of RCC is often determined by using the TNM system of the American Joint Committee on Cancer (AJCC), in which T indicates the size of the primary tumor and extent of invasion; N describes the status of metastasis to regional lymph nodes; and M indicates whether there is distant metastasis.

23.7 Treatment

For localized RCC, primary treatment is nephrectomy (radical nephrectomy, simple nephrectomy, or partial nephrectomy) [5,8–10].

For advanced or disseminated RCC, locoregional therapy (e.g., anti-vascular endothelial growth factor receptor [VEGFR] tyrosine kinase inhibitors [TKIs] sunitinib and pazopanib, and the mammalian target of rapamycin [mTOR] inhibitor temsirolimus) may help palliate symptoms, and systemic therapy has only limited effectiveness [5,8–10].

For renal tumors in childhood, treatment consists of preoperative chemotherapy, followed by surgery (Europe), or up-front surgery prior to administration of chemotherapy (United States). Both approaches achieve overall survival of nearly 90% [6].

23.8 Prognosis

The 5-year disease-specific survival for patients with TNM stage I RCC is up to 95%; that for patients with stage II RCC is around 80%; that for patients with stages I–II RCC, tumor invasion of the urinary collecting system, is about 60%; that for patients with stage III RCC is about 60%; and that for patients with stage IV RCC is <10%.

In patients with localized RCC, metastasis is present in 20%–30% of cases. Patients with metastatic RCC have a 5-year survival rates of <10%. While 30% of patients after treatment may experience recurrence, <15% of children with WT may relapse within 2 years of diagnosis.

References

1. Eble JN (ed.). *Pathology and Genetics of Tumours of the Urinary System and Male Genital Organs*. Lyon, France: IARC Press, 2004.
2. Muglia VF, Prando A. Renal cell carcinoma: Histological classification and correlation with imaging findings. *Radiol Bras* 2015;48(3):166–74.
3. Jonasch E, Gao J, Rathmell WK. Renal cell carcinoma. *BMJ* 2014;349:g4797.
4. Koul H, Huh JS, Rove KO, et al. Molecular aspects of renal cell carcinoma: A review. *Am J Cancer Res* 2011;1(2):240–54.
5. PDQ Adult Treatment Editorial Board. *Renal Cell Cancer Treatment (PDQ®): Health Professional Version. PDQ Cancer Information Summaries*. Bethesda, MD: National Cancer Institute, 2002.
6. PDQ Pediatric Treatment Editorial Board. *Wilms Tumor and Other Childhood Kidney Tumors Treatment (PDQ®): Health Professional Version. PDQ Cancer Information Summaries*. Bethesda, MD: National Cancer Institute, 2002.
7. van den Heuvel-Eibrink MM (ed.). *Wilms Tumor*. Brisbane: Codon Publications, 2016.
8. Doehn C, Grünwald V, Steiner T, Follmann M, Rexer H, Krege S. The diagnosis, treatment, and follow-up of renal cell carcinoma. *Dtsch Arztebl Int* 2016;113(35–36):590–6.
9. Brok J, Treger TD, Gooskens SL, van den Heuvel-Eibrink MM, Pritchard-Jones K. Biology and treatment of renal tumours in childhood. *Eur J Cancer* 2016;68:179–95.
10. Greef B, Eisen T. Medical treatment of renal cancer: new horizons. *Br J Cancer* 2016;115(5):505–16.

24
Testicular Cancer

24.1 Definition

Tumors of the testis consist of the following histological categories: (1) *germ cell tumors* (intratubular germ cell neoplasia [IGCNU] unclassified, seminoma—seminoma with syncytiotrophoblastic cells, spermatocytic seminoma—spermatocytic seminoma with sarcoma, embryonal carcinoma, yolk sac tumor, trophoblastic tumors [choriocarcinoma, trophoblastic neoplasms other than choriocarcinoma—monophasic choriocarcinoma or placental site trophoblastic tumor, teratoma, dermoid cyst, monodermal teratoma, teratoma with somatic-type malignancies], mixed embryonal carcinoma and teratoma, mixed teratoma and seminoma, choriocarcinoma, and teratoma or embryonal carcinoma), (2) *sex cord or gonadal stromal tumors—pure forms* (Leydig cell tumor, malignant Leydig cell tumor, Sertoli cell tumor [Sertoli cell tumor lipid-rich variant, sclerosing Sertoli cell tumor, large cell calcifying Sertoli cell tumor], malignant Sertoli cell tumor, granulosa cell tumor [adult-type granulosa cell tumor, juvenile type granulosa cell tumor], tumors of the thecoma or fibroma group [thecoma, fibroma], sex cord or gonadal stromal tumor, incompletely differentiated, sex cord or gonadal stromal tumors—mixed forms, malignant sex cord or gonadal stromal tumors, tumors containing both germ cell and sex cord or gonadal stromal elements [gonadoblastoma, germ cell–sex cord or gonadal stromal tumor—unclassified]), (3) *miscellaneous tumors of the testis* (carcinoid tumor, tumors of ovarian epithelial types [serous tumor of borderline malignancy, serous carcinoma, well-differentiated endometrioid carcinoma, mucinous cystadenoma, mucinous cystadenocarcinoma, Brenner tumor, nephroblastoma, paraganglioma]), (4) *hematopoietic tumors*, (5) *tumors of collecting ducts and rete* (adenoma, carcinoma), (6) *tumors of paratesticular structures* (adenomatoid tumor, malignant mesothelioma, benign mesothelioma [well-differentiated papillary mesothelioma, cystic mesothelioma], adenocarcinoma of the epididymis, papillary cystadenoma of the epididymis, melanotic neuroectodermal tumor, desmoplastic small round cell tumor), (7) *mesenchymal tumors of the spermatic cord and testicular adnexae*, and (8) *secondary tumors of the testis* [1].

Testicular germ cell tumors represent 95% of all testicular malignancies in adults, and up to 77% in children. In adults, seminoma alone is

responsible for approximately 60% of testicular germ cell tumor cases; while in children, teratoma (62%) is the most common testicular germ cell tumor subtype [2].

24.2 Biology

Suspended outside the body in a fleshy sac called the scrotum, which attaches to the body between the base of the penis and the anus, the testes (singular: testis; commonly known as testicles) are a pair of ovoid glandular organs of 5 × 3 × 2.5 cm in dimension and 10–15 g in weight. In addition to the male sex hormone testosterone, the testes produce as many as 12 trillion sperm in a male's lifetime, about 400 million of which are released in a single ejaculation.

Held in the scrotum by the spermatic cord composed of tough connective tissue, muscle, the vas deferens, blood vessels, lymph vessels, and nerves, the testes are wrapped by the tunica vaginalis (an extension of the peritoneum of the abdomen), the tunica albuginea (a tough, protective sheath of dense irregular connective tissue), and the tunica vasculosa (consisting of blood vessels and connective tissue).

The testes are partitioned by extensions of the tunica albuginea (the middle layers of the testes' covering) into lobules which contain the seminiferous tubules. The seminiferous tubules (150–250 μm in diameter, totaling 800 in each testis) open into a series of uncoiled, interconnected channels called the rete testis, which is connected via ducts (or tubes) to a tightly coiled tube called the epididymis. The epididymis joins to a long, large duct called the vas deferens (which exits via the spermatic cord).

The testes consist of two main types of cells: germ cells and stromal cells. Germ cells line the seminiferous tubules, and are capable of multiplication and differentiation into spermatocytes and spermatids, which move from the lining to the epididymis, where they mature into spermatozoa or sperm cells. Mature sperm cells travel through the vas deferens, and combine with fluids made by the seminal vesicles and the prostate gland to create semen, which is pushed out of body through the urethra during ejaculation.

Stromal cells are supportive cells that are separated into Sertoli cells (or sustentacular cells, which are located in the seminiferous tubules and help make and transport sperm) and Leydig cells (or interstitial cells, which are located between the seminiferous tubules and secrete male sex hormones,

mostly testosterone, contributing to sperm production and maintenance of sex drive, or libido, and other male features).

The testicular collecting system, testicular tunics, and spermatic cord, as well as the rete testis (principally of the intratesticular location), are sometimes referred to as the paratesticular region.

A majority of testicular tumors (77%–95%) arise from germ cells in the seminiferous epithelium (mostly of purely gonadal and rarely of extragonadal origin and presentation). Extragonadal germ cell tumor occurs preferentially along the body midline in children of both genders. Tumors of the paratesticular region are rare (7%–10%) and often benign (e.g., leiomyoma, lipoma, hemangioma, and lymphangioma).

24.3 Epidemiology

Testicular cancer constitutes about 1.5% of male neoplasms and 5% of urologic malignancies globally, with an annual incidence of 3–8 cases per 100,000 men in Western countries and 0.5–3 cases per 100,000 men in Africa, Asia, and South America. The peak incidence of testicular cancer occurs in the 30- to 40-year age group for pure seminoma and in the 15- to 30-year age group for nonseminoma.

Testicular germ cell tumor is the most frequent solid tumor of Caucasian adolescents and young adult males, with a median age at diagnosis of 34. The incidence of pediatric testicular tumors peaks in children aged 2–4 years.

24.4 Pathogenesis

Risk factors for testicular cancer include a history of cryptorchidism or undescended testis, Klinefelter syndrome (testicular dysgenesis), a familial history of testicular cancer among first-degree relatives (father or brothers; isochromosome of the short arm of chromosome 12), Down syndrome, Li–Fraumeni syndrome, the presence of a contralateral tumor or IGCNU (also known as carcinoma *in situ*, which is considered to originate from developmentally arrested immature germ cells/gonocytes that persisted outside of fetal life), subfertility or infertility, tallness, previous marijuana exposure, vasectomy, trauma, mumps, and HIV infection [3].

Testicular cancer is linked to gene mutations on chromosomes 4, 5, 6, and 12 (i.e., expressing SPRY4, kit-ligand, and synaptopodin) [3].

24.5 Clinical features

Symptoms of testicular cancer may include painless unilateral lump or swelling, pain or discomfort in a testicle or the scrotum, dull ache in the lower abdomen or groin, sudden buildup of fluid in the scrotum (hydrocele), breast tenderness or growth (gynecomastia), lower back pain, shortness of breath, chest pain, bloody sputum or phlegm (later-stage testicular cancer), swelling of one or both legs, and blood clot.

24.6 Diagnosis

Diagnosis of testicular germ cell tumors involves physical examination (for intrascrotal mass), biochemical tests for elevated serum tumor markers (e.g., lpha-fetoprotein [AFP], human chorionic gonadotropin [hCG], and lactate dehydrogenase [LDH]), ultrasonography (for any doubtful case), CT/MRI (for retroperitoneal, mediastinal and thorax nodal enlargement), and histological confirmation via testicular biopsy.

Seminoma (called dysgerminoma in the ovary) is the most common testicular germ cell tumor, and usually presents as a bulky, painless, gray-white, lobulated mass of the testis in men of 40 years (compared with nonseminomatous germ cell tumor in men of 25 years). Seminoma may sometimes occur in the mediastinum, pineal gland (germinoma), and retroperitoneum. Histologically, seminoma shows sheets of relatively uniform, large, round-polyhedral cells divided into poorly demarcated lobules by delicate fibrous septa with T lymphocytes and plasma cells; abundant clear or watery cytoplasm (glycogen), large central nuclei, and one or two prominent, elongated, and irregular nucleoli; minimal mitotic figures, occasional infarction and edema; rare fibrosis; and osseous metaplasia. Immunohistochemically, seminoma is positive for placental alkaline phosphatase (PLAP), ferritin, periodic acid-Schiff (PAS) (with and without diastase), vimentin, and angiotensin-1-converting enzyme, but negative for cytokeratin (syncytiotrophoblastic giant cells are positive), α-fetoprotein (AFP), human chorionic gonadotropin (hCG) (syncytiotrophoblastic giant cells are positive), CD30, and EMA [4,5].

Variants of seminoma include seminoma with early carcinomatous transformation (similar to embryonal carcinoma), seminoma with syncytiotrophoblastic giant cells (10%–20% of seminomas, giant cells often related to blood vessels, no cytotrophoblasts, and hCG serum level <1000 IU/L), tubular seminoma (<10 cases reported, tumor cells form tubular structures of various sizes and shapes, areas of classic seminoma present, otherwise similar to classic seminoma) [5].

Teratoma is the most common pediatric germ cell tumor. While teratoma is uniformly benign in children, it is often malignant in postpubertal and adult cases. Macroscopically, teratoma is a large (5–10 cm), multinodular, heterogenous (solid, cartilaginous, cystic) lesion. Histologically, mature teratoma displays a mixture of elements of ectoderm, endoderm, and mesoderm; immature teratoma shows foci resembling embryonic or fetal structures, primitive neuroectoderm, poorly formed cartilage, neuroblasts, loose mesenchyme, and primitive glandular structures. Teratoma with malignant transformation has focal malignancy of the somatic type (e.g., squamous cell carcinoma, adenocarcinoma, and sarcoma). Dermoid cysts may resemble pilomatrixoma with shadow squamous epithelium, calcification, and ossification; show smooth muscle bundles, sweat glands (eccrine or apocrine), and glands lined by ciliated epithelium; and have no atypia, no mitoses, and no IGCNU [6].

The stages of testicular cancer are determined as stages I–III according to the TNM system, which incorporates information relating to the tumor size and location, lymph node involvement, presence/absence of metastasis and levels of serum tumor markers (e.g., hCG, AFP, and LDH) after orchiectomy. Specifically, stage I cancer is confined to the testicle; stage II cancer has spread to the retroperitoneal lymph nodes; stage III cancer has spread beyond the lymph nodes to remote sites in the body, including the lungs, brain, liver and bones. Unlike most other cancer, testicular cancer does not have stage IV. It is notable that 75%–80% of patients with seminoma and about 55% of patients with nonseminoma have stage I disease at diagnosis [2].

24.7 Treatment

Treatment for seminoma and teratoma primarily involves orchiectomy with high ligation of the spermatic cord. In life-threatening metastases in the lung with pulmonary insufficiency, chemotherapy (e.g., bleomycin + etoposide + cisplatin) should be initiated upfront until clinical stabilization prior to orchiectomy. Additional treatments for advanced seminoma consist of radiation of the retroperitoneum and chemotherapy if [2,7]. Mature teratoma in children is usually treated by surgery.

24.8 Prognosis

The expected 5-year survival for seminoma and nonseminomatous GCT is 96% and 90%, respectively.

Further, localized (stage I), regional (stage II and some stage IIIB and IIIC) and distant (stage IIIA, IIIB, or IIIC) testicular cancer have 5 year relative survival rates of 99%, 96% and 73%, respectively.

References

1. Eble JN (ed.). *Pathology and Genetics of Tumours of the Urinary System and Male Genital Organs.* Lyon, France: IARC Press, 2004.
2. Vasdev N, Moon A, Thorpe AC. Classification, epidemiology and therapies for testicular germ cell tumours. *Int J Dev Biol* 2013;57(2–4):133–9.
3. Rajpert-De Meyts E, Skakkebaek NE, Toppari J, et al. Testicular cancer pathogenesis, diagnosis and endocrine aspects. In *Endotext*, ed. LJ De Groot, G Chrousos, K Dungan, et al. South Dartmouth, MA: MDText.com, 2000.
4. Young RH. Sex cord-stromal tumors of the ovary and testis: Their similarities and differences with consideration of selected problems. *Mod Pathol* 2005;18(Suppl 2):S81–98.
5. PathologyOutlines.com. *Seminoma.* http://www.pathologyoutlines.com/topic/testisseminomas.html. Accessed March 8, 2017.
6. Pathologyoutlines.com. *Teratoma.* http://www.pathologyoutlines.com/topic/testisteratoma.html. Accessed March 8, 2017.
7. Ehrlich Y, Margel D, Lubin MA, Baniel J. Advances in the treatment of testicular cancer. *Transl Androl Urol* 2015;4(3):381–90.

25
Urethral Cancer

Anthony Kodzo-Grey Venyo

25.1 Definition

Tumors of the urethra can be either primary or secondary metastatic lesions. The most common histologic types of primary urethral tumors consist of urothelial carcinoma (or transitional cell carcinoma, 55%), squamous cell carcinoma (SCC) (21.5%), and adenocarcinoma carcinoma (16.4%), which are malignant neoplasms of the epithelial layer of the urethra. Other histologic types (e.g., clear cell carcinoma, primary enteric-type mucinous adenocarcinoma of the urethra, transitional cell *in situ* carcinoma, melanoma, and plasmacytoma) are rarely seen. Metastatic carcinomas of the urethra include urothelial carcinoma, clear cell carcinoma, SCC, and adenocarcinoma from elsewhere [1].

25.2 Biology

The male urethra is a narrow, S-curved, fibromuscular tube of 15–25 cm in length that conducts urine and semen from the bladder and ejaculatory ducts, respectively, to the exterior of the body. Although the male urethra is a single structure, it is composed of a heterogeneous series of segments: prostatic, membranous, and spongy.

The male urethra is divided into distal and proximal portions. Extending from the tip of the penis to before the prostate, the distal urethra comprises the meatus, fossa navicularis, penile (pendulous) urethra, and bulbar (bulbous) urethra. Extending from the bulbar urethra to the bladder neck, the proximal urethra includes the membranous urethra and prostatic urethra.

The membranous urethra and prostatic urethra are lined by transitional cells, while the bulbous urethra and penile urethra are covered by the stratified columnar epithelium to the stratified squamous epithelium. The submucosa of the urethra contains numerous glands. These structural features allow development of various histologic types of urethral cancer in males, such as transitional cell carcinoma (especially in the prostatic urethra), SCC (especially in the bulbous and membranous urethra), or adenocarcinoma.

Largely contained within the anterior vaginal wall, the female urethra is about 4 cm in length in adults. The female urethra is lined by transitional cell mucosa proximally and stratified squamous cells distally, which facilitate the development of transitional cell carcinoma in the proximal urethra and SCC in the distal urethra. Arising from metaplasia of the periurethral glands, adenocarcinoma may affect both the proximal and distal portions of the female urethra.

25.3 Epidemiology

Primary urethral cancer represents <1% of urinary tract malignancies, and has an annual incidence of 4.3 cases per 1 million men and 1.5 cases per 1 million women. The incidence of urethral cancer is twice as high in African-Americans as in whites (5 cases per 1 million vs. 2.5 cases per 1 million), and appears to increase with age (averaging 32 cases per 1 million men and 9.5 per 1 million women aged 75–84 years). There is a strong male predominance among urethral cancer patients (male-to-female ratio of 3:1).

25.4 Pathogenesis

A number of postulates have been promulgated as factors responsible for the development of carcinoma of the urethra, including (1) iatrogenic chronic irritation of the urethra from a number of causes, including chronic catheterization and urethroplasty; (2) urethral strictures; (3) previous radiotherapy treatment to areas around the urethra; (4) chronic urethritis after contracting a sexually transmitted disease; and (5) recurrent infections of the urinary tract [2].

Molecularly, urothelial carcinoma is linked to frequent activating mutations in *FGFR3* and several downstream targets of FGFR3 (e.g., *PIK3CA*, *AKT1*, and *RAS*).

25.5 Clinical features

Patients with urethral cancer may display lower urinary tract symptoms (bladder outlet obstruction or retention of urine), visible hematuria or non-visible hematuria, a lump in the urethra, retention of urine, pelvic pain, urethrocutaneous fistula, periurethral abscess formation, dyspareunia, bloody urethral discharge, extraurethral mass, and regional lymph node enlargement (33% of cases) [2].

25.6 Diagnosis

Patients suspected of urethral cancer should undergo (1) examination of the abdomen and inguinal and femoral regions for evidence of lymphadenopathy, and (2) examination with palpation of external genitalia for evidence of suspicious indurations, and in women, a thorough pelvic examination is undertaken [2,3].

Further laboratory investigations include urinalysis, urine microscopy or cytology, full blood count and coagulation, serum urea and electrolytes, liver function tests, prostate-specific antigen (PSA) screen (older males), and *in vitro* culture.

Radiographic investigations include ultrasound, computerized tomography (CT), and MRI (of the abdomen and pelvis, including the genitalia); isotope bone scan (for suspected bony metastasis); retrograde urethrogram (to assess the urethral lesion and the urethra); and urethrocystoscopy (examination under anesthesia and biopsy of the urethral lesion) [2].

Microscopic investigations of tumor tissues and biopsies provide an ultimate means of urethral cancer confirmation.

Urothelial carcinoma of the urethra demonstrates cytologically malignant cells with visible cell membranes, irregular and enlarged nuclei, very prominent nucleoli, dark-looking chromatin, and abundant mitosis [2]. It is worth noting that variants of urothelial carcinoma can develop in the urethra, including a sarcomatoid variant of urothelial carcinoma. CT imaging and histopathological and immunohistochemical features of the sarcomatoid variant of urothelial carcinoma are illustrated in Figure 25.1 [4]. Being a malignant neoplasm, urothelial carcinoma (or transitional cell carcinoma) of the urethra has the ability to metastasize to other sites, including the lung, bone, liver, and brain. Figure 25.2 presents the CT, MRI, and histological findings of such a case [5].

Urethral SCC contains sheets of large prominent tumor cells with focal or abundant keratinization (depending on the grade of differentiation of the tumor), ample cytoplasm, intercellular bridges, high mitotic activity, and prominent nuclear atypia [2].

Urethral adenocarcinoma reveals simple or pseudostratified columnar epithelium, apically located cytoplasm, hyperchromatic nuclei located basically, and occasionally vacuolated cytoplasm with mucin or mucin pools in a true mucinous tumor [2].

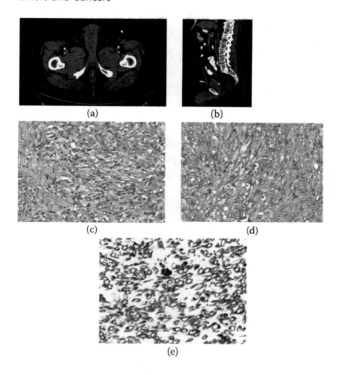

(a)   (b)

(c)   (d)

(e)

Figure 25.1   CT scan of abdomen and pelvis with contrast showing the primary tumor in a 57-year-old man who presented with urinary retention and who was found to have sarcomatoid carcinoma of the urethra: (a) transverse view and (b) sagittal view. Histological features of this sarcomatoid urothelial carcinoma: (c) sarcomatoid changes with areas of necrosis/hemorrhage, H&E 200×, (d) more carcinomatoid features of the tumor cells, H&E 200×, and (e) cytokeratin immunohistochemistry showing strong positivity of the tumor cells, 200×, and carcinoma origin instead of true sarcoma. (Reproduced from Badhiwala, N., et al., *Case Rep. Urol*, 2013, 931893, 2013. Copyright © 2013 Badhiwala, N. et al. This is an open access article distributed under the Creative Commons Attribution License, which permits unrestricted use, distribution, and reproduction in any medium, provided the original work is properly cited).

Clear cell adenocarcinoma of the urethra may show glandular, tubulocystic, solid or diffuse, and papillary or micropapillary growth patterns; variably sized, cuboidal cells in association with abundant clear or eosinophilic cytoplasm and cytoplasmic vacuoles; hyperchromatic and pleomorphic nuclei, prominent nucleoli; the presence of hobnail changes and extracellular material; mitoses; and necrosis [2].

Immunohistochemically, urothelial carcinoma is positive for P53 (80%), CK20 (transurothelial in carcinoma *in situ*, variable in invasive urothelial carcinoma),

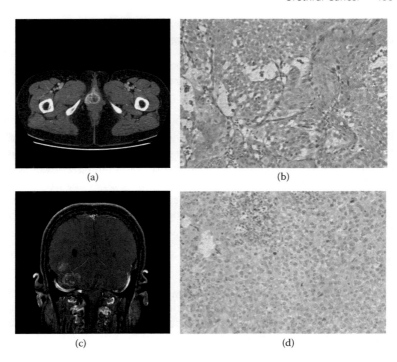

(a)   (b)

(c)   (d)

Figure 25.2   Enhanced pelvic CT scan demonstrates a heterogeneously enhanced mass on female urethra (a). Microphotograph of the urethral mass reveals typical features of transitional cell carcinoma including ramifying papillae, high nuclear/cytoplasmic ratio, and brisk mitotic activity, hematoxylin–eosin 200× (b). Brain MR imaging shows multiple homogeneously enhanced lesions. The largest lesion, seen in the right cerebellar hemisphere, was later surgically resected (c). Microphotograph of the brain mass reveals metastatic carcinoma with increased pleomorphism and mitotic activity which is nearly identical to the urethral mass, hematoxylin–eosin 200× (d). (Reproduced from Moon, K.-S., et al., *BMC Cancer*, 11, 23, 2011. Copyright © 2011 Moon, K.-S. et al. This is an open access article distributed under the Creative Commons Attribution License, which permits unrestricted use, distribution, and reproduction in any medium, provided the original work is properly cited).

CK7, GATA3, high-molecular-weight cytokeratin (CK903 and CK5/6), and p63, but negative for E-cadherin, CD44s, PAX8, PSA, and PSAP. SCC is positive for high-molecular-weight cytokeratin (CK903 and CK 5/6) and p63 (human papillomavirus [HPV] related), but negative for GATA3. Adenocarcinoma is positive for CK20 (variable), CDX2 (variable), and cytoplasmic β-catenin, but negative for nuclear β-catenin, PSA, PSMA, AMACR (distinguishable from prostate), PAX2, and PAX8. Clear cell adenocarcinoma is positive for AMACR

(racemace, 75%), vimentin (75%), p53 (100%), PAX8 (50%), and CK7 (50%), but negative for CK20, p63, GATA3, PSA, and PAP [2].

The stage of urethral cancer is determined according to the TNM system outlined by the American Joint Committee on Cancer as primary tumor (Tx, T0, Ta, Tis, T1, T2, T3, T4), regional lymph nodes (Nx, N0, N1, N2, N3), and distant metastases (Mx, M0, M1) [1].

Differential diagnoses for urethral cancer include (1) urothelial carcinoma of the urinary bladder with extension into the urethra, which tends to be more common than primary urothelial carcinoma of the urethra; (2) extension into the urethra of SCC of the penis, vulva, and cervix; (3) extension into the urethra of adenocarcinoma of nearby organs, including the prostate gland and the colon; and (4) nephrogenic adenoma of the urethra (mimicking clear cell adenocarcinoma).

25.7 Treatment

Localized disease in men is treated by surgery, including (1) conservative or local excision, (2) partial penectomy, (3) radical penectomy (amputation of the penis), (4) pelvic lymphadenectomy, and (5) en bloc resection (including penectomy and cystoprostatectomy, along with the removal of the anterior pubis) [6].

Localized disease in females is treated by radical urethrectomy or urethral-sparing surgery (local excision of the tumor with partial urethrectomy if it is possible to achieve good clear surgical margins). Local radiotherapy provides an alternative treatment instead of surgical treatment.

Advanced disease in both males and females relies on systemic chemotherapy with cisplatin-based agents, as complete excision and eradication of tumor is difficult, and lymph node dissection is of uncertain value in advanced disease [2].

Combination chemotherapy consisting of 5-fluorouracil (100 mg/m^2) and mitomycin C (15 mg/m^2), as well as external beam radiotherapy (30–50 Gy), is useful for treating advanced urethral SCC. Topical instillations of mitomycin C (40 mg/80 µL) weekly for 6 weeks are also effective for treating rapid recurrences of low-grade papillary transitional cell carcinoma of the bulbar urethra pursuant to primary excisions of the urethral tumor. Multimodality treatment consisting of chemotherapy (cisplatin and 5-fluorouracil), external beam radiotherapy (45 Gy), and surgical resection (distal urethrectomy) may be used for late-stage distal urethral tumors.

25.8 Prognosis

Carcinomas of the distal urethra are generally of lower grades than those of the proximal urethra, and can be managed by means of monotherapy with external beam radiotherapy, local resection, partial penectomy, or radical penectomy. Their reported 5-year survival rates range from 83% for early-stage tumors to 45% for advanced tumors, with an overall 5-year survival rate of 60%.

Carcinomas of the proximal urethra require a more aggressive, multi-modality therapeutic approach consisting of chemotherapy, radiotherapy, and surgical resection. Nonetheless, despite this aggressive approach to treatment, carcinomas of the proximal urethra still demonstrate a recurrence rate of >50%.

As bulbar urethral SCC is often diagnosed at the advanced local and late stage, it has a 5-year disease-free survival of up to 15% despite surgery and adjunct chemotherapy.

Unfavorable prognostic factors for urethral cancer include older age (>65 years), larger tumor size, later-stage or higher grade tumor, nodal involvement, lympho-vascular permeation, peri-neural involvement, metastasis, invasion of the prostate gland by prostatic urethral carcinoma, and presence of concomitant urinary bladder tumor.

References

1. Venyo AK. Clear cell adenocarcinoma of the urethra: Review of the literature. *Int J Surg Oncol* 2015;2015:790235.
2. Chavez J A, Zynger D L (eds.) Pathology Outlines - Urethral carcinoma – general. www.pathologyoutlines.com/topic/urethracarcinoma.html; accessed September 09, 2016.
3. Gakis G, Witjes JA, Comperat E, et al. European Association of Urology EAU guidelines on primary urethral carcinoma. *Eur Urol* 2013; 64(5):823–30.
4. Badhiwala N, Chan R, Zhou H-J, Shen S, Coburn M. Sarcomatoid carcinoma of male urethra with bone and lung metastases presenting as urethral stricture. *Case Rep Urol* 2013;2013:931893.
5. Moon K-S, Jung S, Lee K-H, Hwang EC, Kim I-Y. Intracranial metastases from primary transitional cell carcinoma of female urethra: Case report & review of the literature. *BMC Cancer* 2011;11:23.
6. PDQ Adult Treatment Editorial Board. *Urethral Cancer Treatment (PDQ®): Health Professional Version. PDQ Cancer Information Summaries.* Bethesda, MD: National Cancer Institute, 2002.

26

Uterine Corpus Cancer

26.1 Definition

Tumors of the uterine corpus consist of (1) *epithelial tumors and related lesions* (endometrial carcinoma [endometrioid adenocarcinoma—variant with squamous differentiation, villoglandular variant, secretory variant, ciliated cell variant; mucinous adenocarcinoma; serous adenocarcinoma; clear cell adenocarcinoma; mixed cell adenocarcinoma; squamous cell carcinoma; transitional cell carcinoma; small cell carcinoma; undifferentiated carcinoma], endometrial hyperplasia [nonatypical hyperplasia—simple or complex or adenomatous, atypical hyperplasia—simple or complex], endometrial polyp, tamoxifen-related lesions), (2) *mesenchymal tumors* (endometrial stromal and related tumors [endometrial stromal sarcoma—low grade, endometrial stromal nodule, undifferentiated endometrial sarcoma], smooth muscle tumors [leiomyosarcoma—epithelioid variant or myxoid variant; smooth muscle tumor of uncertain malignant potential; leiomyoma NOS—histological variants, mitotically active variant, cellular variant; hemorrhagic cellular variant, epithelioid variant, myxoid, atypical variant, lipoleiomyoma variant; growth pattern variants—diffuse leiomyomatosis, dissecting leiomyoma, intravenous leiomyomatosis, metastasizing leiomyoma], miscellaneous mesenchymal tumors [mixed endometrial stromal and smooth muscle tumor, perivascular epithelioid cell tumor, adenomatoid tumor, other malignant mesenchymal tumors, other benign mesenchymal tumors]), (3) *mixed epithelial and mesenchymal tumors* (carcinosarcoma [malignant Müllerian mixed tumor, metaplastic carcinoma], adenosarcoma, carcinofibroma, adenofibroma, adenomyoma—atypical polypoid variant), (4) *gestational trophoblastic disease* (trophoblastic neoplasms [choriocarcinoma, placental site trophoblastic tumor, epithelioid trophoblastic tumor]; molar pregnancies [hydatidiform mole—complete, partial, invasive, metastatic]; nonneoplastic, nonmolar trophoblastic lesions [placental site nodule and plaque, exaggerated placental site]), (5) *miscellaneous tumors* (sex cord–like tumors, neuroectodermal tumors, melanotic paraganglioma, tumors of germ cell type), (6) *lymphoid and hematopoietic tumors* (malignant lymphoma, leukemia), and (7) *secondary tumors* [1].

Covering neoplasms that occur in the uterus, with the exception of cervical cancer, uterine corpus cancer is predominated by endometrial carcinoma, the incidence of which is greater than that of all other female genital tract malignancies combined. Another relatively infrequent group of uterine corpus cancer is uterine sarcoma, which accounts for 9% of uterine corpus cancer cases.

Based on histopathological criteria, endometrial carcinoma is divided into types I and II. Type I represents 80% of endometrial carcinoma, includes endometrioid adenocarcinoma grades I and II, is estrogen responsive, and has a favorable prognosis. Type II comprises about 20% of endometrial carcinoma, includes endometrioid adenocarcinoma grade 3 and other rare histologies (e.g., serous carcinoma, clear cell carcinoma, and carcinosarcoma or mixed Müllerian tumors), undergoes a rapid progression, and has a poor prognosis [2,3].

26.2 Biology

Located between the urinary bladder anteriorly and the rectum posteriorly in the female pelvis, the uterus is a hollow pear-shaped organ of 8 × 5 × 4 cm in dimension, 40–80 g in weight, and 80–200 mL in volume. The uterus is divided into three main parts: the fundus (connecting with the fallopian tubes), corpus (upper two-thirds of the uterus above the level of the internal cervical os), and cervix (lower one-third of the uterus).

Structurally, the inner surface of the uterus is lined by endometrial mucosa or endometrium, myometrium (consisting of a smooth muscle layer and a network of blood vessels), and serosa or perimetrium (consisting of epithelial cells that envelop the uterus).

The endometrium is composed of the basalis layer, functionalis layer (superficial one-half to two-thirds), and stroma (made up of stromal cells, vessels, stromal granulocytes [e.g., T cells and macrophages], and foamy cells). The functionalis layer is divided into spongiosum (near the basalis) and compactum (near the surface) and is shed monthly. However, before puberty, the endometrium is inactive and contains tubular glands, dense fibroblastic stroma, and thin blood vessels; after menopause, the endometrium is inactive and contains thin, often cystic cavities lined by flat or cuboidal cells and fibrotic stroma.

The uterus is responsive to the hormonal milieu within the body and is responsible for menses, implantation, gestation, labor, and delivery. During the prepubertal and postmenopausal years, the uterus is in a relatively quiescent state.

Endometrial carcinoma arises in a small area (e.g., within an endometrial polyp) or a diffuse multifocal pattern on the lining of the uterus (i.e., the endometrium). Early tumor growth through an exophytic and spreading pattern is characterized by friability and spontaneous bleeding. Later tumor growth shows myometrial invasion and spreading toward the cervix. Endometrial carcinoma type I is attributable to unopposed estrogen stimulation, and often associated with atypical endometrial hyperplasia. Endometrial carcinoma type II is estrogen independent, and develops from the atrophic endometrium [3].

26.3 Epidemiology

Endometrial carcinoma is the fourth most common cancer of women, after breast, lung, and colorectal cancer; the leading malignant cancer of the female genital tract; and the third most common cause of death among gynecological malignancies after ovarian and cervical cancer.

The annual incidence of endometrial carcinoma is 3–25 per 100,000 women, including 3.3 per 100,000 in Uganda; 3.8 per 100,000 in China; >10 per 100,000 in Europe, Australia, New Zealand, South America, and the Pacific Island nations; and 23.3 per 100,000 in the United States. The mortality rate of endometrial carcinoma is 1.7–2.4 per 100,000 women.

About 75% of endometrial carcinoma is found in postmenopausal women (mean age of 61 years), with <20% of cases occurring before menopause.

26.4 Pathogenesis

Risk factors for endometrial carcinoma include obesity, nulliparity, infertility, anovulatory menstrual cycles, tamoxifen therapy, diabetes, hypertension, previous radiotherapy, high-fat diet, and Lynch II syndrome (hereditary nonpolyposis colorectal cancer [HNPCC], with germline mutation in DNA mismatch repair genes MSH1, MSH2, and MSH6).

Endometrial carcinoma type I harbors mutations in the KRAS, PIK3CA, and β-catenin oncogenes; PTEN tumor suppressor gene; and defects in DNA mismatch repair. Endometrial carcinoma type II demonstrates mutations in the P53 gene, E-cadherin tumor suppressor gene, HER-2/neu, and p16 expression and nondiploidy [4].

26.5 Clinical features

Endometrial carcinoma typically manifests abnormal vaginal or uterine bleeding, spotting, or discharge (from a watery and blood-streaked flow to a flow with more blood during or after menopause); difficulty or pain when urinating; pain during sexual intercourse; and pain in the pelvic area. Other presenting symptoms may include purulent genital discharge, weight loss, and a change in bladder or bowel habits. About 5% of patients may be asymptomatic.

26.6 Diagnosis

Endometrial carcinoma tends to form round, polypoid, expansile, friable masses in the corpus proper with an indurated-appearing surface, necrosis, and hemorrhage.

Histologically, endometrial carcinoma often displays back-to-back endometrial-type glands of varying differentiation or atypia, occasional villoglandular pattern, desmoplastic stroma, foamy cells (due to tumor necrosis), adjacent endometrium with endometrial intraepithelial neoplasia (EIN) or atypical hyperplasia, vascular invasion (with chronic inflammation around lymphatics), trophoblastic differentiation with hCG+ cells, and common squamous metaplasia.

Endometrial carcinoma is distinguished into a number of cell types, including endometrioid (also known as typical endometrioid adenocarcinoma, 75%), mixed (10%), uterine papillary serous (<10%), clear cell (4%), carcinosarcoma (3%), mucinous (1%), squamous (<1%), and undifferentiated (<1%).

Endometrioid adenocarcinoma is further separated into well-, moderately, and poorly differentiated subtypes. The well-differentiated subtype (International Federation of Gynecology and Obstetrics [FIGO] grade 1, 45%) shows small, round back-to-back glands; papillary structures (due to budding and branching of large glands); uniform and oval to cylindrical nuclei, minimal nuclear atypia, small discrete nucleoli; and <5% of solid or nonglandular areas. The moderately differentiated subtype (FIGO grade 2, 35%) shows sheet-like tumor cells without glandular features; moderate nuclear atypia, prominent nucleoli; and 6%–50% of solid or nonglandular areas. The poorly differentiated subtype (FIGO grade 3, 20%) shows sheet-like tumor cells without glandular features, high-grade features, poorly formed glands, malignant squamous cells, commonly angiolymphatic invasion, and >50% of solid or nonglandular areas [2,5,6].

Variants of endometrioid adenocarcinoma include villoglandular (papillary) adenocarcinoma (displaying a papillary architecture with delicate fibro-vascular stalks lined by cuboidal to columnar cells with minimal cellular stratification and mild nuclear pleomorphism), secretory adenocarcinoma (displaying abundantly vacuolated columnar cytoplasm and minimal cellular atypia), ciliated adenocarcinoma (showing glands lined predominately by ciliated cells with abundant eosinophilic cytoplasm, mildly irregular nuclear contours and prominent nucleoli) and adenocarcinoma with squamous differentiation (5% or greater of the carcinoma exhibiting keratin pearl formation, defined intercellular bridges, or morules consisting of ovoid or spindled cells) [4].

The stages of endometrial cancer are often determined as I-IV according to the Féderation Internationale de Gynécologie et d'Obstétrique (FIGO) system [2].

26.7 Treatment

Treatments for FIGO stage I and II (type I) endometrial carcinoma are surgery (e.g., total abdominal hysterectomy [TAH] and bilateral salpingo-oophorectomy [BSO]) with or without lymphadenectomy, postoperative vaginal brachytherapy, or radiotherapy alone; treatments for FIGO stage I and II (type II) endometrial carcinoma are surgery and postoperative chemo-therapy with or without radiotherapy [7,8].

Treatments for FIGO stage III and IV endometrial carcinoma are surgery followed by chemotherapy or radiotherapy (operable disease), chemotherapy and radiotherapy (inoperable disease), and chemo-therapy and hormone therapy or biologic therapy (if unsuitable for radiotherapy) [7,8].

26.8 Prognosis

Women with uterine cancer have a 5-year survival rate of 82% and a 10-year survival rate of 79%. For early-stage and localized cancer, a 5-year survival rate of about 95% is expected. For late-stage cancer or cancer showing regional spread, a 5-year survival rate of about 68% is expected. For cancer that has spread more distantly to other areas of the body, a 5-year survival rate of 17% is expected.

Endometrial adenocarcinoma grades 1, 2, and 3 have 5-year survival rates of 94%, 88%, and 79%, respectively.

References

1. Kurman RJ (ed.). *WHO Classification of Tumours of Female Reproductive Organs*. Lyon, France: IARC Press, 2014.

2. Meissnitzer M, Forstner R. MRI of endometrium cancer—How we do it. *Cancer Imaging* 2016;16:11.

3. Amant F, Mirza MR, Koskas M, Creutzberg CL. Cancer of the corpus uteri. *Int J Gynecol Obstetr.* 2015; 131 (suppl 2): S96–104.

4. Dobrzycka B, Terlikowski SJ. Biomarkers as prognostic factors in endometrial cancer. *Folia Histochem Cytobiol.* 2010;48(3):319–22.

5. PathologyOutlines.com. *Endometrioid Carcinoma.* http://www.pathologyoutlines.com/topic/uterusendometrioid.html; accessed March 10, 2017.

6. Masuyama H, Haraga J, Nishida T, et al. Three histologically distinct cancers of the uterine corpus: A case report and review of the literature. *Mol Clin Oncol.* 2016;4(4):563–6.

7. Lee SW, Lee TS, Hong DG, et al. Practice guidelines for management of uterine corpus cancer in Korea: A Korean Society of Gynecologic Oncology Consensus Statement. *J Gynecol Oncol* 2017;28(1):e12.

8. PDQ Adult Treatment Editorial Board. *Endometrial Cancer Treatment (PDQ®): Health Professional Version. PDQ Cancer Information Summaries.* Bethesda, MD: National Cancer Institute, 2002.

27
Vaginal Cancer

27.1 Definition

Tumors of the vagina include (1) *epithelial tumors* (squamous tumors and precursors [squamous cell carcinoma not otherwise specified, or SCC NOS—keratinizing, nonkeratinizing, basaloid, verrucous, warty; squamous intraepithelial neoplasia—vaginal intraepithelial neoplasia (VAIN) 3 or SCC *in situ*; benign squamous lesions—condyloma acuminatum, squamous papilloma or vaginal micropapillomatosis, fibroepithelial polyp]; glandular tumors [clear cell adenocarcinoma, endometrioid adenocarcinoma, mucinous adenocarcinoma, mesonephric adenocarcinoma, Müllerian papilloma, adenoma NOS—tubular, tubulovillous, villous], other epithelial tumors [adenosquamous carcinoma, adenoid cystic carcinoma, adenoid basal carcinoma, carcinoid, small cell carcinoma, undifferentiated carcinoma]), (2) *mesenchymal tumors and tumorlike conditions* (sarcoma botryoides, leiomyosarcoma, endometrioid stromal sarcoma—low grade, undifferentiated vaginal sarcoma, leiomyoma, genital rhabdomyoma, deep angiomyxoma, postoperative spindle cell nodule), (3) *mixed epithelial and mesenchymal tumors* (carcinosarcoma [malignant Müllerian mixed tumor, metaplastic carcinoma], adenosarcoma, malignant mixed tumor resembling synovial sarcoma, benign mixed tumor), (4) *melanocytic tumors* (malignant melanoma, blue nevus, melanocytic nevus), (5) *miscellaneous tumors* (tumors of germ cell type [yolk sac tumor, dermoid cyst], others [peripheral primitive neuroectodermal tumor or Ewing tumor, adenomatoid tumor]), (6) *lymphoid and hematopoietic tumors* (malignant lymphoma, leukemia), and (7) *secondary tumors* [1].

Common primary vaginal tumors are SCC (83%), adenocarcinoma (9%), melanoma (2%), sarcoma (2%), small cell carcinoma (<1%), and lymphoma (<%) [2].

Secondary tumors of the vagina may occur by direct extension (e.g., the cervix, vulva, and endometrium) or by lymphatic or hematogenous spread (e.g., the breast, ovary, or kidney).

27.2 Biology

Situated behind the urethra and bladder and in front of the rectum, the vagina is an elastic, fibromuscular tube of 7.5 cm in length and about 2.5 cm in diameter that extends from the neck of the uterus (the cervix) to the external genitals (the vulva). The vagina is divided into three segments, with the lower third below the level of the bladder base and the urethra anteriorly, the middle third adjacent to the bladder base, and the upper third at the level of the vaginal fornices (which are denoted as anterior, posterior, and lateral with respect to the cervix).

The inner surface of the vagina is lined by the mucosa (basal, intermediate, and superficial squamous epithelium), lamina propria (subepithelial stroma, consisting of elastic fibers, a rich venous and lymphatic network, and multinucleated stellate stromal cells), muscularis (consisting of smooth muscle tissue), and tunica externa (adventitia, consisting of dense irregular connective tissue).

SCC arises from the squamous cells in the mucosa, frequently involving the upper third of the vagina. SCC may be multifocal when developing in a background of VAIN. A primary SCC of the vagina does not usually involve surrounding structures (e.g., cervix or vulva). A tumor involving both the vagina and the cervix is classified as a cervical carcinoma, while a tumor involving both the vagina and the vulva is considered a vulvar carcinoma.

Adenocarcinoma originates in the mucus-producing glandular or secretory cells anywhere in the vagina, and is associated with vaginal adenosis and hysterectomy, but rarely human papillomavirus infection.

Melanoma develops in the pigment-producing cells (melanocytes) of the vagina, and is commonly seen in older women.

Sarcoma develops in the connective tissue cells or smooth muscle cells in the vaginal wall. Sarcoma botryoides is a rare type, occurring mainly in infancy and early childhood.

27.3 Epidemiology

Primary tumors of the vagina comprise approximately 3% of all gynecological malignancies, with an annual incidence of 1 case per 100,000 women.

SCC commonly affects women of >50 years (median age of 60 years; 50% in women aged >70 years and 20% in women aged >80 years), and

occasionally women in their twenties and thirties (often HPV dependent). The age-adjusted incidence rate of invasive SCC in white women is higher than that in black women and Asian or Pacific Islander women.

Adenocarcinoma mainly occurs in postmenopausal women (mean age of 65 years, range 54–78 years). Clear cell carcinoma (a rare form of adeno-carcinoma) is found in women <30 years of age (median age of 19 years) who had exposure during fetal life to diethylstilbestrol (DES) (which was prescribed to pregnant woman between 1938 and 1971 to prevent miscarriage).

27.4 Pathogenesis

Risk factors for vaginal cancer include advancing age (>50 years), prior his-tory of gynecological cancer (e.g., cancer of the cervix or vulva), previous treatment for dysplasia (abnormal cells on the cervix, vagina, and vulva in 30% of cases), genital warts or VAIN (human papillomavirus types 16, 18, and 31 infection), cigarette smoking, prior hysterectomy, prenatal expo-sure to the synthetic hormone DES, and vaginal adenosis (almost all "DES daughters" have vaginal adenosis).

27.5 Clinical features

The most common symptoms of vaginal cancer are abnormal vaginal bleed-ing, difficulty or pain when urinating, pain during sexual intercourse, pain in the pelvic area (the lower part of the abdomen between the hip bones), pain in the back or legs, swelling in the legs, a lump in the vagina, lower extremity edema, fever, fatigue, and weight loss. Some patients may be asymptomatic.

27.6 Diagnosis

Primary SCC of the vagina is typically exophytic or ulcerative with necrosis, and is graded histologically as well differentiated (grade 1), moderately differentiated (grade 2), and poorly differentiated or undifferentiated (grade 3). Well-differentiated SCC shows polygonal squamous cells with ample eosinophilic cytoplasm, abundant keratin pearls, intercellular bridges, nuclear pleomorphism, and mitotic activity. Poorly differenti-ated SCC shows small cells with scant cytoplasm, hyperchromatic nuclei, nuclear pleomorphism, and mitotic activity. Moderately differentiated SCC demonstrates histological features intermediate between well- and

poorly differentiated SCC. HPV+ SCC is often of the nonkeratinizing, basaloid, or warty type and has diffuse p16 immunoreactivity in comparison with HPV− SCC. Immunohistochemically, SCC stains positive for cytokeratin and p16 [3].

Primary adenocarcinoma of the vagina is a polypoid, fungating, exophytic solid mass of 2–5.5 cm. Histologically, the tumor shows cuboidal or columnar tumor cells arranged in glandular structures without specific features (e.g., clear cells, goblet cells, mucinous cells, foci of the endometriosis, or mesonephric remnants), which are typical of non-DES-associated adenocarcinoma, along with markedly enlarged nuclei, marked atypia, and hyperchromasia. The tumor is positive for CK7 and CAM5.2, but negative for CK20, CDX2, ER, PR, PAP, and PSA [4].

Differential diagnoses for vaginal cancer include metastases from other sites (e.g., cervical, endometria, or ovarian cancer, breast cancer, gestational trophoblastic disease, colorectal cancer, urogenital or vulvar cancer, and Skene gland adenocarcinoma).

The stages of vaginal cancer are determined in accordance with the guidelines of the International Federation of Gynecology and Obstetrics (FIGO) and the American Joint Committee on Cancer (AJCC). The FIGO system involves extensive clinical evaluation, including chest radiographs, bimanual and rectovaginal examination under anesthesia, cystoscopy and/or proctoscopy, intravenous pyelogram to evaluate for hydronephrosis, and imaging modalities (e.g., CT, MRI and PET). Stage I indicates that the carcinoma is confined to the vaginal wall, stage II indicates that the carcinoma has involved the paravaginal tissues but not the pelvic wall, stage III indicates that the carcinoma has extended to the pelvic wall, and stage IV indicates that the carcinoma has extended beyond the true pelvis or has involved the mucosa of the bladder or rectum (with stage IVa referring to tumor extension beyond the true pelvis or invasion of the bladder or rectum, and stage IVb referring to distant metastasis) [5,6].

27.7 Treatment

Treatment options for vaginal cancer include surgery, radiotherapy, and chemotherapy.

Surgery (e.g., radical hysterectomy, upper vaginectomy, pelvic lymphadenectomy, trachelectomy, laparotomy and laser surgery) is considered for small (<2 cm), minimally invasive exophytic stage I tumor involving

the upper posterior vagina, for recurrent tumor after radiotherapy, and for stage IVA tumor with a rectovaginal or vesicovaginal fistula. In some cases, neoadjuvant chemotherapy may be used to shrink the tumor prior to removal.

Radiotherapy represents the treatment of choice for invasive vaginal cancer (stage II and higher), implemented via external beam radiation therapy (EBRT) and intracavitary or interstitial brachytherapy (BT).

Chemotherapy (e.g., cisplatin, paclitaxel, carboplatin, fluorouracil, and docetaxel) with palliative EBRT may be applied to treat/control stage IVb, metastatic or recurrent tumors [5,6].

As radiotherapy may reduce hemoglobin level, leading to poor survival for vaginal cancer, transfusion is recommended 8-9 weeks after radiotherapy to maintain hemoglobin levels of >10–11 g/dL and promote tumor oxygenation.

The follow-up for vaginal cancer involves a clinical examination every 3 months for 2 years, then less frequent intervals thereafter.

27.8 Prognosis

If diagnosed and staged early, a combination of surgical resection and radiation is curative in vaginal cancer.

The relative 5-year survival rates for vaginal cancer are 96% for stage 0 (carcinoma in situ), 84% for stage I, 75% for stage II, 57% for stage III, and 36% for stage IV.

For patients with vaginal SCC, the 5-year survival rate is approximately 54%; for patients with vaginal adenocarcinoma, the 5-year survival rate is 60%; and for patients with non-DES-associated adenocarcinoma, the 5-year survival rate is 34%.

Unfavorable prognostic factors for primary vaginal cancer include advanced stage (>60 years), larger tumor size (>4 cm), tumor ulceration, tumor infiltration into the rectovaginal septum, and lower and middle vaginal tumors.

Vaginal SCC has the potential to spread to the vulva, cervix, bladder, and rectum, and metastasize to the obturator, hypogastric, external iliac, groin nodes, liver, lungs, bones, and brain. Vaginal adenocarcinoma tends to metastasize to the lungs and lymph nodes.

References

1. Kurman RJ (ed.). *WHO Classification of Tumours of Female Reproductive Organs.* Lyon, France: IARC Press, 2014.
2. Gardner CS, Sunil J, Klopp AH, et al. Primary vaginal cancer: Role of MRI in diagnosis, staging and treatment. *Br J Radiol* 2015;88(1052):20150033.
3. Pathologyoutlines.com. *Squamous Cell Carcinoma.* http://www.pathologyoutlines.com/topic/vaginascc.html. Accessed March 9, 2017.
4. Pathologyoutlines.com. *Adenocarcinoma.* http://www.pathologyoutlines.com/topic/vaginaadenoNOS.html. Accessed March 9, 2017.
5. PDQ Adult Treatment Editorial Board. *Vaginal Cancer Treatment (PDQ®): Health Professional Version. PDQ Cancer Information Summaries.* Bethesda, MD: National Cancer Institute, 2002.
6. Hacker NF, Eifel PJ, van der Velden J. Cancer of the vagina. *Int J Gynecol Obstetr* 2015; 131 (suppl 2): S84–7.

28
Vulvar Cancer

28.1 Definition

Tumors of the vulva comprise (1) *epithelial tumors* (squamous and related tumors and precursors [squamous cell carcinoma not otherwise specified, or SCC NOS—keratinizing, nonkeratinizing, basaloid, warty, verrucous, keratoacanthoma-like, variant with tumor giant cells; basal cell carcinoma; squamous intraepithelial neoplasia—vulvar intraepithelial neoplasia [VIN] 3 or SCC *in situ*; benign squamous lesions—condyloma acuminatum, vestibular papilloma or micropapillomatosis, fibroepithelial polyp, seborrheic and inverted follicular keratosis, keratoacanthoma], glandular tumors [Paget disease; Bartholin gland tumors—adenocarcinoma, SCC, adenoid cystic carcinoma, adenosquamous carcinoma, transitional cell carcinoma, small cell carcinoma, adenoma, adenomyoma; tumors arising from specialized anogenital mammary-like glands—adenocarcinoma of mammary gland type, papillary hidradenoma; adenocarcinoma of Skene gland origin; adenocarcinomas of other types; adenoma of minor vestibular glands; mixed tumor of the vulva], tumors of skin appendage origin [malignant sweat gland tumors, sebaceous carcinoma, syringoma, nodular hidradenoma, trichoepithelioma, trichilemmoma]), (2) *soft tissue tumors* (sarcoma botryoides, leiomyosarcoma, proximal epithelioid sarcoma, alveolar soft part sarcoma, liposarcoma, dermatofibrosarcoma protuberans, deep angiomyxoma, superficial angiomyxoma, angiomyofibroblastoma, cellular angiofibroma, leiomyoma, granular cell tumor), (3) *melanocytic tumors* (malignant melanoma, congenital melanocytic nevus, acquired melanocytic nevus, blue nevus, atypical melanocytic nevus of the genital type, dysplastic melanocytic nevus), (4) *miscellaneous tumors* (yolk sac tumor, Merkel cell tumor, peripheral primitive neuroectodermal tumor or Ewing tumor), (5) *hematopoietic and lymphoid tumors* (malignant lymphoma, leukemia), and (6) *secondary tumors* [1].

SCC is a predominating malignancy of the vulva, accounting for about 90% of vulvar neoplasms, whereas malignant melanoma (2%–9%), Bartholin gland carcinoma, adenocarcinoma, invasive Paget disease, basal cell carcinoma, verrucous carcinoma, and sarcoma are relatively uncommon [2].

28.2 Biology

Immediately external to the vagina, the vulva is the collective name for the external female genitalia in the pubic region, which include the mons pubis, clitoris, labia minora (inner lips of the vulva, consisting of nonkeratinized stratified squamous epithelium, usually no adnexae), labia majora (outer lips of the vulva, consisting of keratinized stratified squamous epithelium with hair follicles and eccrine, apocrine, and sebaceous glands), vulvar vestibule, vestibulovaginal bulbs, urethral meatus, hymen (nonkeratinized stratified squamous epithelium), Bartholin glands, Skene glands (mucus-secreting columnar epithelium merges with duct urothelium, and then stratified squamous epithelium of the vestibule), and vaginal introitus (vaginal opening).

Vulvar cancer may be distinguished into two types according to their likely tumorigenesis. Vulvar cancer type 1 (human papillomavirus [HPV] dependent, 20%–40% of cases) often develops following HPV infection, leading to VIN (including usual high-grade 2 and 3 and differentiated high-grade 3) and predisposition to vulvar cancer. Vulvar cancer type 2 (HPV independent) evolves from vulvar nonneoplastic epithelial disorders (VNEDs) (including lichen sclerosus, squamous cell hyperplasia, and other dermatoses), leading to cellular atypia and cancer [3].

The most common sites of vulvar cancer involvement are the labia majora (50% of cases), labia minora (15%–20%), clitoris, and perineum, as well as multifocal lesions (5%). A neoplasm affecting both the vagina and vulva (i.e., crossing the hymenal ring) is also considered vulvar tumor [4].

28.3 Epidemiology

Vulvar cancer represents 3%–5% of female genital tract malignancies, and has an annual incidence of 1–2 per 100,000 women, and a peak incidence in the seventh decade (median age of 69.5 years, range of 36–85 years). In younger women, vulvar cancer is associated with warty (HPV infection) or basaloid VIN.

28.4 Pathogenesis

Risk factors for vulvar cancer include hypertension, diabetes mellitus, obesity, HPV types 16 and 18 infection, VIN (including low-grade squamous intraepithelial lesion [LSIL], high-grade squamous intraepithelial lesion [HSIL], and differentiated VIN or dVIN), and VNED (e.g., lichen sclerosus) [5].

Vulvar cancer type 1 (e.g., basaloid and warty carcinomas) tends to occur in younger women (35–65 years), has poorly differentiated intraepithelial-like (basaloid) histology, and is strongly associated with cervical neoplasia, HPV infection, past history of condyloma, sexually transmitted disease (STD), smoking, and preexisting VIN lesion. In contrast, vulvar cancer type 2 mainly affects older women (55–65 years), shows a histology of well-differentiated, keratinizing SCC, and is not associated with cervical neoplasia, HPV infection, past history of condyloma, STD, or smoking. However, it is linked to preexisting vulvar inflammation, lichen sclerosus, squamous cell hyperplasia, and p53 mutations [5].

28.5 Clinical features

Vulvar cancer often has a long history of pruritus, and presents with a vulvar lump or mass, vulvar bleeding, discharge, dysuria, and pain.

Vulvar lesion is usually raised, fleshy, ulcerated, leukoplakic, or warty in appearance. Most SCC lesions are unifocal and occur on the labia majora, while some cases (5%) are multifocal, and involve the labia minora, clitoris, or perineum.

28.6 Diagnosis

Vulvar cancer often presents as a warty tumor, long-standing ulcer, groin mass, plaque, or erythematous rash. Vulvar cancer may spread to adjacent organs (e.g., the vagina, urethra, and anus), or the inguinal and femoral lymph nodes, and the deep pelvic nodes. While well-differentiated lesions spread along the surface with minimal invasion, anaplastic lesions are deeply invasive.

Histologically, vulvar SCC may be well differentiated (abundant eosinophilic cytoplasm, low nuclear-to-cytoplasmic ratio), moderately differentiated (nuclear pleomorphism), or poorly differentiated (basophilic cytoplasm and increased nuclear-to-cytoplasmic ratio). HPV+ tumors often show clusters or single cells with an increased nuclear-to-cytoplasmic ratio, irregular nuclear contours, coarse chromatin, and perinuclear clearing. HPV– tumors contain large cells with round or ovoid nuclei, prominent nucleoli, open or vesicular chromatin, and no perinuclear halo. Immunohistochemically, vulvar SCC is positive for cytokeratins (AE1/AE3, CK5/6, MNF116, 34betaE12), EMA, p63, and p40, but negative for S100, BerEP4, CD34, and CD31. Molecularly, the tumor displays allelic gains at 3q and 12q (HPV+ SCC), a gain of 8q (HPV– SCC), and losses at 3p and 4p (HPV16+ and HPV– SCC). In addition,

hypermethylation of the RAS association domain family gene is observed in lichen sclerosis, and somatic mutations of p53 are found in up to 80% of vulvar SCC [6].

Besides SCC, other histologic types occasionally found on the vulva include malignant melanoma (which makes up 2-9% of malignant tumors of the vulva and 3-5% percent of skin melanomas), verrucous carcinoma (which is frequently misdiagnosed as a condyloma acuminatum), and basal cell carcinoma.

Differential diagnoses of vulvar cancer include VIN, condyloma (condyloma acuminate, condyloma plana or flat, and genital warts; due mainly to low-risk HPV subtype 6 or 11), extramammary Paget disease (characteristic large pale cells in the epithelium and skin adnexa) dysplastic nevi (also called atypical nevus, nevus with architectural disorder, and Clark nevus), autoimmune and bullous disorders, Bowen disease, hypertrophic herpes simplex genitalis/pseudoepitheliomatous hyperplasia, hypertrophic/erosive lichen planus, prurigo nodularis, and other neoplasms of the vulva [3].

VIN is an SCC precursor lesion of the vulva, which is classified into LSIL (vulvar LSIL, flat condyloma, or HPV effect), HSIL (vulvar HSIL, usual-type VIN [uVIN]), and differentiated-type VIN (dVIN) [3].

uVIN (vulvar HSIL) occurs in women of 30–50 years, and is associated with high-risk HPV subtypes 16 and 18. Clinically, it forms multifocal and multicentric lesions (white or erythematous macules or papules), and causes pruritus, dysuria, pain, and ulceration. It can progress to basaloid or warty SCC without treatment. Histologically, it shows hyperkeratosis, parakeratosis, acanthosis with club-shaped rete ridges, disorientation of the cells above the basal cell layer, nuclear clumping with mitotic figures, and variable extension into skin appendages. It stains positive for p16 and Ki67, but negative for p53 [3].

dVIN is less common that uVIN, occurs mainly in postmenopausal women (60–80 years), and often arises in a background of lichen sclerosus and chronic inflammatory dermatoses, but is unrelated to HPV infection. Clinically, it forms unifocal or unicentric gray-white discoloration with a roughened surface, white plaques, white elevated nodules, ulcerative red lesion, or erythematous red lesion. About 33% of dVIN may progress to keratinizing SCC. Histologically, it shows acanthosis, variable parakeratosis, irregular elongation and anastomoses of the rete ridges, nuclear atypia confined to basal and parabasal layers, nuclear enlargement with coarse chromatin, open vesicular nuclei and prominent nucleoli, ample eosinophilic

cytoplasm, scattered mitoses, prominent intercellular bridges, and chronic inflammatory infiltrate in underlying or adjacent papillary dermis, but no extension into skin appendages. It stains positive for p53 and Ki-67, but negative for p16 [3].

The stage of vulvar cancer is usually determined according to the International Federation of Gynecology and Obstetrics (FIGO) staging system. Specifically, stage Ia refers to tumor of ≤2 cm with stromal invasion ≤1 mm, confined to the vulva or perineum, no nodal metastasis; stage Ib contains tumor of >2 cm with stromal invasion >1 mm; stage II refers to tumor of any size with extension to adjacent perineal structures (lower urethra, lower vagina, anus), no nodal metastasis; stage IIIa refers to tumor of any size with or without extension to adjacent perineal structures (lower urethra, lower vagina, anus), with inguino-femoral nodal metastasis [including 1 node metastasis (≥5 mm) or 1–2 node metastasis(es) (<5 mm)]; stage IIIb has ≥2 node metastases (≥5 mm) or ≥3 node metastases (<5 mm); stage IIIc shows node metastases with extra-capsular spread; stage IVa tumor invades upper urethra and/or vaginal mucosa, bladder mucosa, rectal mucosa, or fixed to pelvic bone, or fixed or ulcerated inguinofemoral nodes; and stage IVb indicates distant metastasis including pelvic nodes [2,4].

28.7 Treatment

Surgical resection is the mainstay treatment for vulvar cancer. For localized lesions (T1), a radical or modified radical vulvectomy en bloc with full uni- or bilateral inguinofemoral lymphadenectomy (IFL) or sentinel lymph node (SLN) biopsy may be performed. Pelvic exenteration is usually reserved for locally advanced disease. Primary radiotherapy is often used for unresectable tumors, while adjuvant radiotherapy is indicated for high-risk primary tumors (tumor size of >4 cm, lymphovascular invasion, close or positive surgical margins, metastasis to lymph nodes). In addition, primary chemoradiation can be utilized for unresectable tumors (e.g., when the urethra or anus is involved) to avoid significant postoperative morbidity, and neoadjuvant chemotherapy may be considered when appropriate [5,7].

28.8 Prognosis

The 5-year survival rate is 98% for patients with stage I vulvar cancer, 82% for patients with stage II vulvar cancer, 74% for patients with stage III vulvar cancer, and 31% for patients with stage IV vulvar cancer. Recurrence of vulvar cancer in the groin is in the range of 12%–37%.

Adverse prognostic factors for vulvar cancer include older age, advanced stage at presentation, smoking, ulcerated/matted inguinal lymph nodes, clitoral involvement (due probably to larger tumor size), increased depth of stromal invasion (>4 mm), lymphovascular space involvement (resulting from metastasis to inguinofemoral lymph nodes). Further, the number of positive groin nodes, diameter of nodal metastases, tumor size (>6 cm), and tumor ploidy may also infuence the prognosis. Patients with aneuploid tumors (i.e., an abnormal number of chromosome pairs) have a poorer 5-year survival rate than patients with diploid tumors (23% vs. 62%).

References

1. Kurman RJ (ed.). *WHO Classification of Tumours of Female Reproductive Organs*. Lyon, France: IARC Press, 2014.
2. Canavan TP, Cohen D. Vulvar cancer. *Am Fam Physician* 2002;66(7):1269–74.
3. Sideri M, Jones RW, Wilkinson EJ, et al. Squamous vulvar intraepithelial neoplasia: 2004 modified terminology, ISSVD Vulvar Oncology Subcommittee. *J Reprod Med* 2005;50(11):807–10.
4. Hacker NF, Eifel PJ, van der Velden J. Cancer of the vulva. *Int J Gynecol Obstetr.* 2015; 131 (suppl 2): S76–83.
5. Nooij LS, Brand FA, Gaarenstroom KN, Creutzberg CL, de Hullu JA, van Poelgeest MI. Risk factors and treatment for recurrent vulvar squamous cell carcinoma. *Crit Rev Oncol Hematol* 2016;106:1–13.
6. PathologyOutlines.com. *Squamous Cell Carcinoma.* http://www.pathologyoutlines.com/topic/vulvascc.html. Accessed March 10, 2017.
7. PDQ Adult Treatment Editorial Board. *Vulvar Cancer Treatment (PDQ®): Health Professional Version. PDQ Cancer Information Summaries.* Bethesda, MD: National Cancer Institute, 2002.

Glossary

anaplasia: A term used to describe cancer cells with a total lack of differentiation and with resemblance to original cells either in functions or structures, or both; also known as dedifferentiation (backward differentiation).

aneuploidy: The presence of an abnormal number of chromosomes in a cell. *See* diploidy.

angiogenesis: The growth of new capillary blood vessels from preexisting vessels in the body (especially around a developing neoplasm).

apoptosis: Programmed cell death, with the cells that are damaged beyond repair typically dying as they swell and burst, spilling their contents over their neighbors.

atypia: The state of being not typical or normal. In medicine, atypia is an abnormality in cells in tissue, which may or may not be a precancerous indication associated with later malignancy.

benign tumor: A slow-growing, noncancerous tumor that does not invade nearby tissue or spread to other parts of the body. In most cases, a benign tumor has a favorable outcome, with or without surgical removal. However, a benign tumor in vital structures such as nerves and blood vessels, or undergoing malignant transformation, often has serious consequences. *See* hamartoma.

biopsy: A procedure to remove tumor tissue or cells or tissues for microscopic examination. This is usually conducted through excisional biopsy (removal of an entire lump of tissue), incisional biopsy (removal of part of a lump or a sample of tissue), core biopsy (removal of tissue using a wide needle), or fine-needle aspiration (FNA) biopsy (removal of tissue or fluid using a thin needle).

calcification: The accumulation of calcium salts (e.g., calcium phosphate) in body tissues such as tumor mass, where they do not usually appear. This leads to tissue hardening and produces a dense opacity on a radiographic image.

cancer: Plural cancers or cancer, a group of diseases involving abnormal or uncontrolled expansion of cells that have the potential to invade nearby tissue and/or spread to other parts of the body. *See* tumor, neoplasm, lesion.

carcinoma: A type of cancer that begins in a tissue (called epithelium) that lines the inner or outer surfaces of the body.

CT: Computerized tomography (also known as computed tomography scan [CT scan] or computerized axial tomography [CAT]) utilizes an X-ray machine linked to a computer, together with a dye injected into a vein or swallowed, to take a series of detailed pictures of affected organs or tissues in the body from different angles, in order to reveal the precise location and dimension of tumor.

cyst: A closed capsule or saclike structure usually filled with liquid, semi-solid, or gaseous material (but not pus, which is considered an abscess). As an abnormal formation, a cyst on the skin, mucous membranes, or inside palpable organs can be felt as a lump or bump, which may be painless or painful. While cysts due to infectious causes are preventable, those due to genetic and other causes are not. Most cysts are benign (noncancerous).

dedifferentiation: *See* anaplasia.

desmoplasia: The growth of fibrous or connective tissue around a neoplasm, causing dense fibrosis; it is considered a hallmark of invasion and malignancy.

differentiated: A term used to describe how much or how little tumor tissue looks like the normal tissue it came from. Well-differentiated cancer cells look more like normal cells and tend to grow and spread more slowly than poorly differentiated or undifferentiated cancer cells.

diploidy: The presence of a normal number (two sets) of chromosomes in a cell. *See* aneuploidy.

dysplasia: The overgrowth of immature cells at the location where the number of mature cells is decreasing. This term is particularly used for when cellular abnormality is restricted to the new tissues.

endoscopy: A thin, tubelike instrument with a light and a lens for checking for abnormal areas inside the body.

FISH: Fluorescence *in situ* hybridization for determining the positions of particular genes, for identifying chromosomal abnormalities, and for mapping genes of interest.

grade, grading: A measure of cell anaplasia (reversion of differentiation) in tumor; based on the resemblance of the tumor to the tissue of origin. Depending on the amount of abnormality, a tumor is graded as 1 (well differentiated; low grade), 2 (moderately differentiated; intermediate grade), 3 (poorly differentiated; high grade), or 4 (undifferentiated; high grade). *See* stage, TNM.

H&E stain: Combined use of hematoxylin (positively charged) and eosin (negative charged) to stain nucleic acids (negatively charged) in blue and amino groups in proteins (negatively charged) in pink, respectively. *See* IHC.

hamartoma: a benign, tumor-like, focal malformation resulting possibly from a developmental error. Composed of abnormal or disorganized mixture of cells and tissues, hamartoma grows at the same rate at the surrounding tissues, and rarely invades or compresses nearby structures significantly. In contrast, a true benign tumor may grow faster than surrounding tissues and compresses nearby structures. Despite its benign histology, hamartoma may be implicated in some rare, but life-threatening clinical issues such as those associated with neurofibromatosis type 1 and tuberous sclerosis. A nonneoplastic mass (eg, hemangioma) can also arise in this way, contributing to misdiagnosis. *See* benign tumor.

hyperplasia: A disease associated with the increase in number of normal-looking cells, leading to an enlarged organ. It is also called hypergenesis.

IHC: Immunohistochemistry, a technique that exploits the principle of antibodies binding specifically to antigens in biological tissues to visualize the distribution and localization of specific cellular components within cells and in the proper tissue context. Similar to H&E stain, IHC helps detect cellular abnormalities and verify whether tumor or cancer cells are present at the edge of the material removed (positive margins) or not (negative, not involved, clear, or free margins), or neither negative nor positive (close margins).

Ki-67: A nuclear protein (also known as KI-67 or MKI67) associated with cellular proliferation. and ribosomal RNA transcription. Ki-67 protein is present during all active phases of the cell cycle [G1 (pre-DNA synthesis), S (DNA synthesis), G2 (post-synthesis), and M (mitosis)], but absent in resting phase (G0). The fraction of Ki-67-positive tumor cells detected by MIB-1 (the Ki-67 labeling index or MIB-1 labeling index) often correlates to the aggressiveness and thus the clinical course of cancer. *See* MIB-1.

lesion: A term in medicine to describe all the abnormal biological tissue changes, such as a cut, a burn, a wound, or a tumor. In cancer, *lesion* is used interchangeably with *tumor, cancer,* or *neoplasm*. *See* cancer, tumor, neoplasm.

LOH: Loss of heterozygosity is a gross chromosomal event that results in loss of the entire gene and the surrounding chromosomal region.

malignancy: The state or presence of a malignant tumor.

malignant tumor: A tumor with the capability of invading surrounding tissues, producing metastases, and recurring after attempted removal.

metaplasia: The reversible replacement of one differentiated cell type with another mature differentiated cell type.

MIB-1: a monoclonal antibody raised against Ki-67 protein that allows accurate immunohistochemical detection of active or growing cells. *See* Ki-67.

mitotic figure: Microscopic detection of the chromosomes as tangled, dark-staining threads in cells undergoing mitosis. It is often expressed as mitotic figures per 10 high power fields (hpf, usually 400-fold magnification) (mitotic activity index) or per 1000 tumor cells (mitotic index). As mitotic cell count per 10 hpf equals to an area 0.183 mm2, the American Joint Committee on Cancer specifies that mitotic rate (the proportion of cells in a tissue undergoing mitosis) be reported as mitoses per mm², with a conversion factor of 1 mm² equaling 4 hpf.

MRI: Magnetic resonance imaging (also called nuclear magnetic resonance imaging [NMRI]) relies on a magnet, radio waves, and a computer to take a series of detailed pictures of affected areas inside the body, that help pinpoint the location and dimension of a tumor mass, if present. MRI has better image resolution than CT. It includes T1-weighted, T2 weighted, fluid attenuated inversion recovery (FLAIR, also called dark fluid technique), and diffusion weighted imaging (DWI) sequences. T1 weighted images reveal anatomical details and information about venous sinus permeability or pathologic blush (e.g., water and CSF appear dark; fat and calcification appear white/gray). Use of intravenous contrast gadolinium in T1-weighted sequences further enhances and improves the quality of the images. T2 weighted images provide information about edema, arteries and sinus permeability (e.g., water appears white/hyperintense; fat and calcification appear gray/dark). FLAIR sequences remove the effects of fluid (which normally covers a lesion) from the resulting images (e.g., CSF appears dark; edema appears enhanced). DWI sequences help visualize acute infarction and other inflammatory lesions.

mutation: A change in the structure of a gene caused by the alteration of single base units in DNA, or the deletion, insertion, or rearrangement of larger sections of genes or chromosomes, leading to the formation of a variant that may be transmitted to subsequent generations.

necrosis: A form of cell injury leading to the premature or unprogrammed death of cells and living tissue caused by autolysis (due to too little blood flowing to the tissue). *See* apoptosis.

neoplasia: A term that describes the abnormal growth or proliferation of cells, resulting in a tumor that can be cancerous.

neoplasm: A new and abnormal growth of tissue in a part of the body; it is used interchangeably with *tumor* or *cancer.*

oncogene: A gene whose mutation or abnormally high expression can transform a normal cell into a tumor cell.

parenchyma: The functional tissue of an organ as distinguished from the connective and supporting tissue. *See* stroma.

PCR (polymerase chain reaction): a procedure for rapid, in vitro production of multiple copies of particular DNA sequences relevant to diagnosis.

PET: Positron emission tomography scan combines a computer-based scan with a radioactive glucose (sugar) injected into a vein to generate a rotating picture of the affected area, with malignant tumor cells showing up brighter due to their more active taking up of glucose than normal cells.

pleomorphism: A term used in histology and cytopathology to describe variability in the size, shape, and staining of cells and/or their nuclei; it is a feature characteristic of malignant neoplasia and dysplasia.

radiography: A term used to collectively describe electromagnetic radiation (especially X-rays)–based procedures to visualize the internal structure of a nonuniformly composed and opaque object, such as the human body. *See* MRI, CT.

radiotherapy: Also called radiation, radiation therapy, or X-ray therapy, involving delivery of radiation externally through the skin or internally (brachytherapy) for destruction of cancer cells or inhibition of their growth.

stage, staging: as a measurement of the extent to which a tumor has spread, stage is commonly expressed in two ways: clinical and pathological stages. A clinical stage(ranging from 0, I, II, III, to IV) is an estimate of the extent of the tumor after physical exam, imaging studies (e.g., X-rays, CT, MRI) and tumor biopsies as well as other tests (e.g., blood tests), and provides a vital means for deciding the best treatment to use and for comparing the tumor response to treatment. A pathological stage relies on the results of the exam and tests mentioned above, in addition to information obtained during surgery (e.g., the actual degree of tumor spread, nearby lymph node involvement), and thus offers a more precise guide than clinical stage in helping predict treatment response and outcome/prognosis. For example, in the case of penile cancer, clinical stage 0 correlates to pathological stage Tis or Ta/N0/M0; stage I to T1a/N0/M0; stage II to T1b, T2, or T3/N0/M0; stage IIIa to T1, T2, or T3/N1/M0; stage IIIb to T1, T2, or T3/N2/M0; and

stage IV to T4/any N/M0, or any T/N3/M0, or any T/any N/M1. *See* grade, TNM.

stroma: The parts of a tissue or organ that have a connective and structural role, and that do not conduct the specific functions of the organ (e.g., connective tissue, blood vessels, nerves, ducts). *See* parenchyma.

TNM: A system (i.e., the TNM system) designed by the American Joint Committee on Cancer (AJCC) to pathologically stage a solid tumor. T (tumor; TX, T0, T1, T2, T3, T4) indicates the depth of the tumor invasion—the higher the number, the further the cancer has invaded. N (nodes; NX, N0, N1, N2, N3) indicates whether the lymph nodes are affected, and how much the tumor has spread to lymph nodes near the original site. M (metastasis; MX, M0, M1) indicates whether the tumor has spread to other parts of the body. Thus, the pathological stage of a given tumor may be designated as T1N0MX or T3N1M0 (with numbers after each letter providing further details about the tumor). Knowing a tumor's pathological stage helps select the most appropriate treatments and gives a more accurate guidance on its prognosis. *See* grade, stage.

translocation: a segment from one chromosome is transferred to a nonhomologous chromosome (interchromosomal transloaction) or to a new site on the same chromosome (intrachromosomal translocation). A reciprocal translocation occurs when two nonhomologous chromosomes swap parts; whereas a non-reciprocal translocation occurs when the transfer of chromosomal material is one way, i.e., another segment does not exchange places with the first segment.

tumor: A swelling of a part of the body, generally without inflammation, caused by an abnormal growth of tissue, whether benign or malignant. *See* cancer, neoplasm, lesion.

tumor suppressor gene: A gene (also known as antioncogene) that regulates cell division, repairs DNA mistakes, or instructs cells when to die. *See* apoptosis. When a tumor suppressor gene is mutated, cell growth gets out of control.

ultrasound: A device for delivering sound waves that bounce off tissues inside the body like an echo, and recording the echoes to create a picture (sonogram) of areas inside the body.

undifferentiated: The presence of very immature and primitive cells that do not look like cells in the tissue of their origin. Undifferentiated cells are said to be anaplastic, and an undifferentiated cancer is more malignant than a cancer of that type that is well differentiated. *See* anaplasia, differentiated.

Index